$<$	Is less than
$\leq$	Is less than or equal to
$\lim\limits_{x \to a}$	Limit as x approaches a
$\overleftrightarrow{AB}$	Line containing points A and B
$\overline{AB}$	Line segment between A and B
$\log_b$	Logarithm base b
ln	Logarithm base e
e	2.718 ... (base of the natural logarithms)
$\lvert \mathbf{r} \rvert$	Magnitude of the vector $\mathbf{r}$
$\begin{pmatrix} a_1 \; a_2 \; a_3 \ldots \\ b_1 \; b_2 \ldots\ldots \\ \ldots\ldots\ldots\ldots \end{pmatrix}$	Matrix
$\neq$	Does not equal
$\square$	Numerical or literal expression to be filled in by student
$\bigcirc$	Operation, that is, $+$, $-$, $\times$, $\div$, to be filled in by student
π	Pi (Greek letter used to represent the ratio of the circumference of a circle to the diameter)
$\pm$	Plus or minus
∟	Right angle
(x, y)	Rectangular coordinates of a point in a plane
R	Relation, that is, $<$, $>$, $=$, $>$, $<$ to be filled in by student
sec	Secant
sin	Sine
$\sum$	Sum of certain terms
$\sum\limits_{i=1}^{n}$	Sum of n terms, one for each integer beginning with 1 and ending with n
tan	Tangent
$\mathbf{r}$	Vector $\mathbf{r}$

Mathematics for Technicians with an Introduction to Calculus

Mathematics for Technicians with an Introduction to Calculus

Leland R. Halberg
Howard E. Zink

Lane Community College

Wadsworth Publishing Company, Inc.
Belmont, California

ISBN-0-534-00144-0
L. C. Cat. Card No.-72-160206

Printed in the United States of America

1 2 3 4 5 6 7 8 9 10—75 74 73 72

Preface

Mathematics plays an important part in preparing people for technical occupations. The emphasis in this text is upon developing general mathematical understanding and computational skills that assist technicians in their training and on their jobs.

The content differs from the usual sequence in college mathematics courses in three essential ways:

1. The emphasis is on problem solving.
2. The topics included are particularly relevant to technical application.
3. The presentation is informal. This is a student-oriented text.

The material is designed for a 36-week course, meeting from three to five times weekly. Students should have completed successfully a one-year course of algebra and a one-half-year course of plane geometry.

The first two chapters, labelled Chapters A and B, are provided for readers who need to review some of the essential ideas and manipulative skills of elementary algebra and plane geometry. These chapters are not part of the regular technical mathematics course; rather, they are prerequisites to the course. The material is presented in a format suitable for self-study so that class time need not be devoted to the contents of these chapters. Well-prepared students will be able to complete the review in a very short time. Those who are not well-prepared will find Chapters A and B convenient for occasional reference during the entire course.

Chapter 1 is the beginning of the principal subject matter of the course. Considerable space is given to examples and complete solutions to exercises. Each student is encouraged to have a pencil and paper ready to test his understanding of the exposition. The text clearly specifies places where the reader should make a response before viewing the correct answer. The student will profit as much from being an active reader as he will from doing exercises.

Exercises and problems appear separately at the end of each section. The *exercises* are chosen to emphasize mathematical ideas, relationships, and computational techniques. Exercises performed in an improper or illogical manner strengthen undesirable processes and lead to incorrect results. For this reason, complete solutions to exercises appear in parallel columns with the exercises. The correct solutions are shielded by a gray screen so that the student can avoid viewing the answer until after he has carefully written his own solution. The *problems* illustrate mathematical applications chosen from many technical fields. They require a greater degree of interpretive skill and organization than do the exercises. Complete solutions appear in the Appendix.

This text provides the fundamental tools for understanding specific mathematical descriptions and solutions of problems that technicians are ordinarily required to understand and solve. An instructor working with a class that is limiting its attention to a specific technical area is encouraged to adapt and amplify the applications to his particular class' needs.

Concepts of calculus appear early in the text in the form of area and slope functions. They are presented on the same level as the algebraic concepts and computational skills of the context in which they appear. It is not until Chapters 11 and 12 that the student may realize that preliminaries to calculus have paralleled the development of many topics.

The order of the chapters as they appear in the Contents is important to the continuous development of the topics. Instructors are encouraged to use the text sequence, but those who wish to vary the order will find the following table helpful.

Chapter	Prerequisite
1	(A-Review)
2	1
3	1, 2
4	1, 3
5	1–4
6	1, (B-Review)
7	1, 6
8	1–3, 6, 7
9	4–8
10	1–4, 6
11	1–3, 7, 8, 10
12	1–3, 7, 8, 11

We would like to express our thanks to Louis J. Dunham, Jr., Director of Franklin Institute of Boston, Ralph Mansfield of City College of Chicago, James Mettler of Pennsylvania State University, Schuylkill Campus, Leo Meyer of Chabot College, and Kenneth C. Skeen of Diablo Valley College for their helpful comments in reviewing the manuscript.

We would also like to express our sincere appreciation to the mathematics instructors and students who class tested the preliminary editions of this text, and to those who have given assistance to the details of its production. We give special thanks to Jan Ford and Pamela Laptad for their careful reading and typing of the manuscript. We are grateful to the publisher for constructive reviews and editorial assistance.

Leland R. Halberg
Howard E. Zink

Contents

1 Linear Functions

2 Systems of Linear Equations

3 Quadratic Functions

4 Exponents and Radicals

5 Logarithmic and Exponential Functions

8 Trigonometric Identities and Conditional Equations

9 Complex Numbers

10 Plane Analytic Geometry

11 Differential Calculus

12 Integral Calculus

Appendix

Index

Mathematics for Technicians with an Introduction to Calculus

Chapter A Review of Elementary Algebra Prerequisites (Programmed)

This chapter will help the reader who needs a *brief review* of the elementary topics of algebra considered *prerequisites* to a course in technical mathematics. Some of the topics of beginning algebra that do not appear here are considered briefly in later chapters.

The format of this chapter is especially suitable for review and self-study. Topics are broken down into a sequence of small steps called *frames*, each of which asks for some reasoning and recall response from the reader. The desired response is given at the end of each frame so that you will know immediately whether or not you understand the concept.

To obtain maximum benefit from the use of this program, *it is absolutely necessary that you shield from view the answers at the end of each frame until you have carefully thought out your response and written it down.*

There are special frames to accelerate the review for some readers. These, distinguished by a line surrounding the frame, *pretest* a small area of computational skill or understanding. If you complete these frames without error, you can omit the instructional frames immediately following and proceed to the next *pretest frame*. If the pretest frame is answered incorrectly, the instructional frames between pretests will help you recall the rationale and techniques needed to correct your pretest errors.

Special symbols used to indicate the type of response asked for are as follows:

▭	numerical or literal expression
○	operation, such as $+$, $-$, $\times$, $\div$
R	relation, such as $>$, $<$, $=$, $\geqq$, $\leqq$
______	words or phrases

A-1 Operations with Signed Numbers

If you make no errors with this pretest frame, proceed to the next pretest frame. If you make an error, start with Frame (1).

1. $7 + (-9) = \square$,
2. $(-8) + (-5) = \square$,
3. $(-15) + 18 = \square$,
4. $(-3) + 7 + (-2) + 4 = \square$.

■ 1. (-2), 2. (-13), 3. 3, 4. 6.

(1) Consider the sum of $3 + 4$ and the accompanying diagram.

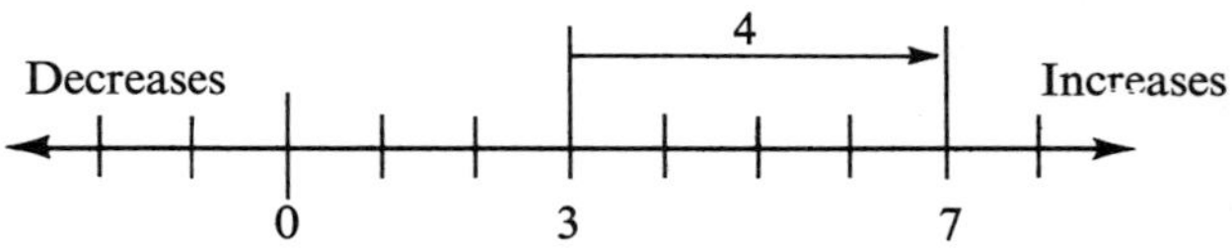

Adding 4 to 3 (increases, decreases) the 3 to $\square$.

■ increases; 7.

(2) Consider the sum of $(-8) + 6$ and the accompanying diagram.

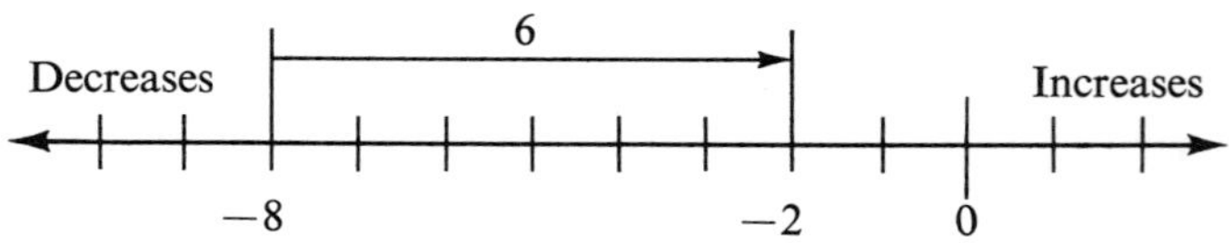

Adding 6 to -8 (increases, decreases) the -8 to $\square$.

■ increases; -2.

(3) Consider the sum of $9 + (-3)$ and the accompanying diagram.

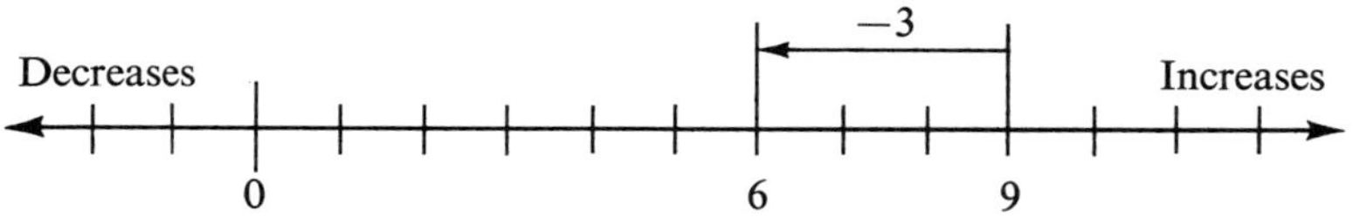

Adding (-3) to 9 (increases, decreases) the 9 to $\square$.

■ decreases; 6.

(4) $8 + (-12) = \square$.

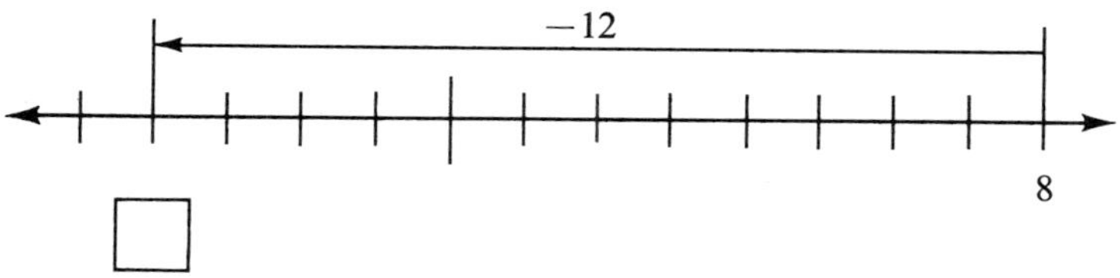

■ (-4)

(5) $(-3) + (-5) = \square$.

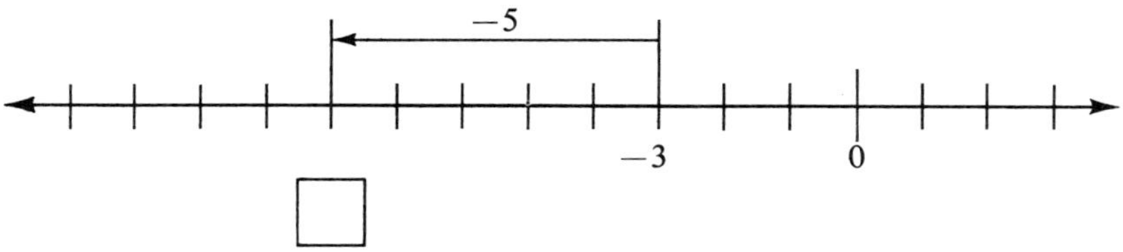

■ (-8)

(6) 1. $4 + (-9) = \square$,

2. $(-8) + 21 = \square$,

3. $(-13) + (-12) = \square$,

4. $(-17) + 5 = \square$,

5. $13 + (-5) = \square$.

■ 1. (-5), 2. 13, 3. (-25), 4. (-12), 5. 8.

(7) $12 + (-7) + 2 + (-6) = \square$.

■ $14 + (-13) = 1$.

---

1. $5 - 8 = \square$,
2. $13 - (-3) = \square$,
3. $(-8) - (-15) = \square$,
4. $(-7) - (-3) + (-4) - 6 = \square$,
5. $4 - (3 + 2) = \square$,
6. $21 - (4 - 6) = \square$.

■ 1. (-3), 2. 16, 3. 7, 4. (-14), 5. (-1), 6. 23.

---

(8) Subtraction is accomplished by changing the problem to an equivalent addition problem:

$$8 - 3 = 8 + (-3) = 5,$$
$$12 - (-8) = 12 + 8 = 20,$$
$$(-4) - 7 = (-4) + (-7) = -11.$$

The operation is changed from subtraction to ______________ and the sign of the subtrahend is changed.

■ addition

(9) Illustrated in a general manner, $a - b$ is changed to [ ].

■ $a + (-b)$

(10) 1. $23 - (-3) =$ [ ],

2. $(-8) - 4 + (-2) - (-7) =$ [ ],

3. $15 - (-15) =$ [ ],

4. $19 - (4 - 9) =$ [ ].

■ 1. $23 + 3 = 26$, 2. $-8 + (-4) + (-2) + 7 = -7$,
3. $15 + 15 = 30$, 4. $19 + (-4) + 9 = 24$.

---

1. $4(-6) =$ [ ], 2. $(-9)8 =$ [ ],
3. $(-7)(-7) =$ [ ], 4. $(-12) \div 4 =$ [ ],
5. $(-6) \div (-12) =$ [ ], 6. $(-8)(3)(-5) =$ [ ].

■ 1. -24, 2. -72, 3. 49, 4. -3, 5. $\frac{1}{2}$, 6. 120.

---

(11) The only effect that signs have on products and quotients is the sign of the answer; in other words, the numerical value of the answer without

regard to sign is unchanged from arithmetic facts. In the following problems assume the signs are correct.

$$(-8)(-6) = (+\boxed{}),$$
$$(-54) \div 9 = (-\boxed{}).$$

■ 48; 6.

(12) The rules for multiplying and dividing signed numbers are alike.
1. If the two factors of multiplication (or division) are alike in sign, the sign of the product (or quotient) is ______.
2. If the two factors of multiplication (or division) are unlike in sign, the sign of the product (or quotient) is ______.

■ 1. positive, 2. negative.

(13) 1. $4(-12) = \boxed{}$,

2. $(-6)(-5) = \boxed{}$,

3. $(-27) \div 9 = \boxed{}$.

■ 1. -48, 2. 30, 3. -3.

(14) When more than two factors are present, (1) an *even* number of negative factors will result in a ______ product, and (2) an *odd* number of factors will result in a ______ product.

■ 1. positive, 2. negative.

(15) 1. $(-8)(-2)(3)(-1) = \boxed{}$,

2. $(-2)(-2)(-3)(-1)(4) = \boxed{}$.

■ 1. -48, 2. 48.

(16) Summary—Operations with Signed Numbers

1. $13 + (-9) = \boxed{}$,
2. $-4 - (-7) = \boxed{}$,
3. $9(-7) = \boxed{}$,
4. $(-3)(-5) = \boxed{}$,
5. $25 \div (-5) = \boxed{}$.

■ 1. 4, 2. 3, 3. -63, 4. 15, 5. -5.

A-2 Integer Exponents

1. $3^2 = \boxed{}$, 2. $\left(\frac{1}{2}\right)^3 = \boxed{}$, 3. $(4a)^2 = \boxed{}$.

■ 1. 9, 2. $\frac{1}{8}$, 3. $16a^2$.

(1) 3^2 is an abbreviation for 3×3.

5^3 is an abbreviation for $5 \times 5 \times 5$.

3^4 is an abbreviation for $\boxed{}$.

■ $3 \times 3 \times 3 \times 3$

(2) In the expression 5^2, the 5 is the *base* and the 2 is the *exponent*. The expression 5^2 means two factors of 5. In the expression 7^3, (1) three is the ______________ and (2) seven is the ______________, and means (3) ______________ factors of (4) ______________.

■ 1. exponent, 2. base, 3. three, 4. seven.

(3) $\left(\frac{2}{3}\right)^3$ is an abbreviation of $\frac{2}{3} \times \frac{2}{3} \times \frac{2}{3}$ and is equal to $\frac{8}{27}$.

$\left(\frac{4}{5}\right)^2 = \boxed{}$.

■ $\frac{4}{5} \times \frac{4}{5} = \frac{16}{25}$.

(4) $(3m)^4$ means $(3m)\,(3m)\,(3m)\,(3m) = 81m^4$.

$(2b)^3 = \boxed{}$.

■ $(2b)(2b)(2b) = 8b^3$.

(5) 1. $5^3 = \boxed{}$, 2. $\left(\frac{2}{5}\right)^2 = \boxed{}$, 3. $(-3a)^2 = \boxed{}$.

■ 1. 125, 2. $\frac{4}{25}$, 3. $9a^2$.

---

1. $3^2 \times 3^5 = 3^{\square}$, 2. $a^4 \cdot a^2 = \boxed{}$, 3. $x^n \cdot x^m = x^{\square}$.

■ 1. ^{7}, 2. a^6, 3. $^{n+m}$.

---

(6) $2^3 \cdot 2^4 = (2 \cdot 2 \cdot 2)(2 \cdot 2 \cdot 2 \cdot 2) = 2^7$. Notice that 7 is the sum of the exponents 3 and 4.

$$\left(\frac{1}{3}\right)^2 \left(\frac{1}{3}\right) = \left(\frac{1}{3}\right)^{\square}.$$

■ 3

Note: $(1/3)^1 = (1/3)$: When no exponent is written, we assume the exponent is *one*.

(7) $a^2 \cdot a^3 = (a \cdot a)(a \cdot a \cdot a) = a^{\square}$.

■ 5

(8) $a^2 b^2 \cdot a^3 b = a^{\square} b^{\square}$.

■ ^{5}; ^{3}.

(9) $x^n \cdot x^{n+1} = x^{\square}$.

■ $x^{n+n+1} = x^{2n+1}$.

(10) $(2n + 1)^2(2n + 1) = (2n + 1)^{\square}$.

■ 3

1. $\dfrac{a^3}{a} = \boxed{}$, 2. $\dfrac{x^n}{x^m} = x^{\square}$, 3. $x^{-2} = \boxed{}^2$.

■ 1. a^2, 2. $^{n-m}$, 3. $\dfrac{1}{x}$.

(11) $$\frac{2^3}{2^2} = \frac{\overset{1}{\cancel{2}} \cdot \overset{1}{\cancel{2}} \cdot 2}{\underset{1}{\cancel{2}} \cdot \underset{1}{\cancel{2}}} = 2.$$

$$\frac{2^3}{2^2} = 2^{\square}.$$

■ 1

(12) $$\frac{a^5}{a^3} = \frac{\overset{1}{a}\,\overset{1}{a}\,\overset{1}{a}\,a\,a}{\underset{1}{a}\,\underset{1}{a}\,\underset{1}{a}} = a^2.$$

$$\frac{a^5}{a^3} = a^{\square}.$$

■ $a^{5-3} = a^2$.

(13) $\frac{x^p}{x^q} = x^{p-q}$. Now suppose that $p = 3$ and $q = 4$. Then $p - q = \boxed{}$.

■ -1

(14) $$\frac{x^3}{x^4} = x^{3 \bigcirc 4} = x^{\square}.$$

■ $-$; -1.

(15) Since $\frac{x^3}{x^4} = \frac{xxx}{xxxx} = \frac{1}{x}$, $x^{-1} = \boxed{}$.

■ $\frac{1}{x^1}$

(16) $\dfrac{y^3}{y^5} = \dfrac{1}{y^{\square}} = y^{\square}.$

■ 2; −2.

(17) $\dfrac{3xy^3}{xy^2} = \boxed{}.$

■ $3y$

(18) $(5)^{-n} = \boxed{}^{\,n}.$

■ $\dfrac{1}{5}$

(19) $\left(\dfrac{2}{3}\right)^{-n} = \boxed{}^{\,n}.$

■ $\dfrac{3}{2}$

1. $(12)^0 = \boxed{}$, 2. $x^0 = \boxed{}$, 3. $(2x^2 + x - 1)^0 = \boxed{}$.

■ 1. 1, 2. 1, 3. 1.

(20) Recall that $\dfrac{x^n}{x^m} = x^{n-m}$. Let $n = 3$ and $m = 3$. $x^{n-m} = x^{\square}$.

■ 0

(21) $\dfrac{2^5}{2^5} = 2^{\square} = 2^{\square} = \boxed{}$.

■ $^{5-5}$; ^{0}; 1.

(22) $\dfrac{a^2b^{-2}}{a^2b^{-2}} = a^{\square}b^{\square} = \boxed{}$.

■ ^{0}; ^{0}; 1.

1. $\{(2)^3\}^2 = \boxed{}$, 2. $(2x^2)^4 = \boxed{}$.

■ 1. 2^6, 2. $16x^8$.

(23) $(x^2)^3 = x^2 \cdot x^2 \cdot x^2 = x^6$.

The product of 2 and 3 is $\boxed{}$.

■ 6

(24) $(y^3)^2 = y^{\square} \cdot y^{\square} = y^{\square}$.

■ ^{3}; ^{3}; ^{6}.

(25) $(a^2b^3)^2 = a^{\square}b^{\square}$.

■ ^{4}; ^{6}.

(26) $(x^n)^2 = x^{\square}$.

■ $x^n x^n = x^{2n}$.

(27) $(a^n b)^m = a^{\square} b^{\square}$.

■ ^{nm}; ^{m}.

(28) Summary—Integer Exponents

1. $a^n \cdot a^m = \boxed{}$, 2. $\dfrac{b^n}{b^m} = \boxed{}$,

3. $x^0 = \boxed{}$, 4. $(x^n)^m = \boxed{}$,

5. $x^{-n} = \boxed{}$.

■ 1. a^{n+m}, 2. b^{n-m}, 3. 1, 4. x^{nm}, 5. $\dfrac{1}{x^n}$.

A-3 Radicals

1. $\sqrt{9} \cdot \sqrt{9} = \boxed{}$, 2. $\sqrt{2} \cdot \sqrt{2} = \boxed{}$, 3. $\sqrt{n} \cdot \sqrt{n} = \boxed{}$.

■ 1. 9, 2. 2, 3. n.

(1) In the example $\sqrt{9} \cdot \sqrt{9} = 9$, the symbol $\sqrt{9}$ represents one of the two *equal* numbers whose product is 9.

$\boxed{} \cdot \boxed{} = (\boxed{})^2 = 9.$

■ 3; 3; 3.

(2) $\sqrt{9} = \square$ because $3 \times 3 = 9$.

■ 3

(3) $\sqrt{36} = \square$ because $\square \times \square = 36$.

■ 6; 6; 6.

(4) The radical ($\sqrt{\ }$) would appear unimportant if there were always a rational equivalent for every radical. This, however, is not the case. For example, find two equal rational (fractional) numbers whose product is 2.

■ No matter to what extent you attempted to manipulate the rational number a/b, $(a/b)^2 \neq 2$.

(5) $\left(\frac{71}{50}\right)^2 = \square$.

■ $\left(\frac{71}{50}\right)^2 = \frac{5041}{2500} = 2.01640.$

(6) Find a closer approximation for $\sqrt{2}$.

■ $\left(\frac{177}{125}\right)^2 = \frac{31329}{15626} = 2.005060$, which is closer to 2 than $\left(\frac{71}{50}\right)^2$.

(7) $(\square)^2 = 2$. (Notice the equal sign.)

■ $\sqrt{2}$ (You may not fully appreciate the fact that $\sqrt{2}$ is a specific point on the number line, and therefore could represent a measurable length.)

(8) 1. $(\boxed{})^2 = 6$, 2. $(\boxed{})^2 = 15$.

■ 1. $\sqrt{6}$, 2. $\sqrt{15}$.

(9) $(\boxed{})^2 = m$.

■ $\sqrt{m}$

1. $\sqrt{12} = 2(\boxed{})$, 2. $\sqrt{4a^3} = \boxed{}\sqrt{a}$,
3. $\sqrt{n} \cdot \sqrt{m} = \boxed{}$.

■ 1. $\sqrt{3}$, 2. $2a$, 3. $\sqrt{nm}$.

(10) $\sqrt{36} = 6$.
$\sqrt{36} = \sqrt{9 \cdot 4} = \sqrt{9} \cdot \sqrt{4} = \boxed{} \cdot \boxed{} = 6$.

■ 3; 2.

(11) $\sqrt{36} = \sqrt{9} \cdot \sqrt{\boxed{}}$.

■ 4

(12) $\sqrt{12} = \sqrt{4 \cdot \boxed{}} = \sqrt{4} \cdot \sqrt{\boxed{}} = 2\sqrt{\boxed{}}$.

■ 3; 3; 3.

(13) $\sqrt{ab^2} = \sqrt{a}\sqrt{b^2} = \sqrt{a} \cdot \square$.

■ b

(14) The square root of a product of two factors is equal to the product of ________________ of each factor.

■ the square root

(15) Write in mathematical symbols the idea expressed in Frame 14: $\square$.

■ $\sqrt{mn} = \sqrt{m}\sqrt{n}$.

(16) $\sqrt{16b^3} = \sqrt{16b^2 b} = \square\sqrt{b}$.

■ $4b$

1. $\sqrt{\dfrac{18a^2}{2}} = \square$, 2. $\sqrt{\dfrac{3}{4}} = \square$, 3. $\sqrt{\dfrac{2}{x^2}} = \square$.

■ 1. $3a$, 2. $\dfrac{\sqrt{3}}{2}$, 3. $\dfrac{\sqrt{2}}{x}$.

(17) $\sqrt{\dfrac{36}{9}} = \sqrt{4} = 2$.

Notice that the same results can be obtained in the following manner:

$$\sqrt{\frac{36}{9}} = \frac{\sqrt{36}}{\sqrt{9}} = \frac{\square}{\square} = \square.$$

■ $\frac{6}{3}$; 2.

(18) $\sqrt{\frac{16}{25}} = \frac{\sqrt{\square}}{\sqrt{\square}} = \square.$

■ $\frac{\sqrt{16}}{\sqrt{25}}$; $\frac{4}{5}$.

(19) $\sqrt{\frac{12}{64}} = \frac{\sqrt{\square}}{\sqrt{\square}} = \frac{\square\sqrt{\square}}{\square} = \frac{\sqrt{\square}}{\square}.$

■ $\frac{\sqrt{12}}{\sqrt{64}} = \frac{2\sqrt{3}}{8} = \frac{\sqrt{3}}{4}.$

(20) $\sqrt{\frac{a^3}{b^2}} = \frac{\square\sqrt{\square}}{\square}.$

■ $\frac{\sqrt{a^2a}}{\sqrt{b^2}} = \frac{a\sqrt{a}}{b}.$

1. $\frac{2}{\sqrt{2}} \boxed{R} \sqrt{2},$ 2. $\frac{1}{\sqrt{3}} = \frac{\square}{3}.$

■ 1. $=$, 2. $\sqrt{3}$.

(21) Frequently expressions such as $2/\sqrt{3}$ must be given a decimal approximation. If $(2\sqrt{3})/3$ equals $2/\sqrt{3}$, which do you think better lends itself to decimal approximation?

■ $\dfrac{2\sqrt{3}}{3}$ can be more accurately and easily calculated than $\dfrac{2}{\sqrt{3}}$.

$$\frac{2\sqrt{3}}{3} \approx \frac{2(1.732)}{3} \quad \text{(Short division);}$$

$$\frac{3}{\sqrt{3}} \approx \frac{2}{1.732} \quad \text{(Long division).}$$

(22) $2/\sqrt{3}$ can be changed to the form $(2\sqrt{3})/3$ by multiplying both numerator and denominator by $\sqrt{3}$.

$$\frac{2}{\sqrt{3}} \cdot \frac{\sqrt{3}}{\sqrt{3}} = \frac{\square}{\square}.$$

■ $\dfrac{2\sqrt{3}}{3}$

(23) To change $1/\sqrt{2}$ to the equivalent form $\sqrt{2}/2$, multiply both numerator and denominator by $\square$.

■ $\sqrt{2}$, because $\dfrac{1}{\sqrt{2}} \cdot \dfrac{\sqrt{2}}{\sqrt{2}} = \dfrac{\sqrt{2}}{2}$.

(24) $\dfrac{2}{\sqrt{5}} = \square$. (Change terms so that no radical appears in the denominator.)

■ $\dfrac{2}{\sqrt{5}} \cdot \dfrac{\sqrt{5}}{\sqrt{5}} = \dfrac{2\sqrt{5}}{5}$.

(25) $\sqrt{\frac{5}{6}} = \frac{\sqrt{5}}{\sqrt{6}} = \frac{\boxed{}}{6}$.

■ $\frac{\sqrt{5}}{\sqrt{6}} \cdot \frac{\sqrt{6}}{\sqrt{6}} = \frac{\sqrt{30}}{6}$.

(26) $\frac{3}{2\sqrt{a}} = \boxed{}$. (Change terms so that no radical appears in the denominator.)

■ $\frac{3}{2\sqrt{a}} \cdot \frac{\sqrt{a}}{\sqrt{a}} = \frac{3\sqrt{a}}{2a}$.

(27) Review—Radicals

1. $\sqrt{2x} \cdot \sqrt{2x} = \boxed{}$, 2. $\sqrt{49x^2y} = \boxed{}$,

3. $\sqrt{\frac{5y^2}{16x^2}} = \boxed{}$, 4. $\frac{2}{3\sqrt{5}} = \boxed{}$.

■ 1. $2x$, 2. $7x\sqrt{y}$, 3. $\frac{\sqrt{5}\,y}{4x}$, 4. $\frac{2\sqrt{5}}{15}$.

(28) Summary—Radicals

1. $\sqrt{a}\sqrt{a} = \boxed{}$, 2. $\sqrt{b^2} = \boxed{}$,

3. $\sqrt{\frac{a}{b}} = \boxed{}$, 4. $\sqrt{mn} = \boxed{}$.

■ 1. a, 2. b, 3. $\frac{\sqrt{a}}{\sqrt{b}} = \frac{\sqrt{a} \cdot \sqrt{b}}{b}$, 4. $\sqrt{m} \cdot \sqrt{n}$.

A-4 Products

Perform the indicated operation:

1. $(2x)(5x) = \square$,
2. $(.5a^2b)(ab^2) = \square$,
3. $(4xy)(2y)(-7z) = \square$.

■ 1. $10x^2$, 2. $.5a^3b^3$, 3. $-56xy^2z$.

(1) $2 \cdot 3 \cdot 3 \cdot 5 = \square$,

$(2 \cdot 3)(3 \cdot 5) = \square$,

$2 \cdot (3 \cdot 3 \cdot 5) = \square$,

$3^2 \cdot 2 \cdot 5 = \square$,

$6 \cdot 15 = \square$.

■ All the expressions are equal to 90. They illustrate how the numbers in multiplication can be indiscriminately grouped (associative principle of multiplication), ordered (commutative principle of multiplication), and partially combined. These fundamental principles remain unchanged with the introduction of literal factors.

(2) $(4y)(3x) = \square$.

■ $4 \cdot 3xy = 12xy$.

(3) $(2ab)(3bc)(3ac) = \square$.

■ $18a^2b^2c^2$

(4) $(-6mn)(-\frac{1}{2}n)(-m^2) = \boxed{}$.

■ $-3m^3n^2$

1. $2x(x-1) = \boxed{}$,
2. $3ab(a^2 - ab + b^2) = \boxed{}$,
3. $5mn\left(\frac{1}{5} + \frac{2}{m} - \frac{m}{n}\right) = \boxed{}$.

■ 1. $2x^2 - 2x$, 2. $3a^3b - 3a^2b^2 + 3ab^3$, 3. $mn + 10n - 5m^2$.

(5) $6(3 + 4)$ (is equal to, is not equal to) $6(3) + 6(4)$.

■ is equal to, because both have the value of 42.

(6) $a(b + c) = ab + ac$ for all real numbers a, b, c. Check the credibility of this statement by assuming $a = 5$, $b = -3$, and $c = 2$.

■ $a(b + c)$; $5(-3 + 2) = 5(-1) = -5$.
$ab + ac$; $5(-3) + 5(2) = -15 + 10 = -5$.

(7) $a(b + c) = ab + ac$ is a generalization that is called the *distributive principle.* Its application is so extensive in mathematics that it is worth identifying by name. $7(2x + 3) = \boxed{} + \boxed{}$, a result that comes from having used the ________________ principle.

■ $14x$; 21; distributive.

(8) 1. $2a(0.4x + 0.5) = \square + \square$,

2. $(-5)(-3a + 4b) = \square + \square$.

■ 1. $0.8ax + 1.0a$, 2. $15a + (-20)b$, or $15a - 20b$.

(9) $\left(\frac{x}{2} - 5\right)4x = \square$.

This is a small variation of the distributive principle. Notice that the result is not affected by a change in order and a reinterpretation of the subtraction; for example,

$$\begin{aligned}(x/2 - 5)4x &= 4x(x/2 + (-5)) \\ &= \square.\end{aligned}$$

■ $2x^2 - 20x$

(10) $3x(4x^2 + 2x - 5) = \square$.

■ This problem illustrates a greater generalization of the distributive principle.

$$\begin{aligned}3x(4x^2 + 2x - 5) &= 3x(4x^2 + (2x - 5)) \\ &= 12x^3 + 3x(2x - 5) \\ &= 12x^3 + 6x^2 - 15x.\end{aligned}$$

Observe the repeated application of the distributive principle. The two intermediate results could easily be omitted.

(11) $5ab(3a - 6 + b) = \square$.

■ $15a^2b - 30ab + 5ab^2$

1. $(2x - 1)(x + 4) = \boxed{}$,
2. $(x - 1)(2x^2 - x + 3) = \boxed{}$.

■ 1. $2x^2 + 7x - 4$, 2. $2x^3 - 3x^2 + 4x - 3$.

(12) Computing the product of $(3x - 2)(x + 5)$ requires a repeated application of the distributive principle. However, in practice, abbreviated methods that give identical results are often used. The following example illustrates the fundamental approach.

$$(3x - 2)(x + 5) = (3x - 2)x + (3x - 2)5.$$

Here $(3x - 2)$ is distributed over $x + 5$. To complete the computation, the distributive principle is again used:

$$(3x - 2)x + (3x - 2)5 = 3x^2 - 2x + 15x - 10.$$

Finally, the like terms $-2x$ and $15x$ can be combined, and we obtain

$$3x^2 - 2x + 15x - 10 = 3x^2 + 13x - 10.$$

Compute the product $(x + 3)(x - 8)$.

■ $$\begin{aligned}(x + 3)(x - 8) &= (x + 3)x - (x + 3)8 \\ &= x^2 + 3x - 8x - 24 \\ &= x^2 - 5x - 24.\end{aligned}$$

(13) Compute the product $(a + b)(c + d)$.

■ $$\begin{aligned}(a + b)(c + d) &= (a + b)c + (a + b)d \\ &= ac + bc + ad + bd.\end{aligned}$$

Notice in this very general problem that the resulting terms of the product represent every possible combination of products, where one factor is from $(a + b)$ and the other from $(c + d)$.

(14) See if the observation made in the previous frame applies to the computation of the product

$$(m + n)(p + q + r).$$

■ Using the distributive principle,

$$\begin{aligned}(m+n)(p+q+r) &= (m+n)p + (m+n)q + (m+n)r \\ &= mp + np + mq + nq + mr + nr.\end{aligned}$$

This demonstrates again that all possible combinations of products are represented in the completed calculation.

(15) Compute the product of the following and add like terms when they occur.

1. $(10x + 3)(x - 2) = \boxed{}$,
2. $(x - 3)^2 = \boxed{}$,
3. $(x - 1)(x^2 - 3x + 1) = \boxed{}$.

■ 1. $(10x + 3)(x - 2) = 10x^2 + 3x - 20x - 6$
$= 10x^2 - 17x - 6;$

2. $(x - 3)^2 = (x - 3)(x - 3) = x^2 - 3x - 3x + 9$
$= x^2 - 6x + 9;$

3. $(x - 1)(x^2 - 3x + 1) = x^3 - 3x^2 + x - x^2 + 3x - 1$
$= x^3 - 4x^2 + 4x - 1.$

Compute the products:

1. $(3x + 4)(3x - 4) = \boxed{}$,
2. $(\sqrt{2x} - 4)(\sqrt{2x} + 4) = \boxed{}$.

■ 1. $9x^2 - 16$, 2. $2x - 16$.

(16) The principles reviewed in the frames preceding the above problem have not changed. The special product of the sum and difference of the same two terms always results in the simplification illustrated below:

$$\begin{aligned}(x+7)(x-7) &= x^2 + 7x - 7x - 49\\ &= x^2 + (7-7)x - 49\\ &= x^2 - 49.\end{aligned}$$

Compute the product of

$$(2x-3)(2x+3) = \square.$$

■ $(2x-3)(2x+3) = 4x^2 - 6x + 6x - 9$
$\qquad\qquad\qquad\quad = 4x^2 - 9.$

(17) $(\sqrt{2}+\sqrt{3})(\sqrt{2}-\sqrt{3}) = \square.$

■ $(\sqrt{2}+\sqrt{3})(\sqrt{2}-\sqrt{3}) = 2 + \sqrt{6} - \sqrt{6} - 3$
$\qquad\qquad\qquad\qquad\quad = 2 - 3$
$\qquad\qquad\qquad\qquad\quad = -1.$

(18) Summary—Products

1. $a \cdot b = b \cdot \square$,
2. $a \cdot (b \cdot c) = (\square) \cdot c$,
3. $a(b+c) = \square + \square$,
4. $(a+b)(c+d) = \square + \square + \square + \square$,
5. $(a+b)(a-b) = \square - \square$.

■ 1. a Commutative principle,
2. $(a \cdot b)$ Associative principle,
3. ab, ac Distributive principle,
4. ac, bc, ad, bd,
5. a^2, b^2.

A-5 Fractions

1. $\dfrac{3a}{2} \cdot \dfrac{4a}{5} = \square$,

2. $2x^2y \cdot \dfrac{4}{xy} = \square$,

3. $\dfrac{5x}{3y} \cdot \dfrac{y}{x} = \square$.

■ 1. $\dfrac{6a^2}{5}$, 2. $8x$, 3. $5/3$.

(1) To find the product of two fractions such as $2/3 \cdot 5/7$, multiply the numerators $(2 \cdot 5)$ to get the numerator of the product and multiply the denominators $(3 \cdot 7)$ to get the denominator of the product.

$$\frac{2}{3} \cdot \frac{5}{7} = \square.$$

■ $\dfrac{2 \cdot 5}{3 \cdot 7} = \dfrac{10}{21}$.

(2) The product of fractions containing literal factors does not change the multiplication procedure. Find the product of

$$\frac{2a}{3b} \cdot \frac{5a}{7b} = \square.$$

■ $\dfrac{10a^2}{21b^2}$

(3) Fractions often can be reduced to lower terms. Consider this possibility in computing the product of

$$\frac{4x}{5y} \cdot \frac{3y^2}{6x^2} = \boxed{}.$$

■ $\dfrac{12xy^2}{30x^2y} = \dfrac{2y}{5x}$.

(4) $\left(\dfrac{0.4ab^2}{0.5c^2}\right) \cdot (10ac^2) = \boxed{}.$

■ Note that $10ac^2 = \dfrac{10ac^2}{1}$, so that

$$\frac{0.4ab^2}{0.5c^2} \cdot \frac{10ac^2}{1} = \frac{4.0a^2b^2c^2}{0.5c^2} = 8.0a^2b^2.$$

Perform the indicated multiplication:

1. $(6ab)\left(\dfrac{1}{2a} + \dfrac{4a}{3b}\right) = \boxed{},$

2. $(3a + 3b)\left(\dfrac{2}{(a+b)} + \dfrac{(a-b)}{3}\right) = \boxed{}.$

■ 1. $3b + 8a^2$, 2. $6 + a^2 - b^2$.

---

(5) The law of distribution, $a(b + c) = ab + ac$, applies to the problems of the previous frame.

$$(10)\left(\frac{a}{2} + \frac{b}{5}\right) = \boxed{}.$$

■ $\dfrac{10a}{2} + \dfrac{10b}{5} = 5a + 2b$.

(6) $6mn\left(\frac{5}{2mn} - \frac{2m}{3}\right) = \boxed{}$.

■ $$6mn\left(\frac{5}{2mn} - \frac{2m}{3}\right) = \frac{30mn}{2mn} - \frac{12m^2n}{3}$$
$$= 15 - 4m^2n.$$

(7) $(a^2 - b^2)\left(\frac{4}{a+b} + \frac{3}{a-b}\right) = \boxed{}$.

■ $$(a^2 - b^2)\left(\frac{4}{a+b} + \frac{3}{a-b}\right) = \frac{4(a^2 - b^2)}{a+b} + \frac{3(a^2 - b^2)}{a-b}$$
$$= \frac{4(a+b)(a-b)}{(a+b)} + \frac{3(a+b)(a-b)}{(a-b)}$$
$$= 4(a-b) + 3(a+b)$$
$$= 4a - 4b + 3a + 3b$$
$$= 7a - b.$$

(8) Summary—Fractions

1. $\frac{a}{b} \cdot \frac{c}{d} = \boxed{}$,

2. $\frac{ax}{ay} = \boxed{}$,

3. $ab\left[\frac{x}{a} + \frac{y}{b}\right] = \boxed{}$.

■ 1. $\frac{ac}{bd}$, 2. $\frac{x}{y}$, 3. $bx + ay$.

A-6 Combining Like Terms

Simplify:

1. $2x + 7x - 3x = \square$,
2. $0.5ax^2 + 0.8ax^2 - 0.2 = \square$,
3. $4(m + n) - \dfrac{3(m + n)}{2} = \square$.

■ 1. $6x$, 2. $1.3ax^2 - 0.2$, 3. $\dfrac{5(m + n)}{2}$.

(1) $4x + 3x = (4 + 3)x = 7x$.
Test the above illustration by letting $x = 2$.

■
$$\begin{aligned} 4x + 3x &= 7x \\ 4(2) + 3(2) &= 7(2) \\ 8 + 6 &= 14 \\ 14 &= 14 \end{aligned}$$

(2) The expression $4x + 3x$ has two terms, $4x$ and $3x$. Since x is a common factor to both terms, $4x + 3x$ may be written as $(4 + 3)x$.
Simplify $1.2y^2 + .7y^2$.

■ $1.2y^2 + .7y^2 = (1.2 + .7)y^2 = 1.9y^2$.

(3) Whenever all the literal factors of two or more terms are identical, the terms may be combined into a single term.

$3m^2n - 9m^2n + 5m^2n = \square$.

■ $(3 - 9 + 5)m^2n = -m^2n$.

(4) Simplify by combining like terms:

$$2ax^2 + 7a - 12ax^2.$$

■ $2ax^2 + 7a - 12ax^2 = (2 - 12)ax^2 + 7a$
$= -10ax^2 + 7a.$

Notice that only two of the terms have identical literal factors.

(5) Simplify $6(m + n) - 4(m + n)$.

■ $6(m + n) - 4(m + n) = (6 - 4)(m + n)$
$= 2(m + n)$ or $2m + 2n$.

Simplify:

1. $5(x + 1) - 4(x + 1) + 7x - 2$,
2. $4(m^2 + m - 1) + 3(m^2 + 4) - m^2$.

■ 1. $8x - 1$, 2. $6m^2 + 4m + 8$.

(6) Frequently parentheses must be removed before like terms are combined. This is done by performing the indicated multiplication. For example,

$$\begin{aligned} 5(x^2 - 2) + 3x^2 &= 5x^2 - 10 + 3x^2 \\ &= (5 + 3)x^2 - 10 \\ &= 8x^2 - 10. \end{aligned}$$

Simplify $4y^2 + 3(y^2 - 2)$.

■ $4y^2 + 3y^2 - 6 = 7y^2 - 6.$

(7) Simplify:

1. $4m + 3(m + n) - n$,
2. $10xy - 4x(y - 2) + 3x$,
3. $0.5z(1.2z - 4) - .3(z^2 - .1) + 2z$.

■ 1. $7m + 2n$, 2. $6xy + 11x$, 3. $0.3z^2 + .03$

(8) Summary—Combining Like Terms

State in your own words the procedures that have been used to combine like terms.

■ 1. Remove parentheses by performing the indicated operations of multiplication.
2. Terms with identical literal factors may be combined by adding their numerical coefficients.

A-7 First-Degree Equations in One Unknown

Find the solution set for each of the following equations.

1. $x + 1.6 = 9.3$,
2. $23 + x = 18$,
3. $x - .05 = .34$.

■ 1. $\{7.7\}$, 2. $\{-5\}$, 3. $\{.39\}$.

(1) Whenever the solution set to an equation is obvious, no more mathematical manipulation is necessary. However, the solution to an equation may not be obvious until the equation is changed to the form $x =$ (a number).

The equation $x + 9.1 = 11.2$ is equivalent to the equation $x = 2.1$ because, if the x is replaced by [ ], both equations will be true.

■ 2.1

(2) $x + 9.1 = 11.2$ can be changed to the basic equation $x = 2.1$ by adding to both sides of the equation the value (-9.1).

$$x + 9.1 + (-9.1) = 11.2 + (-9.1).$$

The result on the left side of the equation is [ ] and on the right side is [ ].

■ x;   2.1.

(3) All equations of the form $x + b = a$, where x represents the only unknown quantity, may be solved in the manner just demonstrated. Solve the following equations:

1. $x + 5 = 12$,   2. $x + \frac{3}{4} = 1$,   3. $x - 7 = 16$.

■ 1. $x + 5 = 12$

$x + 5 + (-5) = 12 + (-5)$

$x = 7$ Solution $\{7\}$

■ 2. $x + \frac{3}{4} = 1$

$x + \frac{3}{4} + \left(-\frac{3}{4}\right) = 1 + \left(-\frac{3}{4}\right)$

$x = \frac{1}{4}$ Solution $\left\{\frac{1}{4}\right\}$

■ 3. $x - 7 = 16$

$x - 7 + 7 = 16 + 7$

$x = 23$ Solution $\{23\}$

Find the solution to each of the following equations:

1. $2.3x = 9.2$,

2. $\frac{2}{3}x = 5$,

3. $-.4x = -.12$.

■ 1. $\{4\}$, 2. $\left\{\frac{15}{2}\right\}$, 3. $\{.3\}$.

(4) The most elementary equivalent equation has the form $x =$ (a number). The equation $1.2x = 4.2$ is equivalent to the equation $x = 3.5$, because both have the solution {☐}.

■ $\{3.5\}$

(5) $1.2x = 4.2$ can be changed to the basic equation $x = 3.5$ by dividing both sides of the equation by 1.2:

$$\frac{1.2}{1.2}x = \frac{4.2}{1.2}.$$

The result on the left side of the equation is ☐ and on the right side is ☐.

■ x; 3.5.

(6) All the equations of the form $ax = b$, where x is the unknown, may be solved by dividing both sides of the equation by the quantity a. The result will be a basic equivalent equation.
Solve the following equations:

1. $5x = 2$, 2. $1.3x = 39$, 3. $2x - 5x = 12$.

■ 1. $5x = 2$

$\frac{5x}{5} = \frac{2}{5}$ Solution $\left\{\frac{2}{5}\right\}$

■ 2. $1.3x = 39$

$\frac{1.3x}{1.3} = \frac{39}{1.3}$

$x = 30$ Solution $\{30\}$

■ 3. $2x - 5x = 12$

$-3x = 12$

$\frac{-3x}{-3} = \frac{12}{-3}$

$x = -4$ Solution $\{-4\}$

Find the solution for each of the following equations:

1. $2x + 7 = x - 8$,
2. $w + 1 + 3w - 4 = 7w - 5$,
3. $1.3t - 4.3t + 2.1 = 2.2t$.

■ 1. $\{-15\}$, 2. $\left\{\frac{2}{3}\right\}$, 3. $\left\{\frac{21}{52}\right\}$.

(7) Finding the solution to the equation $5x + 6 = x - 8$ is usually accomplished by reducing it to the basic equivalent equation of the form $x =$ (a number). We use a combination of the principles of adding (or subtracting) and multiplying (or dividing) both sides of the equation by equal quantities to accomplish this objective. Ordinarily the addition principle must be used first. Add (-6) to both sides of the equation. What is the result?

■ $5x + 6 = x - 8$

$5x + 6 - 6 = x - 8 - 6$

$5x = x - 14.$

(8) Next add $(-x)$ to both sides of the equation. What is the result?

■ $$5x = x - 14$$
$$5x - x = x - x - 14$$
$$4x = -14.$$

(9) The final step is to divide both sides of the equation by [   ]. What is the result?

■ 4

$$\frac{4x}{4} = -\frac{14}{4}$$

$$x = -\frac{7}{2}$$

Solution $\left\{-\frac{7}{2}\right\}$.

(10) Find the solution to the equation

$$12x - 7 = 4 - 3x.$$

■ $12x - 7 = 4 - 3x$ — Add 7 to both sides.

$12x = 11 - 3x$ — Add $3x$ to both sides.

$15x = 11$ — Divide both sides by 15.

$x = \frac{11}{15}$ — Solution $\left\{\frac{11}{15}\right\}$.

(11) Find the solution to the equation

$$14t - 3 + 6t = t + 2t - 6.$$

■ $14t - 3 + 6t = t + 2t - 6$ — First combine like terms on each side of the equation.

$20t - 3 = 3t - 6$ — Add $(-3t)$ to both sides.

$17t - 3 = -6$ — Add 3 to both sides.

$17t = -3$ — Divide both sides by 17.

$t = -\frac{3}{17}$ — Solution $\left\{-\frac{3}{17}\right\}$.

Find the solutions to the following equations.

1. $2(3x + x) = 28$,
2. $3(x - 4) + 7 = (2x + 1) - (5x + 3)$.

■ 1. $\left\{\frac{7}{2}\right\}$, 2. $\left\{\frac{1}{2}\right\}$.

(12) When parentheses appear, they must ordinarily be removed before we can collect like terms. They may be removed by performing the indicated multiplication and reinterpreting the subtraction of expressions in parentheses. Remove the parentheses from the following equations.

1. $3(5x - 4) = 7(2x + 1)$,
2. $4x - 2(x - 1) = 3(x + 2)$.

■ 1. $15x - 12 = 14x + 7$, 2. $4x - 2x + 2 = 3x + 6$.

(13) Solve the following equations. The first step should be to remove the parentheses.

1. $2x - 6(x - 5) = 15$,
2. $2(4x - 8) - (2x + 7) = 85 - 3x$.

■ 1.
$$\begin{aligned} 2x - 6(x - 5) &= 15 \\ 2x - 6x + 30 &= 15 \\ -4x + 30 &= 15 \\ -4x &= -15 \\ x &= \frac{15}{4} \end{aligned}$$
Solution $\left\{\frac{15}{4}\right\}$

■ 2.
$$\begin{aligned} 2(4x - 8) - (2x + 7) &= 85 - 3x \\ 8x - 16 - 2x - 7 &= 85 - 3x \\ 6x - 23 &= 85 - 3x \\ 9x - 23 &= 85 \\ 9x &= 108 \\ x &= 12 \end{aligned}$$
Solution $\{12\}$.

Solve the following equations.

1. $\frac{x}{3} + \frac{1}{2} = 6$,

2. $\frac{4x - 2}{3} + 7 = 2x - \frac{1}{4}$,

3. $2x - \frac{3(x + 1)}{4} = 7 - \frac{2x}{5}$,

4. $\frac{2}{3x - 1} = \frac{5}{x + 2}$.

■ 1. $\left\{\frac{33}{2}\right\}$, 2. $\left\{\frac{79}{8}\right\}$, 3. $\left\{\frac{155}{33}\right\}$, 4. $\left\{\frac{9}{13}\right\}$.

(14) The work of solving equations with fractional expressions can be greatly simplified by first multiplying both sides of the equation by the least common denominator of all the denominators appearing in the equation. For example the first step in solving

$$\frac{2x}{3} + 3 = \frac{1}{6}$$

is to multiply both sides of the equation by 6. What is the resulting equivalent equation?

■ $6\left(\frac{2x}{3} + 3\right) = \left(\frac{1}{6}\right)6$

$4x + 18 = 1.$

(15) The result of having multiplied both sides of a fractional equation by the ________ ________ ________ is an equivalent equation with no fractions.

■ least common denominator

(16) What is the least common denominator of the fractions in the equation

$$\frac{5x}{2} + \frac{3}{4} = 10 - \frac{x}{2}?$$

■ 4

(17) Solve the equation $\frac{5x}{2} + \frac{3}{4} = 10 - \frac{x}{2}$.

■ $\frac{5x}{2} + \frac{3}{4} = 10 - \frac{x}{2}$

$10x + 3 = 40 - 2x$

$12x = 37$

$x = \frac{37}{12}$. Solution $\left\{\frac{37}{12}\right\}$

(18) What is the least common denominator of the fractions in the equation

$$\frac{3x + 2}{x - 1} = \frac{2}{3}?$$

■ $3(x - 1)$

(19) Multiply both sides of the equation by $3(x - 1)$. The result should be an equation with no fractional expressions.

■ $3(x - 1)\left(\frac{3x + 2}{x - 1}\right) = \left(\frac{2}{3}\right) 3(x - 1)$

$9x + 6 = 2x - 2$.

(20) Complete the solution of the preceding equation.

■ $9x + 6 = 2x - 2$

$7x = -8$

$x = -\frac{7}{8}$. Solution $\left\{-\frac{7}{8}\right\}$

(21) Solve the following equations.

1. $\frac{(x-2)}{4} - 1 = 2x + \frac{1}{4}$, 2. $\frac{12}{x+2} = \frac{3}{x}$.

■ 1. $\frac{(x-2)}{4} - 1 = 2x + \frac{1}{4}$

$x - 2 - 4 = 8x + 1$

$x - 6 = 8x + 1$

$-7x = 7$

$x = -1$. Solution $\{-1\}$

■ 2. $\frac{12}{x+2} = \frac{3}{x}$

$12x = 3x + 6$

$9x = 6$

$x = \frac{2}{3}$. Solution $\left\{\frac{2}{3}\right\}$

(22) Summary—First-Degree Equations in One Unknown
To solve the preceding two equations, we must apply all the techniques reviewed in this section. Describe the techniques that result in reducing

$$\frac{x-2}{4} - 1 = 2x + \frac{1}{4}$$

to the equivalent equation $x = -1$.

■ 1. Multiply both sides of the equation by the least common denominator to eliminate fractions.
2. Add equal values to both sides of the equation so that all like terms are combined.
3. Divide both sides of the equation by equal values so that the resulting equation is in the basic form $x =$ (a number).

A-8 Completing the Square

1. $x^2 + 4x + \square = (x + \square)^2$,
2. $9x^2 + 12x + \square = (3x + \square)^2$,
3. $x^2 - 16x + \square = (x - \square)^2$.

■ 1. 4, 2; 2. 4, 2; 3. 64, 8.

(1) Calculate the square of $(a + b)$.

■ $(a + b)^2 = a^2 + ab + ab + b^2$
$= a^2 + 2ab + b^2$.

(2) Compare the following problems to the general results of the previous frame. Supply the missing values.

1. $(x + 2)^2 = x^2 + \square + 4$,
2. $(3x - 1)^2 = 9x^2 - \square + 1$.

■ 1. $4x$, 2. $6x$.

(3) Again compare the general results, $(a + b)^2 = a^2 + 2ab + b^2$, to the following problems and supply the missing values.

1. $(x - 3)^2 = x^2 - 6x + \square$,
2. $(5x + 2)^2 = 25x^2 + 20x + \square$.

■ 1. 9, 2. 4.

(4) If $x^2 - 6x + \square$ is the result of the square of a binomial, the final term must be the square of $\square$.

■ 9; -3.

(5) If the numerical coefficient of x^2 is 1, the last term will be the $\square$ of one-half the numerical coefficient of $\square$.

■ square; x.

(6) Supply the final term if the following are the squares of binomials:

1. $x^2 - 8x + \square$,
2. $w^2 + 10w + \square$,
3. $t^2 + 7t + \square$,
4. $s^2 + 0.1s + \square$.

■ 1. 16, 2. 25, 3. $\frac{49}{4}$, 4. .0025.

(7) The coefficient of x^2 in the example $25x^2 + 20x + 4$ is not 1. The manner in which the 4 is discovered in order to complete the square requires:

1. taking one-half of 20, which is $\square$,
2. dividing the result of Step 1 by the square root of the coefficient of $\square$,
3. squaring the result of Step 2, giving the value $\square$.

■ 1. 10, 2. x^2, 3. 4.

(8) Apply a procedure similar to that used in the preceding frame for discovering the final constant for $9x^2 - 30x + \square$, if it is the result of squaring a binomial.

■ $\left(\frac{-30}{2} \div \sqrt{9}\right)^2 = 25.$

(9) Supply the final term if the following are squares of binomials:

1. $16x^2 + 6x + \square$,
2. $4w^2 - w + \square$,
3. $\frac{s^2}{9} + 5s + \square$.

■ 1. $\frac{9}{16}$, 2. $\frac{1}{16}$, 3. $\frac{225}{4}$.

(10) Summary—Completing the Square

$$a^2x^2 + 2abx + \square = (ax + \square)^2$$

■ b^2; b.

A-9 Inequalities

True or false:

1. $6 < 8$,
2. $7 < -5$,
3. $-8 > 2$,
4. $-9 > -13$,
5. $3.8 \leqq -8.3$,
6. $2(4) \leqq 8$,
7. $9 \geqq -12$,
8. $a \geqq a$.

■ 1. True, 2. False, 3. False, 4. True, 5. False, 6. True, 7. True, 8. True.

(1) For all numbers a and b, a is either less than, equal to, or greater than b. If $a = 7$ and $b = 12$, then a is ____________ b.

■ less than

(2) Statements like "7 is less than 12," or "x is greater than 100" are called inequalities. The statement "y is greater than $2x+5$" is an ____________.

■ inequality

(3) The fundamental symbols used to denote the inequality relationships are "$>$" for greater than and "$<$" for less than. Use the appropriate symbol between the following expressions:

1. $12 \boxed{R} 4$, 2. $9 \boxed{R} 15$, 3. $-9 \boxed{R} -15$,

4. $\frac{1}{2} \boxed{R} \frac{3}{4}$, 5. $-2 \boxed{R} -\frac{4}{3}$.

■ 1. $>$, 2. $<$, 3. $>$, 4, $<$, 5. $<$.

(4) The symbols $<$ and $>$ are combined with the equal sign ($\leqq$ and $\geqq$) to add to the ways of expressing relationships. (The combined relations are usually abbreviated to $\leq$ and $\geq$.)

$\leq$ less than *or* equal to
$\geq$ ____________ than *or* equal to.

■ greater

(5) True or false:

1. $0.75 < 0.075$, 2. $0.33 > 0.3\bar{3}$, 3. $1.0 \leq 0.\bar{9}$,

4. $\frac{2}{3} \leq \frac{14}{21}$, 5. $-5 < -3$, 6. $5 > 3$.

■ 1. False, 2. False, 3. True, 4. True, 5. True, 6. True.

(You may be confused by the third problem. The bar over the 9 means that 9 is repeated an infinite number of times. One is certainly not less than $.\bar{9}$, but it is equal to $.\bar{9}$.)

Graph the solution set for each inequality:

1. $x < 5$, 2. $x \geq 2$, 3. $x + 4 > 9$, 4. $-5 \leq x < -2$.

■ 1. 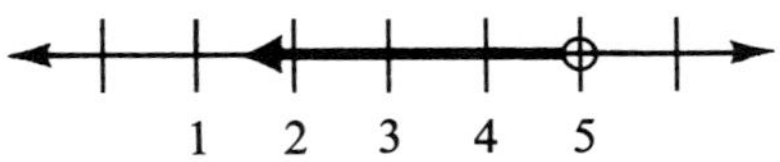

■ 2. 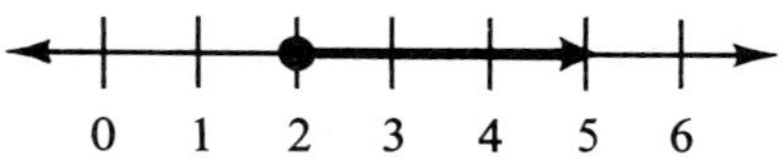

■ 3. 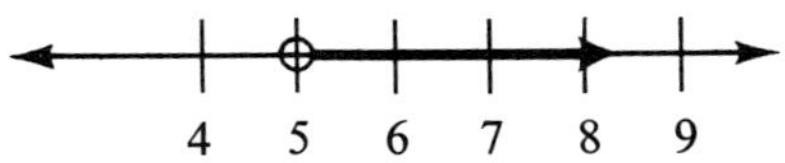

■ 4. 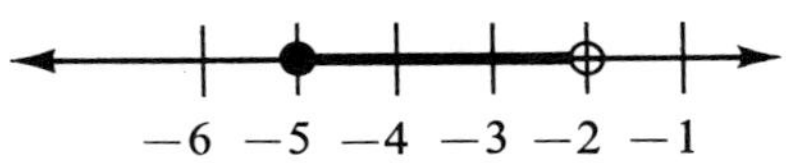

(6) For all a, b, and c, if $a = b$, then $a + c = b + c$. An analogous property exists for inequalities. For all a, b, and c, if $a < b$, then $a + c < b + c$, or if $a > b$, then ☐.

■ $a + c > b + c$.

(7) 1. $6 < 9$ 2. $-4 > -7$

$(6 + 2)$ R $(9 + 2)$, $(-4) + (-10)$ R $(-7) + (-10)$.

■ 1. $<$, 2. $>$.

(8) 1. $5 + 9 > 6$

$5 + 9 - 9 \ \boxed{R}\ 6 - 9,$

2. $3 - 8 < 12$

$3 - 8 + 8 \ \boxed{R}\ 12 + 8.$

■ 1. $>$ $(5 > 6 - 9)$, 2. $<$ $(3 < 12 + 8)$.

(9) 1. $x - 8 < 4$

$x \ \boxed{R}\ 4 + 8,$

2. $p + 5 > 3$

$p \ \boxed{R}\ 3 - 5.$

■ 1. $<$, 2. $>$.

(10) 1. $x - 7 \leq 5$

$x \leq \boxed{},$

2. $p + \dfrac{1}{2} \geq \dfrac{3}{4}$

$\geq \boxed{}.$

■ 1. 12, 2. $\dfrac{1}{4}$.

(11) Basic conditional inequalities like $x < 12$, $p > -2$, $x \leq 12$, and $p \geq 1/4$ may be graphed on the number line.

■

$x < 12$

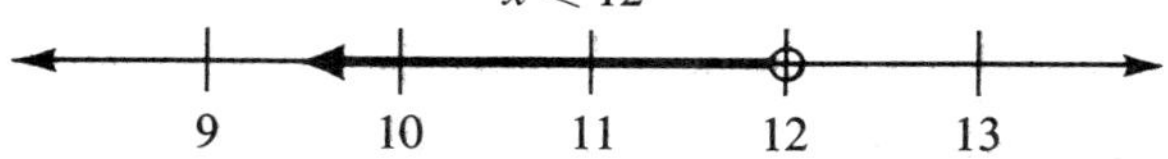

$p > -2$

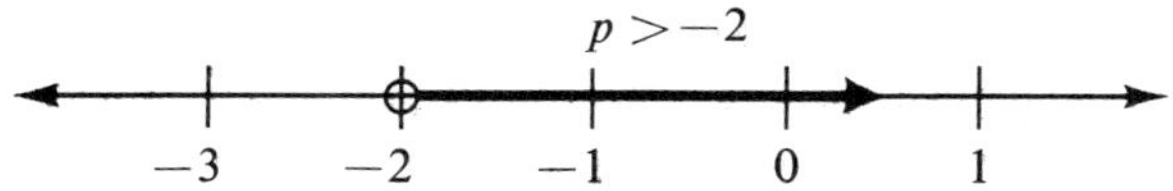

$x \leq 12$

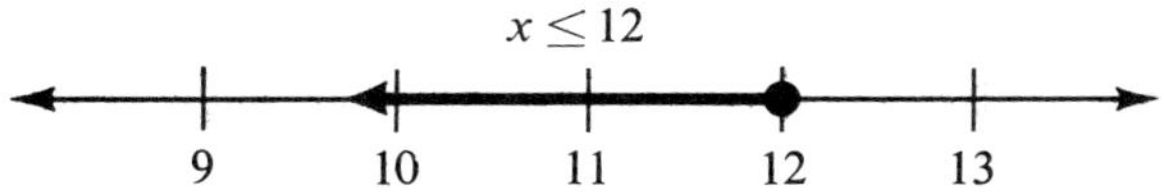

Draw a graph of $p \geq 1/4$.

■ 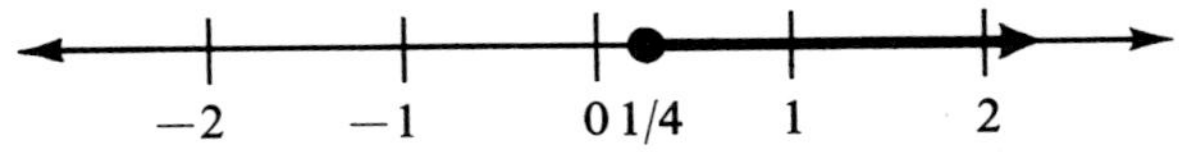

(12) The inequality "5 is less than x and x is less than 10" can be abbreviated $5 < x < 10$. The graph of this inequality is a segment of the number line.

■ 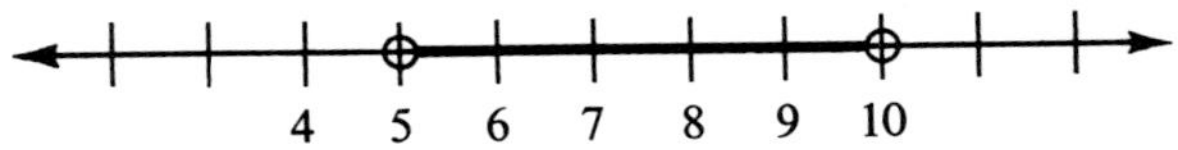

Graph the inequality $-4 < x \leq -2$.

■ 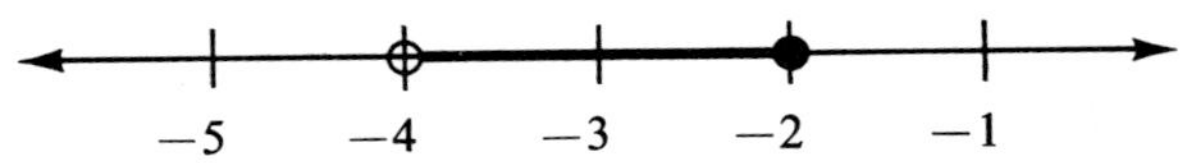

Graph the solution set for each of the inequalities.

1. $2x < 8$, 2. $3x - 7x > 20$.

■ 1. 

2. 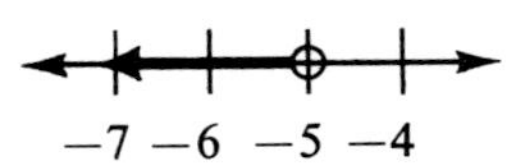

---

(13) For all a, b, and c, if $a = b$ then $ac = bc$. If the relationship is one of $<$ or $>$, rather than $=$, this result is not necessarily true. Complete the following problems and carefully review the results.

1. $3 < 5$ 2. $4 > (-2)$

(2)(3) [R] (2)(5), (3)(4) [R] (3)(−2),

(−2)(3) [R] (−2)(5), (−3)(4) [R] (−3)(−2).

■ 1. $<, >$, 2. $>, <$.

(14) If $a < b$, and $c > 0$, then ac ☐ bc.

■ $<$

(15) If $a < b$, and $c < 0$, then ac ☐ bc.

■ $>$

(16) Multiplying the right and left members of an inequality by a factor less than zero changes the order of the inequality. If $a > b$, then $ac < bc$ if and only if ________________.

■ c is less than zero

(17) Graph the solution set for the inequality $-3x \leq 12$.

$$-3x \leq 12$$
$$x \geq -4.$$

■ 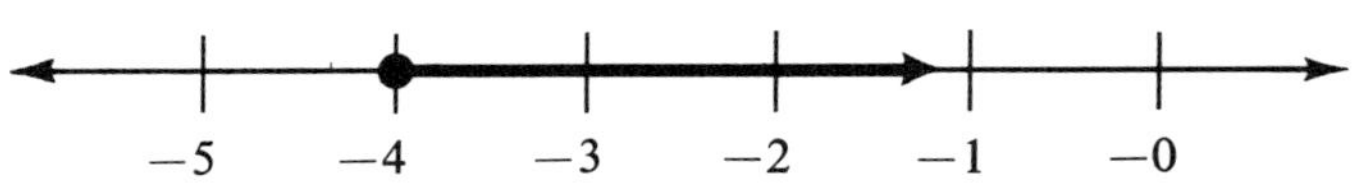

(18) Graph the solution set for the inequality $3x - 5 \leq 7x + 3$.

■ $3x - 5 \leq 7x + 3$

$3x - 7x \leq 3 + 5$

$-4x \leq 8$

$x \geq -2.$

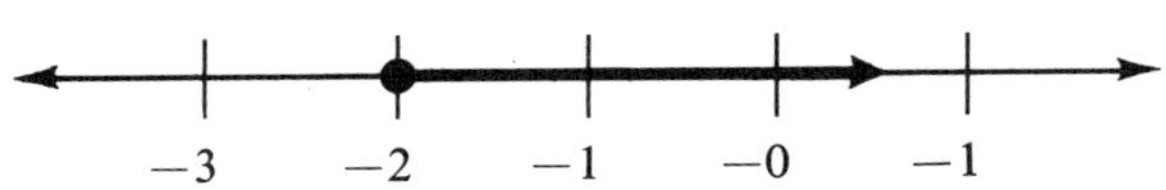

Summary—Inequalities

1. For all a, b, and c, if $a \boxed{R} b$, then $a + c \boxed{R} b + c$.

2. If $a \boxed{R} b$, and $c > 0$, then $ac \boxed{R} bc$.

3. If $a \boxed{R} b$, and $c < 0$, then $ac \boxed{R} bc$.

■ In (1) and (2) both inequality relationships are in the same order. In (3) the order in the inequality relationship must be reversed.

Chapter B Review of Basic Facts in Plane Geometry (Programmed)

A large number of technical problems that are geometric in nature can be analyzed by reducing them to drawings in a plane. Technicians should recognize fundamental geometric relationships so that certain desired results can be measured or calculated.

The purpose of this chapter is to provide a *review* of the more frequently used geometric facts in technical problems. The reader is asked to recall some of the fundamental vocabulary and facts of plane geometry. No attempt is made to reconstruct a body of theorems and their proofs, but some attention is given to establishing the reasonableness of certain results.

A programming technique is used again in this chapter. We reemphasize the value of using the programmed material in the correct manner. Always attempt to *write* conclusions or answers before viewing the answers given at the end of the frame.

B-1 Lines, Angles, and Similar Polygons

If two parallel straight lines are cut by a transversal, (1) alternate interior angles are ________, and (2) corresponding angles are ________.

■ 1. equal, 2. equal.

(1)

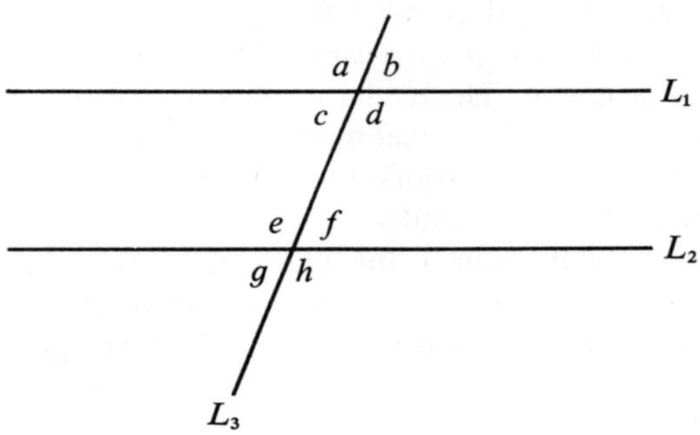

Fig. B-1a

In Fig. B-1a, L_1 and L_2 are parallel. A transversal is a straight line that intersects both L_1 and L_2. In Fig. B-1a, L_3 is a ____________.

■ transversal

(2) In Fig. B-1a, four of the angles designated by lowercase letters are interior angles and the remaining four are exterior angles. If *a*, *b*, *g*, and *h* are exterior angles, angles _____, _____, _____, and _____ are ______________ angles.

■ *c*, *d*, *e*, *f*; interior.

(3) In Fig. B-1a, the pair of angles a and h are referred to as alternate exterior angles. The pair of angles c and f are alternate interior angles. The pair of angles a and e are corresponding angles. All pairs of *equal* angles in Fig. B-1a may be classified as one of these three types. Complete the classification.

■

Alternate exterior	Alternate interior	Corresponding
a, h	c, f	a, e
b, g	d, e	c, g
		b, f
		d, h

1. Two angles whose sides are parallel, right side to right side and left side to left side, are ________.
2. Two angles whose sides are parallel, left side to right side and right side to left side, are ________.

■ 1. equal, 2. supplementary.

(4) Both Figs. B-1b and B-1c show pairs of angles whose respective right sides (R) are parallel and left sides (L) are parallel. Intuition alone should tell you the angles of each pair are ________.

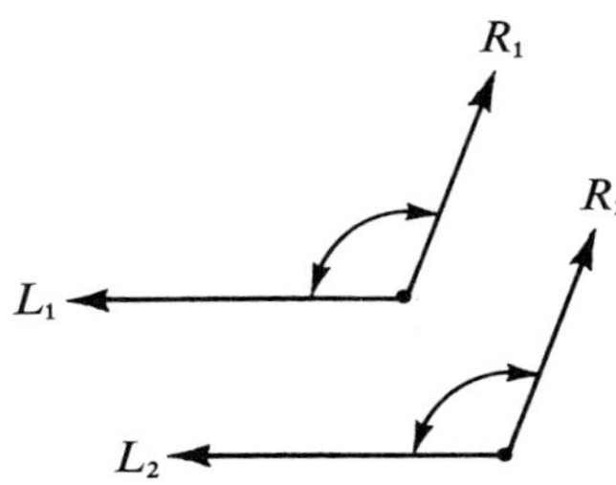

Fig. B-1b

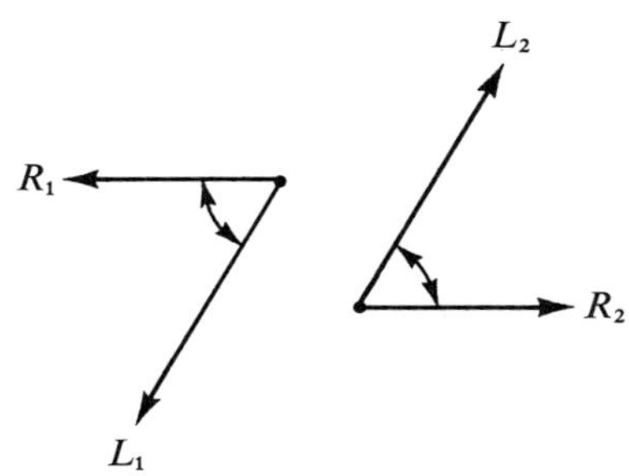

Fig. B-1c

■ equal

(5) Draw a pair of angles so that the left side of one is parallel to the right side of the other.

■ See Fig. B-1d and Fig. B-1e.

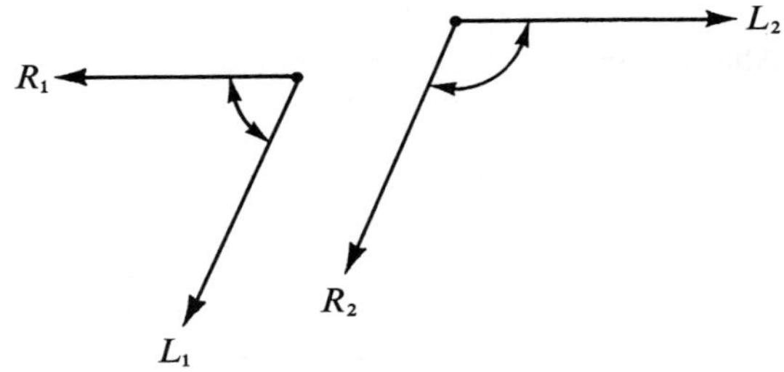

Fig. B-1d

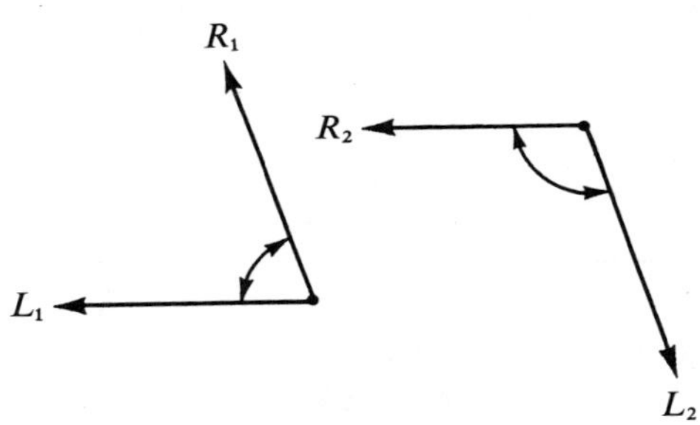

Fig. B-1e

(6) Refer to Fig. B-1d and Fig. B-1e and observe that in each case the pair of angles together form a straight angle. Two angles that together form a _________ _________ are called supplementary angles.

■ straight angle

(7) A straight angle has a measure of ☐ degrees.

■ 180

(8) If angle $A = 72°$ and is the supplement of angle B, what is the measure of angle B?

■ 108° (this is the result of 180° − 72°)

The sum of the measure of the angles of a triangle is equal to ☐ degrees.

■ 180

(9) Consider triangle ABC in Fig. B-1f. Through the vertex C draw L parallel to $\overline{AB}$. Angles b and b' are equal alternate interior angles. Angles a and a' are also equal alternate interior angles. Therefore,

$$\angle a' + \angle c + \angle b' = \angle a + \angle c + \angle b = \boxed{}.$$

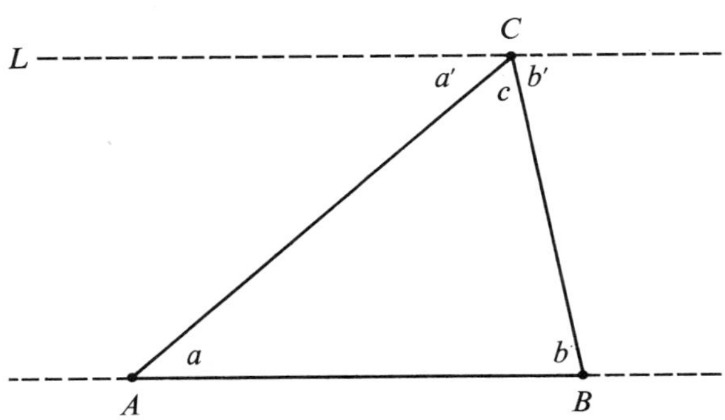

Fig. B-1f

■ 180°

The sum of the interior angles of a polygon of n sides is [ ].

■ $(n-2)180°$

(10) The sum of the measure of the angles of a triangle is [ ].

■ 180°

(11) Consider a quadrilateral. One diagonal of a quadrilateral divides the figure into two triangles. All the angles of both triangles comprise the interior angles of the quadrilateral. Therefore the sum of the angles of the quadrilateral is [ ].

■ 2(180°), or 360°

(12) The technique of dividing a polygon into triangles by drawing all possible diagonals from only one vertex will lead to the solution of the number of degrees in the sum of the interior angles. Complete Table B-1.

Table B–1

Number of sides (n)	Number of triangles	Sum of interior angles
3	1	180°
4	2	2(180°)
5	[ ]	[ ]
6	[ ]	[ ]
⋮	⋮	⋮
10	[ ]	[ ]

■

3	3(180°)
4	4(180°)
⋮	⋮
8	8(180°)

(13) By examining the pattern of the table, a polygon of n sides would be divided into [ ] triangles, and the sum of the measure of the interior angles would be [ ].

■ $n-2$; $(n-2)180°$.

(14) How many degrees are in the sum of the measure of the interior angles of a 20-sided polygon?

■ $(20-2)180° = 3240°$.

Two angles whose sides are mutually perpendicular are either ________ or ____________.

■ equal; supplementary.

(15)

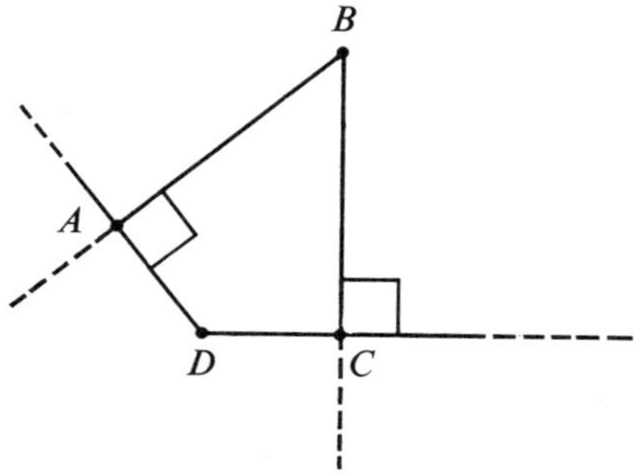

Fig. B-1g

Figure B-1g illustrates $\angle ABC$ and $\angle ADC$ whose sides are mutually ____________.

■ perpendicular

(16) Since the sum of the angles of quadrilateral $ABCD$ is 360°, the sum of $\angle ABC$ and $\angle ADC$ must equal 360° – 2(90°). Two angles whose sum is 180° are ____________ angles.

■ supplementary

(17) Draw two angles such that their corresponding sides, right side to right side and left side to left side, are mutually perpendicular.

■ See Fig. B-1h.

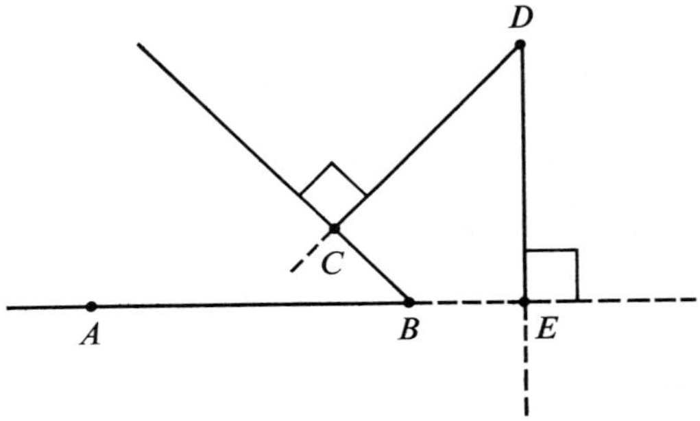

Fig. B-1h

(18) In Fig. B-1h, angles ABC and CDE are equal because they both are supplements of the angle [ ].

■ *CBE*

Geometric figures whose (1) corresponding dimensions are ____________ and whose (2) corresponding angles are ____________ are similar.

■ 1. proportional, 2. equal.

(19) Assume the two triangles of Fig. B-1i are similar. This means that the ratios of corresponding dimensions are equal. Therefore,

$$\frac{\overline{AB}}{\overline{A'B'}} = \frac{\overline{BC}}{\overline{B'C'}} = \frac{\square}{\overline{A'C'}}.$$

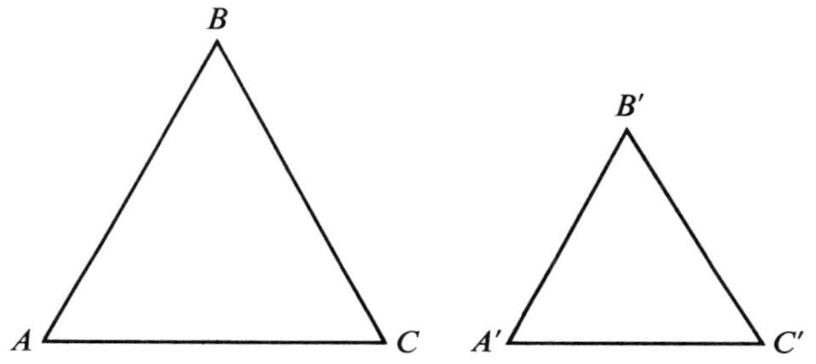

Fig. B-1i

■ $\overline{AC}$

(20) Assume again that the triangles of Fig. B-1i are similar and that $\overline{AB} = 6$, $\overline{A'B'} = 9$, and $\overline{BC} = 8$. Calculate the length of $\overline{B'C'}$.

■ $\frac{6}{9} = \frac{8}{\overline{B'C'}}$

$$\overline{B'C'} = \frac{9(8)}{6} = 12.$$

(21) The ratio of corresponding sides of the two similar polygons in Fig. B-1j is 4/5. If the perimeter of the larger polygon is 28, what is the perimeter of the smaller?

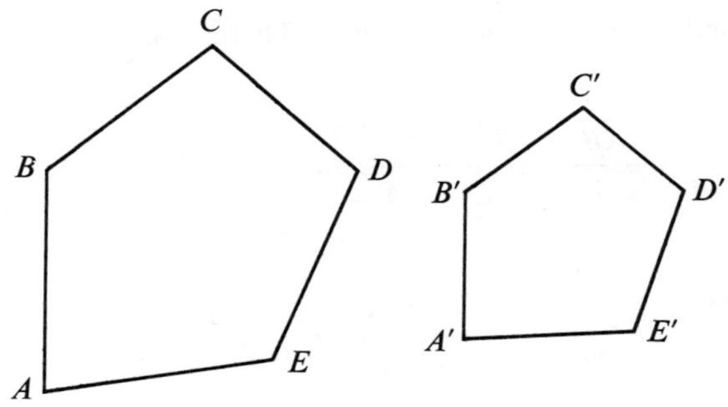

Fig. B-1j

■ $\frac{4}{5}=\frac{p}{28}$; $p=\frac{4 \cdot 28}{5}=22\frac{2}{5}$.

(22) Triangles have the unique characteristic among polygons of being similar if any [ ] corresponding sides are in the same proportion.

■ 2

(23) All ______________ are similar.

■ circles, or regular polygons with equal numbers of sides

(24) If two plane geometric figures are similar and the ratio of their corresponding sides is one, then the figures are ______________.

■ congruent

B-2 Areas

1. The area of a parallelogram is calculated by multiplying ______.

2. The area of a triangle is calculated by multiplying ______.

■ 1. (base) × (altitude), 2. $\frac{1}{2}$ (base) × (altitude).

(1)

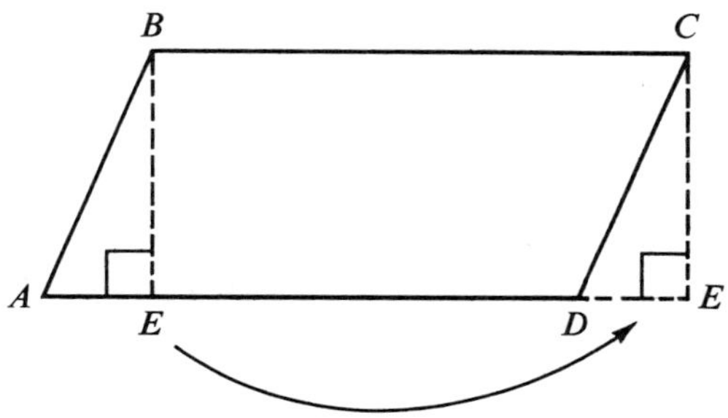

Fig. B-2a

If side $\overline{AD}$ is considered the base and $\overline{BE}$ the altitude, Fig. B-2a should remind you that the area of the parallelogram $ABCD$ may be calculated by finding the product $(\overline{AD})(\overline{BE})$.

Let $\overline{AD} = 12$ and $\overline{BE} = 5$. Find the area of the parallelogram $ABCD$.

■ (Area) = 12(5) = 60.

(2)

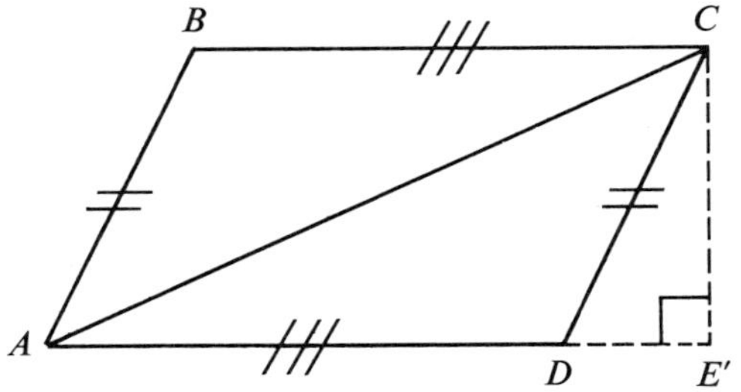

Fig. B-2b

If $ABCD$ is a parallelogram, the diagonal $\overline{AC}$ divides the parallelogram into two congruent triangles. (See Fig. B-2b.) It should be easy to conclude that the area of triangle ADC is equal to ______.

■ $\frac{1}{2}(\overline{AD})(\overline{CE'})$

(3) What is the area of the triangle in Fig. B-2c?

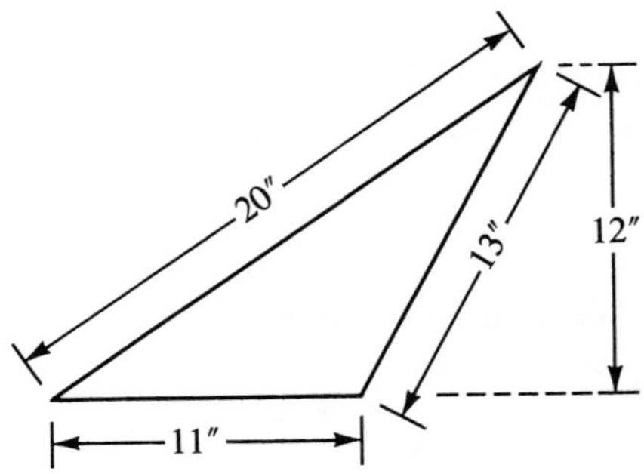

Fig. B-2c

■ $(\text{Area}) = \frac{1}{2}(11)(12) = 66.$

The areas of two similar polygons are to each other as the ________ of any two corresponding dimensions.

■ squares

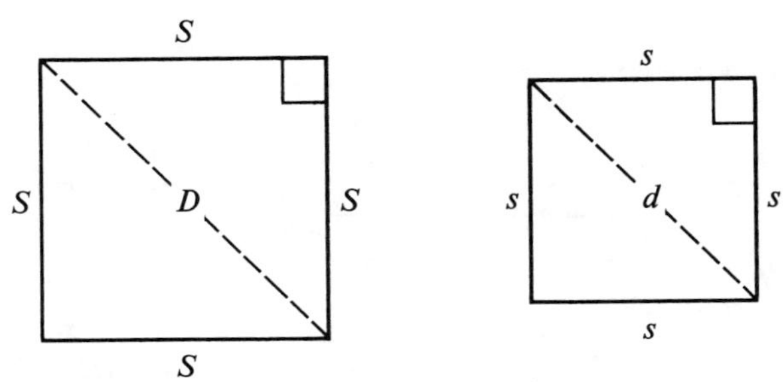

Fig. B-2d

(4) In Fig. B-2d, if $s/S = d/D$, then $s^2/S^2 = d^2/D^2$. The ratio of s^2/S^2 is equal to the ratio of the ________ of the two squares.

■ areas

(5) Assume two similar triangles have altitudes and bases that are related by the factor m. Write the ratio of the areas of the two similar triangles in terms of their bases and altitudes. (See Fig. B-2e.)

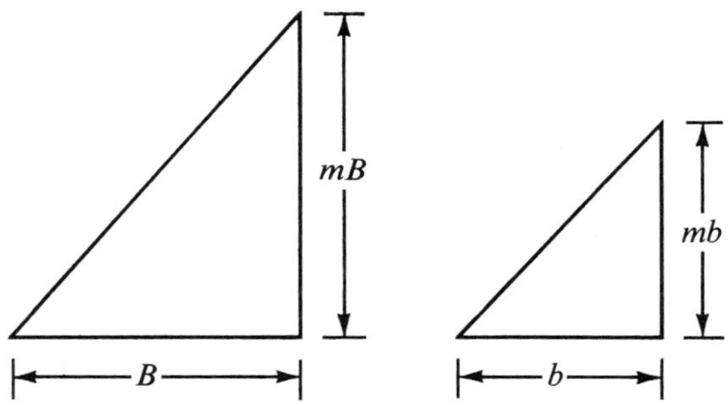

Fig. B-2e

■ $$\frac{(\text{area})}{(\text{Area})} = \frac{\frac{1}{2}(b)(mb)}{\frac{1}{2}(B)(mB)} = \frac{b\,b}{B\,B} = \frac{b^2}{B^2}.$$

(6) Recall the formula for finding the area of a circle. Write the ratio of the areas of the two circles in Fig. B-2f in terms of their radii.

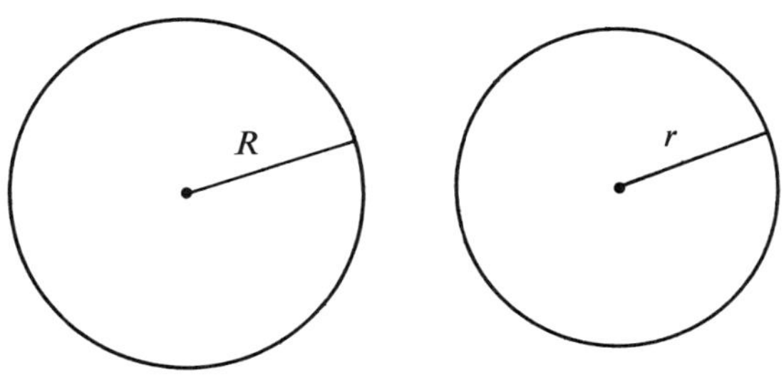

Fig. B-2f

■ $$\frac{(\text{area})}{(\text{Area})} = \frac{\pi r^2}{\pi R^2} = \frac{r^2}{R^2}.$$

B-3 Circles, Chords, and Tangents

An inscribed angle is measured by ______________.

■ one-half its intercepted arc

(1)

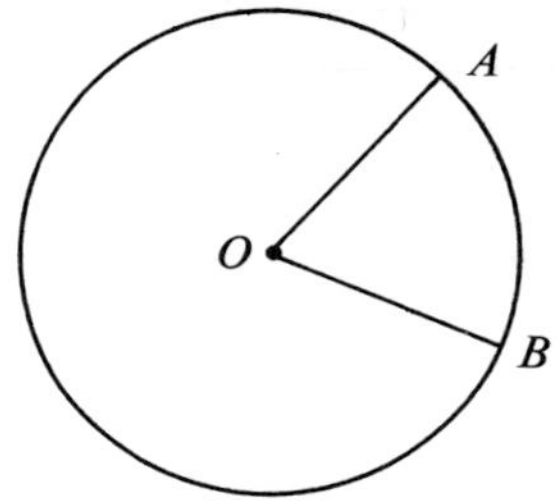

Fig. B-3a

A central angle is an angle whose vertex is on the center of the circle. The number of angle degrees in a central angle equals the number of arc degrees in its intercepted arc. (See Fig. B-3a.)

If $\widehat{AB} = 30$ arc degrees, then $\angle AOB = \square$ angle degrees.

■ 30

(2)

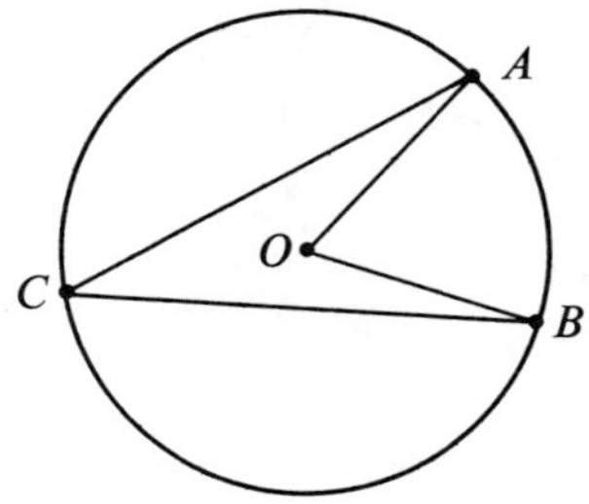

Fig. B-3b

In Fig. B-3b, the inscribed angle ACB intercepts the same arc as the central angle AOB. Two chords intersecting at a common point on the circle form an __________ __________.

■ inscribed angle

(3)

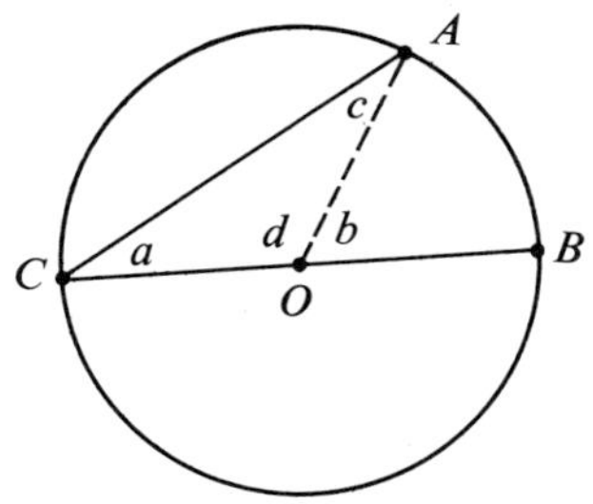

Fig. B-3c

In Fig. B-3c, $\overline{CB}$ is a diameter, and O is the center of the circle.

a. $\angle a = \angle c$,
b. $180° - 2\angle a = \angle d$,
c. $180° - \angle b = \angle d$,
d. $\therefore 2\angle a = \boxed{}$.

■ $\angle b$

(4)

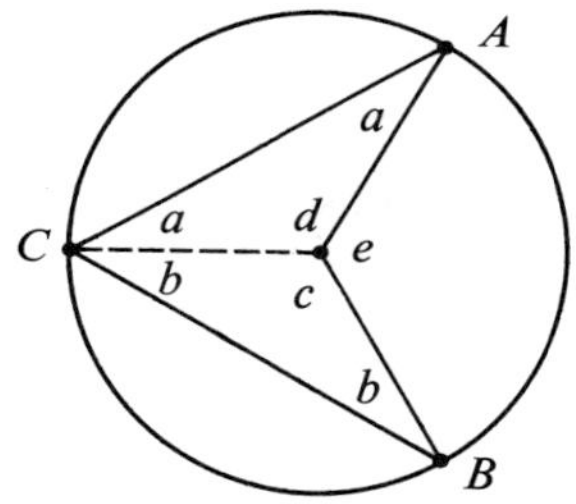

Fig. B-3d

See Fig. B-3d.

a. $2\angle a + 2\angle b + \angle c + \angle d = 360°$,
b. $\angle e + \angle c + \angle d = 360°$,
c. $\therefore 2\angle a + 2\angle b = \angle \boxed{}$,
d. $2(\angle a + \angle b) = \angle \boxed{}$.

■ e; e.

(5)

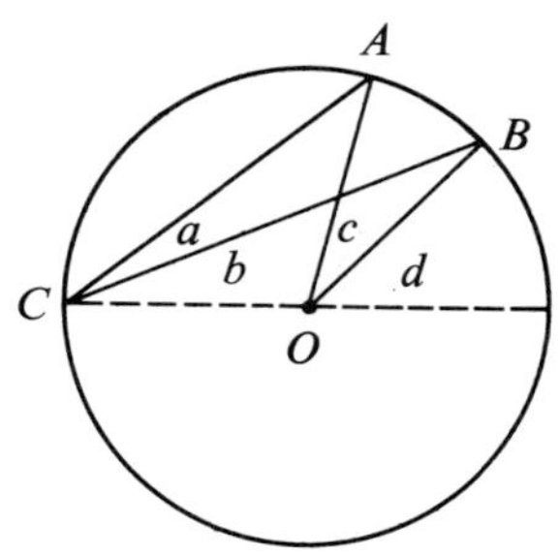

Fig. B-3e

Notice the resemblance of Fig. B-3e to Fig. B-3c.

a. $2(\angle a + \angle b) = \angle c + \angle d$,
b. $2\angle b = \angle d$,
c. $\therefore 2\angle a = \angle \boxed{}$.

■ c

(6)

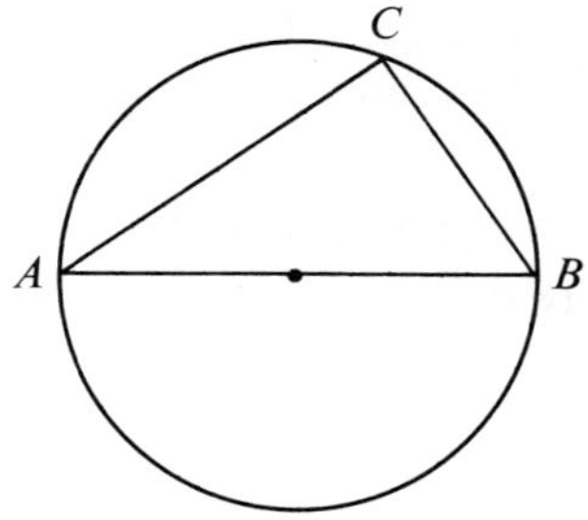

Fig. B-3f

$\overline{AB}$ is a diameter shown in Fig. B-3f. What is the measure of $\angle ACB$?

■ The inscribed $\angle ACB$ is measured by one-half its intercepted arc. Therefore,

$$\angle ACB = 90°.$$

In a right triangle, the square of the hypotenuse is equal to ____________________________.

■ the sum of the squares of the legs

(7) In Fig. B-3g, triangle ABC is a right triangle with height $\overline{CD}$ drawn from the vertex of the right angle to the hypotenuse. Identify all the angles of Fig. B-3g that are complements of $\angle B$.

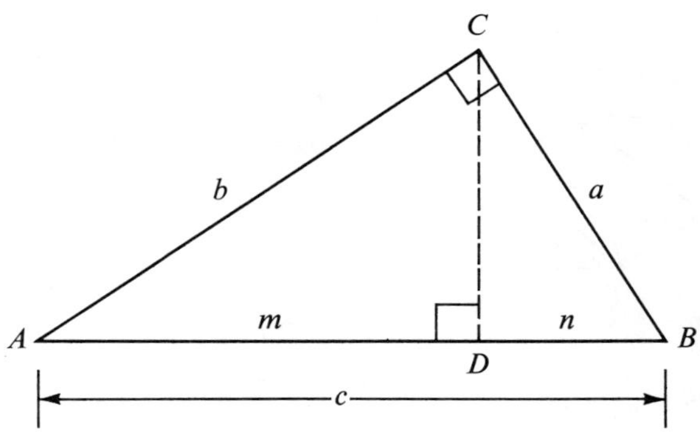

Fig. B-3g

■ $\angle A$ and $\angle DCB$

(8) Since $\angle A$ and $\angle DCB$ are complements of the same angle, they are equal. With this fact, it should be apparent that triangles ABC, ADC, and CDB are ______________.

■ similar

(9) The similar triangles of Fig. B-3g permit us to write the following proportions:

1. $n/a = a/c$ or $a^2 = \square$,
2. $m/b = b/c$ or $b^2 = \square$.

■ 1. cn, 2. cm.

(10) Add together the last results of (1) and (2).

$$\begin{aligned} a^2 + b^2 &= cn + cm \\ &= c\,\square \\ &= c^2. \end{aligned}$$

■ $(n + m)$, which is equal to c

(11) The results of frame 10 are known as the Pythagorean theorem. This theorem is used so frequently in mathematics that it is common to identify it by this name. Stated in words, in a right triangle, the square of the hypotenuse is equal to the sum of the squares of the legs. Or, using the notation of Fig. B-3g, $c^2 = a^2 + b^2$, where c is the (1) ______________ and a and b are (2) ______________.

■ 1. hypotenuse, 2. the legs.

If a diameter of a circle is perpendicular to a chord, it ______ ______.

■ bisects the chord and the major and minor arcs subtended by the chord

(12) Examine Fig. B-3h. $\overline{CE}$ is a diameter perpendicular to chord $\overline{AB}$ at point D. It should be apparent that $\Delta ADO \cong$ [ ].

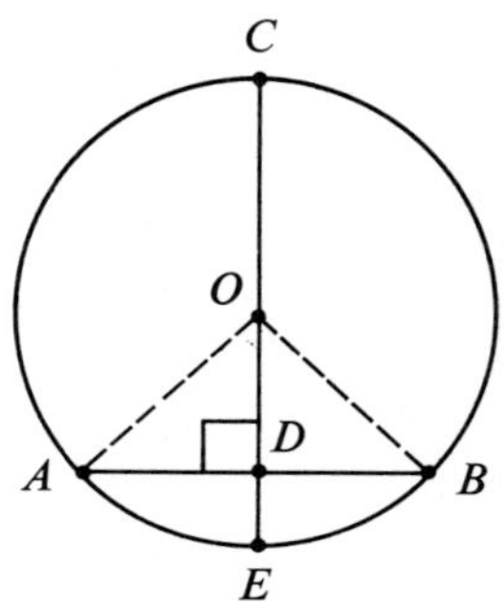

Fig. B-3h

■ ΔBDO

(13) Since $\Delta ADO \cong \Delta BDO$, (1) $\overline{AD} =$ ____.
Therefore, (2) D is the midpoint of ____.

■ 1. $\overline{DB}$, 2. $\overline{AB}$.

(14) Since equal central angles intercept equal arcs, (1) $\widehat{AE} =$ ____ and (2) $\widehat{AC} =$ ____.

■ 1. $\widehat{EB}$, 2. $\widehat{CB}$.

(15)

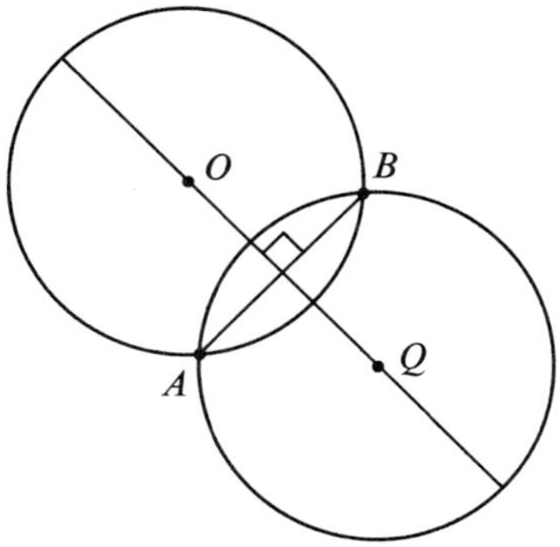

Fig. B-3i

Examine Fig. B-3i. $\overline{AB}$ is a common chord drawn between the intersections of circles O and Q. Since a line segment has only one midpoint, the line of centers ________ the common chord.

■ bisects

If a line is perpendicular to a diameter at one end of the diameter, it is ____________________.

■ tangent to the circle

(16)

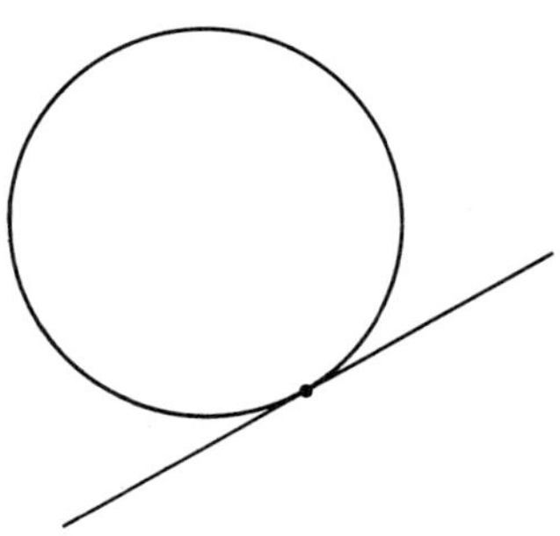

Fig. B-3j

See Fig. B-3j. A tangent to a circle is a straight line that, no matter how far extended, touches the circle at ____________________.

■ only one point

(17) Refer to Fig. B-3k. Diameter $\overline{AB}$ is perpendicular to line L at point B. For any point X on L (other than B), $\overline{OX}$ must be greater than $\overline{OB}$ since the shortest distance from point O to line L is along the line perpendicular to L. Consequently, line L fits the previous definition of a ________________________.

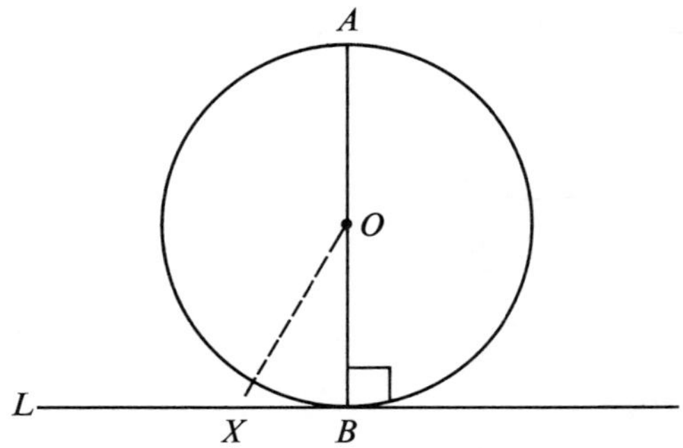

Fig. B-3k

■ tangent to a circle

(18) Look at Fig. B-3l. T is the only point circles O and Q have in common. The radius of circle O is 5 inches and the radius of Q is 6 inches. What is the distance from center to center?

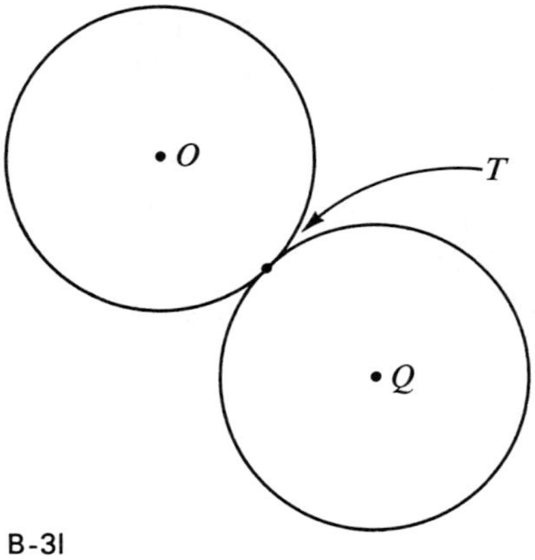

Fig. B-3l

■ If your answer is eleven, you must have concluded that points O, T, and Q are colinear. Can you support this conclusion?

(19) From point A, L_1 and L_2 are drawn tangent to circle O at points B and C respectively. (See Fig. B-3m.) If $\overline{AB} = 12$ and the radius of the circle is 5, find the length of $\overline{AC}$. Justify your answer.

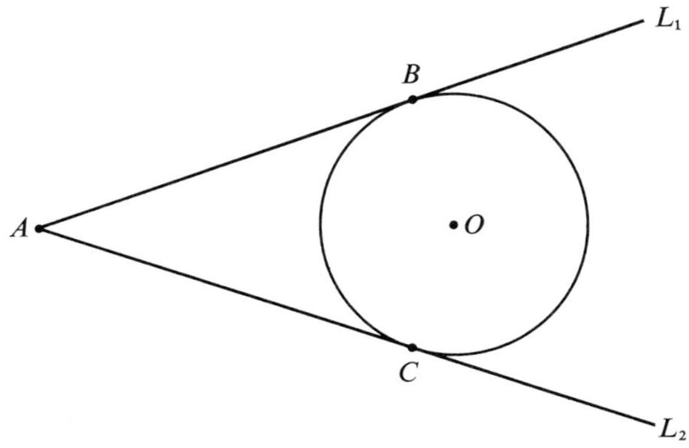

Fig. B-3m

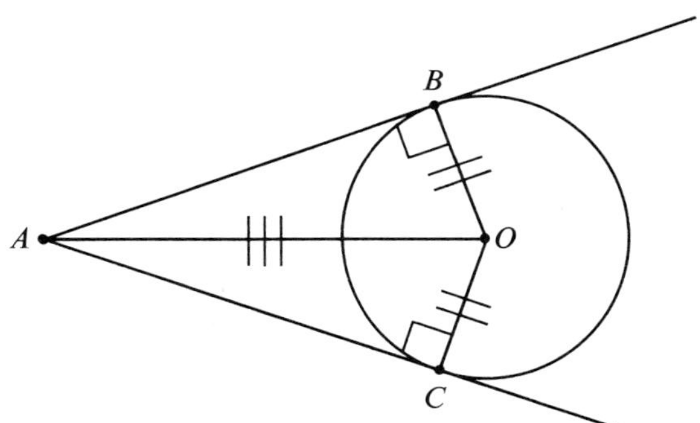

Fig. B-3n

■ 1. $\Delta ABO \cong \Delta ACO$,
2. $\overline{AB} = \overline{AC}$,
3. $\therefore \overline{AC} = 12$.

B-4 Triangles and Auxiliary Lines

The bisectors of the angles of a triangle are concurrent at a point which is ____.

■ equidistant from the sides of the triangle

(1) Let l_A, l_B, and l_C be the bisectors of angles A, B, and C, respectively. (See Fig. B-4a.) The bisector of an angle is the locus of points equidistant from the sides of the angle. From O, the intersection of l_A and l_B, the perpendiculars to the sides of the triangle must be equal in length. Therefore l_C must contain ____________.

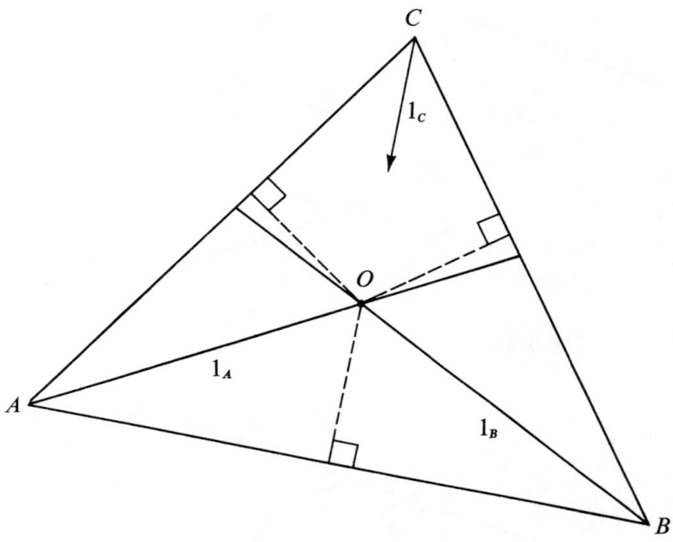

Fig. B-4a

■ point O

(2) A circle is a plane curve, all points of which are equidistant from a fixed point. Therefore the intersection of the bisectors of a triangle is also the ________________________.

■ center of a circle tangent to each of the three sides of the triangle (called the inscribed circle)

The perpendicular bisectors of the sides of a triangle are concurrent at a point which is

____.

■ equidistant from the vertices of the triangle

(3) Examine Fig. B-4b. From the intersection of P_b and P_c, lines are drawn to the vertices of the triangle. As a result of the definition of a perpendicular bisector, we know that line segments $\overline{OC}$, $\overline{OA}$, and $\overline{OB}$ are equal. If point O is equidistant from points B and C, the perpendicular bisector P_a must contain point O. Therefore (1) the ____________________ of the sides of a triangle are concurrent at (2) a point ______________ from the vertices of the triangle.

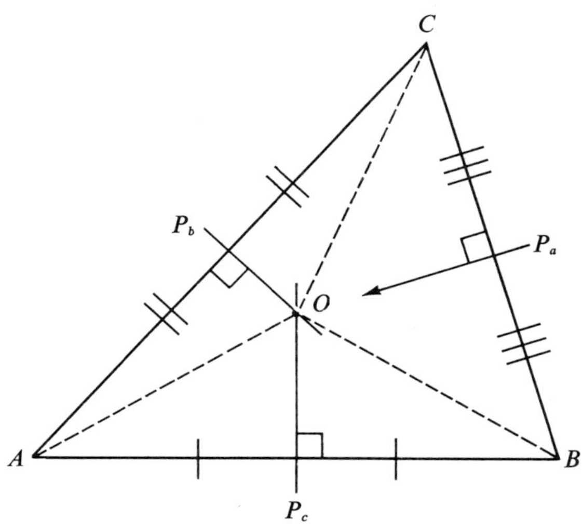

Fig. B-4b

■ 1. perpendicular bisectors, 2. equidistant.

(4) Since point O of Fig. B-4b is equidistant from points A, B, and C, it is the center of the circle that passes through the ________ of the triangle.

■ vertices

(5) Unlike the center of the circle inscribed in a triangle, the center of the circumscribed circle is not necessarily in the interior of the triangle. Sketch a triangle that would not contain the center of a circumscribed circle.

■ 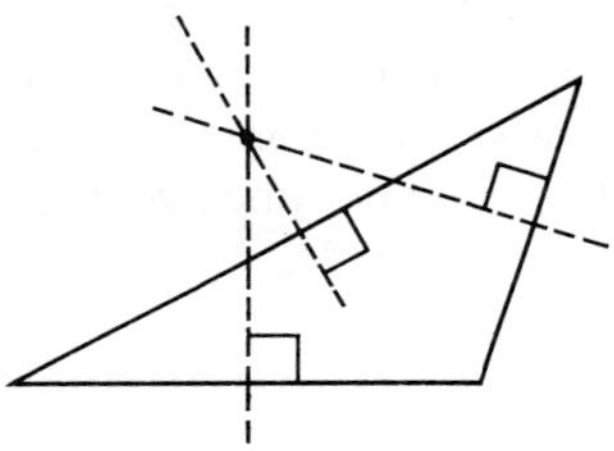

(6) How many circles can be drawn through one point? Two points? Three points?

■ One point – any number
Two points – Any number; however, the circles all will have their centers in the same straight line.
Three points – Only one if the points are not in the same straight line. This can be concluded from the results of Frame 4.

---

1. The centroid of a triangle is the point of intersection of the ________ of the triangle.
2. In a physical triangle the centroid is the ________.

■ 1. medians, 2. center of gravity.

(7) Imagine hanging a physical triangle from one of its vertices with a plumb line continuing beyond the vertex as illustrated in Fig. B-4c. If the triangle is in equilibrium, (1) ________ the area of the triangle must be on each side of (2) ____.

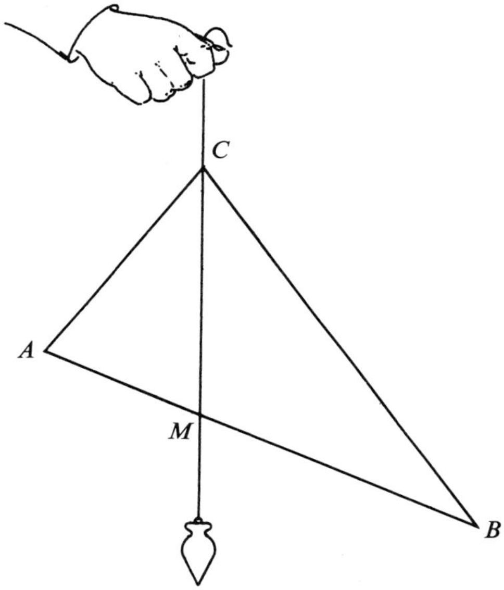

Fig. B-4c

■ 1. one-half, 2. $\overline{CM}$.

(8) Since $\Delta\, ACM$ and $\Delta\, MBC$ have equal areas and equal heights from the base containing point M, they must also have equal bases. Therefore (1) ________ equals (2) ________ and (3) ________ is the midpoint of (4) ________.

■ 1. $\overline{AM}$, 2. $\overline{MB}$, 3. M, 4. $\overline{AB}$.

(9) A line joining the vertex of a triangle to the midpoint of the opposite side is a median. In Fig. B-4c, line segment ________ is a median.

■ $\overline{CM}$

(10) If all the medians of a triangle are drawn, they are concurrent at a point called the centroid of the triangle. This is a point known as the center of gravity in a physical triangle. Sketch a triangle and locate its center of gravity.

■

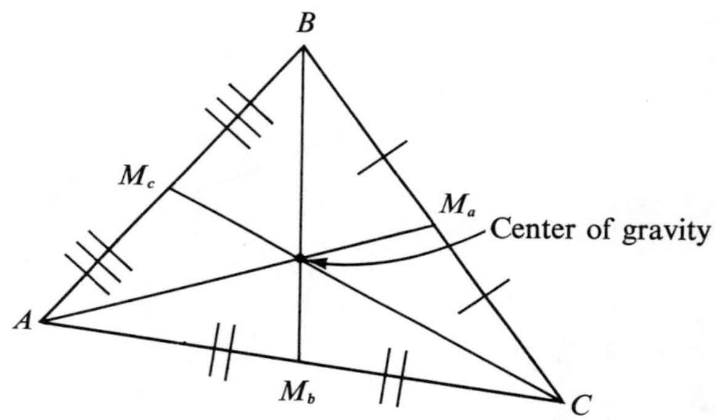

> The medians of a triangle are concurrent at a point which is ________ of the distance from each vertex to the midpoint of the opposite side.
>
> ■ two-thirds

(11) Refer to Fig. B-4d. $\overline{AE}$ and $\overline{BD}$ are medians of triangle ABC. M and N are midpoints of $\overline{AO}$ and $\overline{OB}$, respectively. What relationship exists between $\overline{MN}$ and $\overline{AB}$?

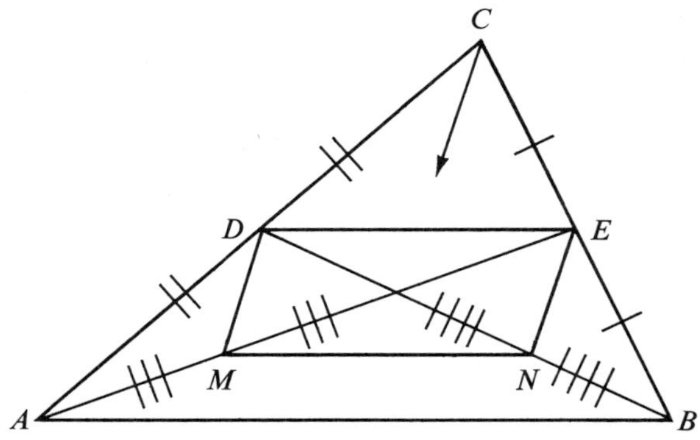

Fig. B-4d

■ $\overline{MN}$ is parallel to $\overline{AB}$ and $\overline{MN} = \frac{1}{2}(\overline{AB})$.

(12) What relationship exists between $\overline{DE}$ and $\overline{AB}$?

■ $\overline{DE}$ is parallel to $\overline{AB}$ and $\overline{DE} = \frac{1}{2}(\overline{AB})$.

(13) The results of Frames 11 and 12 lead to the conclusion that quadrilateral *DENM* in Fig. B-4d is a ______________.

■ parallelogram

(14) In a parallelogram the diagonals bisect each other. How, then, is $\overline{AO}$ related to $\overline{OE}$? (See Fig. B-4d.)

■ $2(\overline{OE}) = \overline{AO}$, or $\overline{AO}$ is $\frac{2}{3}(\overline{AE})$.

The same relationship exists between $\overline{OB}$, $\overline{OD}$, and $\overline{DB}$.

(15) Extending the analysis given medians $\overline{AE}$ and $\overline{DB}$ to the pair of medians coming from C and from either A or B, the conclusions would be equivalent. Thus medians are concurrent and they intersect at a point ________ of the distance from each vertex to the midpoint of the opposite side.

■ two-thirds

Chapter One Linear Functions

1-1 $f(x) = mx$

Engineers and technicians rely heavily upon the use of formulas to help them plan and solve problems pertaining to their jobs. Formulas are concise statements, usually in mathematical form, that clearly define the relationships that exist between two or more variable quantities. For example, bridge builders use formulas that show the relationship between variations in temperature and the expansion of construction materials. Electricians use formulas that show the relationship between electric power requirements and wire size. Formulas also can show the relationship between tensile strength and tension on flexible cables used on construction equipment, and they can describe the relationship between pitch and radius of curvature for freeway construction. Laboratory technicians use formulas to determine the relative amounts of chemical reactants to synthesize specific compounds. Draftsmen use formulas to determine the relationship between certain linear measurements and corresponding areas.

Consider the formula for determining the relationship between the radius of a circle and its circumference, $C = 2\pi r$. For each numerical replacement of r there is *only one* corresponding result C.

r (inches)	C (inches)
1.00	6.28
2.00	12.56
3.00	

■ $C = 2(3.14)(3.00) = 18.84$.

In this example we may refer to r as the **independent variable** and C as the **dependent variable**, since the specific value of C depends upon the independent choice made for the value of r.

In formulas having two variables, it is not always clear which is the independent variable and which is the dependent variable. It is not altogether clear in the example $C = 2\pi r$. Mathematicians have agreed upon a choice of vocabulary and notation that resolves this problem. However, it applies only to formulas or equations with the special property that for each numerical value of the independent variable, there exists only one corresponding value of the dependent variable. Since the formula $C = 2\pi r$ has this characteristic, we shall use it as an illustration.

We say "C is a function of r" or, in notational form, "$C = f(r)$." The expression $C = f(r)$ is read "C equals f of r;" it means that the independent variable C is uniquely determined by the choice of the independent variable r. The formula for the circumference of a circle may be written $f(r) = 2\pi r$, because $C = f(r)$. C and $f(r)$ are both representations of the dependent variable. The dependency of the circumference upon the radius is made specific by this notation.

The mistake of interpreting f and (r) as factors is avoided because parentheses are not ordinarily required in products of literal factors.

Through continued use of *functional* notation and vocabulary, we shall learn to appreciate the convenience and clarity that it gives to the mathematical problems we study.

Many types of **functional** relationships in mathematics are used to represent the relationships between variables common to the engineering world. Let us first look at a simple type in which the dependent variable is equal to a constant multiplied by the independent variable. If x is the independent variable, y is the dependent variable, and m is a **constant** (a number that does not change in value), then we are considering equations in the form $y = mx$, where $y = f(x)$.

Example 1

If $f(x) = 0.5x$, where $y = f(x)$, evaluate $f(1)$, $f(2)$, $f(3)$, and $f(4)$.

i. $f(1) = 0.5(1)$
$= 0.5;$

ii. $f(2) = 0.5(2)$
$= \square;$

iii. $f(3) = 0.5(\square)$
$= \square;$

iv. $f(4) = \square.$

■ ii. 1.0;
iii. 3, 1.5;
iv. 2.0.

x	$f(x)$
1	0.5
2	1.0
3	1.5
4	2.0

As simple as our example may seem, many important relationships may be expressed in this way.

Example 2

$C = pN$. The cost (C) of N items at a constant price per item (p) is a function of N. This may be written $f(N) = pN$, using functional notation.

Example 3

$I = kE$. The current (I) in an electrical circuit of constant conductivity is the product of the constant (k) and the voltage (E). ($k = 1/R$, where R is the resistance.) This equation could be written $f(E) = kE$, where $f(E) = I$.

Example 4

$D = \sqrt{2}\, S$. The diagonal of a square (D) is a function of the length of the side of the square (S).

Example 5

$M_L = kM_B$. The loose-measure of topsoil (M_L) is equal to the product of the swell factor (k) and the bank-measure (M_B). For topsoil, k may be some constant between 110% and 145%; it is a very important consideration in planning excavations.

Example 6

Ordinary proportion is another example. If a 20-foot length of copper wire weighs 1.5 pounds, how much will 92 feet of it weigh? If we let y represent the weight and x represent the length, then

$$\frac{y}{x} = \frac{1.5}{20}, \qquad y = .075x, \quad \text{or } f(x) = .075x.$$

$$f(92) = .075(\boxed{})$$
$$= \boxed{}.$$

■ $f(92) = .075(92)$
$= 6.9$ pounds.

1-1 Exercises

The exercises are given in the left-hand column. The complete solution is given in the right-hand column one frame below the corresponding exercise. *Be sure to shield the solution to each exercise until you have carefully written an independent solution.* Supplementary exercises are given at the end of the chapter.

1. Express the perimeter of a square as a function of its sides using functional notation.

---

2. If $f(s) = 4s$, find $f(3)$, $f(21)$, and $f(.05)$.

■ 1. $f(s) = 4s$, where s is a side and $f(s)$ represents the perimeter.

---

3. Express the circumference of a circle as a function of its diameter using functional notation.

■ 2. $f(3) = 4(3)$
$= 12;$
$f(21) = 4(21)$
$= 84;$
$f(.05) = 4(.05)$
$= .20.$

---

4. If $f(d) = \pi d$, find $f(5)$, $f(13)$, and $f(0.45)$.

■ 3. $f(d) = \pi d$, where d represents the diameter and $f(d)$ represents the circumference.

---

5. Express the value of a sack of quarters as a function of the number of quarters, N. Calculate $f(90)$.

■ 4. $f(5) = \pi(5)$
$= 3.14(5)$
$= 15.7;$

$$f(13) = 3.14(13)$$
$$= 40.8;$$
$$f(.45) = 3.14(0.45)$$
$$= 1.41.$$

---

6. When two or more elements combine chemically, their combining weights will always be in the same ratio. Fifty-six grams of iron and 32 grams of sulfur form exactly 88 grams of iron sulfide. Construct a chart that shows the functional relationship of the combined weights of iron (W_{Fe}) and sulfur (W_S) when the weight of iron is given in 5-gram amounts from 0 to 25 grams.

■ 5. $f(N) = 0.25N;$
$$f(90) = 0.25(90)$$
$$= \$22.50.$$

---

7. A second warehouse is to be a scaled-down version of a larger one in existence. If the ratio is to be 0.85 : 1, determine the corresponding dimensions of the two buildings.

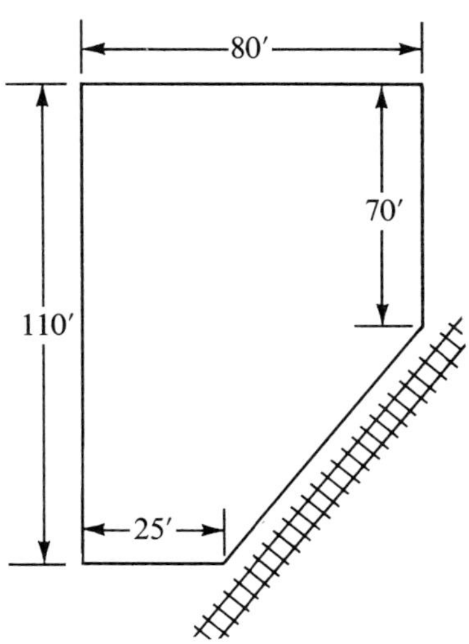

■ 6. $W_S = \frac{32}{56} W_{Fe}.$

W_{Fe}(grams)	W_S(grams)
0	0
5	2.86
10	5.71
15	8.57
20	11.42
25	14.29

■ 7. $d = 0.85D$.

D	d(feet)
25	21.25
70	59.50
80	68.00
110	93.50

1–1 Problems

The problems are not simply an extension of the exercises; rather, they test your ability to interpret, organize, and communicate, in mathematical terms, the *methods* and *solutions* to questions that arise in technical applications of mathematics.

1. Coaxial cable in 200-foot rolls weighs 12.60 pounds. Prepare a table of values that provides the length of cable remaining in a used roll so that the remaining length can be determined by weight within an accuracy of 5 feet. (*Hint:* See Example 6.)
2. Exactly 39.3 grams of sodium combines with 60.7 grams of chlorine to form the chemical compound sodium chloride. Write a formula that defines the functional relationship between the combining weights of sodium and chlorine. Write a second formula that determines the weight of chlorine present in any given amount of sodium chloride. (*Hint:* See number six of the Exercises.)
3. The maximum safe pressure (P_{max}) of a cylindrical boiler is approximated by the formula

$$P_{(max)} = \left(\frac{4\sigma}{d}\right)t.$$

Assume that the longitudinal stress (σ) is 1000 psi and the diameter of the boiler (d) is 48 inches. What is the safe maximum pressure when the thickness (t) is 0.50 inches? What is P_{max} when $t = 0.51$ inches? Find what effect a change of 0.01 inch in thickness (t) has on P_{max}.
4. The chart relates the displacement of a spring and the force applied to it. Can the displacement be expressed in an equation as a function of force? What displacement would result from a force of 3.5 dynes?

Force (dynes)	Displacement (centimeters)
0.0	0.0
2.0	1.0
4.0	2.0
6.0	3.0
15.0	7.5

5. In the equation $C = \pi D$, the variables C and D are said to be in "direct variation." Common ways of expressing this idea are: "C varies directly as D," "C is proportional to D," or simply "C varies as D." In this example π is

referred to as the "constant of proportionality." Express the following direct variations in functional notation.

Resistance (R) of a conductor varies directly as the length (L).
Velocity (v) is proportional to time (t).
Force (F) varies as acceleration (a).

Can you identify any of the constants of proportionality in the above examples?

6. In the United States the usual angular measurement of a right angle is 90 *degrees.* In some European countries the same angle is expressed as 100 *grads.* Write an equation that gives degree measure as a function of grad measure. Make a chart of degree measurements corresponding to every 5 grads from zero to 100 grads.
7. Military surveys use *mils* as the unit of angular measure. A right angle has the value of 1600 mils. Write an equation that gives degree measure as a function of mil measure.

1-2 Slope of $f(x) = mx$ Graphing—Related Function

It is frequently useful to graph a function. To do this we utilize two number lines that intersect at right angles. The horizontal line is called the x-**axis**, and the vertical line is called the y-**axis**. Each point in the plane corresponds to a pair of numbers. For example, in Fig. 1-2a, point A corresponds to the pair of numbers (1, 2). The distance of the point from the y-axis is given first, and the distance from the x-axis is given last. Point B has the **coordinates** $(-2, 1)$, since it is 2 units to the left of the y-axis and one unit above the x-axis.

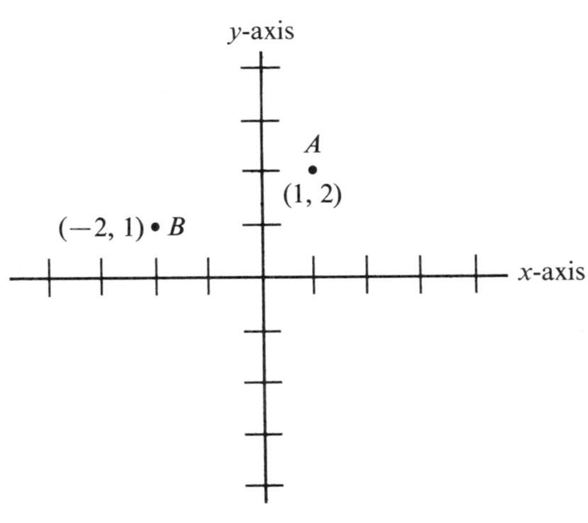

Fig. 1-2a

On Fig. 1-2a plot the points $C(4, 1)$, $D(-3, 4)$, $E(3, -2)$, $F(-4, -2)$, and $G(0, 0)$.

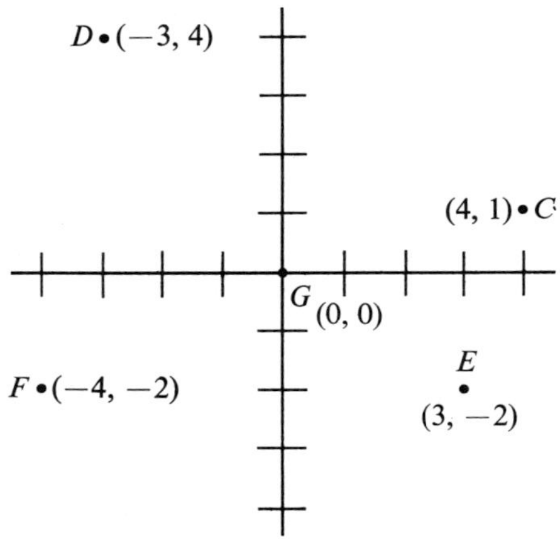

Next let us construct a set of number pairs by using the equation $y = f(x) = 2x$ and place the pairs on a graph.

$$f(-3) = 2(-3) = -6$$
$$f(-2) = -4$$
$$f(-1) = -2$$
$$f(0) = 0$$
$$f(.4) = 0.8$$
$$f(2.7) = 5.4$$

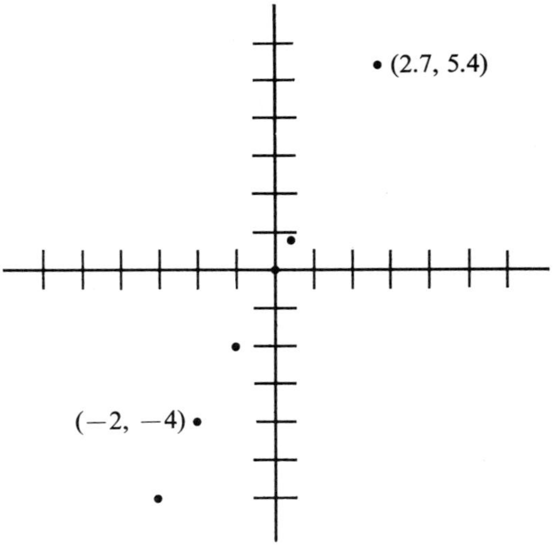

Fig. 1-2b

Each choice of a number to replace x results in a specific value of y. These pairs of numbers may be plotted as in Fig. 1-2b. We should observe that these points lie in a *straight line*.

Consider a second example, $y = f(x) = (-1/3)x$. Plot on the graph in Fig. 1-2c the pairs of numbers that result from x having the values -9, -6, -1.5, 5, and 8.

$f(-9) = \square$ $f(-6) = \square$ $f(-1.5) = \square$

$f(5) = \square$ $f(8) = \square$

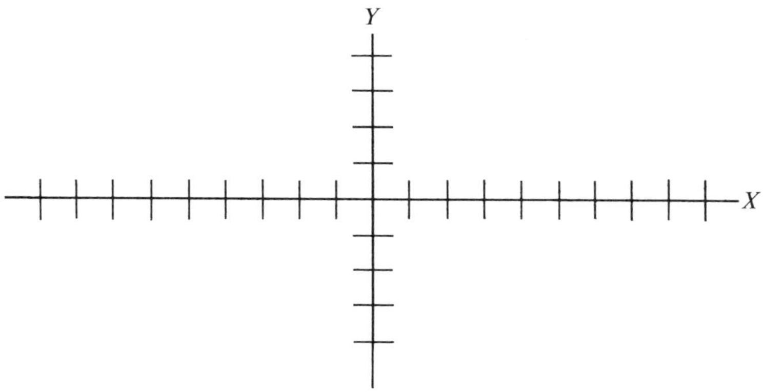

Fig. 1-2c

$f(-9) = 3$ $f(-6) = 2$ $f(-1.5) = \frac{1}{2}$

$f(5) = -\frac{5}{3}$ $f(8) = -\frac{8}{3}$

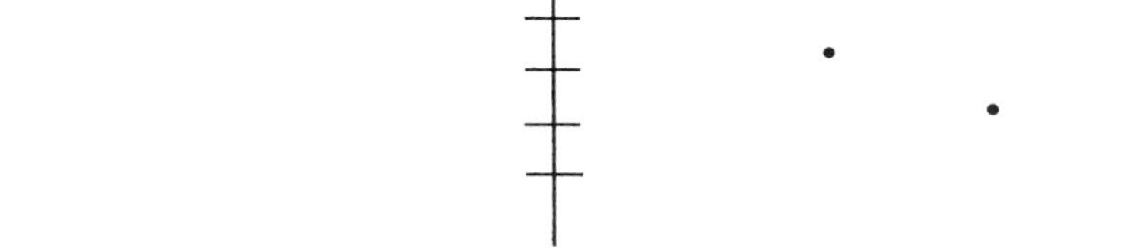

We observe that the points plotted in Fig. 1-2c are also in a straight line. Hence, we call these sets of ordered pairs **linear functions**. Notice that the equations have the general form $f(x) = mx$.

Associated with every linear function is a number we call the **slope**. The slope of a linear function mathematically represents the rate at which the dependent variable changes with respect to corresponding changes of the independent variable. If $P_1(x_1, y_1)$ and $P_2(x_2, y_2)$ are any two ordered pairs of a linear function, then

$$(\text{slope}) = \frac{y_2 - y_1}{x_2 - x_1}. \qquad \text{(Eq. 1-2a)}$$

If we use any two ordered pairs from the graph illustrated in Fig. 1-2b, we may calculate the slope of the functional equation $f(x) = 2x$. For example, using $P_1(-3, -6)$ and $P_2(-1, -2)$,

$$(\text{slope}) = \frac{(-6) - (-2)}{(-3) - (-1)} = \square.$$

■ 2

Experiment with any two other ordered pairs from Fig. 1-2b. What value do you calculate for the slope?

■ Every calculation using ordered pairs from Fig. 1-2b results in the slope equaling two.

Calculate the slope of the graph of the function given in Fig. 1-2c.

■ Since we are permitted to take any two points on the line, your work and the work illustrated here will not necessarily look the same; however, the results should be equal.

$$P_1(-9, 3), \qquad P_2\left(-\frac{3}{2}, \frac{1}{2}\right),$$

$$(\text{slope}) = \frac{(1/2) - 3}{(-3/2) - (-9)},$$

$$= \frac{-5/2}{15/2} = -\frac{1}{3}.$$

A very important observation can be made from the above calculations: the slope of the graph of the functional equation $f(x) = 2x$ is 2, and the slope of $f(x) = -\frac{1}{3}x$ is $-\frac{1}{3}$. Generally, if $f(x) = mx$, m is the slope of the graph. What is the slope of the graph of $y = 5x$?

■ $m = 5$.

From a mathematical point of view, the sign of the slope is very significant. It is important to note that

$$\frac{y_2 - y_1}{x_2 - x_1}$$

is equal to

$$\frac{y_1 - y_2}{x_1 - x_2}$$

but unequal to

$$\frac{y_2 - y_1}{x_1 - x_2}$$

because subtraction is not commutative.

Let us again return to the functional equation $f(x) = 2x$. Wherever the slope of this equation is calculated, the value is two. This fact, stated in another manner, suggests the existence of a second function related to the function defined by $f(x) = 2x$. The related function has the ordered pairs $(x, 2)$. For each x, the corresponding dependent variable is two. If we use functional notation to represent this related equation, we write $f'(x) = 2$.

In order to consider two or more functions in one discussion, we must adapt symbolism to help us distinguish between them. If $f(x) = 2x$, then $f'(x) = 2$ (read "f-prime of x equals 2") designates the related function. The two functions f and f' are related because the second function, f', gives the slope of f. Since f is a straight line, its slope is constant for all values of x (Fig. 1-2d).

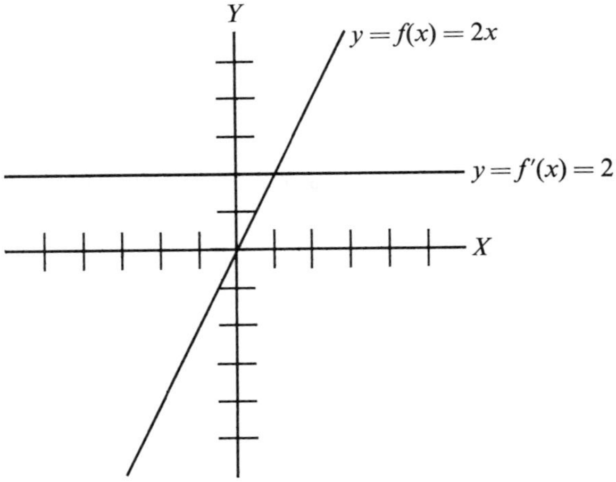

Fig. 1-2d

Graph the f and f' function when $f(x) = -\frac{1}{3}x$.

■ If $f(x) = -\frac{1}{3}x$,

then $f'(x) = -\frac{1}{3}$.

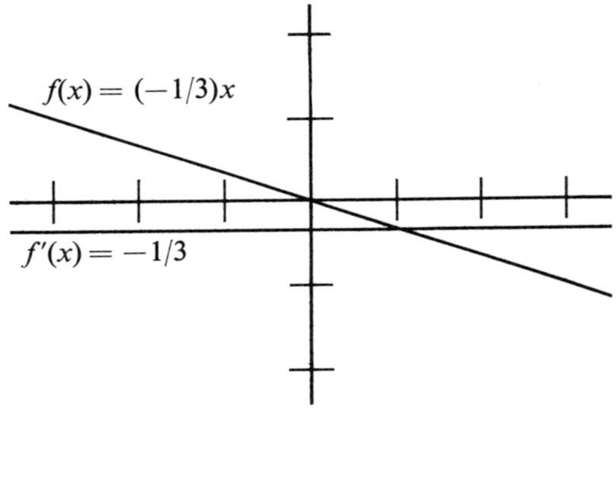

1-2 Exercises

It is important that you write your solution to an exercise before viewing the solution given in the text. Supplementary exercises appear at the end of the chapter.

1. Locate the following points on a graph.

   $P_1(2, 3)$, $P_2(3, 1)$,
   $P_3(-1, 2)$, $P_4(-4, -3)$,
   $P_5(-\frac{2}{3}, -5)$, $P_6(2.5, -4.2)$.

2. a. If $f(x) = 1/2x$, find $f(0)$, $f(1)$, $f(2)$, $f(4)$, and $f(-2)$.
   b. Plot the five points found in (a). Do they lie in a straight line?
   c. Using Eq. 1-2a, find the slope of the line.

■ 1.

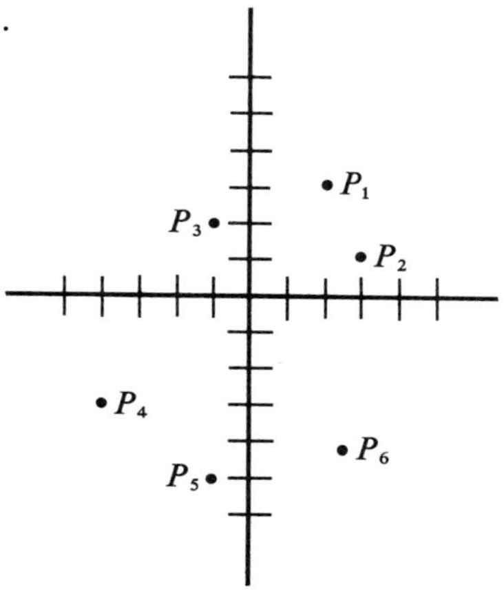

---

3. If f' is used to designate the function which gives the slope of $f(x) = 1/2x$, what is $f'(x)$?

■ 2.

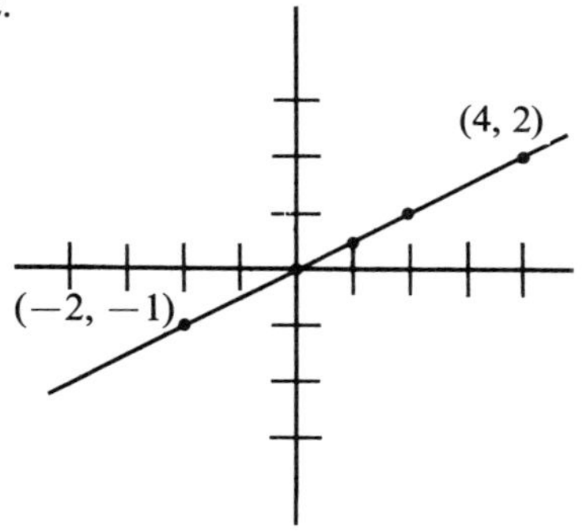

$$m = \frac{2-(-1)}{4-(-2)} = \frac{3}{6} = \frac{1}{2}.$$

---

4. a. If $f(x) = 0.4x$, find and graph $f(1)$, $f(2.8)$, $f(-3.1)$, and $f(-6)$.
   b. Calculate the slope of f.
   c. Write the equation for f'.

■ 3. $f'(x) = 1/2$.

5. The letter f is not the only symbol used to denote a function. In mathematical literature g and h are also rather common.

   a. Graph the function from the equation $g(x) = -1.5x$.

   b. Determine the slope function g'.

■ 4.

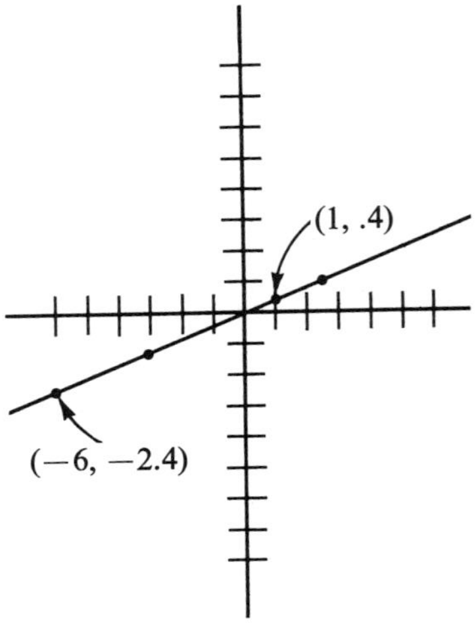

$$m = \frac{(1.4) - (-2.4)}{1 - (-6)}$$
$$= .4$$
$$f'(x) = .4.$$

■ 5.

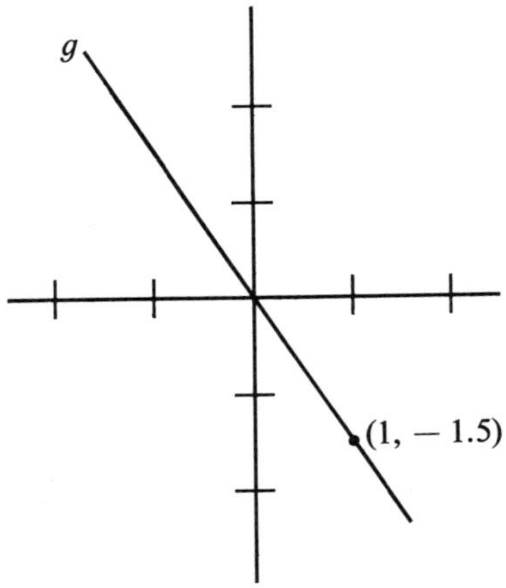

$$g'(x) = -1.5.$$

1–2 Problems

1. Let g' be the function that gives the slope for the function g.
   a. Write an expression for $g(x)$ when $g'(x) = 4$.
   b. Calculate and plot $g(-2)$, $g(0)$, $g(1)$, and $g(3)$.
   c. Calculate m using the slope formula.
2. $P_1(-3, 3)$ and $P_2(2, -2)$ are two points on a linear graph. Calculate the slope of the graph. Can you write an equation for the function f such that $f(-3) = 3$ and $f(2) = -2$?
3. Using a physics or physical science text as a reference, prepare a brief discussion of the mathematics related to the "coefficients of friction."
4. The rate of change of velocity of an experimental electric car under maximum power is a constant 2.2 miles per hour per second between the speeds of zero to 40 mph. Construct a graph that illustrates velocity as a function of time. Express any functional relationships that are apparent from the graph.
5.

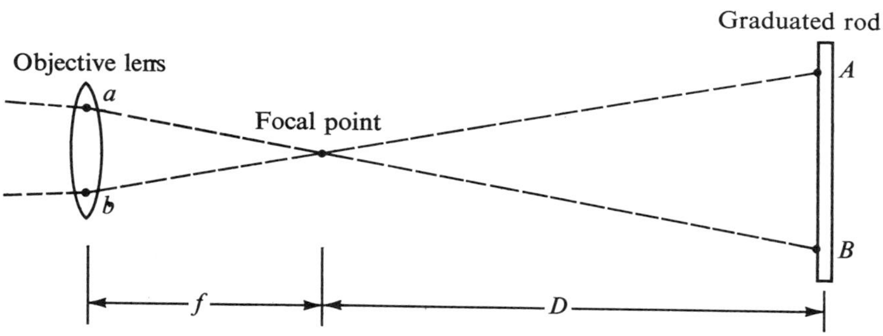

Fig. 1-2e

Transits are equipped to measure distance by using the known distance between "stadia" hairs (a and b) on the objective lens of the transit, the known focal length (f) of the objective lens, and the measured distance between stadia intercepts (A and B) on the graduated rod. As a result of the similar triangles represented in Fig. 1-2e,

$$\frac{D}{f} = \frac{\overline{AB}}{\overline{ab}}.$$

For American instruments

$$\frac{f}{\overline{ab}} = 100.$$

Using this value, write the equation that represents D as a function of $\overline{AB}$. Graph the function as $\overline{AB}$ varies from zero to 10 feet. What is the slope of the function?

1-3 Graphing: $f(x) = mx + b$

The word slope is used in various technical fields and is given meanings somewhat analogous to its meaning in mathematics. If we may borrow some of the vocabulary of surveying, slope is vertical rise divided by horizontal run.

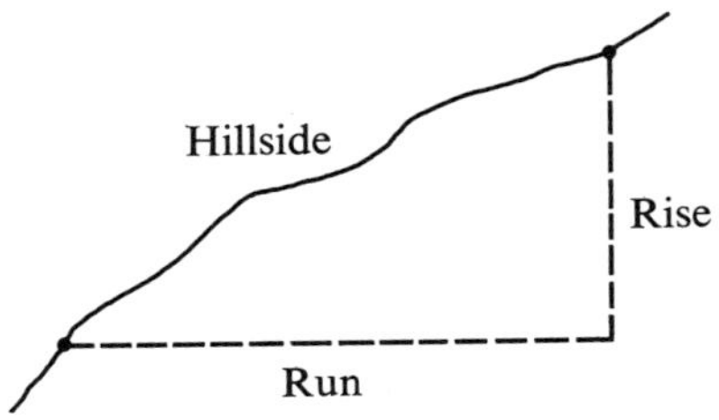

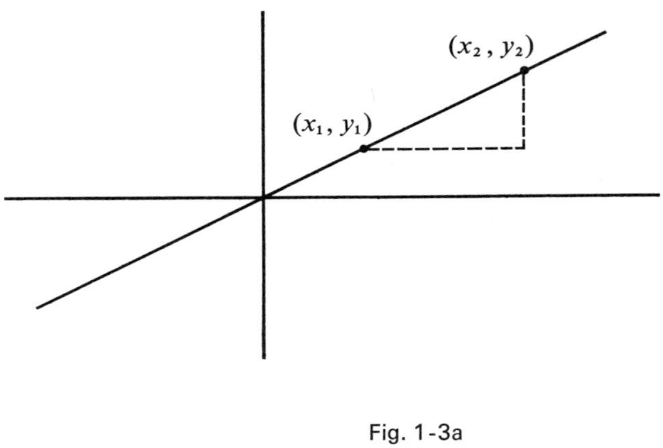

Fig. 1-3a

In mathematics, the difference $y_2 - y_1$ is equivalent to the rise and the difference $x_2 - x_1$, the run (see Fig. 1-3a). In mathematics, there is also a sign associated with slope. The slope is negative if, when the run is in the positive direction, the corresponding "rise" is negative, that is, it decreases (see Fig. 1-3b). This geometric interpretation of slope will enable us to move more meaningfully into a more generalized form of linear functions.

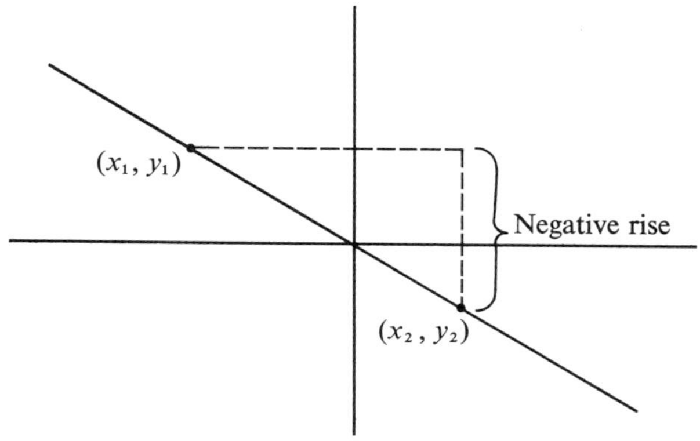

Fig. 1-3b

Consider the graph of $f(x) = 2/3x$ and the graph of a second straight line parallel to f but passing through the y-axis at 5. (See Fig. 1-3c.) Both L_1 and L_2 must have equal slopes and both represent functions of x; however, their corresponding sets of y values differ by a constant 5. In other words, to each $f(x)$

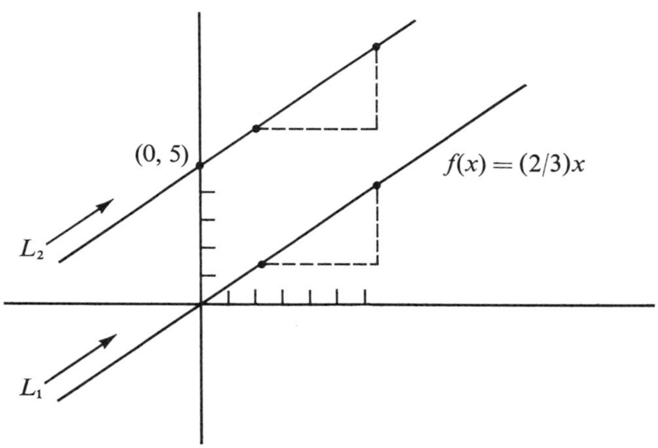

Fig. 1-3c

add the constant 5, and the result should be points on line L_2. If we call the resulting function g, then $g(x) = f(x) + 5$, or $g(x) = (2/3)x + 5$. The functions f and g are both of the general form

$$f(x) = mx + b. \qquad \text{(Eq. 1-3a)}$$

However, in one equation $b = 0$ and in the other $b = 5$. The coefficient m is the slope. The constant b is known as the y-**intercept**, because when $x = 0$, $f(0) = \boxed{}$.

■ b

The formula $F = (9/5)C + 32$ is a very common example of the general form of a linear relationship. This formula may be used to convert temperatures given in the centigrade scale to Fahrenheit scale. If we regard F as a function of C, then 9/5 is the ________ and 32 the ________.

■ slope; F-intercept (F being the vertical axis).

The general equation for linear functions should certainly assist us in graphing. Graph

$$f(x) = -2x + 4 \quad \text{and} \quad g(x) = (1/2)x + 4$$

on the same coordinate system. Use the slope and y-intercept to fix the position of each graph.

■

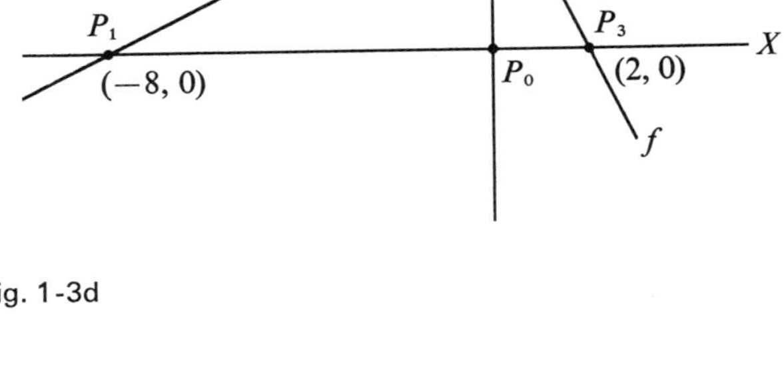

Fig. 1-3d

The functions f and g were chosen to illustrate a special relationship that may exist between the graphs of two or more linear functions. The results illustrated in Fig. 1-3d might suggest that f and g are perpendicular graphs. We can confirm this fact by using the Pythagorean theorem.

For right-triangle $P_1P_2P_0$ calculate the length of $\overline{P_1P_2}$, and for right triangle $P_3P_2P_0$ calculate length $\overline{P_3P_2}$.

■ $\overline{P_1P_2} = \sqrt{8^2 + 4^2} = \sqrt{80}$. $\quad \overline{P_3P_2} = \sqrt{2^2 + 4^2} = \sqrt{20}$.

Triangle $P_1P_2P_3$ is a right triangle if $(\overline{P_1P_3})^2 = (\overline{P_3P_2})^2 + (\overline{P_1P_2})^2$. Confirm this numerically.

■ $(\overline{P_1P_3})^2 = 10^2 = 100$.

$$(\overline{P_3P_2})^2 + (\overline{P_1P_2})^2 = (\sqrt{20})^2 + (\sqrt{80})^2 = 20 + 80 = 100.$$

Thus, the graphs of f and g are perpendicular.

Observe the slopes of f and g. They are *negative reciprocals*. In general, if two linear functions have slopes m_1 and m_2 such that

$$(m_1)(m_2) = -1, \qquad \text{(Eq. 1-3b)}$$

then the graphs of the two functions are perpendicular. Eq. 1-3b is, of course, the relationship that exists between negative reciprocals.

Example 1

If $f(x) = 0.25x + 1.85$ and $g(x) = mx - 9.2$, what value of m makes the graphs of f and g perpendicular?

$$(0.25)(m) = -1$$

$$m = \frac{-1}{0.25} = -4.00.$$

1-3 Exercises

Shield the correct responses until you have written your solution.

1. Graph the following equations on the same coordinate system:

$f(x) = 3x,$

$g(x) = 3x - 2,$

$h(x) = 3x + 3.$

---

2. Graph the following equations on the same coordinate system:

$f(x) = -2x,$

$g(x) = -2x + 4,$

$h(x) = -2x - 1.$

■ 1.

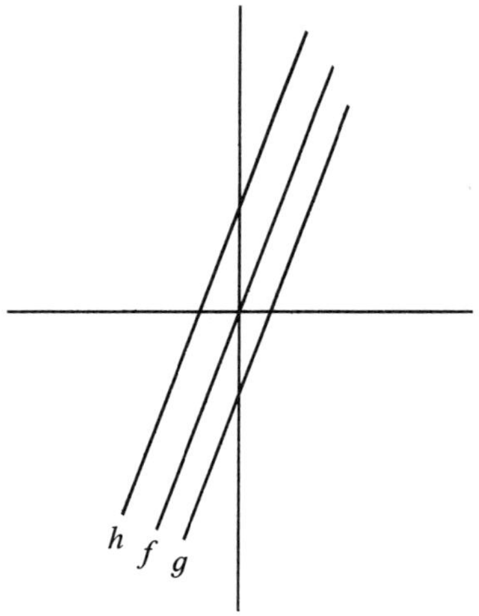

3. Graph $3x - 2y = 5$.

■ 2.

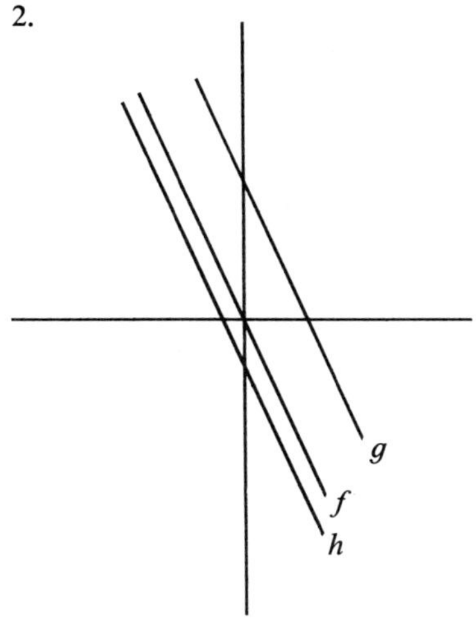

4. Write the equations of the following lines:

a. $m = 3$, $b = 2$,

b. $m = -2$, $b = 4$,

c. $m = 0$, $b = 3$,

d. $m = 0$, $b = 0$.

■ 3.

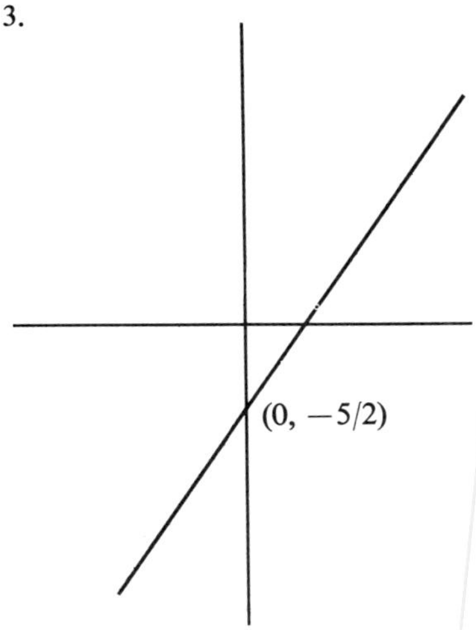

5. If $f(x) = 3x - 5$, find $f(0)$. What is the name given to $f(0)$?

■ 4. a. $y = 3x + 2$,
b. $y = -2x + 4$,
c. $y = 3$,
d. $y = 0$ (notice that this is the x-axis).

6. Write the equation of the line which has a y-intercept of -2 and which is perpendicular to $y = 3x + 3$.

■ 5. $f(0) = 3(0) - 5$
$= -5$;
$f(0)$ is the y-intercept.

■ 6. $m_2 = -\dfrac{1}{m_1}$

$= -\dfrac{1}{3}$.

Therefore, $y = -\dfrac{1}{3}x - 2$.

1–3 Problems

1. The resistance (R_m) used to extend the range (E) of a voltmeter movement is determined by the functional equation

$$R_m = \frac{\Omega}{V} E - R_v,$$

where Ω/V is the meter sensitivity and R_v is the meter resistance. If R_m is a function of E, then the formula has the same form as Eq. 1-3a.
   a. Identify the parts of the formula that play the roles of $f(x)$, x, m, and b.
   b. If $\Omega/V = 10{,}000$ and $R_v = 100$, find R_m when $E = 1, 10, 150$.
2. The formula $v = at + v_0$ represents the functional relationship between final velocity (v) and time (t) when v_0 is the initial velocity and (a) is acceleration. Graph the function when $v_0 = 88$ ft/sec and $a = -32$ ft/sec^2. From the graph, determine t for $v = 0$. Check the results by substitution into the formula.
3. The total inductance (L_t) of two coils (L_1 and L_2) is given by the formula

$$L_t = 2M + (L_1 + L_2),$$

where M is the mutual inductance which depends upon variations in the coupling of the coils. If $L_1 = L_2 = 2.2$ and M ranges from 0 to 0.80, graph L_t as a function of M.

4. Point B in a survey is 300 feet north and 400 feet east of point A. Find, with respect to point A, a point C directly north of point A such that angle ABC is a right angle. (*Hint:* Use an approach that includes Eq. 1-3b.)
5. Sometimes linear equations are obscured by other algebraic factors.
   a. Graph
   $$f(x) = \frac{x^2 + 3x + 2}{x + 2}.$$
   b. Graph $g(x) = x + 1$.
   c. Do you think the functions f and g are equal?
   d. Does $f(-2) = g(-2)$?
6. The set of numbers used to replace the independent variable is called the **domain** of the function. Usually the domain of linear functions is the set of real numbers.
   a. What is the domain of $f(x) = x + 1$?
   b. What is the domain of $g(x) = \dfrac{x^2 + 1}{x - 1}$?

1-4 Determining Linear Functions Given Point and Slope

Often we are given certain information about a function, and from this information we must determine the equation of the function. If we know the slope and one point on the graph of a linear function, then the equation can be found. In Section 1-2 we developed the formula

$$m = \frac{y_2 - y_1}{x_2 - x_1},$$

where m represents the slope. If the values m and (x_1, y_1) are known, then by letting (x_2, y_2) be any *general point* (x, y) on the graph of the function, the equation becomes

$$m = \frac{y - y_1}{x - x_1}.$$

This is usually written in the form

$$y - y_1 = m(x - x_1), \qquad \text{(Eq. 1-4a)}$$

and is called the **point-slope form** of linear equations.

Example 1

What is the equation of the line passing through the point (4, 5) with a slope of 2?

i. $m = 2$, $x_1 = 4$, and $y_1 = 5$.
ii. $y - 5 = 2(x - 4)$ or in functional notation, $f(x) = 2x - 3$, where $y = f(x)$.

1-4 Exercises

Use an answer mask. Find the line passing through the given point and having the given slope.

1. $P(5, 7)$, $m = 2$.

---

2. $P(-2, 4)$, $m = -\left(\frac{3}{4}\right)$.

■ 1. $(y - 7) = 2(x - 5)$
$y = 2x - 3.$

---

3. $P(3/5, -2/3)$, $m = \left(\frac{5}{3}\right)$.

■ 2. $(y - 4) = \left(\frac{-3}{4}\right)(x + 2)$
$y = \frac{-3}{4}x + \frac{5}{2}.$

---

4. $P(\sqrt{2}, \sqrt{3})$, $m = \sqrt{3}$.

■ 3. $\left(y + \frac{2}{3}\right) = \left(\frac{5}{3}\right)\left(x - \frac{3}{5}\right)$
$y = \frac{5}{3}x - \frac{5}{3}.$

---

5. $P(13.2, 17.8)$, $m = 0.340$.

■ 4. $(y - \sqrt{3}) = \sqrt{3}(x - \sqrt{2})$
$y = \sqrt{3}x - \sqrt{6}$
$+ \sqrt{3}.$

■ 5. $(y - 17.8) = 0.340(x - 13.2)$
$y = 0.340x + 13.3.$

1-4 Problems

1. The formula $v = at + v_0$ gives the functional relationship between velocity (v) and time (t) when a is the constant acceleration and v_0 the initial velocity. Write the equation in point-slope form. If $a = 9.8$ meters per second per second, find the initial velocity (v_0) of an object that reaches a velocity of 87 meters per second in 3.2 seconds.
2. The temperature of a boiler rises at a constant rate of 2.3 degrees per minute. If the temperature is 132.5° F. after 35 minutes, what is the equation relating time and temperature?
3. Using the equation of Problem 2, find the temperature after one hour.

1-5 Determining Linear Functions from Two Points

If two points of a linear function are known, then it is easy to find the slope of the line of the function by Eq. 1-2a,

$$m = \frac{y_2 - y_1}{x_2 - x_1}.$$

If we substitute this value for m in $y - y_1 = m(x - x_1)$, we get the two-point formula

$$y - y_1 = \frac{y_2 - y_1}{x_2 - x_1}(x - x_1). \qquad \text{(Eq. 1-5a)}$$

Example 1

Determine the equation of the linear function whose graph passes through $P_1(2, 4)$ and $P_2(4, 10)$.

i. $y - y_1 = \dfrac{y_2 - y_1}{x_2 - x_1}(x - x_1)$.

ii. $y - 4 = \dfrac{10 - 4}{4 - 2}(x - 2)$.

iii. $y - 4 = \boxed{}\,(x - 2)$.

iv. $y = \boxed{}$.

■ iii. 3; iv. $3x - 2$.

Example 2

If $f(-2) = -1$, $f(-1/2) = -4$, and f is a linear function, determine the equation of f.

i. $y - (-1) = \dfrac{(-4) - (-1)}{\boxed{}}(x - (-2))$.

ii. $y + 1 = \boxed{}\,(x + 2)$.

iii. $y = \boxed{}$.

■ i. $\left(-\frac{1}{2}\right) - (-2)$; ii. (-2); iii. $-2x - 5$.

Example 3

If a speed of 15 miles per hour is equal to 22 feet per second, what speed in miles per hour is 40 feet per second? (*Note*: Since zero miles per hour is equal to zero feet per second, we actually have two known, ordered pairs of numbers.)

If we let y represent speed in miles per hour and x speed in feet per second, then

$$y - 0 = \frac{15 - 0}{22 - 0}(x - 0),$$

$$y = \frac{15}{22}x, \quad \text{or} \quad f(x) = \frac{15}{22}x;$$

so if

$$x = 40, \qquad f(40) = \boxed{}$$

■ $f(40) = \frac{15}{22}(40)$

$= 27\frac{3}{11}$ miles per hour.

1-5 Exercises

1. Write the equation of the straight line passing through $P_1(1, 2)$ and $P_2(4, 5)$.

---

2. Write the equation of the straight line passing through $P_1(3, -4)$ and $P_2(-2, 3)$.

■ 1. $y - 2 = \frac{5-2}{4-1}(x - 1)$

$y = x + 1.$

---

3. Write the equation of the straight line passing through $P_1(-4, -6)$ and $P_2(6, -2)$.

■ 2. $y + 4 = \frac{3+4}{-2-3}(x - 3)$

$y + 4 = -\frac{7}{5}x + \frac{21}{5}$

$y = -\frac{7}{5}x + \frac{1}{5}.$

---

4. Write the equation of the straight line passing through $P_1(-7, 5)$ and $P_2(3, 5)$.

■ 3. $y + 6 = \frac{-6+2}{-4-6}(x + 4)$

$y + 6 = \frac{2}{5}x + \frac{8}{5}$

$y = \frac{2}{5}x - \frac{22}{5}.$

---

5. If $f(6) = 5$, $f(0) = 0$, and f is a linear function, write the equation of f.

■ 4. $$y - 5 = \frac{5-5}{3+7}(x+7)$$
$$y - 5 = 0$$
$$y = 5.$$

---

6. If the graph of a linear function passes through $P_1(-4, 2)$ and $P_2(3, 1/2)$, what is the y-intercept?

■ 5. $$y - 0 = \frac{5-0}{6-0}(x-0)$$
$$y = \frac{5}{6}x.$$

■ 6. $$y - 2 = \frac{1/2 - 2}{3+4}(x+4)$$
$$y - 2 = -\frac{3}{14}x - \frac{6}{7}$$
$$y = -\frac{3}{14}x + \frac{8}{7};$$
y-intercept $= 8/7$.

1–5 Problems

The following problems are applications of linear functions. In each problem two points are given from which the equation can be determined. Give attention to which variable is most likely independent.

1. A stroboscope enables a technician to determine that a motor increasing in speed at a constant rate turns at 120 rps at the end of 1.08 seconds and 240 rps at the end of 2.16 seconds. Determine the equation that relates time (t) to rate of rotation (w).
2. Using the equation from Problem 1, find w when $t = 1.50$ seconds.
3. The kinetic energy (KE) of a photoelectron having escaped a metal surface is a linear function of the frequency (f) of the light falling on the metal, minus the energy (w) required to free the electron from the metal surface ($KE = hf - w$). If $KE = 2.00$ ev when $f = 5.75 \times 10^{26}$ and 2.98 ev when $f = 7.25 \times 10^{26}$, determine the equation constants h and w.
4. Using the equation of Problem 3, find KE when $f = 6.00(10^{26})$.
5. Over the linear portion of the operating range of a transistor, I_c ranges from 15 to 30 milliamperes as I_b ranges from .012 to .030 milliamperes. If the quiescent bias current, I_b, equals .021 milliamperes, find the corresponding value of I_c.

6. If 0° C corresponds to 32° F and 100° C corresponds to 212° F, determine the equation relating the two temperature scales.
7. The nominal length of a steel measuring tape is 100.000 feet. At 70° F the actual length to the nearest thousandth is 100.009 feet. At 106° F the actual length is 100.035 feet.
    a. Find the equation that gives actual length as a function of temperature.
    b. What is the actual length at 90° F?
    c. At what temperature are the actual and nominal lengths equal?

1-6 Linear Inequalities

Statements of inequality are very similar to equations. The statement $y > 2x + 3$ can best be studied by first considering $y = 2x + 3$. If we graph $y = 2x + 3$, the coordinates of every point on the line satisfy the equation. (See Fig. 1-6a). The coordinates of every point above the line satisfy the inequality $y > 2x + 3$.

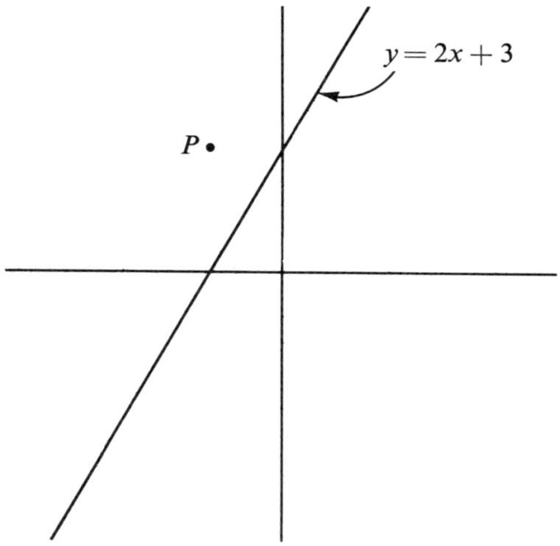

Fig. 1-6a

For example, point P with coordinates $(-2, 4)$ is above the line and $4 > 2(-2) + 3$ is a true statement. The solution set of the inequality is the half-plane illustrated in Fig. 1-6b. The dotted line represents the boundary points, which are not part of the solution set.

Simplifying and solving inequalities is similar to solving equations except that, if each number of an inequality is multiplied or divided by a negative number, the direction of the inequality is reversed.

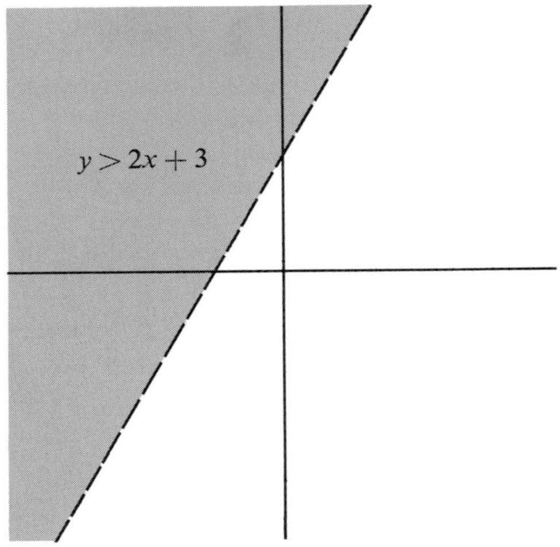

Fig. 1-6b

Example 1

Solve for y and graph the inequality $3x - 5 < 4(x + y)$.
Solution:

i. $3x - 5 < 4x + 4y$.

ii. $-x - 5 < 4y$.

iii. $-4y < x + 5$.

iv. $y \text{ R } -\frac{x}{4} - \frac{5}{4}$.

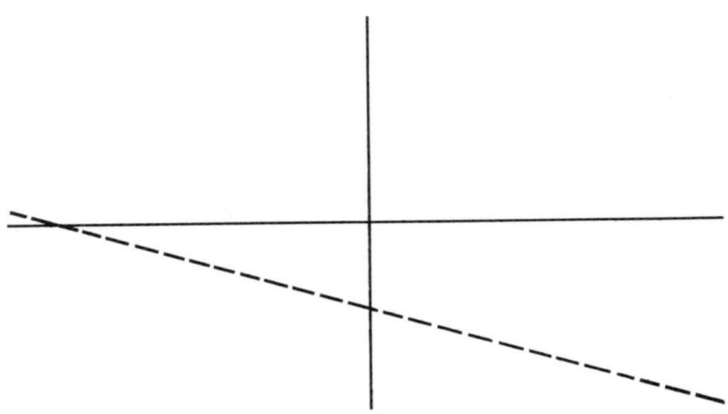

Fig. 1-6c

v. The graph can be drawn readily by first considering the graph of

$$y = -\frac{x}{4} - \frac{5}{4}.$$

Complete Fig. 1-6c.

■ iv. $y > -\frac{x}{4} - \frac{5}{4}.$

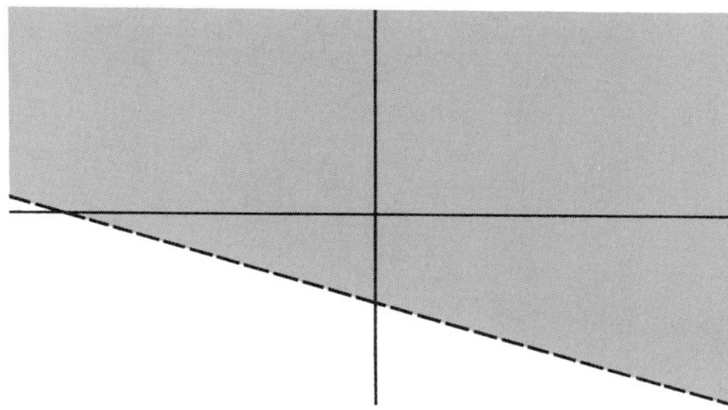

Example 2

Graph the set of ordered pairs (x, y) such that $-3x \geq 9$.
Solution:
If $-3x \geq 9$, then x ☐ -3. See Fig. 1-6d. A solid boundary implies that points on the boundary are also in the solution set.

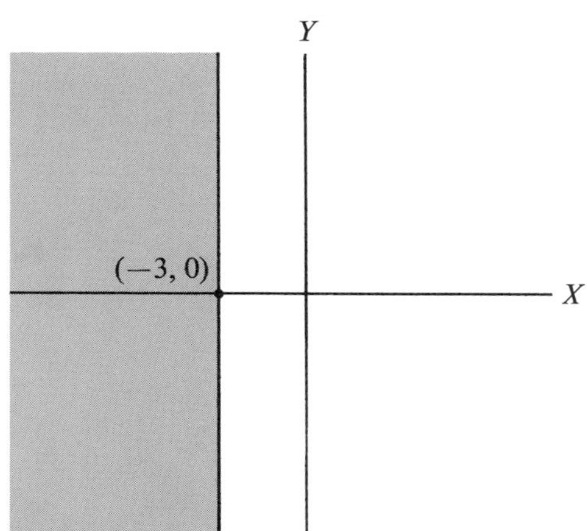

Fig. 1-6d

■ $x \leq -3$.

Notice that all y's are permitted, since there is no restriction on them.

1-6 Exercises

Remember to use an answer shield.

1. Graph $y > \frac{2}{3}x - 4$.

---

2. Graph $y \leq \frac{x+5}{2}$.

■ 1.

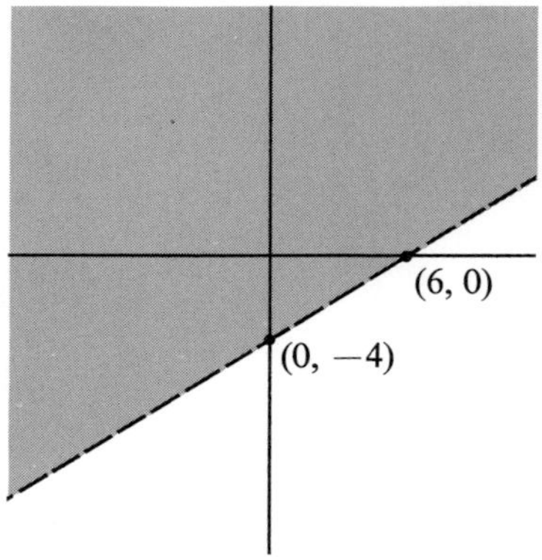

---

3. Simplify the inequality
$3 - 4x < 2x - 15,$
and graph the solution set on the number line.

■ 2.

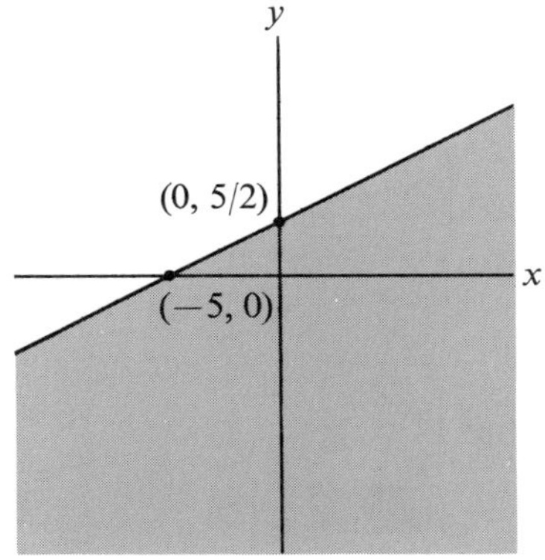

---

4. Graph the set of ordered pairs (x, y) such that $6 - 2x < 10$.

■ 3.

$$\begin{aligned} 3 - 4x &< 2x - 15 \\ -6x &< -18 \\ x &> 3. \end{aligned}$$

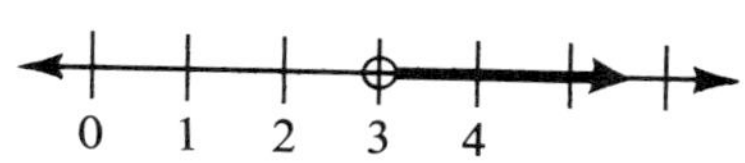

---

5. Graph $\frac{x}{2} - \frac{y}{5} > 1$.

■ 4.

$$\begin{aligned} 6 - 2x &< 10 \\ -2x &< 4 \\ x &> -2. \end{aligned}$$

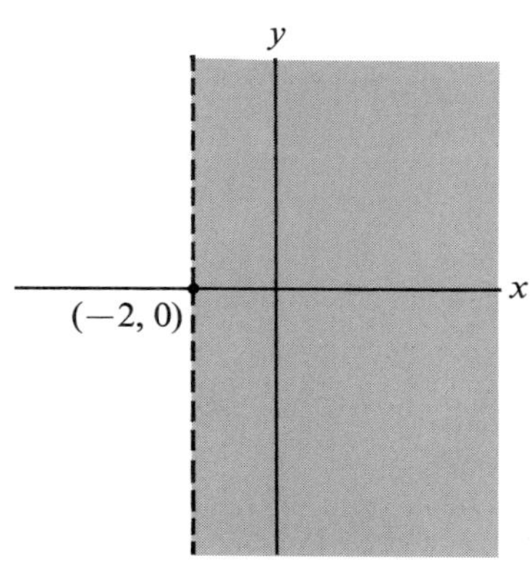

6. Graph $y - 0.50 < 0.25(x + 3.00)$.

■ 5. $\frac{x}{2} - \frac{y}{5} > 1$

$$5x - 2y > 10$$

$$-2y > -5x + 10$$

$$y < \frac{5}{2}x - 5.$$

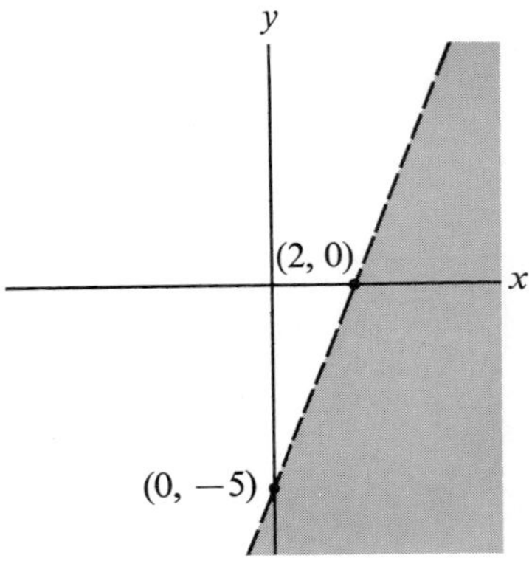

■ 6. $y < .25x + 1.25$.

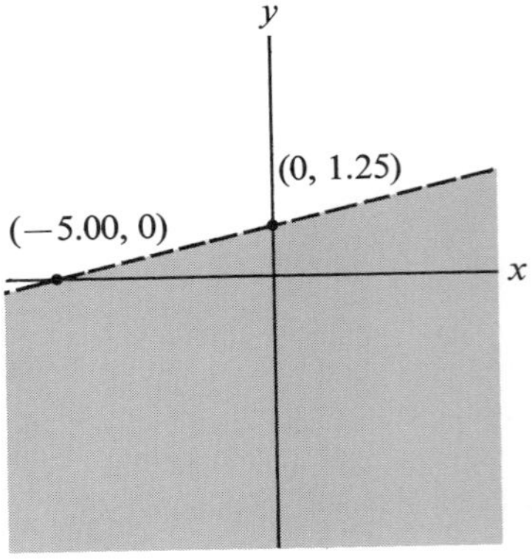

1–6 Problems

1. Products A and B require the same machine operation. If product A takes 12 minutes of machine time per item and product B takes 20 minutes of machine time per item, what is the possible total production of products A and B from a machine in an 8-hour shift? Represent the solution graphically.
2. A tire manufacturer makes two grades of tires, grade A wholesaling at \$20 and grade B at \$12. The annual total sales must be greater than \$500,000 to show a profit. Represent the information graphically.
3. If the specifications call for a steel rod to be 40 inches long with less than 2 percent error in length and if the actual length of the rod is indicated by L, does the compound inequality $39.2 < L < 40.8$ correctly state the situation?

1-7 Related Functions

The motion of a freely falling object near the surface of the earth is described by three very simple formulas that are probably familiar to anyone who has studied some physical science.

$$s = 16t^2 \qquad \text{(Eq. 1-7a)}$$

$$v = 32t \qquad \text{(Eq. 1-7b)}$$

$$a = 32 \qquad \text{(Eq. 1-7c)}$$

The first equation gives the number of feet (s) the object falls from rest in (t) seconds. The second gives the velocity (v) in feet per second at any time (t) during its fall. The third equation says that the acceleration (a) is a constant 32 feet per second per second. We now want to show that these three functional equations—and others like them—have a very unique mathematical relationship: if only one of them is known, the others can be derived.

We studied in Section 1-2 the related functional equations $f(x) = mx + b$ and $f'(x) = m$. Equations 1-7b and 1-7c are illustrations of f and f':

$$v = f(t) = 32t \quad (b = 0),$$

$$a = f'(t) = 32.$$

In order to discover a second important function related to f, we shall first graph $v = f(t) = 32t$. Since the slope equals 32, we shall use a small vertical scale and plot only nonnegative values of time (t).

t (sec)	v (ft/sec)
1	32
2	
3	
t	

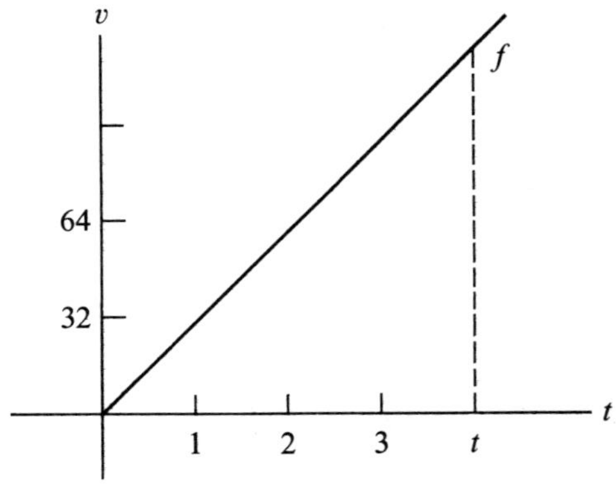

■ (2, 64); (3, 96); (t, $32t$).

Now let us examine the **area** enclosed by the t axis, the abscissa t, and the graph of the function f. The plane geometric figure formed by segments of these lines is a right triangle.

$$\begin{aligned}(\text{Area}) &= \frac{1}{2}(\text{altitude})(\text{base}) \\ &= \frac{1}{2}(32t)(t) \\ &= \boxed{}.\end{aligned}$$

■ $16t^2$

It is very reasonable that the result for this area is the same as the expression for s in Equation 1-7a. Distance (s) is ordinarily the product of average velocity and time. Since in this example the velocity is changing at a constant rate from zero to $32t$, the average value $16t$ multiplied by the time t does in fact equal $16t^2$.

For future reference and convenience, let us use F to represent the function that gives the area between the nonnegative graph of a function f, the horizontal axis, the vertical axis, and any positive abscissa. This notation gives $s = F(t) = 16t^2$.

Let us experiment with a second problem.

Assume we have a functional equation $f(s) = ks$, where $k > 0$. Determine the equations for the related functions f' and F.

■ $f(s) = ks, k > 0$

$f'(s) = k$

$$F(s) = \frac{1}{2}(s)(ks)$$

$$= \frac{1}{2}ks^2 .$$

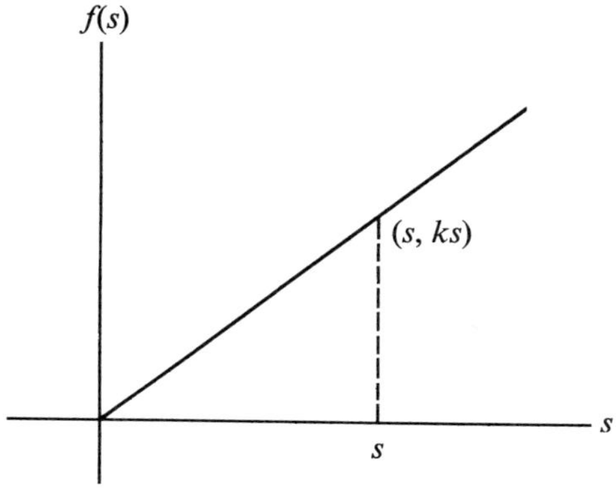

Although it may appear that we have only a function f, its slope k, and the area $(1/2)ks^2$, these three functions in fact describe the characteristics of a simple spring. The function f is known as Hooke's Law and describes the

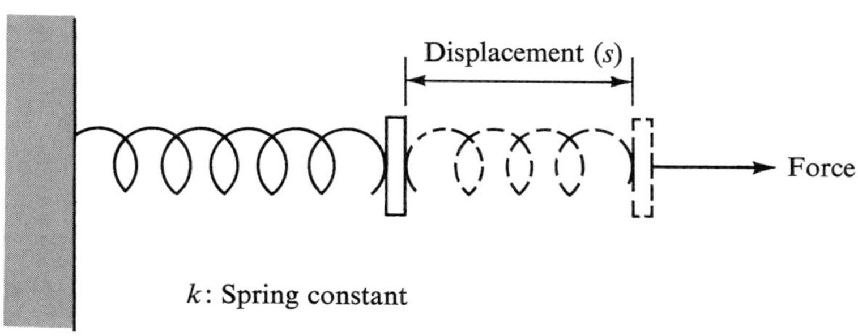

relationship between force and displacement (s). The function f' describes how the force changes with respect to displacement; this is a property determined by the stiffness of the spring, called the spring constant (k). The function F describes the *work* required to stretch the spring. Work is defined in physics as the product of *average* force and displacement.

1-7 Exercises

1. Graph $f(x) = 2x$. Calculate the area enclosed by f, $x = 5$, and the x-axis.

2. Given $f(x) = mx$, determine the equation for F assuming $m > 0$.

■ 1.

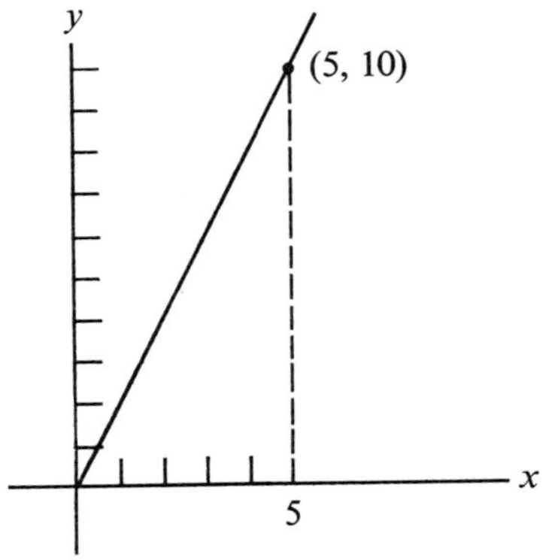

$$\text{Area} = \frac{1}{2}(\text{base})(\text{height})$$
$$= \frac{1}{2}(5)(10)$$
$$= 25.$$

3. Using the equation derived in Exercise 2, calculate $F(5)$ when $m = 2$.

■ 2.

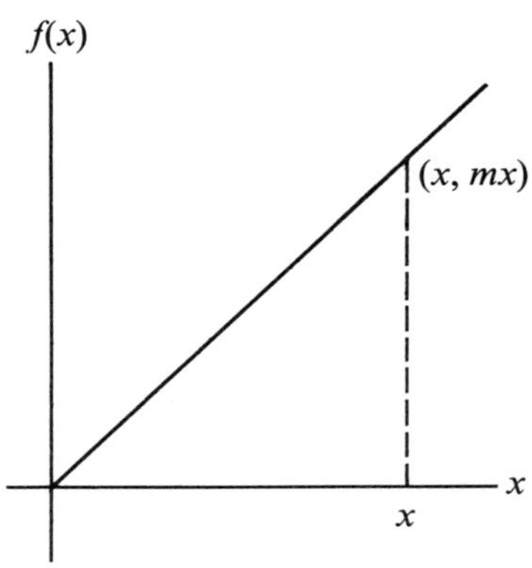

$$F(x) = \frac{1}{2}(x)(mx)$$
$$= \frac{1}{2}mx^2.$$

---

4. If $F(x) = 2x^2$ represents the area designated on the coordinate system, determine the equation of the function.

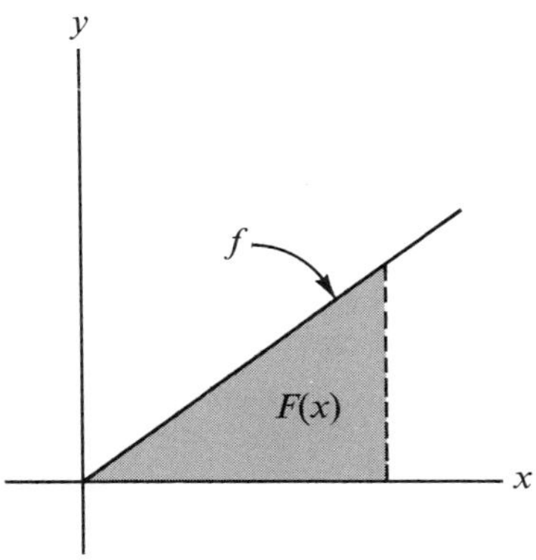

■ 3. $F(x) = \frac{1}{2}(2)x^2$

$$F(5) = \frac{1}{2}(2)(5)^2$$
$$= 25.$$

This result is identical to that calculated in Exercise 1.

---

5. If $f(x) = 3.2x$, calculate $F(0.1)$.

■ 4. $F(x) = 2x^2$

$$F(x) = \frac{1}{2}mx^2.$$

Therefore, $\frac{1}{2}mx^2 = 2x^2$, or $m = 4$. The function f has the form $f(x) = mx$, so $f(x) = 4x$. (See Exercise 2.)

6. If $f(x) = \frac{2}{3}x$, calculate the area designated on the coordinate system.

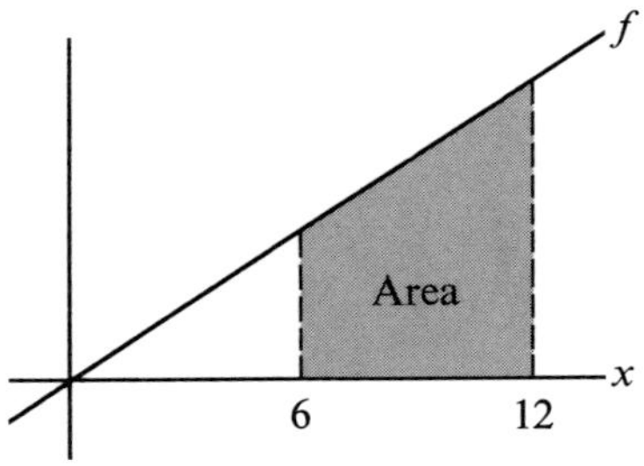

■ 5. $F(x) = \frac{1}{2}mx^2$

$= \frac{1}{2}(3.2)x^2;$

$F(0.1) = \frac{1}{2}(3.2)(0.1)^2$

$= .016.$

■ 6. $F(x) = \frac{1}{2}mx^2$

$= \frac{1}{2}\left(\frac{2}{3}\right)x^2$

$= \frac{x^2}{3};$

$(\text{Area}) = F(12) - F(6)$

$= \frac{(12)^2}{3} - \frac{6^2}{3}$

$= 36.$

1-7 Problems

1. Only the equation $f(x) = mx$ and its related functions are discussed in this section. By using area formulas it is possible to determine the related f' and F functions from the more general linear function $f(x) = mx + b$.

a. Determine the area designated on the coordinate system in Fig. 1-7a where $f(x) = 2x + 4$.

b. Develop the general equation for F when $f(x) = mx + b$, $m > 0$, and $b > 0$.

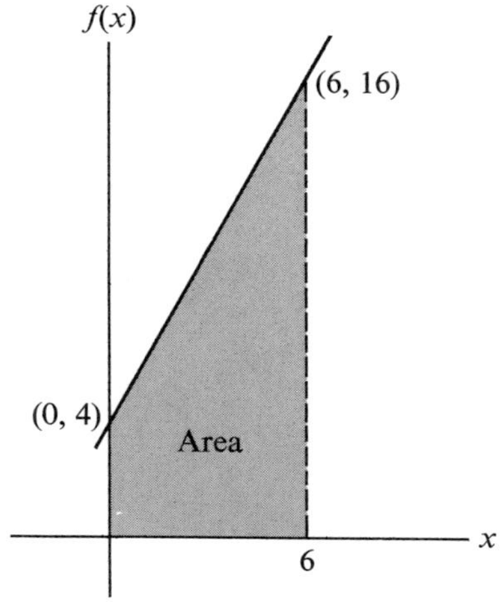

Fig. 1-7a

2. An airplane is designed to accelerate at 15 ft/sec² and is airborne at a velocity of 180 ft/sec. What is the distance it must ordinarily travel on the runway before it is airborne? (*Hint:* This problem requires that you find the three related functions f', f, and F beginning with $f'(t) = 15$.)
3. If a simple spring has a spring constant $k = 6$ pounds per inch, find the work required to stretch it 5 inches.
4. How much work is required to stretch the spring in Problem 3 another inch, that is, from 5 to 6 inches?
5. Assume the current (i) flowing in a circuit is a linear function of time (t), such that $i = 0.05t$. The total charge (q) transmitted in the circuit is the area under the graph of $f(t) = i = 0.05t$ from $t_1 = 0$ to $t_2 = t$. Find the formula for total charge as a function of time.

1 Supplementary Exercises and Problems

1. If $f(t) = 9.8t$, find $f(0)$, $f(.005)$, and $f(7.6)$. (1-1)
2. If $f(s) = ks$, find k when $f(3.2) = 27.6$. (1-1)
3. If $f(w) = kw$ and $f(-2/3) = 11/6$, find $f(1)$, $f(\frac{1}{2})$, and $f(12/5)$. (1-1)
4. Express selling price (P) as a function of cost (c) when the mark-up is 25% of the cost. Find $f(1.00)$, $f(5.00)$, and $f(15.75)$. (1-1)
5. Express selling price (P) as a function of cost (c) when cost is 75% of selling price. Find $f(1.00)$, $f(5.00)$, and $f(15.75)$. (1-1)

6. At 80° Fahrenheit a 100.000-foot steel measuring tape has an actual length of 100.011 feet, measured to the nearest foot.
   a. Find the equation that gives corrected measurements (C) as a function of nominal measurements (N) for measurements taken at 80° F. (*Hint:* Use ordinary proportion.)
   b. Complete the following table using the equation developed in part (a).

80°F

Nominal (N) (feet)	Corrected (C) (feet)
38.049	
94.765	
476.819	

(1-1)

7. Plot the points $P_1(4, 2)$, $P_2(2, 4)$, $P_3(-4, -2)$, $P_4(-2, -4)$, $P_5(-4, 2)$, $P_6(4, -2)$, $P_7(-2, 4)$, and $P_8(2, -4)$ on a coordinate graph. (1-2)
8. Which variable in the ordered pair (t, p) is plotted on the horizontal axis? (1-2)
9. What is the slope of the linear function containing $P_1(-4, 3.2)$ and $P_2(2.5, -2)$? (1-2)
10. Graph $f(x) = 2.5x$ and $g(x) = -2.5x$ on the same coordinate system. (1-2)
11. If $f(1.5) = 2.1$ and $f(3.5) = 4.9$, what is the slope of f if f is linear? (1-2)
12. The graphs of the functions f and g in Fig. 1-7b are parallel. If the equation of f is $f(x) = 1.5x$, what is the equation of g? (1-3)

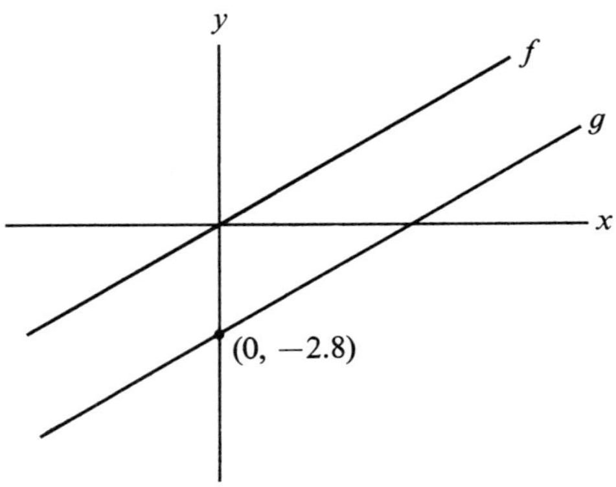

Fig. 1-7b

13. Rewrite the equation $F = (9/5)C + 32$ in functional notation so that C is a function of F. (1-3)

14. Solve the equation

$$a = \frac{v - v_0}{t}$$

for v. If v is a function of t, what is the "v-intercept"?

15. Corrections (C_t) in the length of a 100-foot steel measuring tape (nominal length at 70° F) is a function of temperature (t). Using appropriate scales, graph $C_t = .000645(t - 70°)$. What is the slope? What is the "C_t-intercept"? (1-3)

16. Find the equation for g (perpendicular to f) in Fig. 1-7c. Where does g cross the x-axis? (1-3)

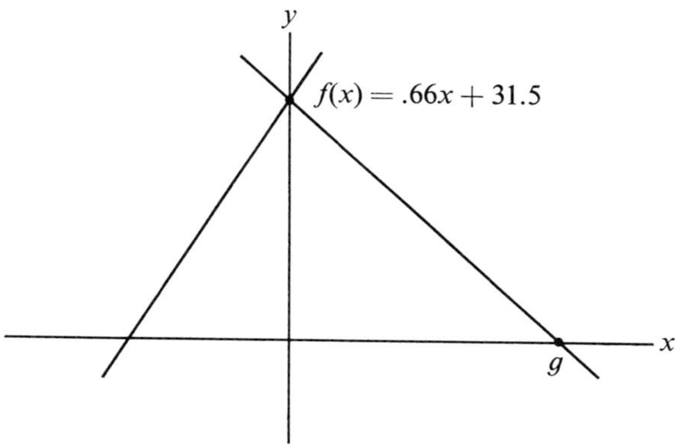

Fig. 1-7c

17. What is the y-intercept of the linear function with slope $-.02$ passing through $(12, -.08)$? (1-4)

18.

Linear function	Slope	Point
g	2/3	(6, 5)
h	2/3	(−3, −4/3)

Are the graphs of g and h colinear? Why? (1-4)

19. Find the equation of the line passing through (12.2, 15.4) and perpendicular to $f(x) = -1.3x + 4.5$. (1-4)

20. Write the equation of the linear function passing through $P_1(2/5, -3/5)$ and $P_2(4, 6\frac{1}{2})$. (1-5)

21. If $g(.2) = 120$, $g(.8) = 75$, and g is linear, find the equation for g. (1-5)

22. Many formulas require temperatures to be given in the absolute Kelvin (K) scale. If 32° Fahrenheit (F) corresponds to 273.16° K and 212° F corresponds to 373.16° K, find the equation that gives temperature on the Kelvin scale as a function of temperature on the Fahrenheit scale. (1-5)

23. Determine the intercepts of the load line in Fig. 1-7d. (1-5)

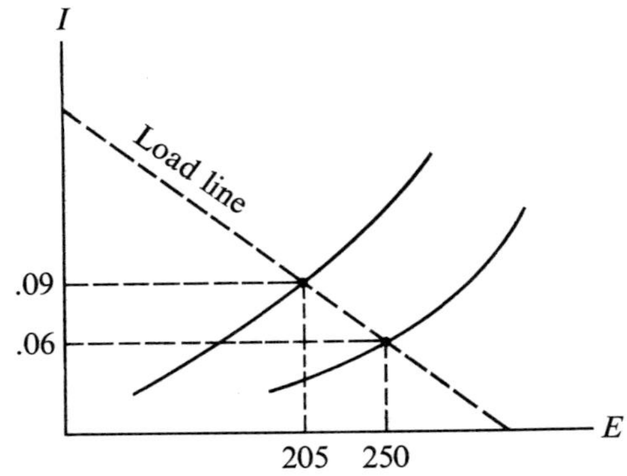

Fig. 1-7d

24. A straight line passes through $P_1(-4.3, -6.2)$ and $P_2(2.1, -3.8)$. Will it pass through P_3 (10.1, 2.7)? (1-5)

25. Graph the inequality $x - 2y < 4$. (1-6)

26. Write the inequality for the solutions illustrated in the graph. (1-6)

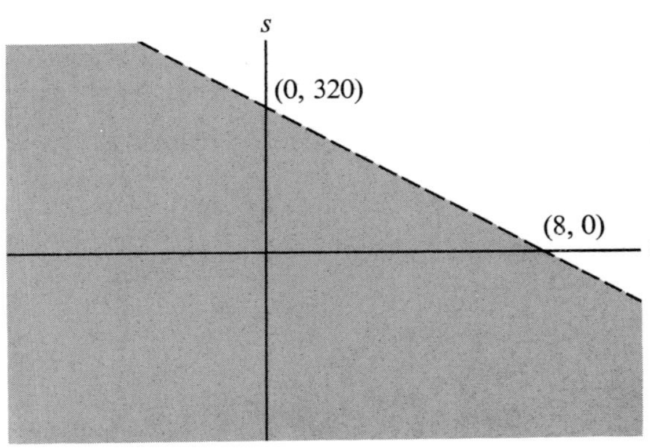

27. Graph the inequality

$$\frac{y-2}{5} \geq \frac{x-4}{3}. \qquad (1\text{-}6)$$

28. Find the designated area. (1-7)

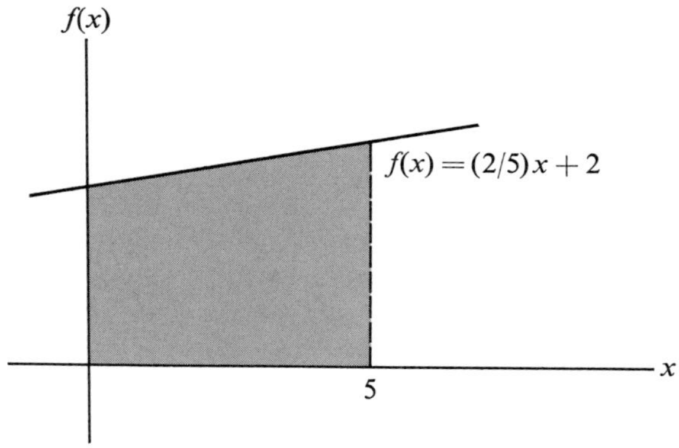

29. Find the designated area. (1-7)

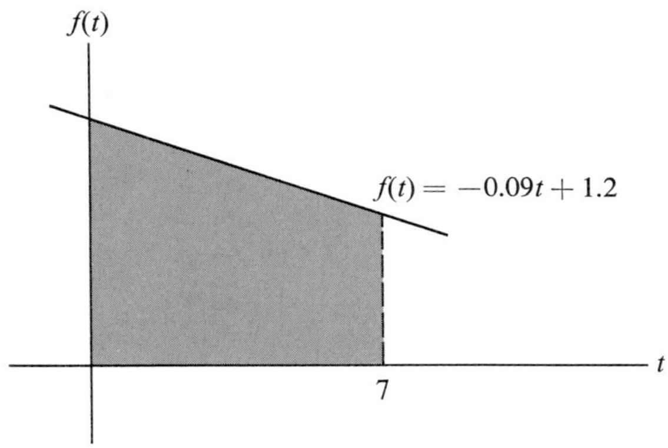

30. Find the shaded area. (1-7)

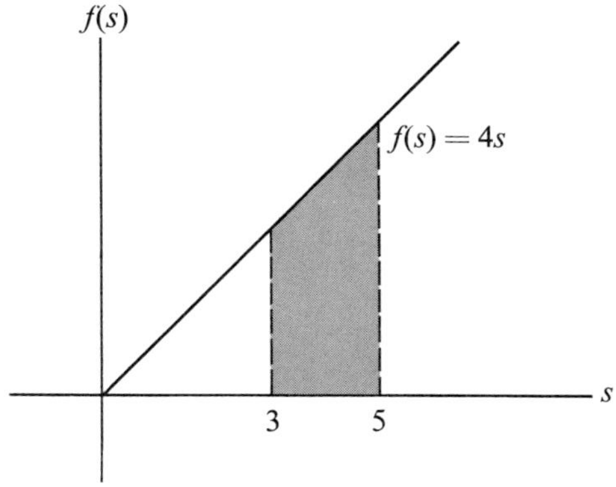

31. Find the shaded area in literal terms. (1-7)

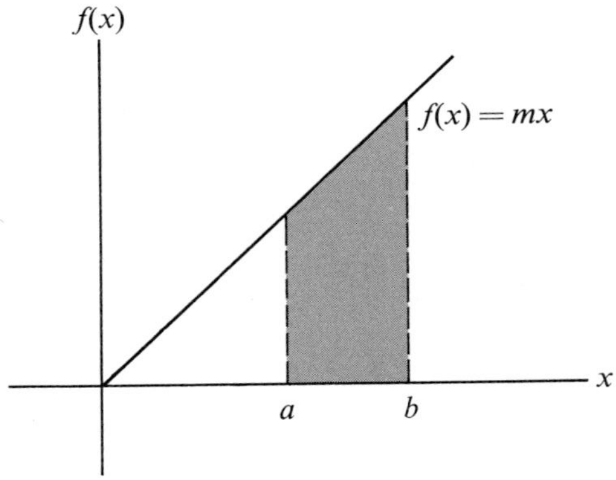

32. Find the shaded area. (1-7)

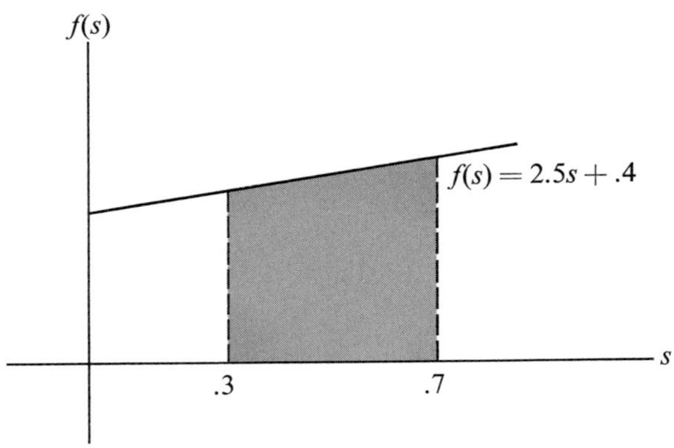

Chapter Two Systems of Linear Equations

2-1 Graphical Solutions—Two Variables

The relationship between sales and expenses in producing salable goods is often analyzed by using mathematical models. The models seek to capture the essential features of real-world situations. For example, assume that b represents a factory's fixed expenses in preparing to manufacture some particular item. Some fractional part (m) of all sales (s) is also expense. Therefore the mathematical model for total expense (t) is

$$t = ms + b. \qquad \text{(Eq. 2-1a)}$$

The business objective is to produce and sell beyond the *break-even* point, where expenses equal sales; in other words, where

$$t = s. \qquad \text{(Eq. 2-1b)}$$

Both Eqs. 2-1a and 2-1b represent linear functions between the variables s and t. Each equation alone has an unlimited number of solutions. However, there is one point represented in Fig. 2-1a that is the intersection of the two graphs, and the ordered pair of numbers corresponding to this point is the only solution common to both solution sets. In our example this is the break-even point.

Many times the two linear equations define independent relationships between the same two variables. If their individual solution sets intersect, the intersections are usually of considerable importance.

How many intersections may occur between the solutions of two different linear equations in two unknowns?

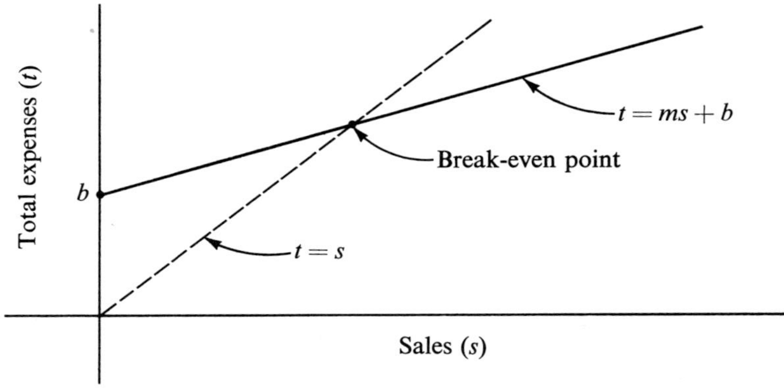

Fig. 2-1a

■ Because the graph of a linear equation in two unknowns is a straight line, there is at most one intersection.

Let us consider another, but quite different, example.

Example 1

Suppose an airplane has enough fuel for 5 hours of flying time cruising at 240 miles per hour (air speed). If a 30-mile-per-hour east wind is blowing, how far east can the plane fly and return before the fuel is gone?

Relative ground speed against the wind is 210 miles per hour. Relative ground speed with the wind is 270 miles per hour. Let t represent the time flying against the wind and $(5 - t)$ the time flying with the wind. If d represents the distance one way, then

$$d = 210t \quad \text{and} \quad d = 270(5 - t).$$

Graph both equations on the same coordinate system.

■

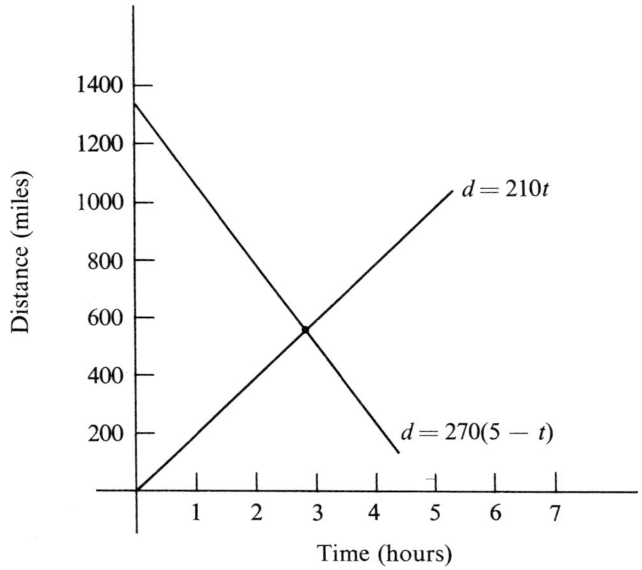

The intersection of the graphs is the point whose coordinates satisfy both equations. From the graph the distance is approximately ______ miles.

■ 590

If the flight is over water, the 590-mile distance in this problem is called the point of no return. Notice that this point does not correspond to one-half the flying time; in fact, it seldom does.

2-1 Exercises

1. Approximate graphically a solution common to both equations in the following system:

$$x + y = 6$$
$$x - y = 4.$$

2. Approximate graphically a solution common to both equations in the following system:

$$x + y = -2$$
$$x - y = 6.$$

■ 1.

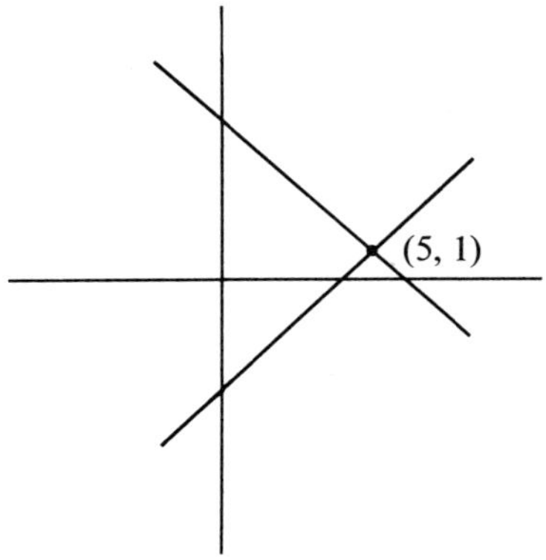

3. Determine graphically the simultaneous solution of the following system:

$$3x - y = 5$$
$$2x + 3y = 4.$$

■ 2.

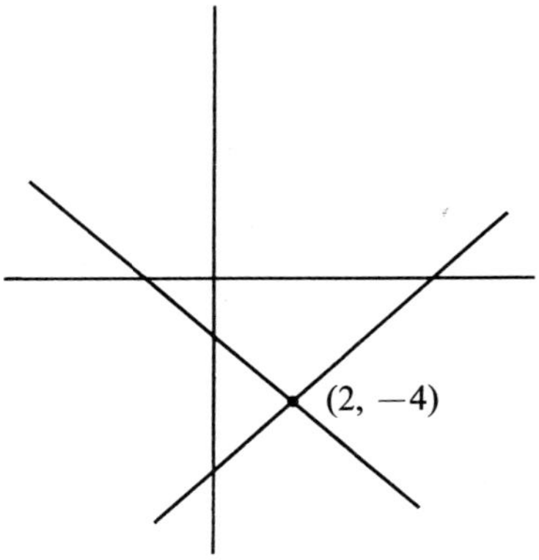

4. Determine graphically the simultaneous solution of the following system:

$$x = 2y$$
$$x - y = 2.$$

■ 3.

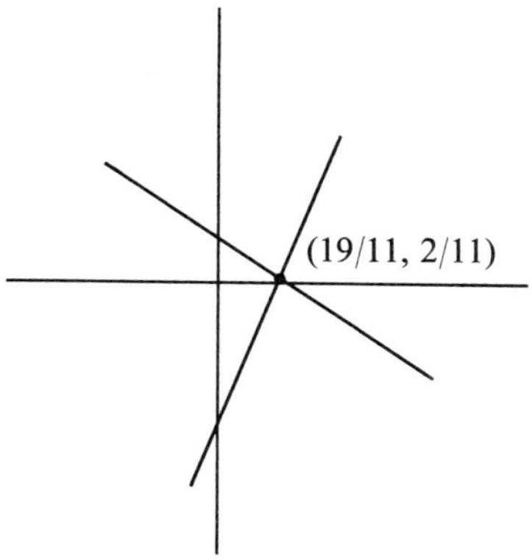

5. The sum of two numbers is 13, and their difference is 3. Let x be the larger and y the smaller unknown number. Write two equations that show their relationship. Graph each and find their intersection.

■ 4.

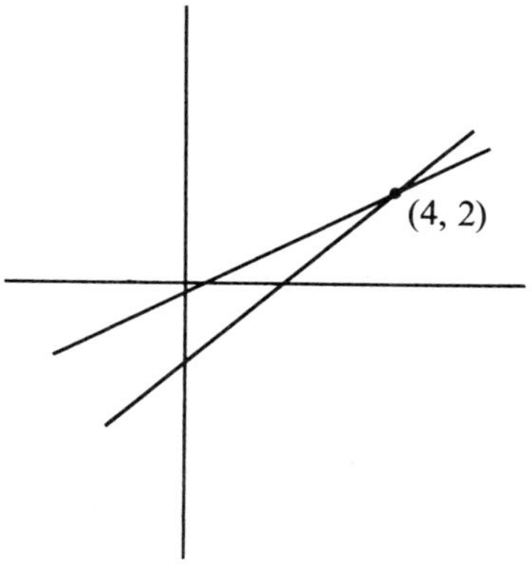

■ 5. $x + y = 13$
$x - y = 3$.

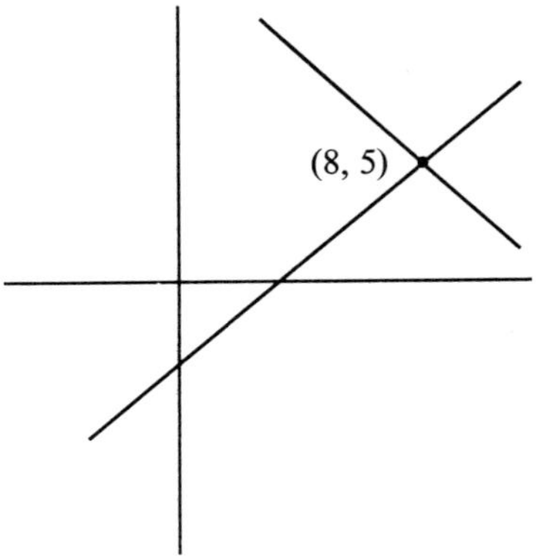

2-1 Problems

1. Suppose in Eq. 2–1*a* that $m = 0.3$ and $b = \$11,200$. Approximate graphically the sales required to break even. What effect does an increase in m have on the break-even point?
2. Suppose the airplane's fuel supply will last 6 hours if the air speed is reduced to 200 miles per hour. How is the point of no return affected under the same flying conditions that existed in Example 1?
3. The unknown values of current (I_1 and I_2) in the branches of the electric circuit in Fig. 2–1b can be determined by finding the simultaneous solution to the two equations

$$60I_2 + 20I_1 = 6$$

and

$$20I_2 + 80I_1 = 12.$$

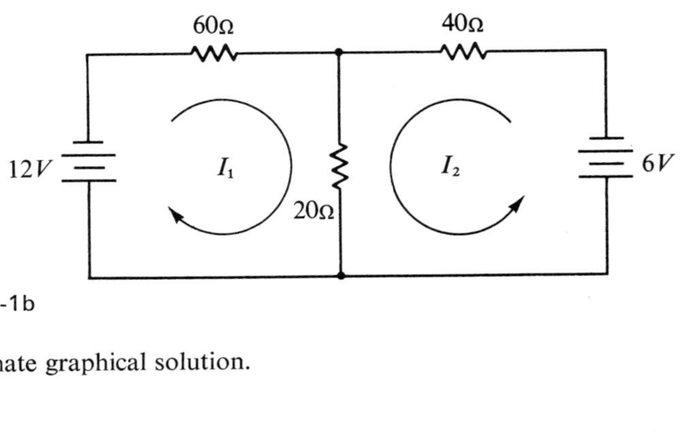

Fig. 2-1b

Find an approximate graphical solution.

4. By graphical methods, show the solutions in the first quadrant that are common to

$$2x + y \leq 18$$

and

$$y - x \leq 6.$$

2-2 Algebraic Solutions—Two Variables

Graphical solutions to systems of equations have the same disadvantages as measurements in that they are approximations. Problem 3 of Section 2-1 is an example of how inadequate a graphical solution may be. In both branches of the circuit (Fig. 2-1b) very small values of current flow, which is characteristic of modern circuits. Even a reasonably good graph results in what appears to be no current in either branch.

Our purpose in this section is to develop algebraic methods for solving systems of linear equations; the accuracy of the results is then limited only by the accuracy of the coefficients in the equations.

Let us consider some of the relationships that exist in a system of linear equations in two unknowns. In order to do this it is necessary for us to have two linear equations where we can assume the coefficients are exact numbers and the exact intersection of the two is known.

Check to see that these conditions are met in the system

$$3x - 2y = -3, \qquad \text{(Eq. 2-2a)}$$

$$5x - y = 2, \qquad \text{(Eq. 2-2b)}$$

where the intersection is (1, 3).

■ i.
$$3x - 2y = -3$$
$$3(1) - 2(3) \stackrel{?}{=} -3$$
$$-3 = -3.$$

ii.
$$5x - y = 2$$
$$5(1) - 3 \stackrel{?}{=} 2$$
$$2 = 2.$$

In the following experiments, we shall for convenience refer to Eq. 2-2a and Eq. 2-2b as only (a) and (b), respectively.

Experiment 1

Multiply the left and right members of Equation (a) by some constant k. Does the resulting equation have in its solution set the ordered pair (1, 3)?

■ i. $3x - 2y = -3$ (a)
$3kx - 2ky = -3k,$ (a) multiplied by k

ii. $3k(1) - 2k(3) \stackrel{?}{=} -3k$
$-3k = -3k.$

Experiment 2

Multiply the left and right members of Equation (b) by some constant k. Does the resulting equation have in its solution set the ordered pair (1, 3)?

■ i. $5x - y = 2$ (b)
$5kx - ky = 2k,$ (b) multiplied by k

ii. $5k(1) - k(3) \stackrel{?}{=} 2k$
$2k = 2k.$

Experiment 3

Subtract Equation (b) from (a), left side from left side and right side from right side. Does the resulting equation have (1, 3) in its solution set?

■ i.

$$\begin{array}{rl} 3x - 2y = -3 & \text{(a)} \\ 5x - y = 2 & \text{(b)} \\ \hline -2x - y = -5 & \text{(a) minus (b)} \end{array}$$

ii. $-2(1) - 3 \stackrel{?}{=} -5$
$-5 = -5.$

Experiment 4

Add Equations (a) and (b), left side to left side and right side to right side. Does the resulting equation have (1, 3) in its solution set?

■ i.

$$\begin{array}{rll} 3x - 2y = & -3 & \text{(a)} \\ 5x - \ \ y = & 2 & \text{(b)} \\ \hline 8x - 3y = & -1 & \text{(a) plus (b)} \end{array}$$

ii.

$$8(1) - 3(3) \stackrel{?}{=} -1$$
$$-1 = -1.$$

Experiment 5

Multiply Equation (b) by (− 2) and add the result to Equation (a). Does the final result have (1, 3) in its solution set?

■ i.

$$\begin{array}{rl} 5x - y = 2 & \text{(b)} \\ -10x + 2y = -4 & \text{(b) multiplied by } (-2) \end{array}$$

ii.

$$\begin{array}{rl} 3x - 2y = -3 & \text{(a)} \\ -10x + 2y = -4 & \text{(b) multiplied by } (-2) \\ \hline \end{array}$$

iii.

$$-7x + 0y = -7 \qquad \text{(sum)}$$

iv.

$$-7(1) + 0(3) \stackrel{?}{=} -7$$
$$-7 = -7.$$

Experiment 6

Multiply Equation (a) by 5 and Equation (b) by (− 3) and add the results. Does the final result have (1, 3) in its solution set?

■ i. $15x - 10y = -15$ (a) multiplied by (5)

ii. $-15x + \ 3y = -6$ (b) multiplied by (−3)

iii. $0x - \ 7y = -21$ (sum)

iv.

$$0(1) - 7(3) \stackrel{?}{=} -21$$
$$-21 = -21.$$

The experiments produced a *family* of equations with the common solution (1, 3). For our purposes, the most important results are in step (iii) of both Experiments 5 and 6. The resulting equation in each case is an equation of only one unknown, $-7x = -7$ and $-7y = -21$. The solutions to these two equations are {1} and {3}, respectively, the coordinates of the simultaneous solution of the original system.

Try to use the techniques suggested by Experiments 5 and 6 to find the simultaneous solution to the following system of equations:

$$2x + 3y = 5, \qquad \text{(Eq. 2-2c)}$$

$$3x - 4y = 6. \qquad \text{(Eq. 2-2d)}$$

■ i.

$$\begin{array}{rl} 6x + 9y = 15 & \text{(c) multiplied by (3)} \\ -6x + 8y = -12 & \text{(d) multiplied by } (-2) \\ \hline 17y = 3 & \text{(sum)} \\ \therefore\ y = \dfrac{3}{17} & \left\{\dfrac{3}{17}\right\} \end{array}$$

ii.

$$\begin{array}{rl} 8x + 12y = 20 & \text{(c) multiplied by (4)} \\ 9x - 12y = 18 & \text{(d) multiplied by (3)} \\ \hline 17x \qquad = 38 & \text{(sum)} \\ x = \dfrac{38}{17} & \left\{\dfrac{38}{17}\right\} \end{array}$$

iii. Check to see if (38/17, 3/17) is the simultaneous solution to the original system:

$$2\left(\frac{38}{17}\right) + 3\left(\frac{3}{17}\right) \stackrel{?}{=} 5 \qquad \text{substitution into (c)}$$

$$\frac{76}{17} + \frac{9}{17} \stackrel{?}{=}$$

$$\frac{85}{17} \stackrel{?}{=}$$

$$5 = 5.$$

$$3\left(\frac{38}{17}\right) - 4\left(\frac{3}{17}\right) \stackrel{?}{=} 6 \qquad \text{substitution into (d)}$$

$$\frac{114}{17} - \frac{12}{17} \stackrel{?}{=}$$

$$\frac{102}{17} \stackrel{?}{=}$$

$$6 = 6.$$

Step (ii) of the last example can be replaced by what is frequently an easier method for obtaining the second variable in a simultaneous solution of two equations in two unknowns. Since for each y (or x) there is only one x (or y), the x

(or y) coordinate of the simultaneous solution can be found by substituting the y (or x) value first obtained into either of the original equations and solving for x (or y).

For example, substitute $y = 3/17$ into Eq. 2-2c.

$$2x + 3\left(\frac{3}{17}\right) = 5,$$

$$2x = \left(\frac{85}{17}\right) - \left(\frac{9}{17}\right),$$

$$x = \frac{38}{17}.$$

The check of the simultaneous solution (38/17, 3/17) *must* be made by substitution into Eq. 2-2d, because Eq. 2-2c was used to find the x-value of the ordered pair.

What happens in an attempt to use an algebraic solution to a system of equations that has no simultaneous solution?

The equations $y = (2/3)\,x - 5$ and $y = (2/3)\,x + 1$ have the same slope. We know the graphs of this pair of equations have no intersection. If we subtract one equation from the other, the result is $0 = -6$, which is *false* for every ordered pair (x, y). We may expect this type of result in every attempt to solve a system that has no simultaneous solution.

2-2 Exercises

1. Determine algebraically the simultaneous solution of the system

$$\begin{aligned} x - 2y &= 2, \quad &\text{(a)}\\ x - y &= 8. \quad &\text{(b)} \end{aligned}$$

---

2. Solve simultaneously:

$$\begin{aligned} 2r + s &= 6, \quad &\text{(a)}\\ 3r + 4s &= 4. \quad &\text{(b)} \end{aligned}$$

■ 1. i.

$$\begin{aligned} x - 2y &= 2 \quad &\text{(a)}\\ x - y &= 8 \quad &\text{(b)}\\ \hline -y &= -6 \quad &\text{(difference)}\\ y &= 6 \quad &\{6\}. \end{aligned}$$

ii.

$$\begin{aligned} x - 2y &= 2 \quad &\text{(a)}\\ -2x + 2y &= -16 \quad &\text{(b) times } (-2)\\ \hline -x &= -14 \quad &\text{(sum)}\\ x &= 14 \quad &\{14\}. \end{aligned}$$

iii. Solution: (14, 6).

3. Solve simultaneously:

$$6v + 3t - 14 = 0, \qquad \text{(a)}$$
$$2v - t = 0. \qquad \text{(b)}$$

■ 2. i.
$$\begin{array}{rl} 8r + 4s = 24 & \text{(a) times 4} \\ 3r + 4s = 4 & \text{(b)} \\ \hline 5r = 20 & \text{(difference)} \\ r = 4 & \{4\}. \end{array}$$

ii.
$$\begin{array}{rl} 6r + 3s = 18 & \text{(a) times 3} \\ -6r - 8s = -8 & \text{(b) times } (-2) \\ \hline -5s = 10 & \text{(sum)} \\ s = -2 & \{-2\}. \end{array}$$

iii. Solution: $(r, s) = (4, -2)$.

4. Solve simultaneously:

$$2y - 3x = 6, \qquad \text{(a)}$$
$$4y - 3(2x + 3) = 0. \qquad \text{(b)}$$

■ 3. i.
$$\begin{array}{rl} 6v + 3t = 14 & \text{(a)} \\ 6v - 3t = 0 & \text{(b) times 3} \\ \hline 12v = 14 & \text{(sum)} \\ v = \frac{7}{6} & \left\{\frac{7}{6}\right\}. \end{array}$$

ii. Since for each v there is one and on one t, in this exercise we will find th corresponding t by substituting 7/6 fo v in equation (a).

$$6\left(\frac{7}{6}\right) + 3t = 14$$
$$3t = 14 - 7$$
$$t = \frac{7}{3} \qquad \left\{\frac{7}{3}\right\}.$$

iii. Solution: $(v, t) = \left(\frac{7}{6}, \frac{7}{3}\right)$.

5. The graphic solutions for the problems in Section 2-1 can now be done algebraically. Solve for I_2 and I_1 in Problem 3.

■ 4. This system has no simultaneous solution. both equations are written so that y is give in terms of x, it becomes very apparent th the graphs have equal slopes and therefo have no intersections.

$$y = \frac{3}{2}x + 3 \qquad \text{(a)}$$

$$y = \frac{3}{2}x + \frac{9}{4} \qquad \text{(b)}$$

A system with no solution is said to be **inconsistent**. Attempts to solve inconsistent systems algebraically always result in *false* equations, such as $0 = 3/4$.

■ 5. i. $60I_2 + 20I_1 = 6$ (a)

$20I_2 + 80I_1 = 12$ (b)

ii. $60I_2 + 20I_1 = 6$ (a)

$5I_2 + 20I_1 = 3$ (b) divided by 4

$55I_2 = 3$ (difference)

$$I_2 = \frac{3}{55}.$$

iii. $30I_2 + 10I_1 = 3$ (a) divided by 2

$30I_2 + 120I_1 = 18$ (b) times $\frac{3}{2}$

$-110I_1 = -15$ (difference)

$$I_1 = \frac{15}{110}$$

$$I_1 = \frac{3}{22}.$$

2-2 Problems

1. An electrical experiment results in the following equations for currents i_1 and i_2 of a certain circuit:

$$i_1 = 3i_2 \quad \text{and} \quad 414i_1 - 2i_2 = 5.$$

Find i_1 and i_2.

2. The sum of the acute angles of any right triangle is 90°. If, for a particular right triangle, the difference is 18°, what are the angles?
3. The line with equation $y = mx + b$ passes through a point (x, y) if and only if the coordinates of the point satisfy the equation. Find the values of m and b so that the line will contain the points (2,3) and (1,4). (The techniques of this section provide a second method for determining the equation of a linear function when two points are known.)
4. A certain wire has a resistance of 0.42 ohms at a temperature of 20° C and a resistance of 0.48 ohms at 70° C. If the resistance R is a linear function of the temperature T, find the equation of the function. (*Hint*: See Problem 3.)

5. One alloy is 45 percent and another 25 percent zinc. These are to be mixed so as to make an alloy that is 40 percent zinc. How many pounds of each are required for a 1000-pound mixture?
6. Two points on a radar scope move at a constant rate around a circle 150 inches in circumference. When they move in opposite directions, they meet every 5 seconds. When they move in the same direction they meet every 25 seconds. Find the rate of revolution for each point.
7. If a linear graph is fitted to the plate-current plate-voltage curve, determine P_v(min) (see Fig. 2.2a).

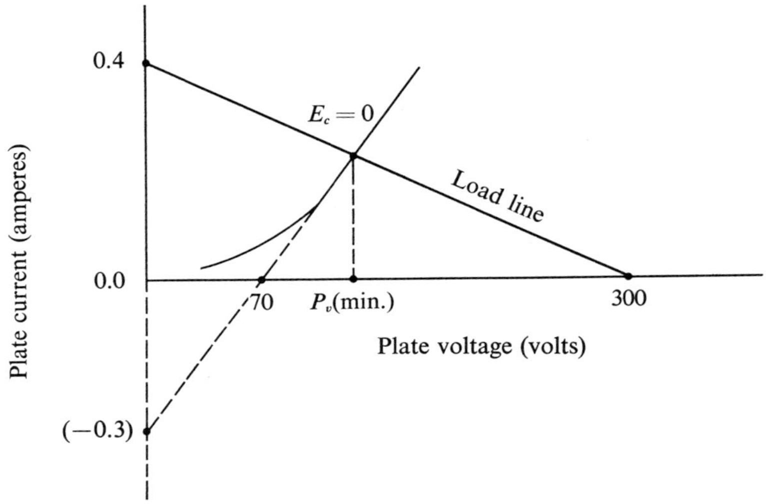

Fig. 2-2a

2-3 Solution with Computational Aids—Two Variables

When specific types of problems are solved in a general manner, the results can frequently be applied to all problems of the same type, abbreviating the time and energy for subsequent computations. Using the methods of Section 2-2, we shall solve a general system of two equations in two unknowns simultaneously. The equations

$$a_1x + b_1y = c_1 \qquad \text{(Eq. 2-3a)}$$

and

$$a_2x + b_2y = c_2 \qquad \text{(Eq. 2-3b)}$$

may be solved for y by multiplying the first equation by a_2, the second by a_1, and then subtracting.

$$a_1a_2x + a_2b_1y = a_2c_1$$

$$\underline{a_1a_2x + a_1b_2y = a_1c_2}$$

$$a_1b_2y - a_2b_1y = a_1c_2 - a_2c_1 \qquad \text{(difference)}$$

$$(\boxed{})y = a_1c_2 - a_2c_1$$

$$y = \boxed{}. \qquad \text{(Eq. 2-3c)}$$

■ $(a_1b_2 - a_2b_1)y$;

■ $$y = \frac{a_1c_2 - a_2c_1}{a_1b_2 - a_2b_1}. \qquad \text{(Eq. 2-3c)}$$

Similarly, we may solve for x by multiplying Eq. 2-3a by _____ and Eq. 2-3b by _____ and subtracting.

■ b_2; b_1.

$$a_1b_2x + b_1b_2y = b_2c_1$$

$$\underline{a_2b_1x + b_1b_2y = b_1c_2}$$

$$\boxed{} \qquad \text{(difference)}$$

$$x = \boxed{}. \qquad \text{(Eq. 2-3d)}$$

■ $(a_1b_2 - a_2b_1)x = b_2c_1 - b_1c_2$;

■ $$x = \frac{b_2c_1 - b_1c_2}{a_1b_2 - a_2b_1}. \qquad \text{(Eq. 2-3d)}$$

Test the results of our general solution on the system

$3x + y = 7,$

$2x - 5y = -1.$

■ $a_1 = 3 \qquad b_1 = 1 \qquad c_1 = 7$
$a_2 = 2 \qquad b_2 = -5 \qquad c_2 = -1$

$$y = \frac{a_1c_2 - a_2c_1}{a_1b_2 - a_2b_1} = \frac{3(-1) - 2(7)}{3(-5) - 2(1)} = \frac{-17}{-17} = 1$$

$$x = \frac{b_2c_1 - b_1c_2}{a_1b_2 - a_2b_1} = \frac{-5(7) - 1(-1)}{3(-5) - 2(1)} = \frac{-34}{-17} = 2$$

Check: $3x + y = 7 \qquad 2x - 5y = -1$
$3(2) + 1 \overset{?}{=} 7 \qquad 2(2) - 5(1) \overset{?}{=} -1$
$7 = 7 \qquad -1 = -1$

Equations 2-3c and 2-3d may be used to solve any linear system of two equations in two unknowns provided $a_1b_2 - a_2b_1 \neq 0$. This expression has the value zero only when the system is inconsistent or when the system contains equivalent equations. The reason for this may be demonstrated by solving Eq. 2-3a and Eq. 2-3b for y.

$$y = -\frac{a_1}{b_1}x + \frac{c_1}{b_1} \qquad \text{(Eq. 2-3e)}$$

$$y = -\frac{a_2}{b_2}x + \frac{c_2}{b_2}. \qquad \text{(Eq. 2-3f)}$$

If

$$\frac{a_1}{b_1} = \frac{a_2}{b_2},$$

then the two equations have graphs that are either identical or parallel and, therefore, solution sets that either are equal or have no intersection. It should be clear that

$$\frac{a_1}{b_1} = \frac{a_2}{b_2}$$

is equivalent to $a_1 b_2 - a_2 b_1 = 0$.

When systems of equations are developed in the process of solving a problem, the equations must be written in a *standard form* comparable to Eqs. 2-3a and 2-3b in order to use the generalized procedure illustrated in this section.

Put the system

$$2x = 3(y - 5) + 2,$$

$$3x - y = 4(x + 1),$$

in standard form.

■ i. $2x = 3y - 15 + 2$
$3x - y = 4x + 4.$

ii. $2x - 3y = -13$
$-x - y = 4$ Standard Form.

It is clear only now that $a_1 = 2, b_1 = -3, c_1 = -13, a_2 = -1, b_2 =$ ☐, $c_2 =$ ☐.

■ -1; 4.

The formulas developed so far, if organized systematically, can make the solution of systems of equations a very routine matter. However, the arithmetic required in using Eqs. 2-3c and 2-3d is frequently a source of error. Modern electronic desk calculators with programming capability and of course computer systems can easily relieve us of that problem. Even hand calculations can be improved by using a systematic procedure. The following chart illustrates a procedural technique used by computer programmers.

Flow Chart

Read the coefficients and constants of the system of equations.

Compute the value of $a_1b_2 - a_2b_1$ and call it d.

If d is zero, the system has no solution and the calculation should stop. However if $d \neq 0$, continue.

Calculate the numerator of the x value and call it d_1.

Calculate the numerator of the y value and call it d_2.

Calculate the values of x and y using the predetermined values of d, d_1, and d_2.

Record the solution.

End the calculation.

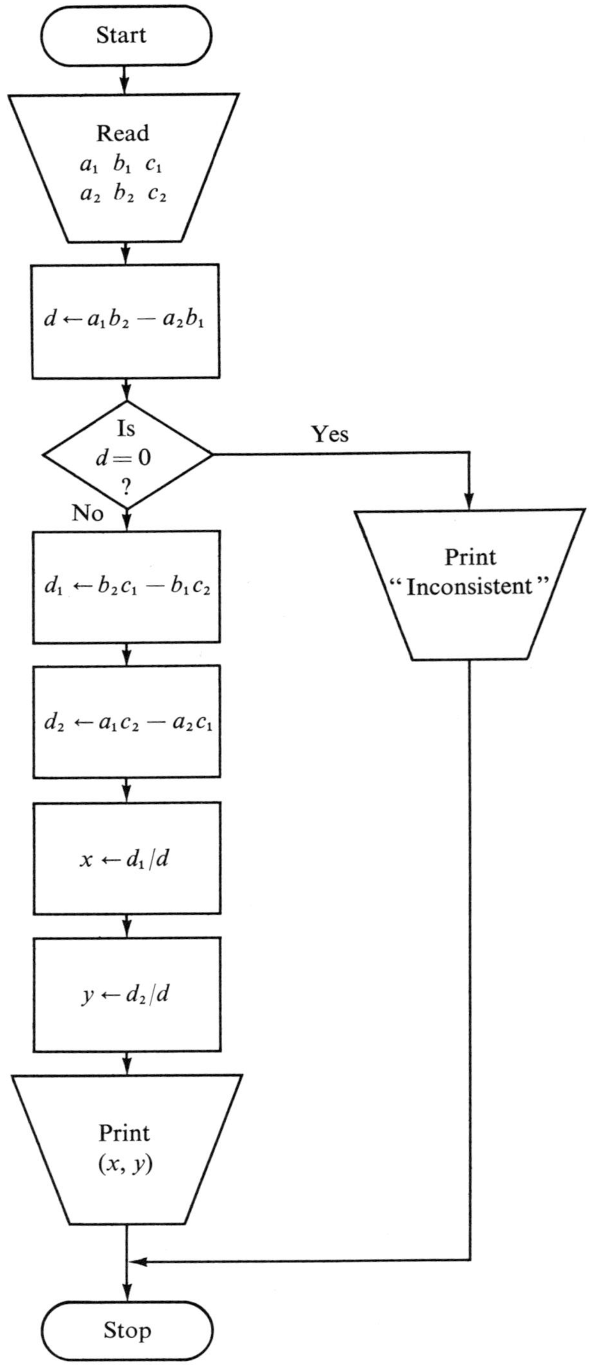

2-3 Exercises

Use the systematic procedure illustrated in the flow chart. If computing devices are available, take this opportunity to use them.

1. Solve the system

$$x - 2y = 11,$$
$$2x - 3y = 18.$$

---

2. Solve the system

$$y - 2 = 3(x - 5),$$
$$y - 3 = 2(x + 4).$$

■ 1. $a_1 = 1 \quad b_1 = -2$
$a_2 = 2 \quad b_2 = -3$
$c_1 = 11$
$c_2 = 18$
Solution: (3, −4).

---

3. Solve the system

$$-2x + 3y = 9,$$
$$8x - 12y = 3.$$

■ 2. First put the equations in standard form.

$$3x - y = 13$$
$$2x - y = -11$$

$a_1 = 3 \quad b_1 = -1$
$a_2 = 2 \quad b_2 = -1$
$c_1 = 13$
$c_2 = -11$
Solution: (24, 59).

---

4. Solve the system

$$0.1R - 3r = 8,$$
$$0.2R - 4r = 24.$$

■ 3. $a_1 = -2 \quad b_1 = 3$
$a_2 = 8 \quad b_2 = -12$
$c_1 = 9$
$c_2 = 3$

$$a_1b_2 - a_2b_1 = (-2)(-12) - (8)(3) = 24 - 24 = 0.$$

This system is inconsistent.

5. Solve the system

$$0.500F_1 + 0.600F_2 = 10.0,$$
$$0.866F_1 - 0.800F_2 = 20.0.$$

■ 4. $a_1 = 0.1$ $b_1 = -3$
$a_2 = 0.2$ $b_2 = -4$
$c_1 = 8$
$c_2 = 24$

Solution: $R = 200$
$r = 4$.

■ 5. $a_1 = 0.500$ $b_1 = 0.600$
$a_2 = 0.866$ $b_1 = -0.800$
$c_1 = 10.0$
$c_2 = 20.0$

Solution: $F_1 \approx 21.7$
$F_2 \approx -1.46$.

2-3 Problems

1. A chemist has a 20 percent solution and an 8 percent solution of sulfuric acid. How many cubic centimeters of each solution should he mix in order to obtain 180 cubic centimeters of a 15 percent solution?
2. An airplane flies 400 miles in 5 hours against the wind. The return trip flying with the wind is made in 3 hours, 20 minutes. What is the air speed of the airplane and the wind velocity?

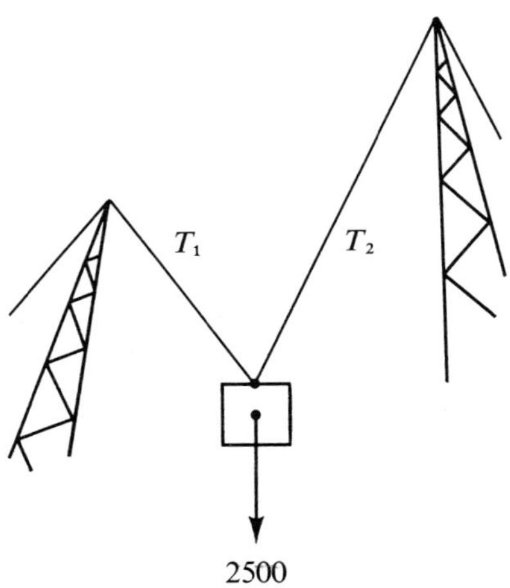

Fig. 2-3a

3. The cables from two cranes support a large weight that is in equilibrium. (See Fig. 2-3a.) In a particular situation, the tension (T) on each cable can be determined by solving the system

$$0.766T_1 = 0.342T_2,$$
$$0.643T_1 + 0.940T_2 = 2500.$$

Find T_1 and T_2.

4. Two 112-kilowatt (150 H.P.) motors are being considered for installation at a pumping station. The first costs $1500 and has an efficiency of 83 percent. The second costs $1200 and has an efficiency of 80 percent. How many hours of full load operation at 1.8 cents per kilowatt hours are necessary to justify the purchase of the more expensive motor?

5. A telephone 150-pair cable can be made with either enameled or tinned wire. There are 550 soldered connections in a given installation. A soldered connection with enameled wire costs 1.85 cents; with tinned wire it costs 1.25 cents. The 150-pair cable with enameled wire costs 65 cents per linear foot and with tinned wire costs 83 cents per linear foot. Calculate the length of cable for which the installation cost would be the same for both types of wire.

2-4 Larger Linear Systems—Augmented Matrix

Our attention has been focused on problems involving systems of equations in two unknowns. In order to analyze the distribution of current in a circuit like the one in Fig. 2-4a, a system of three equations in three unknowns is required.

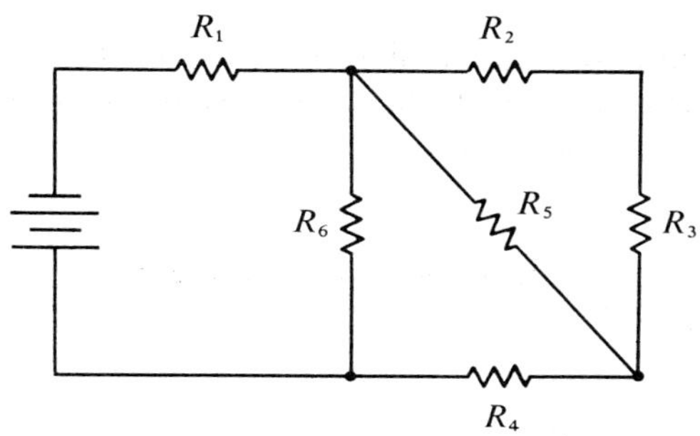

Fig. 2-4a

Suppose the system is

$$6i_1 - 5i_2 - 2i_3 = 2, \quad \text{(Eq. 2-4a)}$$

$$4i_1 + \ i_2 + 3i_3 = 10, \quad \text{(Eq. 2-4b)}$$

$$5i_1 + 3i_2 + 7i_3 = 13. \quad \text{(Eq. 2-4c)}$$

A solution to Eq. 2-4a would be an **ordered triplet** of numbers (i_1, i_2, i_3). For example, $(1, 2, -3)$ is a solution.

■ $$\begin{aligned} 6(1) - 5(2) - 2(-3) &\stackrel{?}{=} 2, \\ 6 - 10 + 6 &\stackrel{?}{=} 2, \\ 2 &= 2. \end{aligned}$$

There are, however, an infinite number of solutions to Eq. 2-4a. Likewise, Eq. 2-4b and Eq. 2-4c each has an infinite solution set, but a particular solution to one of the equations is not necessarily a solution to the others.

The ordered triplet $(1, 2, -3)$ is not a solution to either Eq. 2-4b or Eq. 2-4c.

■ i. $$\begin{aligned} 4(1) + (2) + 3(-3) &\stackrel{?}{=} 10 \\ 4 + 2 - 9 &\stackrel{?}{=} 10 \quad \text{(Eq. 2-4b)} \\ -3 &\neq 10. \end{aligned}$$

ii. $$\begin{aligned} 5(1) + 3(2) + 7(-3) &\stackrel{?}{=} 13 \\ 5 + 6 - 21 &\stackrel{?}{=} 13 \quad \text{(Eq. 2-4c)} \\ -10 &\neq 13. \end{aligned}$$

A graphical treatment in three dimensions would show that each equation has a solution set that is represented by a plane. The intersection of the planes, should there be any, may be only a point.

In the case of the system being illustrated, the point is $(3, 4, -2)$. Verify this statement.

■ i. $$\begin{aligned} 6(3) - 5(4) - 2(-2) &\stackrel{?}{=} 2 \quad \text{(Eq. 2-4a)} \\ 18 - 20 + 4 &= 2. \end{aligned}$$

ii. $4(3) + (4) + 3(-2) \stackrel{?}{=} 10$ (Eq. 2-4b)

$12 + 4 - 6 = 10.$

iii. $5(3) + 3(4) + 7(-2) \stackrel{?}{=} 13$ (Eq. 2-4c)

$15 + 12 - 14 = 13.$

The method we shall develop for determining simultaneous solutions, when they exist, will be analytic and may be used not only for linear systems in three unknowns but for systems of any number of unknowns.

As early as 1729, a brilliant mathematician named Colin Maclaurin observed the relationship between the general solution of systems of linear equations and a branch of mathematics that deals with arrays of numbers in rows and columns called **determinants**. Maclaurin, as well as a later mathematician, Gabriel Cramer, published works that generalized the solution of linear systems of equations by the methods of determinants. In practical applications the classical method known as *Cramer's Rule* becomes prohibitively laborious, even in terms of using computers.

The methods we shall pursue are suited to hand computations for small systems as well as computer-assisted computations for larger systems.

Consider the somewhat special system of three equations in the variables x, y, and z.

$$-x + 4y + 3z = -2, \qquad \text{(Eq. 2-4d)}$$

$$y + 5z = 2, \qquad \text{(Eq. 2-4e)}$$

$$-3z = -3. \qquad \text{(Eq. 2-4f)}$$

The solution of this system is quite easy even without special methods. From Eq. 2-4f, z must equal ______.

■ one

Substituting the value of z into Eq. 2-4e allows us to evaluate y.

■ $y + 5(1) = 2$ (Eq. 2-4e)

$y = -3.$

The final unknown can be evaluated by substituting 1 for z and (-3) for y into Eq. 2-4d.

■ $$\begin{aligned} -x + 4(-3) + 3(1) &= -2 \qquad \text{(Eq. 2-4d)} \\ -x - 12 + 3 &= -2 \\ -x &= 7 \\ x &= -7. \end{aligned}$$

The ordered triplet $(-7, -3, 1)$ is the intersection of the solution set of each equation—in other words, the simultaneous solution of the system.

An analogous system in four unknowns or more is no more difficult. Solve the system

$$\begin{aligned} -x - 3y + z - w &= -6, \\ 12y - 2z + w &= 23, \\ 16z + 7w &= -7, \\ 39w &= -39. \end{aligned}$$

■ i. $$\begin{aligned} 39w &= -39 \\ w &= -1. \end{aligned}$$

ii. $$\begin{aligned} 16z + 7w &= -7 \\ 16z + 7(-1) &= -7 \\ 16z &= 0 \\ z &= 0. \end{aligned}$$

iii. $$\begin{aligned} 12y - 2z + w &= 23 \\ 12y - 2(0) + (-1) &= 23 \\ 12y &= 24 \\ y &= 2. \end{aligned}$$

iv. $$\begin{aligned} -x - 3y + z - w &= -6 \\ -x - 3(2) + 0 - (-1) &= -6 \\ -x &= -1 \\ x &= 1. \end{aligned}$$

Solution: $(1, 2, 0, -1)$.

Let us call systems of the preceding type **triangular systems.**

If you review and summarize the six experiments with linear equations performed in Section 2-2, you may conclude that the simultaneous solution of a linear system of equations is *unchanged* if

1. *the order of the equations in the system is changed,*
2. *any equation is multiplied by a nonzero constant,*
3. *any result of conclusion* (2) *is added to any of the other equations in the system.*

The importance of these conclusions is that they provide a means of changing *every* linear system to an equivalent triangular system.

To demonstrate we return to the first system appearing in this section. However, before we proceed with the application of the three rules, we should realize that they only affect the values of coefficients and constants. Therefore the amount of writing can be considerably reduced by recording only what is absolutely essential to the computation. The system

$$\begin{aligned} 6i_1 - 5i_2 - 2i_3 &= 2, \\ 4i_1 + i_2 + 3i_3 &= 10, \\ 5i_1 + 3i_2 + 7i_3 &= 13, \end{aligned}$$

becomes

$$\left(\begin{array}{ccc|c} 6 & -5 & -2 & 2 \\ 4 & 1 & 3 & 10 \\ 5 & 3 & 7 & 13 \end{array}\right),$$

which is a row-and-column array of numbers called an **augmented matrix**. Be sure to notice the relative position of the coefficients and constants in the system and in the matrix. We shall also need to be able to identify a specific element in the matrix. We shall let

$$\left(\begin{array}{ccc|c} a_1 & b_1 & c_1 & d_1 \\ a_2 & b_2 & c_2 & d_2 \\ a_3 & b_3 & c_3 & d_3 \end{array}\right)$$

code the relative positions of the numbers in our specific example.

To test our understanding, $c_3 = \square$ and $a_2 = \square$.

■ 7; 4.

In order to triangularize the system, we shall arbitrarily produce zeros below the *diagonal* a_1, b_2, c_3. However, in doing so, we can systematize our procedure.

i. Multiplying the first row by $k_1 = -2/3$ and adding the result to the second row produces the equivalent system

$$\left(\begin{array}{ccc|c} 6 & -5 & -2 & 2 \\ 0 & \dfrac{13}{3} & \dfrac{13}{3} & \dfrac{26}{3} \\ 5 & 3 & 7 & 13 \end{array}\right).$$

The number k_1 is the particular number that will produce a zero in position a_2 and is equal to $-a_2/a_1$. Notice that only the second row is changed from the original matrix.

ii. Multiplying the first row by

$$k_2 = -\frac{a_3}{a_1} = -\frac{5}{6}$$

and adding the results to the third row produces the equivalent system

$$\left(\begin{array}{ccc|c} 6 & -5 & -2 & 2 \\ 0 & \frac{13}{3} & \frac{13}{3} & \frac{26}{3} \\ 0 & \frac{43}{6} & \frac{52}{6} & \frac{68}{6} \end{array}\right).$$

iii. Multiplying the second row of the last result by

$$k_3 = -\frac{b_3}{b_2} = -\frac{43}{26}$$

and adding the result to the third row produces the desired triangular matrix

$$\left(\begin{array}{ccc|c} 6 & -5 & -2 & 2 \\ 0 & \frac{13}{3} & \frac{13}{3} & \frac{26}{3} \\ 0 & 0 & \frac{9}{6} & -\frac{18}{6} \end{array}\right).$$

It is important to recognize that the result of each step in the procedure is the matrix for the following step and the code for identifying the elements does not change; in other words, the subscripts 1, 2, and 3 will always identify the first, second, and third row of each equivalent matrix.

Solving the system represented by the last matrix gives the ordered triplet (________), which agrees with the values we expected.

■ (3, 4, −2)

The following flow chart both summarizes and generalizes the procedure.

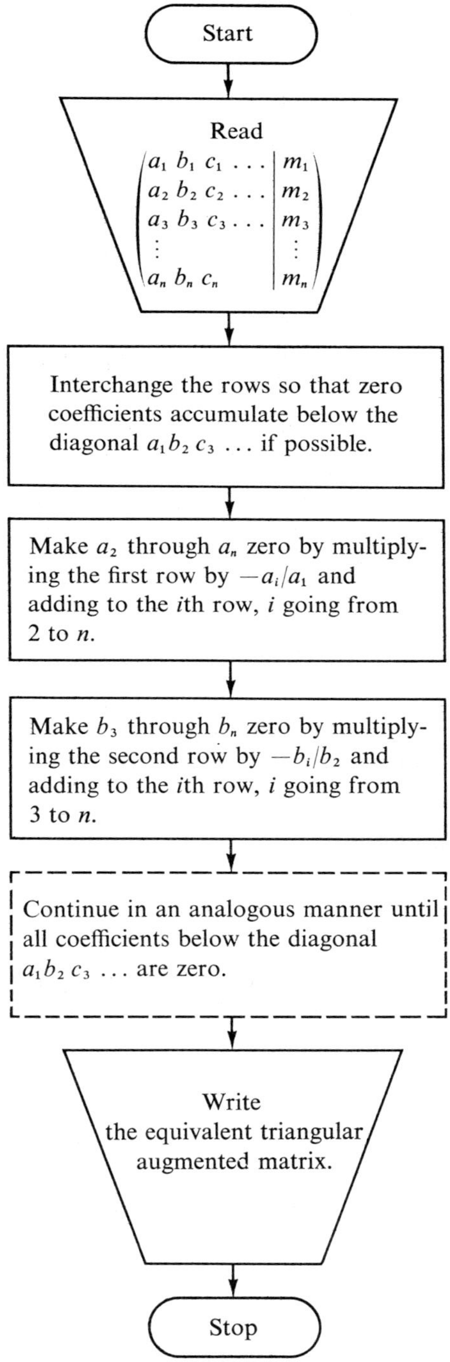

The linear system of coefficients is an n-by-n array, augmented by constants m_i.

Interchanging the rows has no effect on the solution. If no non-zero a_i occurred, the system would, in effect, have $n - 1$ variables.

This generates zeros in the first column below a_1.

This generates zeros in the second column below b_2.

This generates zeros in subsequent columns below the diagonal $a_1 b_2 c_3 d_4 \ldots$.

When the triangular augmented matrix is known, the solution to the corresponding linear system can be found.

To obtain the equivalent triangular system is our immediate objective; with it we can easily solve a system of linear equations. Our treatment of matrices is limited to this same objective.

Example 1

Find the simultaneous solution (x, y, z, w) for the system

$$\begin{aligned} x + y &= 2, \\ y + z &= -1, \\ z + w &= 0, \\ x - y + z - w &= 0. \end{aligned}$$

i. $\left(\begin{array}{cccc|c} 1 & 1 & 0 & 0 & 2 \\ 0 & 1 & 1 & 0 & -1 \\ 0 & 0 & 1 & 1 & 0 \\ 1 & -1 & 1 & -1 & 0 \end{array}\right)$

$\rightarrow \left(\begin{array}{cccc|c} 1 & 1 & 0 & 0 & 2 \\ 0 & 1 & 1 & 0 & -1 \\ 0 & 0 & 1 & 1 & 0 \\ 0 & -2 & 1 & -1 & -2 \end{array}\right)$ (Row 1 multiplied by -1 and added to row 4.)

$\rightarrow \left(\begin{array}{cccc|c} 1 & 1 & 0 & 0 & 2 \\ 0 & 1 & 1 & 0 & -1 \\ 0 & 0 & 1 & 1 & 0 \\ 0 & 0 & 3 & -1 & -4 \end{array}\right)$ (Row 2 multiplied by 2 and added to row 4.)

$\rightarrow \left(\begin{array}{cccc|c} 1 & 1 & 0 & 0 & 2 \\ 0 & 1 & 1 & 0 & -1 \\ 0 & 0 & 1 & 1 & 0 \\ 0 & 0 & 0 & -4 & -4 \end{array}\right).$ (Row 3 multiplied by -3 and added to row 4.)

ii. Equivalent triangular system:

$$\begin{aligned} x + y + 0z + 0w &= 2, \\ y + z + 0w &= -1, \\ z + w &= 0, \\ -4w &= -4. \end{aligned}$$

iii. Solution: $\boxed{}$.

■ (2, 0, −1, 1)

The clue to discovering that a system is *inconsistent* is the appearance of a false equation in the triangulation process. The system

$$\begin{aligned} x - y + 2z &= 1, \\ -x + 2y + z &= 3, \\ -4x + 7y + z &= 2 \end{aligned}$$

is inconsistent and has an equivalent triangular system

$$\begin{aligned} -4x + 7y + z &= 2, \\ y + 3z &= 10, \\ 0 &= 24. \end{aligned}$$

The false equation is $\boxed{}$.

■ $0 = 24$.

A **dependent** system is a system that has an infinite solution set because two or more of the equations in the system are equivalent. If the equivalent equations are not observed before a solution is attempted, the triangulation process will produce a row of zeros. For example,

$$\begin{aligned} 2x + y - 3z &= 4, \\ x - y + z &= 2, \\ 4x + 2y - 6z &= 8 \end{aligned}$$

is a dependent system. An equivalent triangular system is

$$\begin{aligned} 2x + y - 3z &= 4, \\ -3y + 5z &= 0, \\ 0 &= 0. \end{aligned}$$

2-4 Exercises

1. Write the augmented matrix for the system

$$\begin{aligned} 3x + 5y - 2z &= 12, \\ x + 4y + z &= -4, \\ -x + 3y - 7z &= 0. \end{aligned}$$

2. Write the augmented matrix for the following system. (What is the coefficient of y in the first equation?)

$$\begin{aligned} x - z &= 5, \\ 4y + x - 2z &= 3, \\ \frac{x}{2} + z - \frac{y}{3} &= -4. \end{aligned}$$

■ 1. $\left(\begin{array}{rrr|r} 3 & 5 & -2 & 12 \\ 1 & 4 & 1 & -4 \\ -1 & 3 & -7 & 0 \end{array}\right)$.

3. Write the augmented matrix for the system

$$\frac{x+y}{2} - z = 4,$$
$$5x = \frac{4y - z}{2},$$
$$y = 3x - 1.$$

■ 2. First arrange the equations in standard form.

$$x + 0y - z = 5,$$
$$x + 4y - 2z = 3,$$
$$\frac{1}{2}x - \frac{1}{3}y + z = -4.$$

The augmented matrix is

$$\left(\begin{array}{ccc|c} 1 & 0 & -1 & 5 \\ 1 & 4 & -2 & 3 \\ \frac{1}{2} & -\frac{1}{3} & 1 & -4 \end{array}\right).$$

---

4. Solve the system by first finding a triangular form.

$$4x - 3y = 8,$$
$$x + \frac{1}{2}y = 3.$$

■ 3. The standard form and the corresponding matrix are:

$$x + y - 2z = 8,$$
$$10x - 4y + z = 0,$$
$$-3x + y + 0z = -1,$$

and

$$\left(\begin{array}{ccc|c} 1 & 1 & -2 & 8 \\ 10 & -4 & 1 & 0 \\ -3 & 1 & 0 & -1 \end{array}\right).$$

---

5. Solve the system

$$R_1 + 2R_2 + R_3 = 20,$$
$$3R_1 - R_2 + R_3 = 22,$$
$$2R_1 + R_2 - 4R_3 = 7.$$

■ 4. $\left(\begin{array}{cc|c} 4 & -3 & 8 \\ 1 & \frac{1}{2} & 3 \end{array}\right).$

Multiply the first row by $(-1/4)$ and add to the second row to obtain

$$\left(\begin{array}{cc|c} 4 & -3 & 8 \\ 0 & \frac{5}{4} & 1 \end{array}\right).$$

The equivalent triangular system is

$$4x - 3y = 8,$$
$$\frac{5}{4}y = 1.$$

Solution:

i. $y = \frac{4}{5}$.

ii. $4x - 3\left(\frac{4}{5}\right) = 8$

$$4x = \frac{52}{5}$$

$$x = \frac{13}{5} \qquad \left\{\left(\frac{13}{5}, \frac{4}{5}\right)\right\}.$$

---

6. Solve the system

$$\begin{aligned} x + y + 2z &= 0, \\ y + z &= 2, \\ x - z &= 4. \end{aligned}$$

■ 5. $\left(\begin{array}{ccc|c} 1 & 2 & 1 & 20 \\ 3 & -1 & 1 & 22 \\ 2 & 1 & -4 & 7 \end{array}\right)$

$$\to \left(\begin{array}{ccc|c} 1 & 2 & 1 & 20 \\ 0 & -7 & -2 & -38 \\ 2 & 1 & -4 & 7 \end{array}\right)$$

$$\to \left(\begin{array}{ccc|c} 1 & 2 & 1 & 20 \\ 0 & -7 & -2 & -38 \\ 0 & -3 & -6 & -33 \end{array}\right)$$

$$\to \left(\begin{array}{ccc|c} 1 & 2 & 1 & 20 \\ 0 & -7 & -2 & -38 \\ 0 & 0 & -\frac{36}{7} & -\frac{117}{7} \end{array}\right).$$

Solution:

i. $\left(-\frac{36}{7}\right)R_3 = \left(-\frac{117}{7}\right)$

$$R_3 = \frac{13}{4}.$$

ii. $-7R_2 - 2\left(\frac{13}{4}\right) = -38$

$$R_2 = \frac{9}{2}.$$

iii. $R_1 + 2\left(\frac{9}{2}\right) + \left(\frac{13}{4}\right) = 20$

$$R_1 = \frac{31}{4}.$$

7. A nonlinear function is known to have the general form

$$y = f(x) = Ax^2 + Bx + C.$$

If the graph of f passes through the points (6,0), (3,1), and (2,2), find the numerical coefficients A, B, and C.

■ 6. $$\left(\begin{array}{ccc|c} 1 & 1 & 2 & 0 \\ 0 & 1 & 1 & 2 \\ 1 & 0 & -1 & 4 \end{array}\right)$$

$$\rightarrow \left(\begin{array}{ccc|c} 1 & 1 & 2 & 0 \\ 0 & 1 & 1 & 2 \\ 0 & -1 & -3 & 4 \end{array}\right)$$

$$\rightarrow \left(\begin{array}{ccc|c} 1 & 1 & 2 & 2 \\ 0 & 1 & 1 & 2 \\ 0 & 0 & -2 & 6 \end{array}\right).$$

Solution:

i. $-2z = 6$; $z = -3$.

ii. $y - 3 = 2$; $y = 5$.

iii. $x + 5 - 6 = 0$; $x = 1$.

■ 7. i. The system that must be solved is

$$\begin{aligned} 36A + 6B + C &= 0, \\ 9A + 3B + C &= 1, \\ 4A + 2B + C &= 2. \end{aligned}$$

ii. The corresponding augmented matrix is

$$\left(\begin{array}{ccc|c} 36 & 6 & 1 & 0 \\ 9 & 3 & 1 & 1 \\ 4 & 2 & 1 & 2 \end{array}\right)$$

$$\rightarrow \left(\begin{array}{ccc|c} 36 & 6 & 1 & 0 \\ 0 & \frac{3}{2} & \frac{3}{4} & 1 \\ 4 & 2 & 1 & 2 \end{array}\right)$$

$$\rightarrow \left(\begin{array}{ccc|c} 36 & 6 & 1 & 0 \\ 0 & \frac{3}{2} & \frac{3}{4} & 1 \\ 0 & \frac{4}{3} & \frac{8}{9} & 2 \end{array}\right)$$

$$\rightarrow \left(\begin{array}{ccc|c} 36 & 6 & 1 & 0 \\ 0 & \frac{3}{2} & \frac{3}{4} & 1 \\ 0 & 0 & \frac{2}{9} & \frac{10}{9} \end{array}\right).$$

iii. $\frac{2}{9}C = \frac{10}{9}$; $C = 5.$

iv. $\frac{3}{2}B + \frac{3}{4}(5) = 1;$

$$B = -\frac{11}{6}.$$

v. $36A + 6\left(-\frac{11}{6}\right) + 5 = 0;$

$$A = \frac{1}{6}.$$

Therefore,

$$f(x) = \frac{1}{6}x^2 - \frac{11}{6}x + 5.$$

2-4 Problems

1. If $y = f(x) = Ax^2 + Bx + C$ is the general form of an equation whose graph passes through $(-1, -5)$, $(2, 4)$, and $(0, -4)$, find A, B, and C.
2. A, B, and C are centers of the circles in Fig. 2-4b. Each circle is tangent to the other two shown. Find the radius of each circle, if

$$\overline{AB} = 4 \text{ inches},$$
$$\overline{AC} = 5 \text{ inches},$$
$$\overline{BC} = 7 \text{ inches}.$$

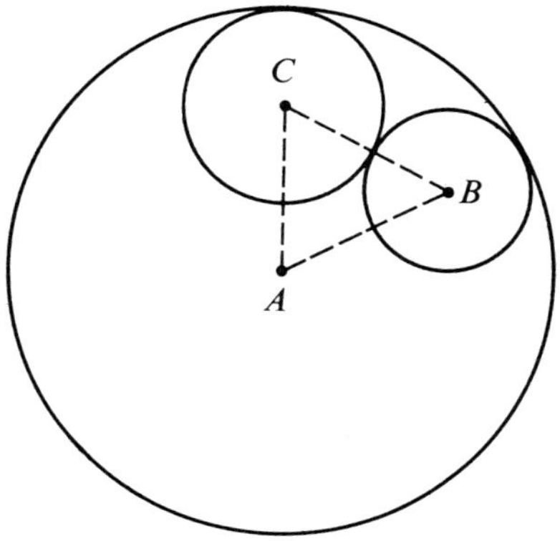

Fig. 2-4b

3. A "delta-network" (Fig. 2-4c) can be replaced by a "y-network" (Fig. 2-4d) by using the following equations:

$$R_a + R_b = \frac{R_1R_2 + R_2R_3}{R_1 + R_2 + R_3},$$

$$R_a + R_c = \frac{R_1R_2 + R_1R_3}{R_1 + R_2 + R_3},$$

$$R_b + R_c = \frac{R_1R_3 + R_2R_3}{R_1 + R_2 + R_3}.$$

Find the values of R_a, R_b, and R_c when $R_1 = 3$, $R_2 = 4$, and $R_3 = 2$.

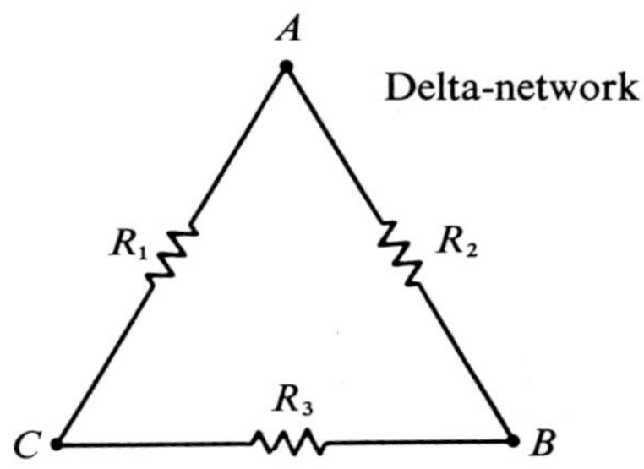

Fig. 2-4c

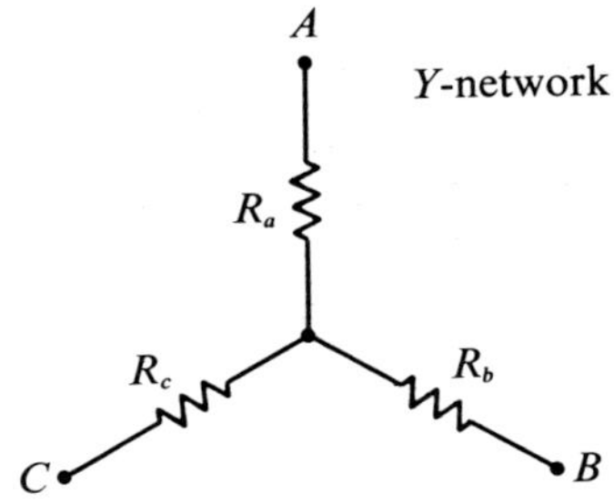

Fig. 2-4d

4. In the gear system illustrated in Fig. 2-4e, point A makes a revolution every 365 milliseconds, and point B makes a revolution every 687 milliseconds. How long is it from the time these points are together until they are next together?

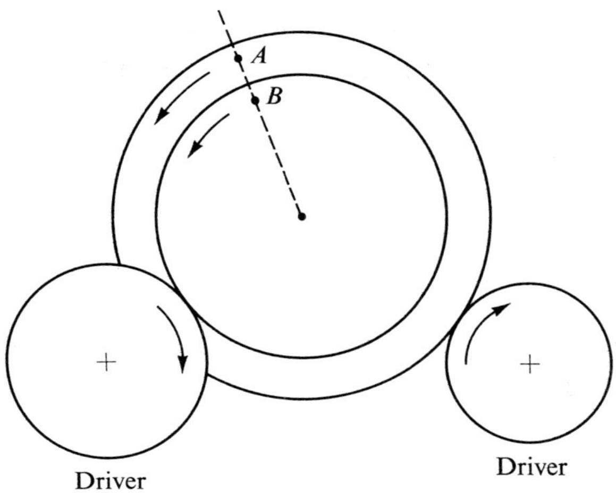

Fig. 2-4e

5. In applying Kirchoff's laws to the electric circuit given in Fig. 2-4f, the following equations result. Determine the indicated currents.

$$I_A + I_B + I_C = 0,$$
$$6I_A - 10I_C = 8,$$
$$6I_A - 2I_B = 5.$$

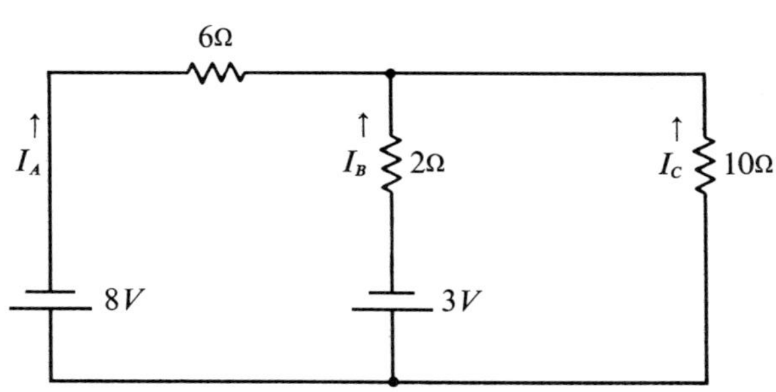

Fig. 2-4f

6. Three pipes empty into a setting tank. If pipes No. 1 and No. 2 fill the tank in 6 hours, or pipes No. 2 and No. 3 in 8 hours, or pipes No. 1 and No. 3 in 10 hours, how long does it take to fill the tank if all three pipes operate? If only pipe No. 1 operates?

---

2 Supplementary Exercises and Problems

1. Approximate graphically the solution common to the system of equations

$$r = \frac{2}{3}t + 12,$$

$$10t = 4r - 3. \qquad (2\text{-}1)$$

2. Write the system of equations to represent the values given on the graph in Fig. 2-4g. (2-1)

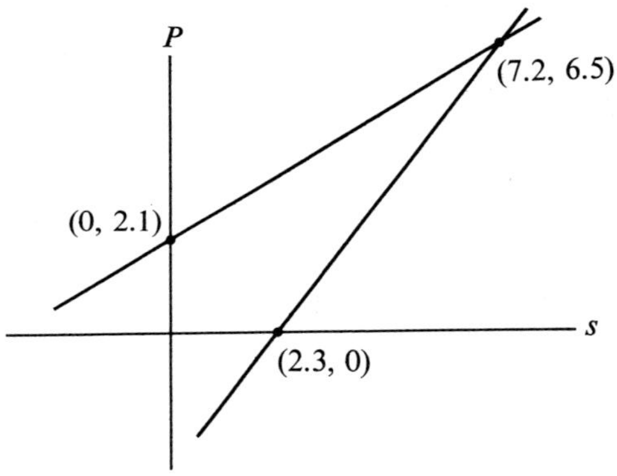

Fig. 2-4g

3. Solve graphically the system

$$8s - \frac{1}{2}w = 60,$$

$$3s + 2w = 40. \qquad (2\text{-}1)$$

4. Solve algebraically

$$2m + n = 7,$$

$$m - 2n = 6. \qquad (2\text{-}2)$$

5. Solve Exercise 1 algebraically. (2-2)
6. Solve algebraically the system

$$2(s - 3t) + t = 19,$$

$$\frac{1}{2}(s + 3t) = -9. \qquad (2\text{-}2)$$

7. Solve simultaneously

$$5(y - 1) = 2x,$$

$$\frac{2x - 5}{5} = y + 3. \qquad (2\text{-}2)$$

8. Solve simultaneously

$$p = -\frac{5}{4}q - 2,$$

$$4(p + 1) = q - 4. \qquad (2\text{-}2)$$

9. Solve for the simultaneous solution $(1/x, 1/y)$ in the system

$$\frac{1}{x} + \frac{1}{y} = \frac{4}{15}, \qquad \frac{3}{x} + \frac{2}{y} = \frac{19}{15}. \qquad (2\text{-}2)$$

10. Put in standard form and solve simultaneously

$$18 + 2k = h,$$

$$h + k = -9. \qquad (2\text{-}3)$$

11. Solve the system

$$\frac{I_a - 2}{5} - \frac{I_b - 10}{4} = \frac{10 - I_a}{3},$$

$$\frac{I_b + 2}{3} + \frac{2I_a + I_b}{16} = \frac{I_a + 13}{8}. \qquad (2\text{-}3)$$

12. Solve the system

$$1.3u - 1.8v = 22,$$

$$2.4u + 1.4v = 20. \qquad (2\text{-}3)$$

13. Solve the triangular system

$$3u - \frac{1}{6}v + w = 4,$$

$$v + \frac{2}{3}w = -\frac{2}{3},$$

$$4w = 2. \qquad (2\text{-}4)$$

14. Find the solution of the system

$$\begin{aligned} u - 4v - 2w &= -7, \\ 2u - 3v + w &= 1, \\ u + v - w &= 4. \end{aligned} \qquad (2\text{-}4)$$

15. Find the solution of the system

$$\begin{aligned} 2A - B + 2D &= 2, \\ 4A - 2B + 3C &= -7, \\ B + C + D &= 6, \\ 6A - 4C + 5D &= 23. \end{aligned} \qquad (2\text{-}4)$$

16. When triangularizing an augmented matrix, how is an inconsistent system identified? How is a dependent system identified? (2-4)

17. Three pollsters in different communities of a large metropolitan area collected responses from a total of 1000 people. Community A reported 10% For, 50% Against, and 25% Undecided. Community B reported 25% For, 55% Against, and 20% Undecided. Community C reported 15% For, 35% Against, and 30% Undecided. Combining the data, the results proved to be 18% For, 52% Against, and 25% Undecided. Approximately how many people were polled in each community? (2-4)

Chapter Three Quadratic Functions

3-1 Graph of a Quadratic Function

Rectangular air ducts with equal perimeters do not necessarily have the same air capacity. If, for example, the width and depth are 5 inches and 7 inches, the cross-sectional area is 35 square inches. If the dimensions are 4 inches by 8 inches, the cross-sectional area is only 32 square inches. Both, of course, have perimeters of 24 inches. Let us examine more of the possibilities suggested in Fig. 3-1a.

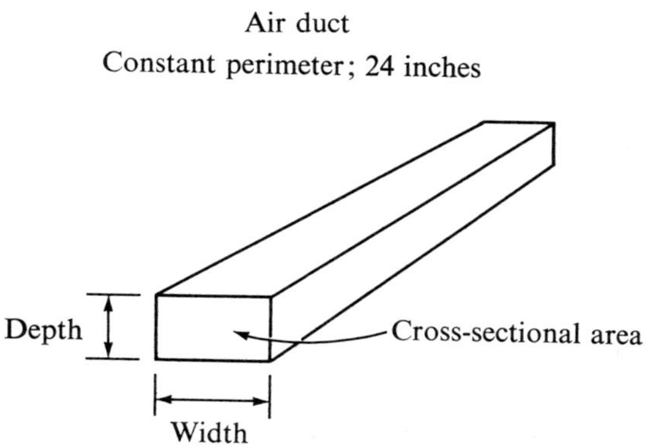

Fig. 3-1a

Fill in the blanks in Table 3-1a. The least area is [ ] square inches. The greatest area is [ ] square inches.

Table 3-1a

Width (inches)	0	1	2	3	4	5	6	7	8	9	10	11	12
Depth (inches)					8	7							
Cross-Sectional Area (sq inches)					32	35							

■ 0; 36.

Table 3-1b

Depth (inches)	12	11	10	9	8	7	6	5	4	3	2	1	0
Area (sq inches)	0	11	20	27	32	35	36	35	32	27	20	11	0

Notice that when the perimeter remains constant, the depth and area are functions of the width. Graph the set of pairs (width, depth) from Table 3-1b and derive the functional equation relating them.

■ $d =$ (depth)
$w =$ (width)
$p =$ (24-inch perimeter)

$$2d + 2w = 24$$
$$d + w = 12$$
$$d = -w + 12.$$

or

$$d = f(w) = -w + 12. \qquad \text{(Eq. 3–1a)}$$

Graph the set of ordered pairs (width, area) from Table 3-1b.

We can derive the equation of the function illustrated in Fig. 3-1c by using well-known formulas from geometry. The area of a rectangle is $A =$ (width) (depth) $= wd$. Eq. 3-1a expresses d in terms of w. Substituting $(-w + 12)$ for d, the area formula becomes

$$A = w(-w + 12)$$
$$= -w^2 + 12w,$$

or

$$A = f(w) = -w^2 + 12w. \qquad \text{(Eq. 3-1b)}$$

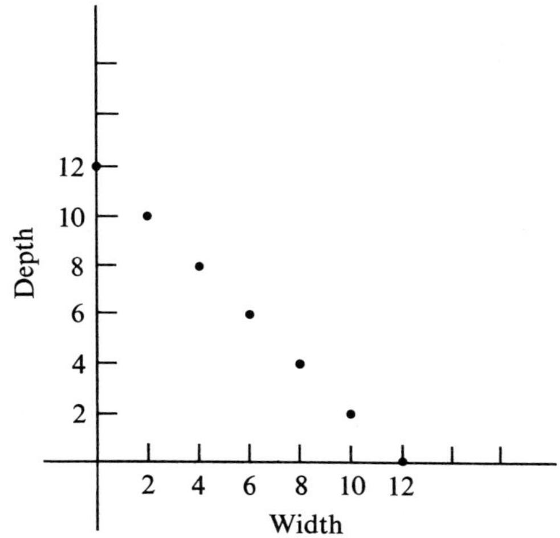

Fig. 3-1b

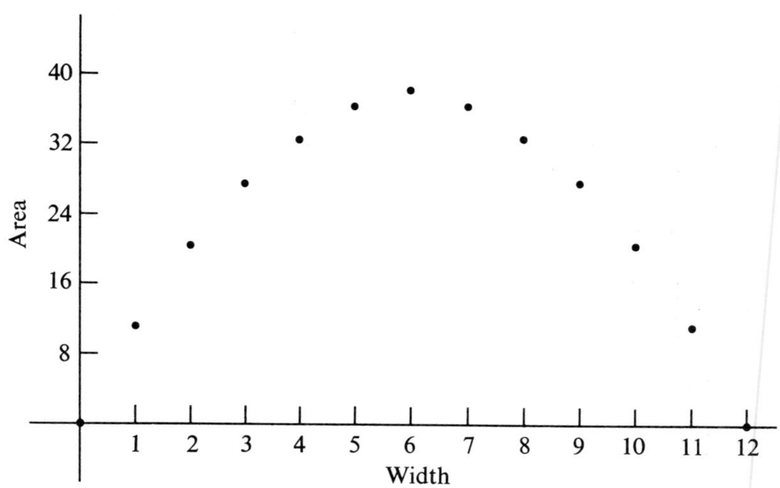

Fig. 3-1c

Verify that Eq. 3-1b has a solution set that contains the ordered pairs (width, area) given in Table 3-1b.

■ i. $f(2) = -(2)^2 + 12(2)$
$= -4 + 24$
$= 20.$

ii. $f(5) = -(5)^2 + 12(5)$
$= -25 + 60$
$= 35.$

iii. etc.

Equation 3-1a is of the general form $f(x) = ax + b$, which was the subject of Chapter 1. Equation 3-1b is a **quadratic equation** and is an example of an equation with the general form

$$f(x) = ax^2 + bx + c, \quad a \neq 0. \qquad \text{(Eq. 3-1c)}$$

The square of the independent variable is the feature that gives it the name quadratic.

The graph of a quadratic function is called a **parabola**, a nonlinear curve that has special characteristics. The characteristics that are not studied in this chapter will be studied in Chapter 10. Because the graph is not linear, more than two points of a quadratic function must be calculated in order to draw a very accurate parabola.

Example 1

Graph $h(x) = x^2 - 4x + 4$. Only a small part of the infinite solution set may be calculated and graphed. However, we have graphed a significant portion of the curve, a portion that best serves our immediate purposes.

Table 3-1c

x	$h(x)$
−2	16
−1	☐
0	4
1	1
2	0
3	1
4	4
5	9
6	☐

■ i. $h(-1) = (-1)^2 - 4(-1) + 4$
$= 1 + 4 + 4$
$= 9.$

ii. $h(6) = (6)^2 - 4(6) + 4$
$= 36 - 24 + 4$
$= 16.$

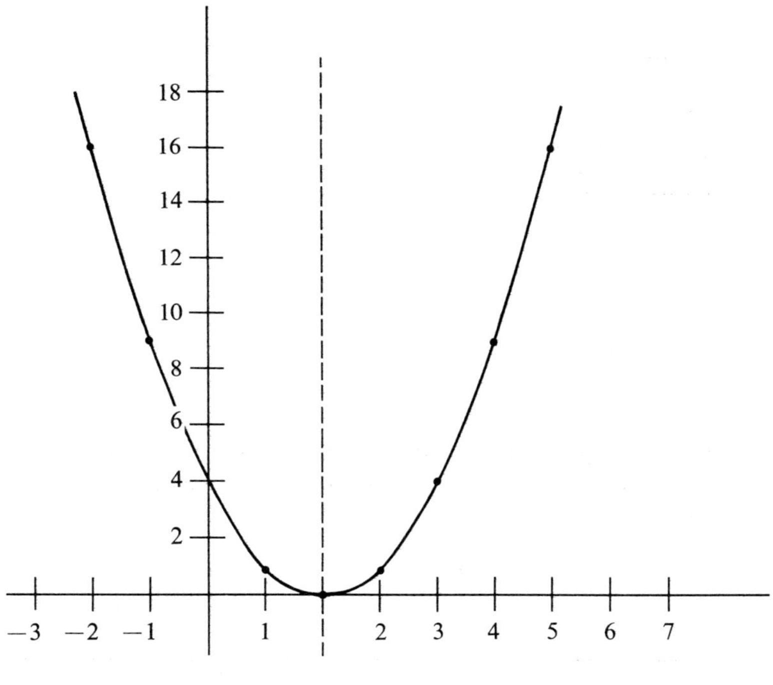

Fig. 3-1d

To gain a further understanding of quadratic functions, let us study the features of the graphs in Figs. 3-1c and 3-1d. Notice first of all that the graph for $f(w) = -w^2 + 2w$ curves in the opposite direction of the graph for $h(x) = x^2 - 4x + 4$. The explanation can be found in the sign of the coefficient of the squared variable. When the absolute value of the independent variable in either equation is large, the squared term dominates the other terms of the equation and clearly has the greatest effect on the value of the dependent variable.

Closer examination of either graph also reveals that parabolas have symmetry about a straight line passing through the highest or lowest point of the graph, as in Figs. 3-1c and 3-1d, respectively. Examine the ordered pairs in Table 3-1c. Zero appears only once in the $h(x)$ column; however, 1, 4, 9, and 16 each appear twice. Consider the *average* of the abscissas for each of the equal values of $h(x)$.

Table 3-1d

$h(x)$	x_1	x_2	$\frac{x_1 + x_2}{2}$	$\left(\frac{x_1 + x_2}{2}, h(x)\right)$
1	1	3	2	(2, 1)
4	0	4	2	(2, 4)
9	−1	5	2	(2, 9)
16	☐	☐	☐	(☐, ☐)

■ −2; 6; 2; (2, 16).

Plotting the ordered pairs in the last column of Table 3-1d onto the graph of $h(x) = x^2 - 4x + 4$ demonstrates the position of the line of symmetry (dotted line of Fig. 3-1d).

3-1 Exercises

1. Graph the quadratic equation

$$f(x) = x^2 + 2x + 1$$

from $x = -4$ to $x = 2$. Does $f(x)$ have a least or greatest value?

2. Graph the quadratic equation

$$f(w) = w^2 - 4w$$

from $w = -1$ to $w = 5$. Describe the position of the line of symmetry. Does $f(w)$ have a least or greatest value?

■ 1.

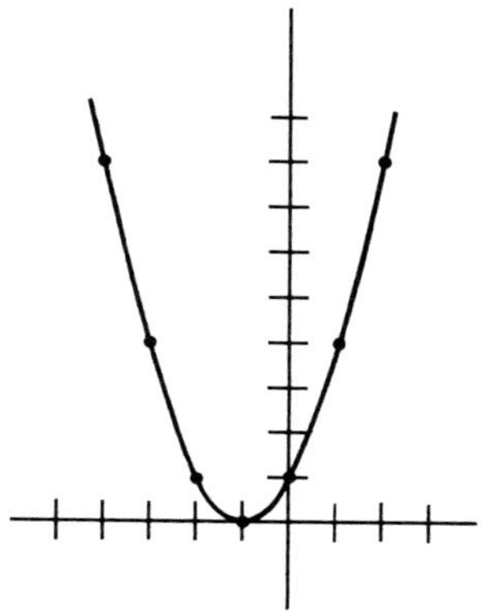

x	$f(x)$
−4	9
−3	4
−2	1
−1	0
0	1
1	4
2	9

$f(x)$ has a least value, $f(-1)$.

---

3. Graph the quadratic equation

$$f(s) = -s^2 - 2s + 4$$

from $s = -4$ to $s = 2$. Describe the position of the line of symmetry. Does $f(s)$ have a least value?

■ 2.

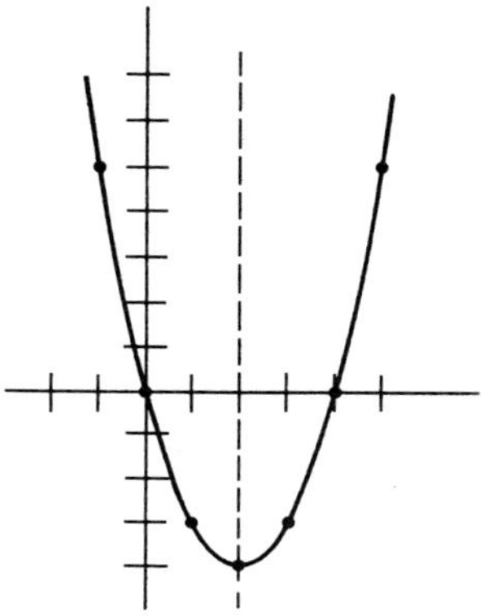

w	$f(w)$
−1	5
0	0
1	−3
2	−4
3	−3
4	0
5	5

Line of symmetry: $w = 2$.
$f(w)$ has a least value, $f(2)$.

---

4. Graph the quadratic equation

$$f(z) = z^2 + z + 1.$$

What is the equation of the line of symmetry?

■ 3.

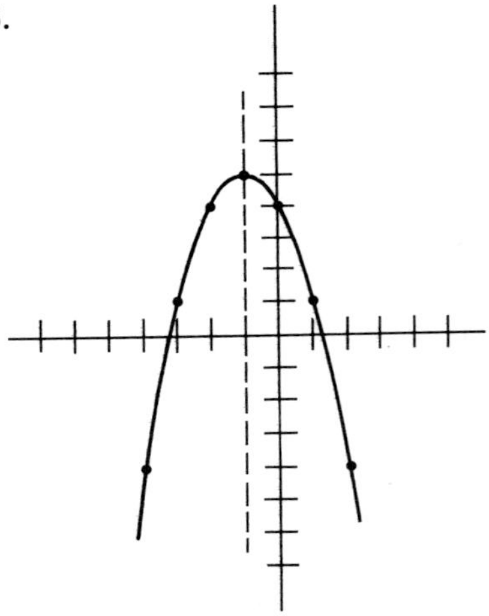

s	$f(s)$
−4	−4
−3	1
−2	4
−1	5
0	4
1	1
2	−4

Line of symmetry: $s = -1$.
No, $f(s)$ has a greatest value $f(-1)$.

■ 4.

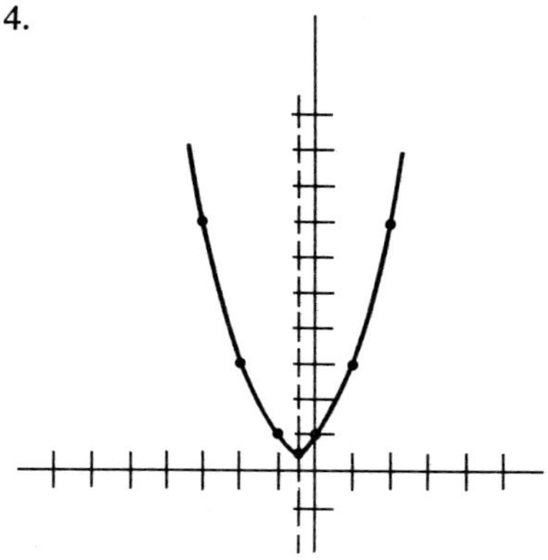

z	$f(z)$
−3	7
−2	3
−1	1
0	1
1	3
2	7

Line of symmetry: $z = -1/2$.

3-1 Problems

1. The equation $P = E^2/R$ relates the power (P) dissipated in a circuit and the applied voltage (E). Graph P as a function of E when $R = 1000$ ohms and E ranges from zero to 10 volts.
2. The kinetic energy of a one-pound mass in free fall is determined by the formula $E_k = 512t^2$. Calculate and graph the results from $t = 0$ seconds to $t = 5$ seconds.
3. The height (s) of a free-flying object when shot vertically at an initial velocity v_0 from a position s_0 is given by the formula

$$s = -16t^2 + v_0t + s_0.$$

If $v_0 = 128$ feet per second and $s_0 = -12$ feet, graph s as a function of t through a range that will permit you to read from the graph (1) the maximum height, (2) the time from $t = 0$ to the time when the object returns to $s = 0$.

4. Find the centrifugal force on a weight placed at position B. The formula is

$$F=\frac{mv^2}{r}.$$

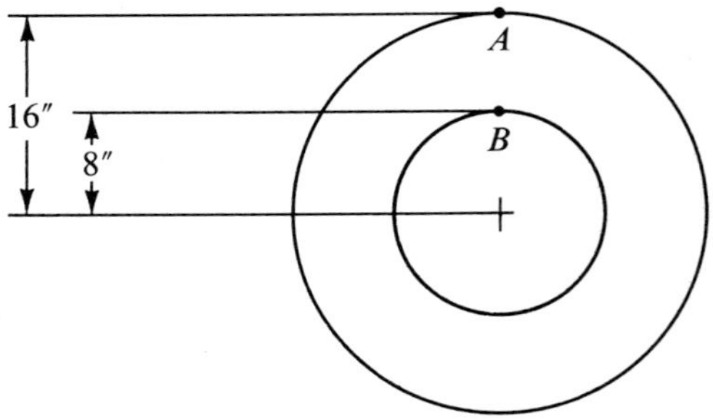

F is in poundals when v is in feet per second, m in pounds, and r in feet. Point A has a linear velocity of 80 miles per hour, and the mass at point B is 4 ounces. What is the effect on centrifugal force when the speed is reduced by one-half?

5. The length (L in centimeters) of a simple pendulum is related to the period (T in seconds) by the formula

$$L=\frac{980\,T^2}{4\pi^2}.$$

Find L for a one-second and two-second period.

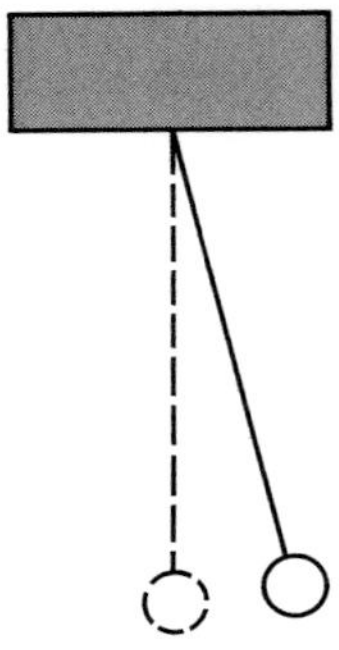

3-2 The Zeros of a Quadratic Function

The **zeros** of the quadratic equation $f(x) = ax^2 + bx + c$ are those values of x for which $f(x) = 0$. For example the *only* zeros of the equation $f(x) = 4x^2 - 5x - 6$ are 2 and $-3/4$.

The fact that these two numbers are zeros for f may be verified by evaluating $f(2)$ and $f(-3/4)$.

■ i. $f(2) = 4(2)^2 - 5(2) - 6$

$= 16 - 10 - 6$

$= 0.$

$$\text{ii. } f\left(-\frac{3}{4}\right) = 4\left(-\frac{3}{4}\right)^2 - 5\left(-\frac{3}{4}\right) - 6$$
$$= \frac{9}{4} + \frac{15}{4} - \frac{24}{4}$$
$$= 0.$$

We can easily show that the two numbers are the *only* zeros by returning to Section 3-1 and reviewing the variety of graphs drawn in the examples and exercises. The zeros of a quadratic function correspond to the points where the graph intersects the x-axis, in other words, the x-intercepts. Consequently, there are at most two zeros. Notice that only one zero and no *real* zeros are also possibilities.

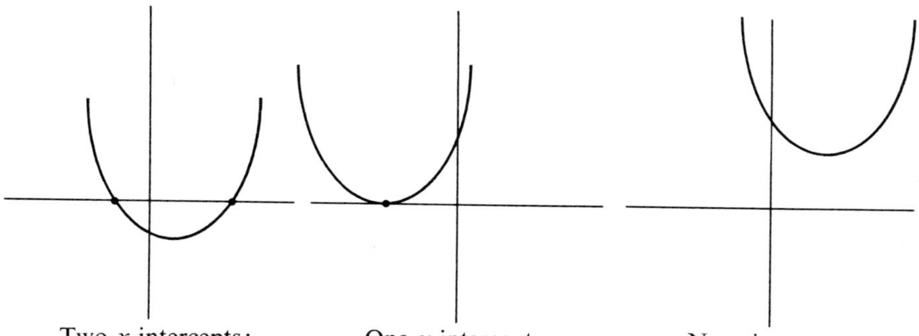

Two x-intercepts; Two real zeros. One x-intercept; One real zero. No x-intercept; No real zeros.

There are several ways of finding the zeros of a quadratic equation. Graphing, an obvious method, gives only approximations. There are several algebraic methods. The choice of method depends upon the completeness of the quadratic equation.

The **pure quadratic** has the general form $f(x) = ax^2 + c$, $a \neq 0$. Finding the zeros is equivalent to finding the roots to the equation $ax^2 + c = 0$.

Example 1

Find the zeros of $f(x) = 4x^2 - 12$.

i. $4x^2 - 12 = 0.$

ii. $4x^2 = 12.$

iii. $x^2 = 3.$

Solution set: $\{\sqrt{3}, -\sqrt{3}\}$.

Check the results.

■ i. $f(\sqrt{3}) = 4(\sqrt{3})^2 - 12$

$= 4(3) - 12$

$= 0.$

ii. $f(-\sqrt{3}) = 4(-\sqrt{3})^2 - 12$

$= 4(3) - 12$

$= 0.$

The **incomplete quadratic** has the form $f(x) = ax^2 + bx$. Finding the zeros is equivalent to finding the roots to the equation $ax^2 + bx = 0$. The simplest technique involves factoring an x from the left member of the equation. The result is two first-degree factors with a zero product. Since every zero product must have one or more zero factors, the x's that produce zero factors are the roots.

Example 2

Find the zeros of $f(x) = .050x^2 + 6.2x$.

i. $.050x^2 + 6.2x = 0$

$x(.050x + 6.2) = 0.$

ii. Either $x = 0$ or $.050x + 6.2 = 0$.

iii. Therefore $x = 0$ and

$$x = -\frac{6.2}{.050}$$

$$= -124 \qquad \{0, -124\}.$$

Check the results.

■ i. $f(0) = .050(0)^2 + 6.2(0)$

$= 0.$

ii. $f(-124) = .050(-124)^2 + 6.2(-124)$

$= 768.8 - 768.8$

$= 0.$

The algebraic method used to determine the zeros of a **complete quadratic** equation in the form $f(x) = ax^2 + bx + c$ is equivalent to finding the roots of the equation $ax^2 + bx + c = 0$ and depends upon the values a, b, and c. The method discussed here is a general one, applicable to every quadratic equation. The technique is called *completing the square.* It involves adjusting the value of the constant term so that the expression

$$x^2 + \frac{b}{a}x + c_1$$

(where c_1 is the adjusted constant) is the square of a first-degree polynomial.

Example 3

Find the zeros of $f(x) = x^2 + 6x + 5$.

i. $x^2 + 6x + 5 = 0.$

ii. $x^2 + 6x + \boxed{9} = \boxed{9} - 5.$ (Nine comes from squaring half the coefficient of x; that is, $(6/2)^2 = 9$)

iii. $(x + 3)(x + 3) = \boxed{9} - 5.$

iv. $(x + 3)^2 = 4.$

v. $x + 3 = 2$ or $x + 3 = -2.$

vi. $x = -1$ or $x = -5$ $\{-1, 5\}.$

Check the results.

■ i. $f(-1) = (-1)^2 + 6(-1) + 5$
$= 1 - 6 + 5$
$= 0.$

ii. $f(-5) = (-5)^2 + 6(-5) + 5$
$= 25 - 30 + 5$
$= 0.$

Example 4

Find the zeros of $f(x) = 2x^2 + 8x - 3$. We simplify the process by first reducing the coefficient of x^2 to one.

i. $2x^2 + 8x - 3 = 0.$

ii. $x^2 + 4x - 1.5 = 0.$

iii. $x^2 + 4x + \boxed{} = 1.5 + \boxed{}.$

iv. $(x + \square)(x + \square) = 5.5.$

v. $x + \square = \pm\sqrt{5.5}.$

vi. $x = -\square \pm \sqrt{5.5}.$

Check the results.

■ iii. 4; 4

iv. 2; 2

v. 2

vi. 2

Check:

i. $f(-2+\sqrt{5.5}) = 2(-2+\sqrt{5.5})^2 + 8(-2+\sqrt{5.5}) - 3$
$= 8 - 8\sqrt{5.5} + 11 - 16 + 8\sqrt{5.5} - 3$
$= 0.$

ii. $f(-2-\sqrt{5.5}) = 2(-2-\sqrt{5.5})^2 + 8(-2-\sqrt{5.5}) - 3$
$= 8 + 8\sqrt{5.5} + 11 - 16 - 8\sqrt{5.5} - 3$
$= 0.$

The solution of many problems requires solving an equation of the form $ax^2 + bx + c = 0$, $a \neq 0$, which, of course, is equivalent to a quadratic $f(x) = 0$.

Example 5

What size squares should be cut from the corners of a 10 inch by 6 inch rectangular piece of sheet metal in order to form an open top rectangular box with a bottom of 24 square inches?

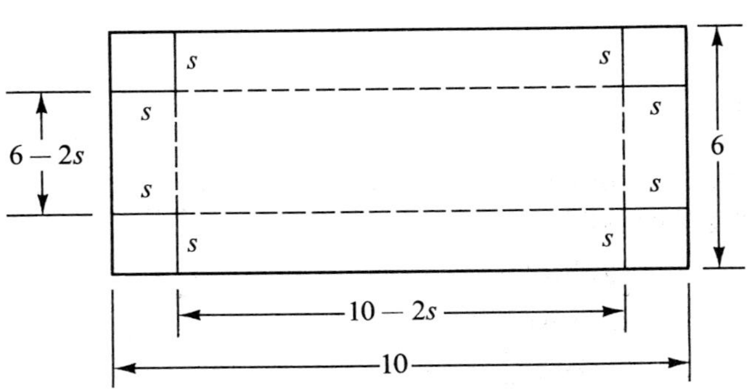

i. $(10 - 2s)(6 - 2s) = 24$
ii. $4s^2 - 32s + 36 = 0$
iii. $s^2 - 8s + 9 = 0$
iv. (Complete the solution)

■ iv. $s^2 - 8s + 16 = 16 - 9$
$(s-4)^2 = 7$
$s = 4 \pm \sqrt{7}.$

Only $s = 4 - \sqrt{7} \approx 1.35$ inches is an appropriate solution to the problem.

3-2 Exercises

1. Find the zeros of the equation

$$f(x) = x^2 - 9.$$

---

2. Find the zeros of the equation

$$h(x) = x^2 + 6x.$$

■ 1. $x^2 - 9 = 0$
$x^2 = 9$
$x = \pm 3.$

Solution set: $\{3, -3\}$.

---

3. Find the roots to the equation

$$4x^2 - 9 = 0.$$

■ 2. $x^2 + 6x = 0$
$x(x+6) = 0$
$x = 0$
$x = -6.$

Solution set: $\{0, -6\}$.

---

4. Find the roots to the equation

$$4x^2 - 3 = -4x.$$

■ 3. $4x^2 - 9 = 0$

$$x^2 = \frac{9}{4}$$

$$x = \pm\frac{3}{2}.$$

Solution set: $\left\{\frac{3}{2}, -\frac{3}{2}\right\}$.

---

5. Find the roots to the equation

$$7x^2 + 3x = 4.$$

■ 4. $4x^2 + 4x - 3 = 0$

$$x^2 + x - \frac{3}{4} = 0$$

$$x^2 + x + \frac{1}{4} = \frac{3}{4} + \frac{1}{4}$$

$$\left(x + \frac{1}{2}\right)^2 = 1$$

$$x = -\frac{1}{2} \pm 1.$$

Solution set: $\left\{\frac{1}{2}, -\frac{3}{2}\right\}$.

---

6. Solve $x^2 + 6x + 7 = 0$.

■ 5. $7x^2 + 3x = 4$

$$x^2 + \frac{3}{7}x = \frac{4}{7}$$

$$x^2 + \frac{3}{7}x + \frac{9}{196} = \frac{4}{7} + \frac{9}{196}$$

$$\left(x + \frac{3}{14}\right)^2 = \frac{121}{196}$$

$$x + \frac{3}{14} = \pm\frac{11}{14}$$

$$x = -\frac{3}{14} \pm \frac{11}{14}.$$

Solution set: $\left\{\frac{4}{7}, -1\right\}$.

---

7. Find the roots to the equation

$$8w = 4 - w^2.$$

■ 6. $x^2 + 6x + 7 = 0$

$$x^2 + 6x + 9 = 9 - 7$$

$$(x + 3)^2 = 2$$

$$x + 3 = \pm\sqrt{2}$$

$$x = -3 \pm \sqrt{2}.$$

Solution set: $\{-1.59, \quad -4.41\}$ approximate.

8. Find the roots to the equation

$$0.3t^2 - 6.0t = 1.2.$$

■ 7.

$$\begin{aligned} w^2 + 8w &= 4 \\ w^2 + 8w + 16 &= 4 + 16 \\ (w+4)^2 &= 20 \\ w + 4 &= \pm 2\sqrt{5} \\ w &= -4 \pm 2\sqrt{5}. \end{aligned}$$

Solution set:
{.47, −8.47} approximate.

---

9. Find the roots to the equation

$$\frac{x^2}{4} + \frac{x}{6} - \frac{1}{8} = 0.$$

■ 8.

$$\begin{aligned} .3t^2 - 6.0t &= 1.2 \\ t^2 - 20t &= 4 \\ t^2 - 20t + 100 &= 4 + 100 \\ (t-10)^2 &= 104 \\ t - 10 &= \pm 2\sqrt{26} \\ t &= 10 \pm 2\sqrt{26}. \end{aligned}$$

Solution set:
{20.2, −.2} approximate.

---

10. Find the roots to the equation

$$x - \frac{5}{x} = 4.$$

■ 9.

$$\frac{x^2}{4} + \frac{x}{6} - \frac{1}{8} = 0$$

$$x^2 + \frac{2}{3}x - \frac{1}{2} = 0$$

$$x^2 + \frac{2}{3}x + \frac{1}{9} = \frac{1}{2} + \frac{1}{9}$$

$$\left(x + \frac{1}{3}\right)^2 = \frac{11}{18}$$

$$x + \frac{1}{3} = \pm\frac{\sqrt{22}}{6}$$

$$x = -\frac{1}{3} \pm \frac{\sqrt{22}}{6}.$$

Solution set:
{.449, −1.125} approximate.

■ 10.

$$x - \frac{5}{x} = 4$$

$$\begin{aligned} x^2 - 5 &= 4x \\ x^2 - 4x - 5 &= 0 \\ x^2 - 4x + 4 &= 5 + 4 \\ (x-2)^2 &= 9 \\ x - 2 &= \pm 3 \\ x &= 2 \pm 3. \end{aligned}$$

Solution set: $\{5, -1\}$.

3-2 Problems

1. The height of a projectile above the ground (in feet), as a function of time (in seconds), is

$$h = 160t - 16t^2.$$

When will the projectile hit the ground? (*Hint*: Find t when $h = 0$.)

2. A strip of metal 11 inches wide is to be folded into an open-top rectangular trough having a cross-sectional area of 12 square inches. Find the width and depth of the trough.

3. The combined resistance R_s of resistors R_1 and R_2 in series is determined by the formula

$$R_s = R_1 + R_2 .$$

The same two resistors in parallel have a combined resistance R_p, which is determined by the formula

$$R_p = \frac{R_1 R_2}{R_1 + R_2}.$$

Calculate the value of R_1 and R_2 when $R_s = 300$ and $R_p = 60$.

4. The cross-section of a brass bar is a regular hexagon that measures 5/8 inches across the flats. Find its tensile strength if a bar with one square-inch cross-sectional area can support 32,000 pounds. (The ratio of the cross-sectional area of the two rods of the same material is equal to the ratio of the respective tensile strengths.)

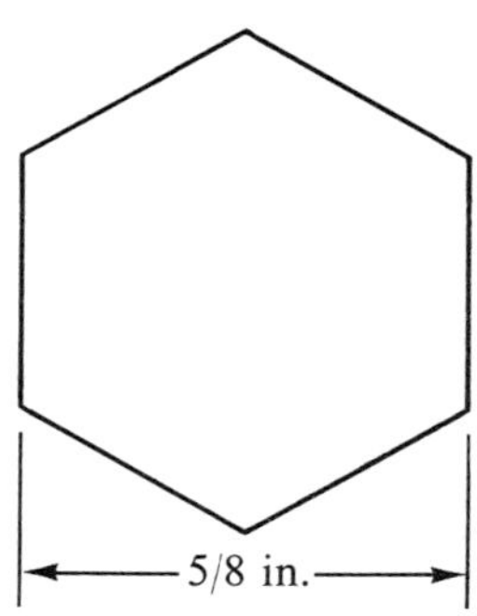

5. Calculate the distance x so that the material from the conveyer belt drops into the middle of the gondola. (The formula $s = 16t^2$ represents the distance s in feet that a freely falling object falls in t seconds.) The velocity of the conveyer is 6 ft/second.

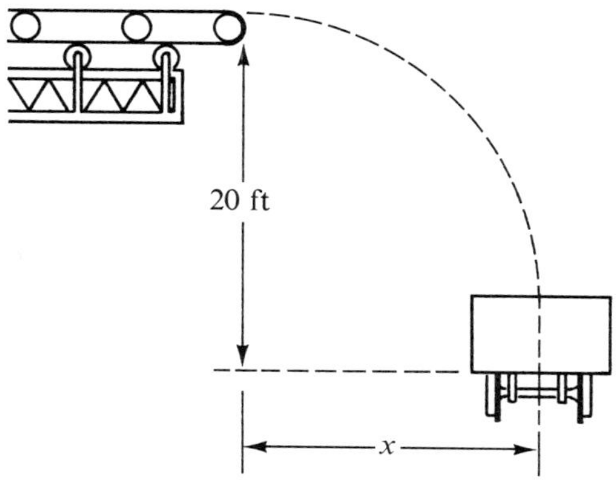

3-3 Quadratic Formula

Although it is always possible to find the zeros of a quadratic equation by completing the square, we shall save time and reduce errors if we solve the general quadratic $ax^2 + bx + c = 0$ for x in terms of a, b, and c, and then use the result as a formula to solve for the zeros of all other quadratic equations.

Solve for the roots of $ax^2 + bx + c = 0$ by completing the square.

■ i. $ax^2 + bx + c = 0$ (Divide by a)

ii. $x^2 + \frac{b}{a}x + \frac{c}{a} = 0$ (Subtract c/a from both sides and complete the square)

iii. $x^2 + \frac{b}{a}x + \frac{b^2}{4a^2} = \frac{b^2}{4a^2} - \frac{c}{a}$ (Add the fractions on the right)

iv. $\left(x + \frac{b}{2a}\right)^2 = \frac{b^2 - 4ac}{4a^2}$ (Take the square root of both sides of the equation)

v. $x + \frac{b}{2a} = \frac{\pm\sqrt{b^2 - 4ac}}{2a}$ (Subtract $b/2a$ from both sides and combine the like fractions)

vi. $\mathbf{x = \frac{-b \pm \sqrt{b^2 - 4ac}}{2a}}$ (Eq. 3-3a)

Equation 3-3a, known as the **quadratic formula**, is so useful that it is worth memorizing.

Example 1

Solve for the roots of the equation $32x^2 - 12x - 9 = 0$.
The coefficient $a = 32$, $b = -12$, and $c = -9$. Substitute the values into the quadratic formula.

$$x = \frac{-b \pm \sqrt{b^2 - 4ac}}{2a}$$

$$= \frac{-(-12) \pm \sqrt{(-12)^2 - 4(32)(-9)}}{2(32)}$$

$$= \frac{12 \pm \sqrt{\boxed{}}}{64}$$

$$= \frac{12 \pm \boxed{}}{64} \qquad \{\boxed{}, \boxed{}\}.$$

■ 1,296; 36; $\left\{\frac{3}{4}, -\frac{3}{8}\right\}$.

Example 2

Solve $x(x + 1.8) = 3.6$ for its roots. First we must put the equation in the form $ax^2 + bx + c = 0$ so that a, b, and c can be determined.

$$x(x + 1.8) = 3.6$$

$$x^2 + 1.8x - 3.6 = 0$$

Therefore, $a = \boxed{}$, $b = \boxed{}$, and $c = \boxed{}$.

■ 1; 1.8; −3.6.

Complete the solution using the quadratic formula.

■ $$x = \frac{-1.8 \pm \sqrt{(1.8)^2 - 4(1)(-3.6)}}{2(1)}$$

$$= \frac{-1.8 \pm \sqrt{17.64}}{2}$$

$$= \frac{-1.8 \pm 4.2}{2}.$$

Solution set: $\{-3.0, 1.2\}$.

Example 3

Find the zeros of the quadratic equation $f(x) = x^2 + 2x + 3$. If we graph $f(x) = x^2 + 2x + 3$, we discover the graph does not intersect the x-axis. The function has no zeros for any real value of x. However, if we apply the quadratic formula, we get the following results:

$$x = \frac{-2 \pm \sqrt{(2)^2 - 4(1)3}}{2(1)}$$

$$= \frac{-2 \pm \sqrt{-8}}{2}$$

$$= \frac{-2 \pm 2\sqrt{-2}}{2}$$

$$= -1 \pm \sqrt{-2}.$$

If we evaluate $f(-1 + \sqrt{-2})$, it is indeed equal to zero, as is $f(-1 - \sqrt{-2})$.

■ $$f(-1 + \sqrt{-2}) = (-1 + \sqrt{-2})^2 + 2(-1 + \sqrt{-2}) + 3$$

$$= 1 - 2\sqrt{-2} - 2 - 2 + 2\sqrt{-2} + 3$$

$$= 0.$$

There are, of course, no two *equal real* numbers whose product is (-2); in other words, $\sqrt{-2}$ is not a *real* number. Mathematicians have chosen to call numbers of this type *imaginary numbers*. The name might imply that they are of little value, but this is not the case as we shall see in Chapter 9.

Calculating the value of $b^2 - 4ac$ from the quadratic formula will indicate immediately the type of zeros that may be expected for a specific quadratic function. For this reason $b^2 - 4ac$ is called the **discriminant**. If $b^2 - 4ac > 0$, there are two real zeros. If $b^2 - 4ac = 0$, there is only one real root. The third and only other possibility is $b^2 - 4ac < 0$.

In this case the zeros are ______________________.

■ not *real* numbers

A flow chart can be used to summarize and organize the general procedure for calculating the zeros of quadratic functions.

Record the constants from the quadratic in standard form.

Calculate the value of the discriminant (d).

If $d > 0$, calculate the two real roots.

If $d = 0$, calculate the one real root.

If $d < 0$, there are no real roots.

Write out the results.

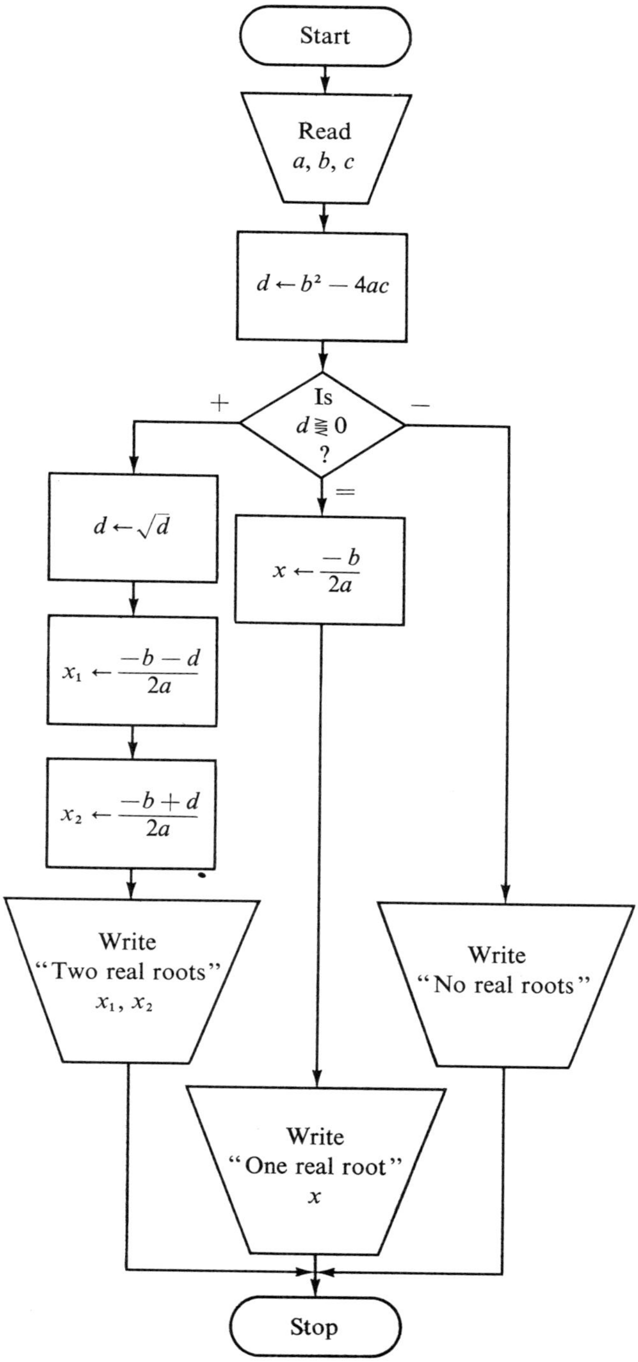
Start
Read
a, b, c
$d \leftarrow b^2 - 4ac$
Is
$d \gtreqless 0$
?
+
=
−
$d \leftarrow \sqrt{d}$
$x \leftarrow \frac{-b}{2a}$
$x_1 \leftarrow \frac{-b-d}{2a}$
$x_2 \leftarrow \frac{-b+d}{2a}$
Write
"Two real roots"
x_1, x_2
Write
"No real roots"
Write
"One real root"
x
Stop

3-3 Exercises

1. Use the quadratic formula to find the zeros of

$$f(x) = x^2 - x - 32.$$

2. Calculate the roots of

$$12x^2 + x = 20,$$

using the quadratic formula.

■ 1.
$$x = \frac{-(-1) \pm \sqrt{(-1)^2 - 4(1)(-32)}}{2(1)}$$
$$= \frac{1 \pm \sqrt{129}}{2}.$$

Solution set: $\{-5.17, 6.17\}$ approximate.

3. Solve for w when

$$25w^2 + 12 = 40w.$$

■ 2.
$$x = \frac{-1 \pm \sqrt{(1)^2 - 4(12)(-20)}}{2(12)}$$
$$= \frac{-1 \pm \sqrt{961}}{24}.$$

Solution set: $\left\{-\frac{4}{3}, \frac{5}{4}\right\}.$

4. Solve for t when

$$2t^2 + 3t - 4 = 0.$$

■ 3.
$$w = \frac{-(-40) \pm \sqrt{(-40)^2 - 4(25)(12)}}{2(25)}$$
$$= \frac{40 \pm \sqrt{400}}{50}.$$

Solution set: $\left\{\frac{6}{5}, \frac{2}{5}\right\}.$

5. Find the zeros of the equation

$$f(x) = 0.24x^2 - 0.31x - 0.75.$$

■ 4.
$$t = \frac{-3 \pm \sqrt{(3)^2 - 4(2)(-4)}}{2(2)}$$
$$= \frac{-3 \pm \sqrt{41}}{4}.$$

Solution set: $\{.85, -2.35\}$ approximate.

6. Solve for x when

$$x^2 + 6x + 9 = 0.$$

■ 5.
$$x = \frac{-(-.31) \pm \sqrt{(-.31)^2 - 4(.24)(-.75)}}{2(.24)}$$
$$= \frac{.31 \pm \sqrt{.8161}}{.48}.$$

Solution set: {2.52, 1.23} approximate.

7. Find the zeros of the quadratic equation

$$f(x) = x^2 - 3x + 3.$$

■ 6.
$$(\text{Discriminant}) = (6)^2 - 4(1)9 = 0.$$

The only real root is

$$x = \frac{-b}{2a} = \frac{-6}{2(1)} = -3.$$

Solution set: $\{-3\}$.

8. Find the zeros of the quadratic equation

$$f(s) = 2.0s^2 - 0.6,$$

using the quadratic formula.

■ 7.
$$(\text{Discriminant}) = (-3)^2 - 4(1)3 = -3.$$

There are no *real* roots;

however, $f\left(\frac{3 \pm \sqrt{-3}}{2}\right) = 0.$

■ 8.
$$s = \frac{-0 \pm \sqrt{0^2 - 4(2.0)(-0.6)}}{2(2.0)}$$
$$= \frac{\pm\sqrt{4.8}}{4}.$$

Solution set: $\{.547, -.547\}$.

3-3 Problems

1. The sum of a number and its square is 56. Find the number.
2. A projectile is fired vertically into the air. The distance (in feet) above the ground as a function of time (in seconds) is given by

$$S = 300 - 100t - 16t^2.$$

When will the projectile hit the ground?

3. Find a value for x so that the shaded area is equal to 96. (Refer to Section 1–7.)

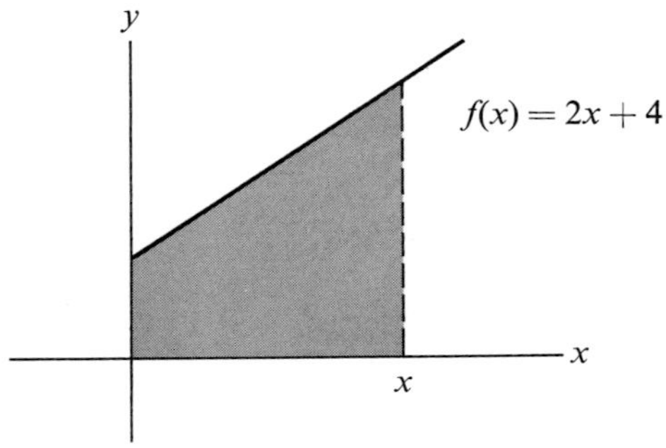

4. If 6.25 foot-pounds of work is done when a spring is extended, how far is it extended if the spring constant is 6 pounds per inch?
5. If the combined area of the inner circles is 4/5 the area of the outer circle, find the diameters of the inner circles.

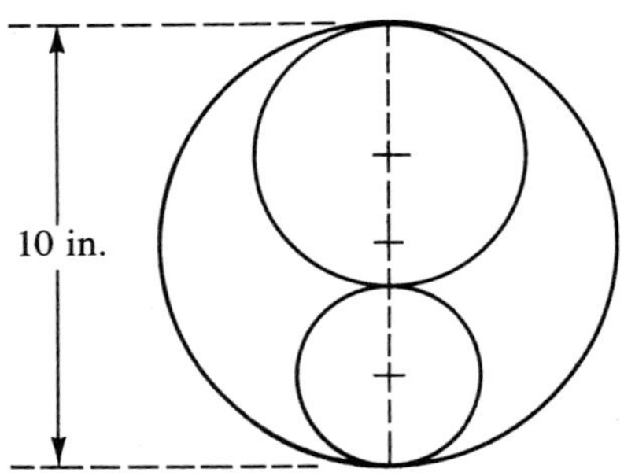

6. Find the radius of curvature passing through point A and tangent to the horizontal and vertical lines c_1 and c_2.

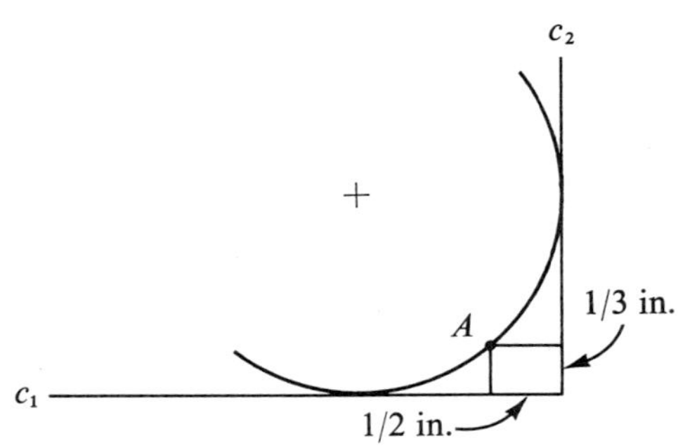

7. A cube of metal alloy is placed in a cooling bath, with the result that the volume shrinkage is 2.00 cubic centimeters and the linear shrinkage is .125 centimeters on each edge. Find the dimensions before cooling.
8. Find the distance x so that the area of the triangle is divided exactly in half by a line drawn parallel to the base.

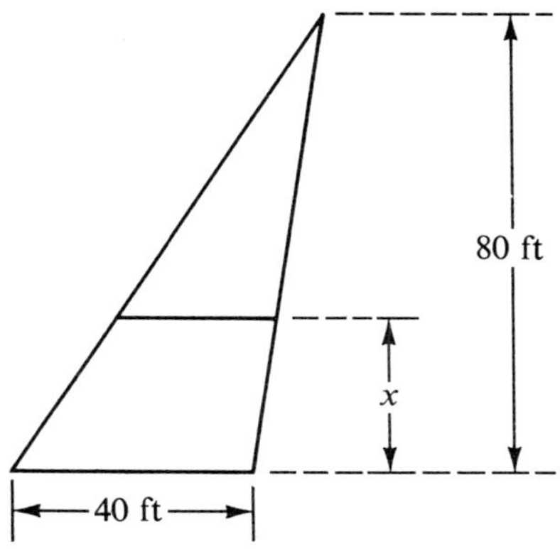

3-4 Maxima and Minima of a Quadratic Function

In addition to the zeros of a quadratic function, there is usually another ordered pair in the solution set that is of special interest and importance. Every quadratic function f has either a greatest or least $f(x)$. In Section 3-1, which discusses graphing quadratic functions, we discovered that the line of symmetry contains this particular ordered pair. Let us use a principle of electronics to illustrate the importance of the greatest $f(x)$ in a quadratic function.

Assume we have a six-volt power supply with an internal resistance of eight ohms that is hooked to a speaker. The power (P) that is delivered to the speaker is a function of the current (I). The constants in this particular problem require that

$$P = f(I) = 6I - 8I^2. \qquad \text{(Eq.3-4a)}$$

This is a quadratic equation and has one particular value I that corresponds to maximum power. Our problem is to find that particular I.

A graphical approach is one way to solve the problem. It involves approximations which may in many cases be satisfactory. There are other methods that give precise results and are easier than graphing.

Recall from Section 3-1 that the line of symmetry contains the maximum or minimum point of a parabola. Also recall that the average of every two abscissas with equal ordinates identifies the position of the line of symmetry. Therefore the average of the zeros of a quadratic would correspond to the abscissa of the maximum or minimum point. If

$$x_1 = \frac{-b + \sqrt{b^2 - 4ac}}{2a}$$

is one zero, and

$$x_2 = \frac{-b - \sqrt{b^2 - 4ac}}{2a}$$

is the other zero, the average is

$$\begin{aligned} \bar{x} &= \frac{1}{2}(x_1 + x_2) \\ &= \frac{1}{2}\left(\frac{-b + \sqrt{b^2 - 4ac} - b - \sqrt{b^2 - 4ac}}{2a}\right) \\ &= \frac{1}{2}\left(\frac{-2b}{2a}\right) \\ &= \frac{-b}{2a}. \qquad \text{(Eq. 3-4b)} \end{aligned}$$

This very simple result is the abscissa of the maximum or minimum point on a parabola. See Fig. 3-4a.

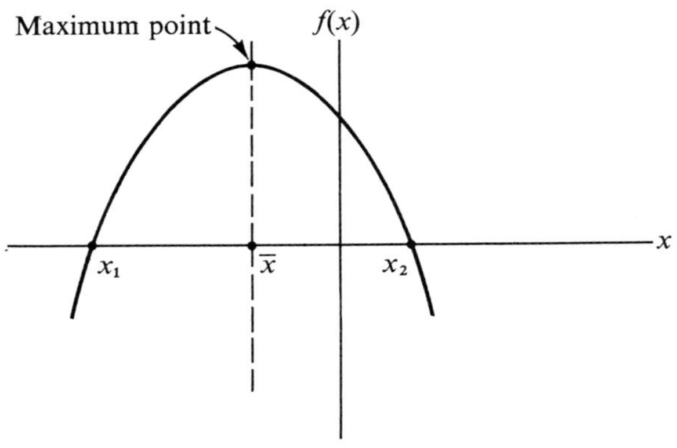

Fig. 3-4a

Applying Eq. 3-4b to Eq. 3-4a, the I for maximum power (P) can be readily calculated.

$$P = -8I^2 + 6I$$

$$\bar{I} = -\frac{6}{2(-8)} = \frac{3}{8} \text{ amperes}$$

$$P_{max} = f\left(\frac{3}{8}\right) = \boxed{\qquad\qquad}.$$

■ $P_{max} = f\left(\frac{3}{8}\right) = (-8)\left(\frac{3}{8}\right)^2 + 6\left(\frac{3}{8}\right)$

$= \frac{9}{8}$ watts.

Note: Those with a special interest in electronics can see that with the values of power and current for the speaker, the correct resistance for the speaker can be calculated.

$$R = \frac{P}{I^2} = \frac{9}{8} \Big/ \left(\frac{3}{8}\right)^2 = 8\Omega.$$

Notice that the resistance of the source and load are equal.

3-4 Exercises

1. Find the maximum or minimum point of

$$f(x) = -x^2 + 3x.$$

2. Find the maximum or minimum point of

$$h(x) = -3x^2 + 5x - 1.$$

■ 1. $\bar{x} = -\frac{b}{2a} = -\frac{3}{2(-1)} = \frac{3}{2}$

$$f\left(\frac{3}{2}\right) = -\left(\frac{3}{2}\right)^2 + 3\left(\frac{3}{2}\right)$$

$$= -\left(\frac{9}{4}\right) + \left(\frac{9}{2}\right) = \frac{9}{4}.$$

Maximum point: $\left(\frac{3}{2}, \frac{9}{4}\right)$.

---

3. Find the maximum or minimum point of

$$f(t) = 0.8t^2 - 4.0t + 1.0.$$

■ 2. $\bar{x} = -\frac{b}{2a} = -\frac{5}{2(-3)} = \frac{5}{6}$

$$f\left(\frac{5}{6}\right) = -3\left(\frac{5}{6}\right)^2 + 5\left(\frac{5}{6}\right) - 1$$

$$= -\left(\frac{25}{12}\right) + \left(\frac{50}{12}\right) - \left(\frac{12}{12}\right)$$

$$= \frac{13}{12}.$$

Maximum point: $\left(\frac{5}{6}, \frac{13}{12}\right)$.

---

4. Find the maximum power (P) if

$$P = 12I - 300I^2.$$

■ 3. $\bar{t} = -\frac{b}{2a} = -\frac{-4.0}{2(0.8)}$

$$= 2.5;$$

$$f(2.5) = 0.8(2.5)^2 - 4.0(2.5) + 1.0$$

$$= -4.0.$$

Minimum point: $(2.5, -4.0)$.

■ 4. $P = f(I) = -300I^2 + 12I$

$$\bar{I} = -\frac{b}{2a} = -\frac{12}{2(-300)} = .02$$

$$P_{\text{max}} = f(.02) = -300(.02)^2 + 12(.02)$$

$$= -.12 + .24$$

$$= .12$$

3-4 Problems

1. The height of an object above the ground is given by the equation

$$S = -16t^2 + 80t + 50.$$

At about what time will the height be a maximum?

2. A rectangular field with a river along one side is to be fenced along the other three sides. If the total length of the fence is 600 feet, what is the maximum possible area of the field?

3. A Norman window consists of the combination of a rectangle and semicircle. If the perimeter is to be limited to 12 feet of material, what dimensions provide the greatest area?

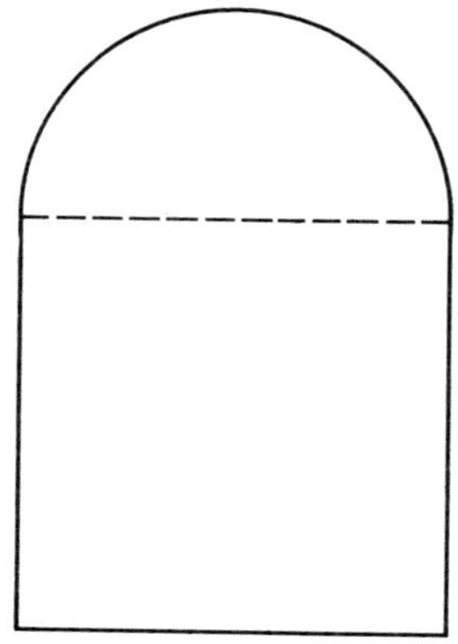

4. The bending moment (M) of a uniformly loaded beam of length (L) at any distance (s) from one end is

$$M = \frac{wLs}{2} - \frac{ws^2}{2},$$

where w is a constant weight per unit length. Find the point at which the bending moment is the greatest.

5. If a farmer harvests his crop July 15, he will have 30,000 pounds of bush beans worth $12 per hundred pounds. Every day he waits, the crop increases by 3,000 pounds, but the quality of the crop is affected, and each day the price drops 80 cents per hundred pounds. When should the crop be harvested in order to make the most money?

6. Find the dimensions of the rectangle $CDEF$ with the greatest area.

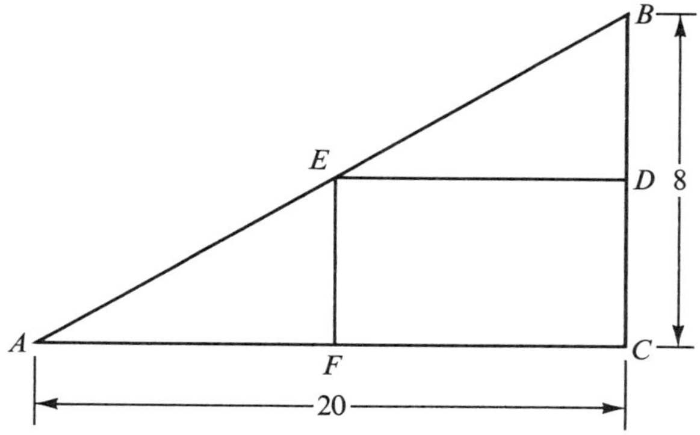

3-5 Related Function—Slope

In Chapter 1 we studied two functions, f' and F, related to each linear function f. You will recall that f' gives the slope or rate of change of f, and F gives the first quadrant area under the graph of f. In this section we *begin* to consider similar functions related to quadratics.

As we consider the rate of change or slope of a quadratic function, we notice that the definition of slope for linear functions cannot be applied to non-linear functions without some modification.

Choose any two points P_1 and P_2 on the graph of a quadratic function (Fig. 3-5a). The value of

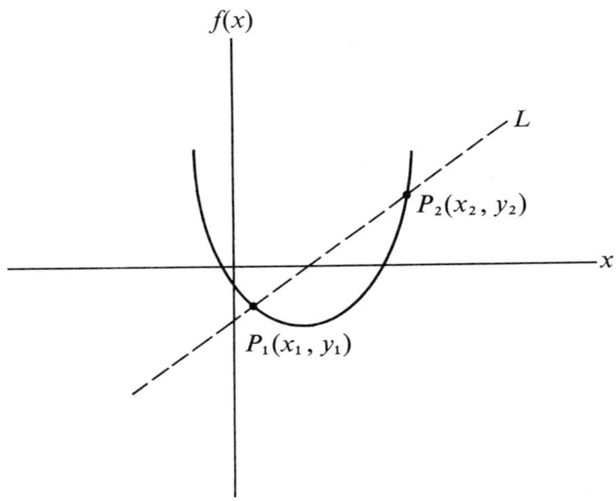

Fig. 3-5a

$$\frac{y_2 - y_1}{x_2 - x_1}$$

is the slope of the secant L, and it changes as either P_1 or P_2 moves along the parabola. If, however, P_2 moves closer to P_1 until, in fact, P_2 and P_1 become the same point, there is a unique line associated with P_1 that graphically defines the slope. This line is equivalent to the *tangent line* to the curve at P_1 (Fig. 3-5b).

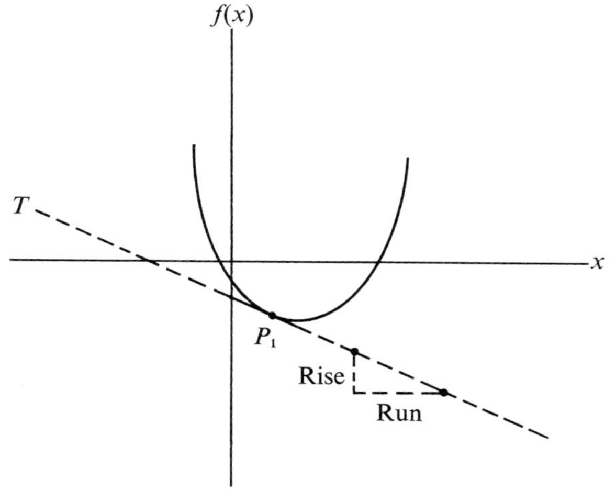

Fig. 3-5b

Every point on the parabola has a different slope. The magnitude of the slope at each point can be approximated by the rise and run of the tangent line. In Fig. 3-5b the slope at P_1 is approximately [ ].

■ $\left(\frac{\text{rise}}{\text{run}}\right) \approx -\frac{1}{2}$.

It should not be difficult to see that the slopes of a quadratic function are themselves a function of x. The more difficult problem is to determine whether or not there is a reasonable equation that relates the x's and corresponding values of slope.

Experiment

Figure 3-5c shows a carefully prepared graph of

$$f(x) = \frac{x^2}{3} + \frac{x}{2} - 2.$$

x	$f(x)$
5.00	8.83
4.00	5.33
3.00	2.50
2.00	0.33
1.00	−1.17
0	−2.00
−1.00	−2.17
−2.00	−1.67
−3.00	−0.50
−4.00	1.33
−5.00	3.83

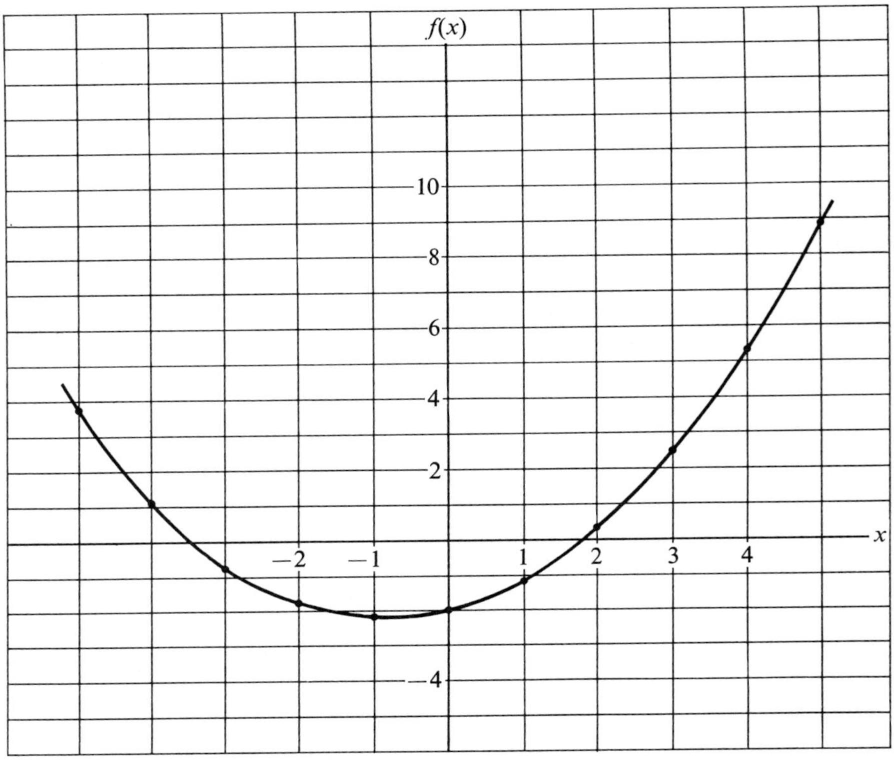

Fig. 3-5c

At every integer value of x from -5 to 5 draw a line tangent to the parabola and estimate as nearly as possible the value of its slope. Record in a table the corresponding values of x and estimated slope. Graph the results.

■

x	Slope of tangent line
5	23/6
4	19/6
3	15/6
2	11/6
1	7/6
0	1/2
−1	− 1/6
−2	− 5/6
−3	− 3/2
−4	−13/6
−5	−17/6

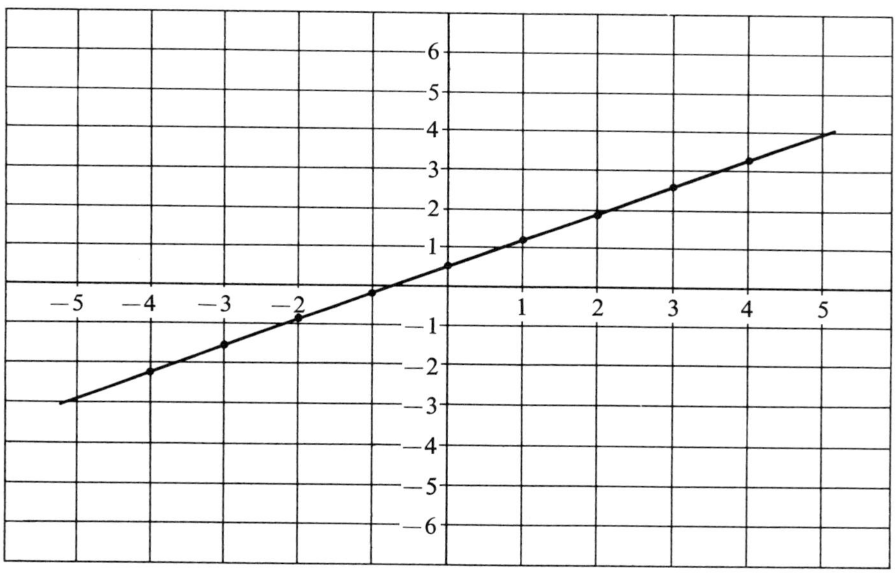

Fig. 3-5d

The experimental results indicate that the functional relationship between the x's and the corresponding values of the slope appears to be linear. If we make this assumption and fit a straight line to the points in Fig. 3-5d, we can then determine the related equation.

If we approximate the y-intercept by 1/2, and the slope by 2/3, the equation of the graph is [ ].

■ $y = \frac{2}{3}x + \frac{1}{2}$.

For convenience we shall write $y = f'(x)$ and identify the related function as f', notation equivalent to that used in Chapter 1.

One way of checking the validity of our experimental result for f' is to test it on a portion of the curve where we know the value of the slope.

The minimum point of the parabola in Fig 3-5c should have a slope of ______.

■ zero

The value of x for the minimum point can be found by using Eq. 3-4b,

$$f(x) = \frac{x^2}{3} + \frac{x}{2} - 2$$

$$\bar{x} = -\frac{b}{2a} = -\frac{1/2}{2(1/3)} = -\frac{3}{4}.$$

Next, evaluate $f'(-3/4)$.

■ $$f'\left(-\frac{3}{4}\right) = \left(\frac{2}{3}\right)\left(-\frac{3}{4}\right) + \left(\frac{1}{2}\right)$$
$$= \left(-\frac{1}{2}\right) + \left(\frac{1}{2}\right)$$
$$= 0.$$

In Chapter 11, we develop exact methods for showing that if

$$f(x) = \frac{x^2}{3} + \frac{x}{2} - 2,$$

then the related equation that characterizes the slope at every x is exactly

$$f'(x) = \frac{2}{3}x + \frac{1}{2}.$$

3-5 Exercises

1. Carefully graph the equation

$$f(x) = x^2 - 4x + 4$$

from $x = -1$ to $x = 5$. At every integer value of x estimate the slope of the line tangent to the curve. Graph the results and estimate the equation of the graph.

2. If $f'(x) = 2x - 4$ is the equation of the slopes of

$$f(x) = x^2 - 4x + 4,$$

(a) what is the slope at $x = 7$?
(b) for what x is the slope zero?

■ 1.

x	-1	0	1	2	3	4	5
Slope	-6	-4	-2	0	2	4	6

The actual related slope function is

$$f'(x) = 2x - 4.$$

3. Graphically determine the related slope function, f', for the equation

$$f(x) = -\frac{1}{3}x^2.$$

■ 2. (a) $f'(x) = 2x - 4$

$$f'(7) = 2(7) - 4 = 10.$$

(b) Finding where the slope is zero is equivalent to finding the zero of the f' function or the root of $2x - 4 = 0$. Therefore, the slope of f is zero at $x = 2$.

4. Using the results of Exercise 3, find:
(a) the slope of f at $x = 3$.
(b) the x where the slope is zero.

■ 3. The exact result is

$$f'(x) = -\frac{2}{3}x.$$

5. If $h(x) = -16t^2 + 40t$ and $h'(x) = -32t + 40$, find the maximum point of h using the fact that the slope at the maximum point is zero.

■ 4. (a) $f'(x) = -\frac{2}{3}x$

$f'(3) = -\frac{2}{3}(3) = -2.$

(b) $-\frac{2}{3}x = 0$

$\therefore x = 0.$

■ 5. i. $h'(x) = -32t + 40$

$-32t + 40 = 0$

$-32t = -40$

$t = \frac{5}{4};$

ii. $h\left(\frac{5}{4}\right) = -16\left(\frac{5}{4}\right)^2 + 40\left(\frac{5}{4}\right)$

$= 25;$

$\therefore$ max point is $\left(\frac{5}{4}, 25\right)$.

3-5 Problems

1. The elevation of a road as it passes over a hill is given by

$$h = 100 - .001s^2.$$

Find the slope and elevation when $s = 0$, $s = 50$, and $s = 100$. ($h = f(s)$ and $f'(s) = -.002s$.)

2. The cables of a suspension bridge hang in the approximate shape of a parabola. If the height of the cable above the road bed is given by

$$h = .01s^2 + 10,$$

find h and the slope in the middle of the bridge and at each end of the cable. ($h = f(s)$ and $f'(s) = .02s$.)

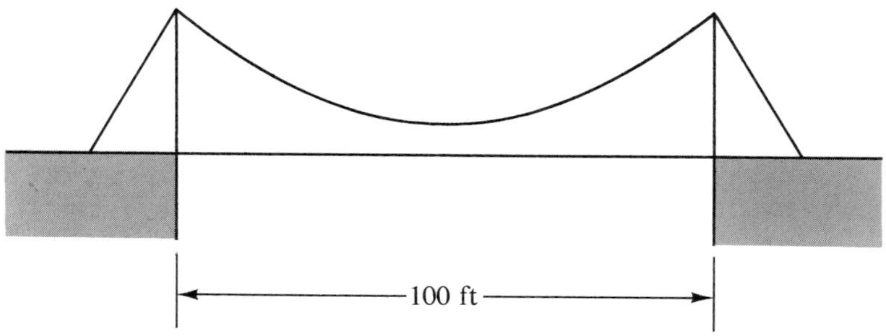

3. No current in a capacitive circuit will be flowing when the rate of change of voltage (equivalent to slope) is zero. If voltage (v) is a function of time (t) defined by the equation

$$v = -\frac{1}{2}t^2 + 3t,$$

at what instant of time will the current be zero? ($v = f(t)$ and $f'(t) = -t + 3$.)

4. Carefully prepare the graph of

$$f(x) = -\frac{1}{4}x^2 + x - 1,$$

and for at least five points along the x-axis determine the slope of the lines tangent to the graph of f. Each value thus determined should be graphed. The points should approximate points on the graph of f'. Try to determine the equation of f' from these points.

3-6 Related Function—Area

In Chapter 1 we used geometry to find the functional equation for F, which gives the first quadrant area under the graph of the linear function f. We have not yet developed the necessary background to discover the equation of a similar function F when the f function is quadratic. This is done in Chapter 12. However, there are many techniques of approximating an area in the first quadrant under a curve without knowing any functional equations. Engineers frequently find it necessary to use approximate techniques in beam design, circuit analysis, motion problems, hydraulics, and other problems.

To cite a specific case, Fig. 3-6a shows the graph of a transient current as a function of time. The shaded *area* is equivalent to the total flow of *charge* from t_1 to t_2.

If each vertical unit represents .02 amperes and each horizontal unit .01 seconds, then each square of the grid represents (.02)(.01) coulombs of charge. The total charge from t_1 to t_2 is therefore approximately [ ] coulombs.

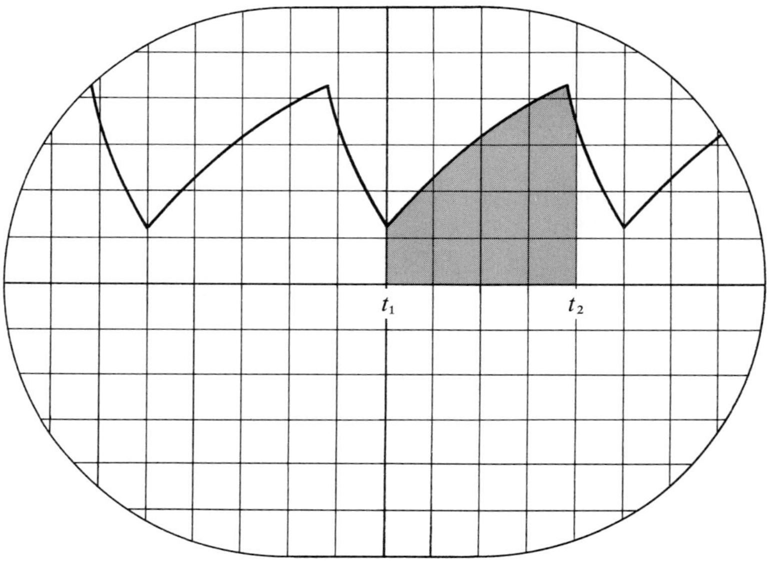

Fig. 3-6a

■ 11.5 (.02)(.01) = .0023.

3-6 Exercises

1. Graph and estimate the area in the first quadrant under the curve

$$f(x) = 2x^2 + x,$$

from $x = 0$ to $x = 3$.

2. Graph and estimate the area in the first quadrant under the curve

$$h(x) = -\frac{x^2}{4} + 6x.$$

■ 1.

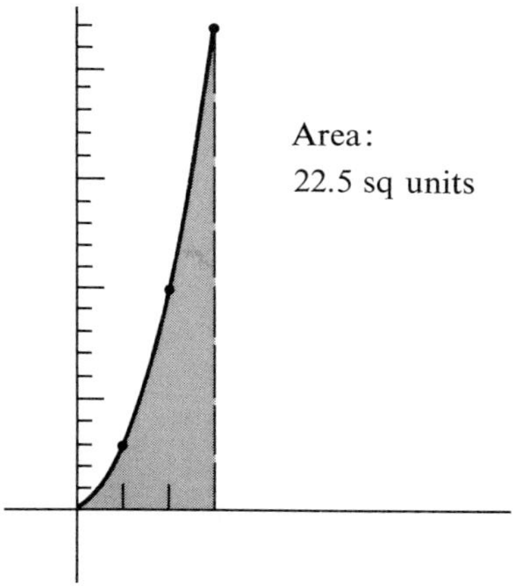

Area: 22.5 sq units

x	$f(x)$
0	0
1	3
2	10
3	21

■ 2.

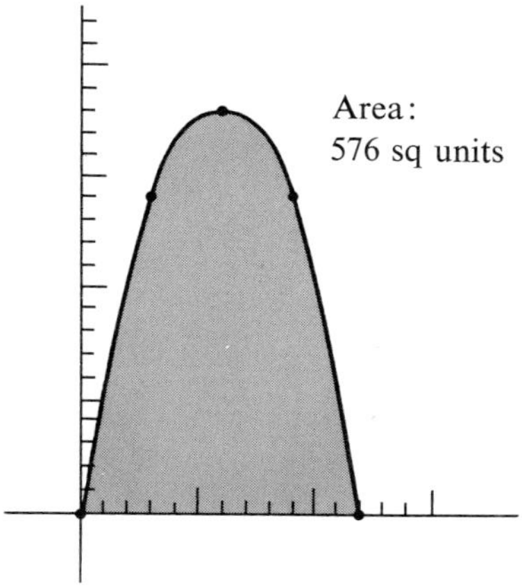

Area: 576 sq units

x	$h(x)$
0	0
6	27
12	36
18	27
24	0

3-6 Problems

1. Fig. 3-6b is a graph of rocket velocity as a function of time. The area in the first quadrant under the curve is equivalent to the *distance* the rocket travels as a function of time. Estimate the distance traveled in 8 seconds.

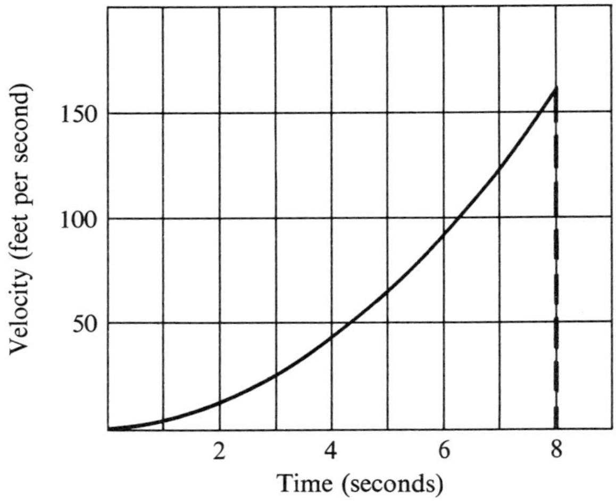

Fig. 3-6b

2. The *bending moment* of a loaded beam is equivalent to the area under the curve of Fig. 3-6c. Find the bending moment 18 feet from the end of the beam.
3. The force required to reduce the volume of a compression chamber increases as the piston moves toward the top of its stroke. The area under the curve is equivalent to the *energy* required to compress the gas in the chamber. Estimate the energy from Fig. 3-6d.
4. The area under the curve of Fig. 3-6e is equivalent to the *voltage* across a capacitor when divided by the capacitance (c). Find the approximate voltage at $t = .02$ seconds when $c = .006$ farads.
5. What is a planimeter?

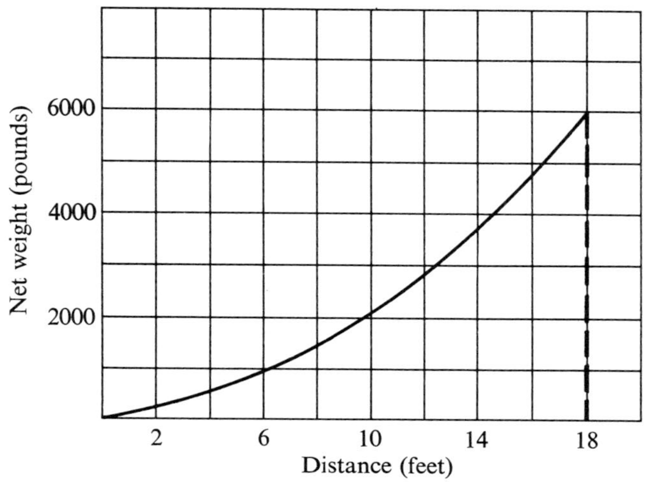

Fig. 3-6c

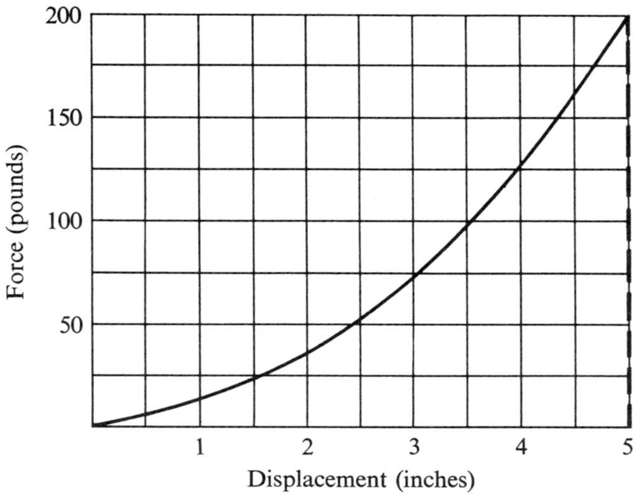

Fig. 3-6d

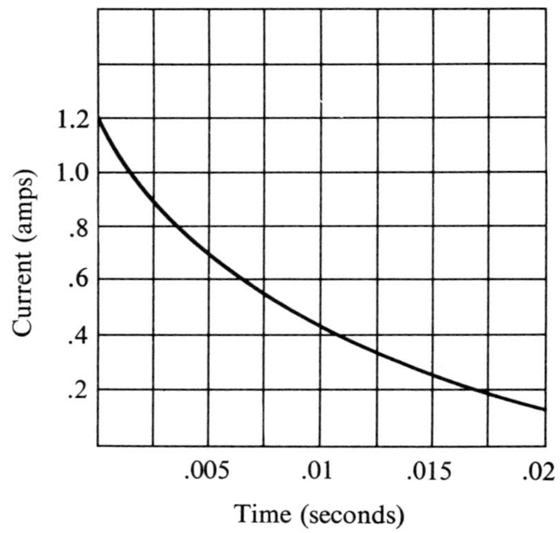

Fig. 3-6e

---

3 Supplementary Exercises and Problems

1. Compare the graphs of $f(x) = x^2 + 3x$ and $g(x) = x^2 + 3x - 6$ by plotting them on the same coordinate system. (3-1)
2. Compare the graphs of $f(s) = 1/2s^2$ and $g(s) = 6s^2$ by plotting them on the same coordinate system. (3-1)
3. Compare the graphs of $f(t) = t^2 + 5t$ and $g(t) = t^2 - 5t$ by plotting them on the same coordinate system. (3-1)
4. Graph $f(r) = -1.6r^2 + 4.5r$ and approximate the equation of the line of symmetry. (3-1)
5. Solve the equation $1.1t^2 + 6.3t = 12.3t$. (3-2)
6. $h = \dfrac{V^2}{2g}$. Solve for V in terms of the other variables. (3-2)
7. Find the zeros of the equation $f(s) = 7 + 7s + s^2$ by completing the square. (3-2)
8. $H = \dfrac{RI^2t}{4.18}$. Solve for I in terms of the other variables. (3-2)
9. Find the zeros of

$$f(W) = \frac{KV}{A}(W_f W - W^2). \qquad (3\text{-}2)$$

10. Find the zeros of $f(t) = 2\pi \text{ft} - \pi \text{ft}^2$. (3-2)
11. Solve $(x + 2)(x - 2) = 4$ for x. (3-2)

12. Solve $m^2 + Rm = \frac{1}{c}$ for m in terms of the other variables. (3-2)

13. Find the dimensions of the area B if the uniform border A is equal in area to B. (3-2)

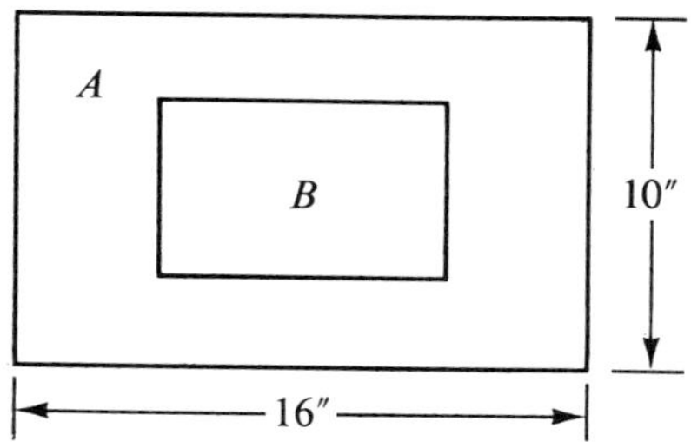

14. Using the discriminant of the quadratic formula, find how many times the graphs of the following quadratics intersect the x-axis.

(a) $f(x) = x^2 - x - 1$,
(b) $g(x) = 2 + x(1 - 5x)$,
(c) $h(x) = x^2 - 2.60x + 1.69$,
(d) $k(x) = -\frac{x^2}{2} + \frac{x}{3} + \frac{2}{3}$. (3-3)

15. Find the k that causes $f(w) = w^2 + kw + 4$ to have only one real zero.

16. Find the values of c that cause $f(z) = z^2 + 9z + c$ to have two real zeros. (3-3)

17. If the legs of the triangle are in the ratio of 3 to 1, find the length of each. (3-3)

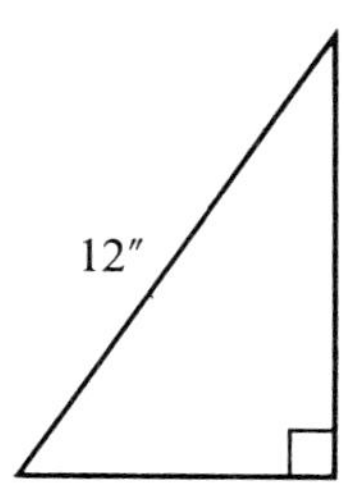

18. The illumination P of a source of light and the intensity of illumination I at a distance of r meters from the source are given by the equation $P = Ir^2$. If two sources, $P_1 = 30$ and $P_2 = 50$, are separated by 20 meters, what point between them has equal illumination from both sources? (3-3)

19. Find the value of x so that the line parallel to the base of the trapezoid divides the area into equal parts. (3-3)

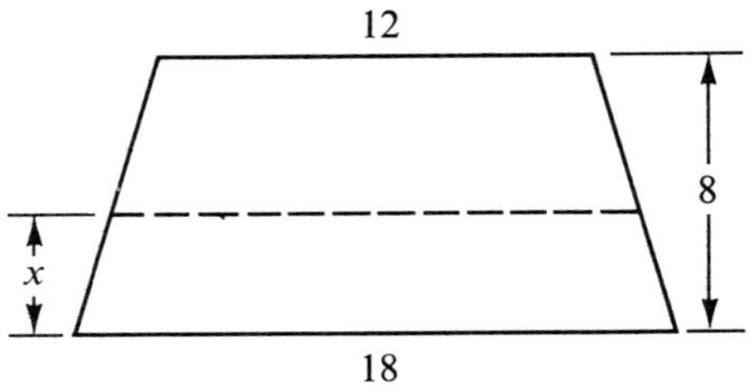

20. Solve for f in the formula $x = 2\pi fL - \dfrac{1}{2\pi fC}$. (3-3)

21. Find the maximum or minimum value of $P = .8I - 6.5I^2$, when $P = f(I)$. (3-4)

22. Use the minimum or maximum value $\bar{x} = \dfrac{-b}{2a}$ of the equation

$$f(x) = ax^2 + bx + c$$

to calculate $f(\bar{x})$. Simplify. (3-4)

23. What is the largest rectangular area that can be enclosed by 250 feet of fencing? (3-4)

24. A television cable company has 2040 monthly subscribers at \$5.00 per month. Experience has shown that for each decrease of \$.20 per month the company receives 50 additional subscribers. What monthly charge will result in the largest monthly income? (3-4)

25. Use a graphical approach for approximating the related f' equation for the equation

$$f(x) = -\frac{x^2}{2} + 4x.$$

26. If $g(x) = 4.2x^2 + 3.5x - 7.1$ and $g'(x) = 8.4x + 3.5$, find the coordinates of the point on g where the slope is 0.5. (3-5)

27. If $h(x) = 22x^2 + 100x + 51$, what is the x-intercept for the related slope function h'? (3-5)

28. The velocity (v) of an object in meters per second is changing with respect to time (t) such that $v = 9.8t^2 + 100t$. Beginning with $t = 0$, how long will it take for the object to travel approximately 1000 meters? (3-6)

Chapter Four Exponents and Radicals

4-1 Laws of Exponents—Integer Exponents

This section demonstrates how the notation of exponents is extended to include zero and negative integers.

For convenience, we state the definition and laws pertaining to positive-integer exponents. If a and b are any numbers other than zero, and n and m are positive integers, then

$$a^n = aaa \ldots a, \ (n \text{ factors of } a), \qquad \text{(Def. 4-1a)}$$

$$a^m a^n = a^{m+n}, \qquad \text{(Law 4-1a)}$$

$$\frac{a^m}{a^n} = a^{m-n}, \qquad \text{(Law 4-1b)}$$

$$(a^m)^n = a^{mn}, \qquad \text{(Law 4-1c)}$$

$$(ab)^n = a^n b^n, \qquad \text{(Law 4-1d)}$$

$$\left(\frac{a}{b}\right)^n = \frac{a^n}{b^n}. \qquad \text{(Law 4-1e)}$$

If these laws are not familiar, you should review Section A-2.

Consider next the value of a number raised to the zero power, a^0. In terms of Law 4-1a,

$$a^m \cdot a^0 = a^{m+0} = a^m.$$

Thus, it is logical that $a^0 = \boxed{}$, provided $a \neq 0$.

■ one

This result is equally justifiable as a consequence of Law 4-1b. The value of a fraction with equal numerator and denominator is one.

$$1 = \frac{a^m}{a^m} = a^{m-m} = a^0.$$

Therefore $12^0 = 1$, $\pi^0 = 1$, $(x + y)^0 = 1$, and so on. Strange as these results may seem, the laws of positive integer exponents are preserved as we extend them to include zero.

What meaning is given to negative integers as exponents? They arise quite naturally from the laws just stated. If in the case of Law 4-1b, $n > m$, then the difference, $m - n$, is less than 0. For example,

$$\frac{a^2}{a^5} = a^{2-5} = a^{-3}.$$

Since

$$\frac{a^2}{a^5} = \frac{\overset{1}{\cancel{a}}\,\overset{1}{\cancel{a}}}{\underset{1}{\cancel{a}}\,\underset{1}{\cancel{a}}\,a\,a\,a} = \frac{1}{a^3},$$

we can see that a^{-3} is equivalent to $1/a^3$. Generally,

$$a^{-n} = \frac{1}{a^n}. \qquad \text{(Law 4-1f)}$$

Example 1

Evaluate 5^{-3}.

$$5^{-3} = \frac{1}{5^3} = \frac{1}{125}.$$

Example 2

Evaluate $\left(\frac{2}{3}\right)^{-2}$.

$$\left(\frac{2}{3}\right)^{-2} = \frac{1}{(2/3)^2} = \frac{1}{4/9} = \frac{9}{4}.$$

Observe that the reciprocal of $\frac{2}{3}$ is $\frac{1}{2/3}$, or $\frac{3}{2}$. It is therefore more convenient to proceed in the following manner:

$$\left(\frac{2}{3}\right)^{-2} = \left(\frac{3}{2}\right)^{2} = \frac{9}{4}.$$

Example 3

Evaluate $\frac{1}{1.3^{-2}}$.

$$\frac{1}{(1.3)^{-2}} = (1.3)^2 = 1.69.$$

4-1 Exercises

1. Rewrite the following expressions without using zero or negative exponents and simplify.

(a) $3x^{-2}$ (e) $6x^{-2}(2x^2)^{-2}$

(b) $\frac{2x^{-3}}{3y^{-2}}$ (f) $a^{-1} + b^{-1}$

(c) $(3ax^2)^0$ (g) $\left(\frac{2x^2}{3x^{-3}}\right)^{-1}$

(d) $\frac{a^2b^{-3}}{a^{-2}b^{-4}}$ (h) $(6x^3y)^0 5^{-2}$

---

2. Evaluate the following functions:

(a) $f(x) = \dfrac{3x^3}{2x^2}$,
$f(3) = \square$,

(b) $f(x) = (27x^{-3})^0 + (2x)^{-1}$,
$f(4) = \square$,

(c) $f(x) = (x + x^2)^{-2}$,
$f(1.5) = \square$,

(d) $f(x) = (x^2 + x^3)^2(x^3 + x^2)^{-2}$,
$f(3.75) = \square$.

■ 1. (a) $3x^{-2} = \dfrac{3}{x^2}$.

(b) $\dfrac{2x^{-3}}{3y^{-2}} = \dfrac{2y^2}{3x^3}$.

(c) $(3ax^2)^0 = 1$.

(d) $\dfrac{a^2b^{-3}}{a^{-2}b^{-4}} = a^4b$.

(e) $6x^{-2}(2x^2)^{-2}$

$$= \frac{6}{x^2}\left(\frac{1}{4x^4}\right)$$

$$= \frac{6}{4x^6} = \frac{3}{2x^6}.$$

(f) $a^{-1} + b^{-1} = \dfrac{1}{a} + \dfrac{1}{b}$

$$= \frac{b + a}{ab}.$$

(g) $\left(\dfrac{2x^2}{3x^{-3}}\right)^{-1} = \dfrac{3x^{-3}}{2x^2} = \dfrac{3}{2x^5}$.

(h) $(6x^3y)^0 5^{-2} = \dfrac{1}{25}$.

---

3. Graph $y = x^2$.

■ 2. (a) $f(x) = \dfrac{3x^3}{2x^2} = \dfrac{3x}{2}$

$$f(3) = \frac{3(3)}{2} = \frac{9}{2}.$$

(b) $f(x) = (27x^{-3})^0 + (2x)^{-1}$

$$= 1 + \frac{1}{2x}$$

$$f(4) = 1 + \frac{1}{2(4)} = \frac{9}{8}.$$

(c) $f(x) = (x + x^2)^{-2}$

$$= \frac{1}{(x + x^2)^2}$$

$$f(1.5) = \frac{1}{(1.5 + (1.5)^2)^2}$$

$$= \frac{1}{(1.5 + 2.25)^2}$$

$$= \frac{1}{(3.75)^2} \approx .0711.$$

(d) $f(x) = (x^2 + x^3)^2$

$$\times (x^3 + x^2)^{-2}$$

$$= \left(\frac{x^3 + x^2}{x^3 + x^2}\right)^2 = 1$$

$f(3.75) = 1.$

---

4. Graph $y = x^3$.

■ 3.

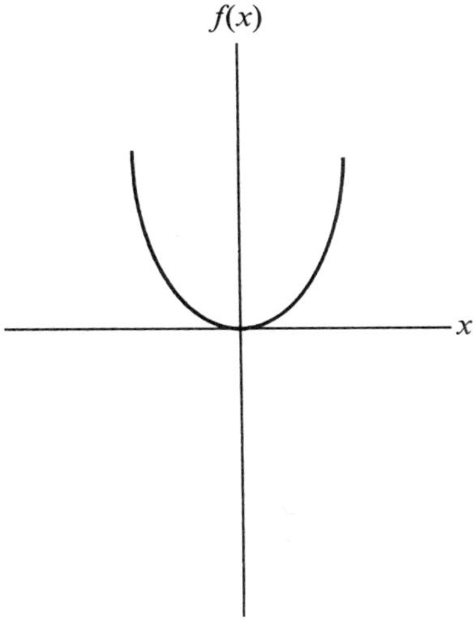

5. Graph $y = x^{-1}$.

■ 4.

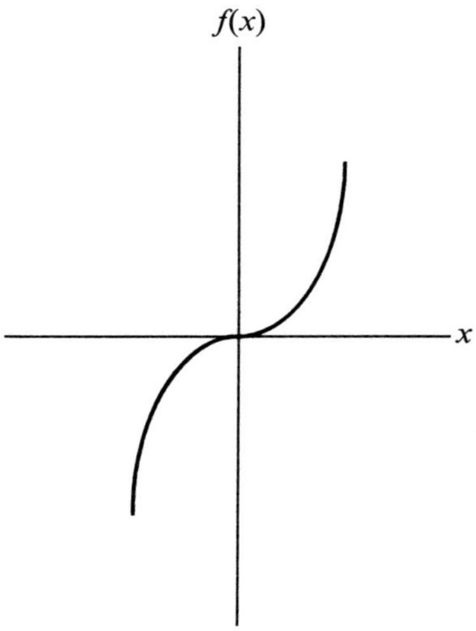

6. Graph $y = x^{-2}$.

■ 5.

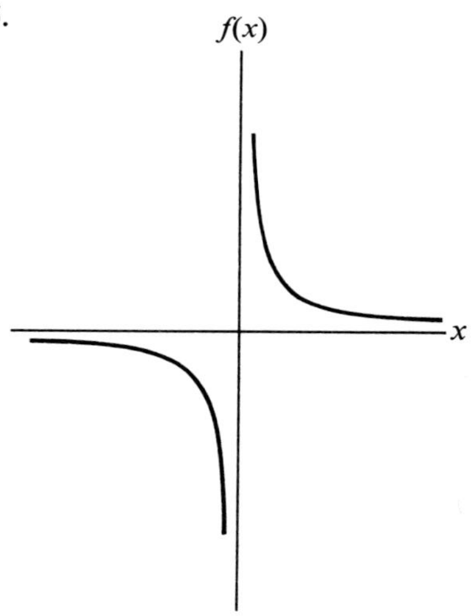

■ 6.

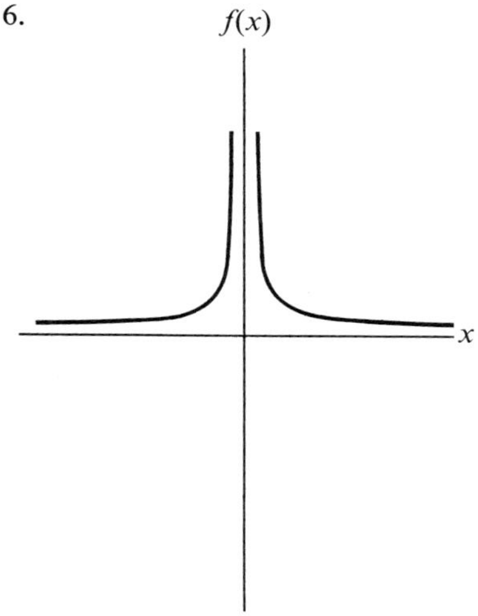

4-1 Problems

1. Calculate V (voltage) using the formula

$$eV = \frac{1}{2} mv^2,$$

if

$$e = 1.602 \times 10^{-19},$$
$$m = 9.108 \times 10^{-28},$$
$$v = 1.23 \times 10^5.$$

2. Calculate the radius in centimeters of the first Bohr orbit for hydrogen using the formula

$$r = \epsilon_0 \frac{n^2 h^2}{\pi m e^2},$$

if

$$\epsilon_0 = 8.85 \times 10^{-12},$$
$$h = 6.62 \times 10^{-34},$$
$$m = 9.11 \times 10^{-31},$$
$$e = 1.60 \times 10^{-19},$$
$$n = 1.$$

3. Calculate the capacitance of a two-plate parallel capacitor using the formula

$$C = \frac{2.3(10^{-13})KA}{s},$$

if

$$K = 78.5,$$
$$A = 4.00,$$
$$s = 2.50(10^{-3}).$$

Calculate the rate of flow (Q) of liquid through a capillary tube using the formula

$$Q = \frac{\pi R^4(\Delta P)}{8\mu L},$$

if

$$R = 2.0(10^{-2}),$$
$$\Delta P = 15.0,$$
$$\mu = 5.60(10^{-3}),$$
$$L = 12.0.$$

4-2 Rational Exponents and Radicals

The appearance of fractional exponents in technical literature is quite common. For example, the flow (Q) of water over a rectangular weir is described by the empirical formula

$$Q = 3.33bH^{3/2}.$$

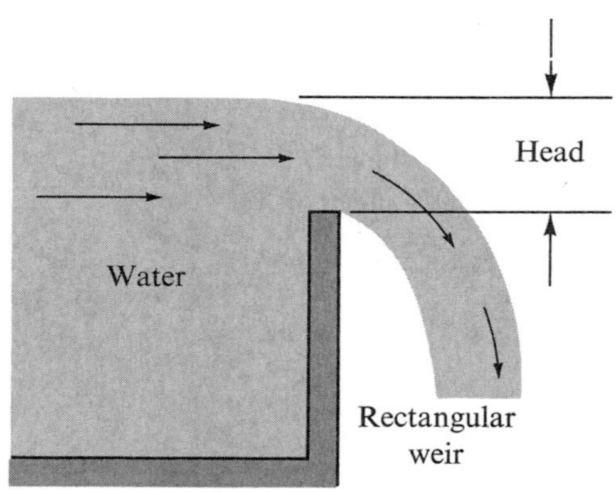

Fig. 4-2a

The factor $H^{3/2}$ is to be interpreted as the square root of H cubed, or as the cube of the square root of H, both of which are equivalent. Let us illustrate by letting $H = 4$.

$$4^{3/2} = (\sqrt[2]{4^3}) = \sqrt[2]{64} = 8,$$

or

$$4^{3/2} = (\sqrt[2]{4})^3 = 2^3 = 8.$$

As a matter of fact, if we agree that

$$a^{m/n} = \sqrt[n]{a^m}, \quad \text{or} \quad (\sqrt[n]{a})^m, \qquad \text{(Def 4-2a)}$$

then the six laws of exponents appearing in Section 4-1 hold for rational exponents. The equivalence between expressions in radical and exponential form may be more fully appreciated if we make a few fundamental comparisons. The meaning of radical notation enables us to write

$$\begin{aligned}
&\sqrt[2]{a} \cdot \sqrt[2]{a} = (\sqrt[2]{a})^2 = a, \\
&\sqrt[3]{a} \cdot \sqrt[3]{a} \cdot \sqrt[3]{a} = (\sqrt[3]{a})^3 = a, \\
&\vdots \\
&\underbrace{\sqrt[n]{a} \cdot \sqrt[n]{a} \cdots \sqrt[n]{a}}_{n \text{ factors}} = (\sqrt[n]{a})^n = a.
\end{aligned}$$

Using Law 4-1c, we can write

$$\begin{aligned}
&a^{1/2} \cdot a^{1/2} = (a^{1/2})^2 = a, \\
&a^{1/3} \cdot a^{1/3} \cdot a^{1/3} = (a^{1/3})^3 = a, \\
&\vdots \\
&\underbrace{a^{1/n} \cdot a^{1/n} \cdots a^{1/n}}_{n \text{ factors}} = (a^{1/n})^n = a.
\end{aligned}$$

Definition 4-2a permits us to interpret the meaning of any positive real number taken to any rational power.

Example 1

If $f(x) = x^{.75}$, find $f(16)$.

$$f(16) = 16^{.75} = 16^{75/100} = 16^{3/4} = (\sqrt[4]{16})^3 = 2^3 = 8.$$

Example 2

If $f(x) = x^{-2/3}$, find $f(0.125)$.

$$f(0.125) = (0.125)^{-2/3} = \frac{1}{(0.125)^{2/3}} = \frac{1}{(\sqrt[3]{0.125})^2} = \frac{1}{(.5)^2} = \frac{1}{.25} = 4.$$

Example 3

If $f(x) = x^{2.5}$, find $f(8)$.

$$f(8) = 8^{2.} = 8^2 \cdot 8^{1/2} = 64\sqrt{8} = 128\sqrt{2}.$$

The ability to interchange radical and exponential forms enables us to perform simplifications that cannot be accomplished in both forms with equal ease.

Example 4

If $f(x) = \sqrt[4]{(x-2)^2}$, find $f(27)$.

$$f(27) = \sqrt[4]{(27-2)^2} = 25^{2/4} = 25^{1/2} = \sqrt{25} = 5.$$

Example 5

If $f(x) = \sqrt[4]{x}$, find $f(12)$.

$$f(12) = \sqrt[4]{12} = (12)^{1/4} = ((12)^{1/2})^{1/2} = \sqrt{\sqrt{12}}$$
$$\approx \sqrt{3.46} \approx 1.86.$$

Under certain circumstances the base of an exponential expression may be negative. An *even* number of negative factors has a positive product. An *odd* number of negative factors has a negative product. For this reason, an even-numbered root (for example, a square root) of a negative number does not exist among the *real* numbers; but an odd-numbered root (for example, a cube root) of a negative number does exist among the *real* numbers.

Example 6

If $f(x) = x^{3/2}$, find $f(12)$ and $f(-16)$.

i. $f(12) = 12^{3/2} = (\sqrt[2]{12})^3 \approx (3.464)^3 \approx 41.57.$

ii. $f(-16)$ is not a real number. Numbers like $\sqrt{-16}$ are treated in Chapter 9, where we discuss the so-called imaginary numbers.

Example 7

If $h(x) = x^{4/3}$, find $h(-9.261)$.

$$h(-9.261) = (-9.261)^{4/3} = (\sqrt[3]{-9.261})^4 \approx (-2.1)^4 \approx 19.45.$$

Almost all technical problems require that numerical expressions with exponents or radicals be replaced with decimal approximations. The interpretation of rational exponents is one problem, and the numerical approximation of exponential expressions is another. Tables of roots, calculators with automatic square root, and the A and K scales of slide rules are usually the most convenient aids for calculating square and cube roots. Techniques for evaluating other rational powers are best understood in terms of the use of logarithms, the subject of Chapter 5.

Extending the range of a table is ordinarily done by using modified *scientific notation*, representing a number as a product of a number appearing in the table and a power of ten. The power of ten must be such that its root is known by inspection.

Example 8

Find the square root of 67,300.

Both 67.3 and 6.73 occur in the range of most square root tables. If we elect to write

$\sqrt{67300} = \sqrt{67.3 \times 10^3}$,

we discover that the square root of 10^3 is not known by inspection. If, instead, we write

$\sqrt{67300} = \sqrt{6.73 \times 10^4}$,

then the approximate root is 2.59×10^2, or 259.

Example 9

Find the cube root of .000482.

$\sqrt[3]{.000482} = \sqrt[3]{482 \times 10^{-6}} \approx 7.84 \times 10^{-2}$.

Notice that the exponent of ten is divisible by three.

4-2 Exercises

Try to acquire experience in the use of tables of roots, the slide rule, and calculators in determining the decimal approximation of each exercise.

1. $20.0^{1/2}$

---

2. $200^{0.5}$

■ 1. $20.0^{1/2} = \sqrt{20.0} \approx 4.47$.

---

3. $\sqrt{18.5}$

■ 2. $200^{0.5} = \sqrt{2 \times 10^2} \approx 14.1.$

---

4. $\sqrt[3]{21.0}$

■ 3. $\sqrt{18.5} \approx 4.30.$

---

5. $\sqrt[3]{21{,}000}$

■ 4. $\sqrt[3]{21.0} \approx 2.76.$

---

6. $\sqrt[3]{210{,}000}$

■ 5. $\sqrt[3]{21{,}000} = \sqrt[3]{21 \times 10^3} \approx 27.6.$

---

7. $192^{0.25}$

■ 6. $\sqrt[3]{210{,}000} = \sqrt[3]{210 \times 10^3}$
$\approx 59.4.$

---

8. $\sqrt[4]{(.0125)^2}$

■ 7. $192^{.25} = 192^{1/4} = (192^{1/2})^{1/2}$
$= \sqrt{\sqrt{192}} \approx \sqrt{13.8}$
$\approx 3.72.$

---

9. $\sqrt{4.25 \times 10^7}$

■ 8. $\sqrt[4]{(.0125)^2} = .0125^{2/4}$
$= \sqrt{.0125}$
$= \sqrt{1.25 \times 10^{-2}}$
$\approx .1118.$

---

10. $(\sqrt[3]{6.42})^6$

■ 9. $\sqrt{4.25 \times 10^7} = \sqrt{42.5 \times 10^6}$
$\approx 6{,}520.$

---

11. If $f(x) = x^{1/2}$, find $f(0.04)$, $f(0.40)$, and $f(4.00)$.

■ 10. $(\sqrt[3]{6.42})^6 = (6.42)^{6/3}$
$= (6.42)^2$
$\approx 41.22.$

12. If $g(x) = x^{1/3}$, find $g(-27)$, $g(-.008)$, and $g(6400)$.

■ 11.
$$\begin{aligned} f(0.04) &= (0.04)^{1/2} \\ &= \sqrt{4(10^{-2})} \\ &= 2(10^{-1}) = 0.20 \\ f(0.40) &= (0.40)^{1/2} \\ &= \sqrt{40(10^{-2})} \\ &= 6.32(10^{-1}) \\ &= 0.632 \\ f(4.00) &= (4.00)^{1/2} \\ &= \sqrt{4.00} = 2. \end{aligned}$$

13. Graph $f(x) = x^{1/2}$.

■ 12.
$$\begin{aligned} g(-27) &= (-27)^{1/3} \\ &= \sqrt[3]{-27} = -3 \\ g(-.008) &= (-.008)^{1/3} \\ &= \sqrt[3]{-8(10^{-3})} \\ &= -2(10^{-1}) \\ &= -.2 \\ g(6400) &= (6400)^{1/3} \\ &= \sqrt[3]{6.4(10^{3})} \\ &\approx 1.86(10^{1}) = 18.6. \end{aligned}$$

14. Graph $f(x) = x^{1/3}$.

■ 13.

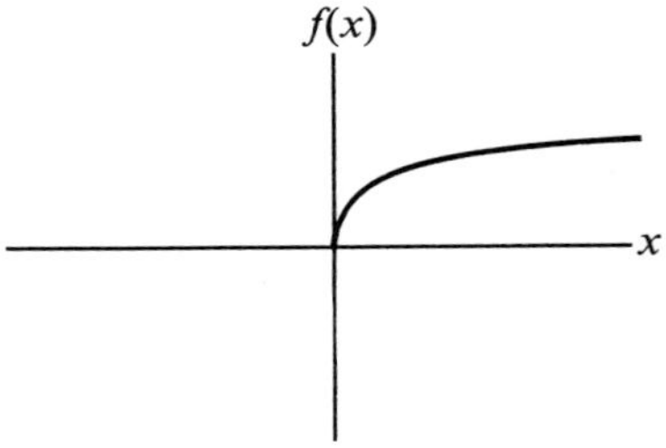

■ 14.

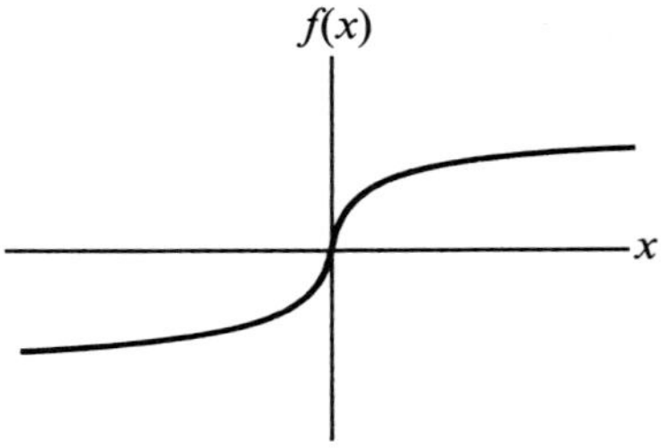

4-2 Problems

1. Current (i) in an RC circuit is found by using the formula

$$i = \frac{E}{R} e^{-t/RC}.$$

Calculate the value of i, when

$$\begin{aligned} E &= 10, \\ R &= 150, \\ C &= .01, \\ t &= 2, \\ e &= 2.72. \end{aligned}$$

2. The Manning formula for the velocity (V) of flowing liquid in pipes is

$$V = \frac{1.49}{n} R^{2/3} S^{1/2}.$$

Calculate V, when

$$\begin{aligned} n &= .01, \\ R &= 3.0, \\ S &= .75. \end{aligned}$$

3. Plot Q as a function of H from $H = 0$ feet to $H = 2$ feet, given that

$$Q = 3.33bH^{1.5},$$

where

Q represents cubic feet of water per second;
$b = 10$, width of weir in feet;
H represents head of water in feet.

4. Engine horsepower (hp) requirements for operating air compressors can be determined by using the formula

$$\text{hp} = \frac{.0643\,V\,n}{n-1}\left[\left(\frac{P_2}{P_1}\right)^{\frac{n-1}{n}} - 1\right].$$

Calculate the horsepower required when

$V = 8.5$ cubic feet per minute,
$n = 1.5$ (for nonisothermic adiabatic compression),
$P_2 = 65$ pounds per square inch,
$P_1 = 15$ pounds per square inch.

4-3 Operations with Radicals

By utilizing the laws of exponents, we can show that

$$\text{i.}\quad \sqrt[n]{a}\cdot\sqrt[n]{b}\equiv\sqrt[n]{ab},$$

$$\text{ii.}\quad \frac{\sqrt[n]{a}}{\sqrt[n]{b}}\equiv\sqrt[n]{\frac{a}{b}}.$$

Example 1

$$\sqrt[n]{a}\cdot\sqrt[n]{b}=a^{1/n}\cdot b^{1/n}=(ab)^{1/n}=\sqrt[n]{ab}.$$

Example 2

$$\frac{\sqrt[n]{a}}{\sqrt[n]{b}}=\frac{a^{1/n}}{b^{1/n}}=\left(\frac{a}{b}\right)^{1/n}=\boxed{}.$$

■ $\sqrt[n]{\frac{a}{b}}$

Occasionally it is necessary to multiply or divide polynomial expressions involving radicals. This can be easily done by applying the results just illustrated. You must be careful to observe that the indexes of the radicals are equal or are made equal before performing either the multiplication or division operation.

Example 3

$$\begin{aligned}(\sqrt{a}+\sqrt{b})^2&=(\sqrt{a})^2+2\sqrt{a}\sqrt{b}+(\sqrt{b})^2\\&=a+2\sqrt{ab}+b.\end{aligned}$$

Example 4

$$\begin{aligned}(\sqrt[3]{2x^2y})(\sqrt[3]{4xy^2})&=\sqrt[3]{(2x^2y)(4xy^2)}\\&=\sqrt[3]{8x^3y^3}=2xy.\end{aligned}$$

Example 5

$$\frac{\sqrt{x^2-1}}{\sqrt{x+1}} = \sqrt{\frac{x^2-1}{x+1}} = \sqrt{\frac{(x+1)(x-1)}{x+1}} = \boxed{}.$$

■ $\sqrt{x-1}$

Observe that in both the initial and final expressions of Example 5, x must be greater than or equal to *one* in order for the expressions to have meaning in the context of real numbers.

If a radical is the divisor in a division operation, then it is often easier and more accurate to *rationalize* the divisor before proceeding with the division operation. Examples 6 and 7 illustrate this process.

Example 6

Evaluate $\frac{5}{\sqrt{3}}$ to four significant figures.

The disadvantage of evaluating the square root of 3 first is that it must be an approximation of not less than five significant figures, which, of course, becomes the divisor in the division process. An alternative is to proceed in the following manner:

$$\frac{5}{\sqrt{3}} = \frac{5}{\sqrt{3}} \cdot \frac{\sqrt{3}}{\sqrt{3}} = \frac{5\sqrt{3}}{3} \approx \frac{5(1.732)}{3} \approx 2.887.$$

The denominator is rationalized by multiplying by an identical radical. Of course the numerator must also be multiplied by the same quantity. Because division is usually the most complicated arithmetic operation, dividing by three is much simpler than dividing by 1.7321.

Example 7

Evaluate $\frac{3}{2+\sqrt{5}}$ to four significant figures.

$$\frac{3}{2+\sqrt{5}} \cdot \frac{2-\sqrt{5}}{2-\sqrt{5}} = \frac{6-3\sqrt{5}}{4-5} \approx \frac{6-6.7083}{-1} = .7083.$$

The simplest way to rationalize binomials where one or both terms are radicals is to multiply by the **conjugate**. Two binomials are said to be conjugates if they have the property $(n + m)(n - m) = n^2 - m^2$; in other words, the binomials are the sum and difference of the same two terms. The conjugate of $2\sqrt{3} - \sqrt{7}$ is [   ], because their product is [   ].

■ $2\sqrt{3} + \sqrt{7}$; $12 - 7$ or 5.

4-3 Exercises

Retain three significant figures in all numerical solutions.

1. Evaluate $\sqrt{2} \cdot \sqrt{10}$.

2. Evaluate $\sqrt[3]{4} \cdot \sqrt[3]{10}$.

■ 1. $\sqrt{2} \cdot \sqrt{10} = \sqrt{20} = 2\sqrt{5} \approx 4.47$.

3. Simplify $\sqrt{ab} \cdot \sqrt{bc}$.

■ 2. $\sqrt[3]{4} . \sqrt[3]{10} = \sqrt[3]{8(5)} = 2\sqrt[3]{5} \approx 3.42$.

4. Simplify $\sqrt{8ab^3} \cdot \sqrt{2a^3b}$.

■ 3. $\sqrt{ab} \cdot \sqrt{bc} = \sqrt{ab^2c} = b\sqrt{ac}$.

5. $(\sqrt{a} - \sqrt{2b})^2 = ?$

■ 4. $\sqrt{8ab^3} \cdot \sqrt{2a^3b} = \sqrt{16a^4b^4} = 4a^2b^2$.

6. $(2\sqrt{3} + \sqrt{6})(2\sqrt{3} - \sqrt{6}) = ?$

■ 5. $(\sqrt{a} - \sqrt{2b})^2$
$= (\sqrt{a})^2 - 2\sqrt{a}\sqrt{2b} + (\sqrt{2b})^2$
$= a - 2\sqrt{2ab} + 2b$.

7. Use the quadratic formula to find the radical roots of

$$x^2 + 5x + 3 = 0$$

and check the solution by substituting into the original equation.

■ 6. $(2\sqrt{3}+\sqrt{6})(2\sqrt{3}-\sqrt{6})$
$= 4(3) - 6$
$= 6.$

8. Find the average of the roots of

$$x^2 - 4x - 2 = 0.$$

■ 7. $x^2 + 5x + 3 = 0$

$$x = \frac{-5 \pm \sqrt{25 - 12}}{2}$$

$$x = \frac{-5 \pm \sqrt{13}}{2}.$$

Check (only one check is shown):

$$\left(\frac{-5+\sqrt{13}}{2}\right)^2 + 5\left(\frac{-5+\sqrt{13}}{2}\right) + 3 \stackrel{?}{=} 0$$

$$\frac{25 - 10\sqrt{13} + 13}{4} + \frac{-25 + 5\sqrt{13}}{2} + 3 \stackrel{?}{=}$$

$$\frac{25}{4} - \frac{5\sqrt{13}}{2} + \frac{13}{4} - \frac{50}{4} + \frac{5\sqrt{13}}{2} + \frac{12}{4} \stackrel{?}{=}$$

$$\frac{-25}{4} + \frac{25}{4} \stackrel{?}{=}$$

$$0 = 0.$$

9. Evaluate $\dfrac{\sqrt{6}}{\sqrt{3}}$.

■ 8. $x_1 = \dfrac{4+\sqrt{24}}{2} = 2+\sqrt{6}$

$$x_2 = \frac{4-\sqrt{24}}{2} = 2-\sqrt{6}$$

$$\frac{x_1 + x_2}{2} = \frac{2+\sqrt{6}+2-\sqrt{6}}{2}$$

$$= 2.$$

---

10. Evaluate $\dfrac{\sqrt{45}}{\sqrt{15}}$.

■ 9. $\dfrac{\sqrt{6}}{\sqrt{3}}=\sqrt{\dfrac{6}{3}}=\sqrt{2}\approx 1.41.$

---

11. Simplify $\dfrac{\sqrt{ax}}{\sqrt{bx}}$.

■ 10. $\dfrac{\sqrt{45}}{\sqrt{15}}=\sqrt{\dfrac{45}{15}}=\sqrt{3}\approx 1.73.$

---

12. Evaluate $\dfrac{5}{\sqrt{6}-4}$.

■ 11. $\dfrac{\sqrt{ax}}{\sqrt{bx}}=\sqrt{\dfrac{ax}{bx}\cdot\dfrac{b}{b}}=\sqrt{\dfrac{ab}{b^2}}$

$$=\frac{\sqrt{ab}}{b}.$$

---

13. Evaluate $\dfrac{\sqrt{2}-\sqrt{3}}{\sqrt{2}+\sqrt{3}}$.

■ 12. $\left(\dfrac{5}{\sqrt{6}-4}\right)\cdot\left(\dfrac{\sqrt{6}+4}{\sqrt{6}+4}\right)$

$$=\frac{5\sqrt{6}+20}{6-16}$$

$$=\frac{-20-5\sqrt{6}}{10}$$

$$=-2-\frac{\sqrt{6}}{2}\approx -3.22.$$

---

14. Simplify $\dfrac{\sqrt{a}+\sqrt{6}}{\sqrt{a}-\sqrt{6}}$.

(Rationalize the denominator.)

■ 13. $\left(\dfrac{\sqrt{2}-\sqrt{3}}{\sqrt{2}+\sqrt{3}}\right)\cdot\left(\dfrac{\sqrt{2}-\sqrt{3}}{\sqrt{2}-\sqrt{3}}\right)$

$$=\frac{2-2\sqrt{6}+3}{2-3}$$

$$=\frac{5-2\sqrt{6}}{-1}$$

$$=2\sqrt{6}-5\approx -.101.$$

■ 14. $$\left(\frac{\sqrt{a}+\sqrt{6}}{\sqrt{a}-\sqrt{6}}\right)\cdot\left(\frac{\sqrt{a}+\sqrt{6}}{\sqrt{a}+\sqrt{6}}\right)$$
$$=\frac{a+2\sqrt{6a}+6}{a-6}.$$

4-3 Problems

1. Calculate F_0 when $L = .02$ henrys and $C = .006$ farads using the formula

$$F_0=\frac{1}{2\pi\sqrt{LC}}.$$

2. Determine the velocity (V) of a deep-water wave using the formula

$$V=\frac{\sqrt{g\lambda}}{\sqrt{2\pi}},$$

when $g = 32$ ft/sec^2 and $\lambda = 56$ feet.

3. Calculate the current (I) using the formula

$$I=\frac{E}{\sqrt{R^2+X^2}},$$

when $E = 6$ volts, $R = 100$ ohms, and $X = 60$ ohms.

4. Calculate the corrected horsepower (H_c) of an internal combustion engine when $H_0 = 86.4$, $P_0 = 30.40$ and $T_0 = 32.0$ using the formula

$$H_c=\frac{29.92\ H_0}{P_0}\sqrt{\frac{T_0+460}{520}}.$$

5. Calculate Z when

$$Z=\sqrt{R^2+\left(2\pi fL-\frac{1}{2\pi fC}\right)^2},$$

where

$$R=40{,}000,\ f=455(10^3),\ L=.006,\ C=360\ (10^{-12}).$$

4-4 Equations with Radicals

Solving equations with radicals requires no special technique unless the unknown is a factor or term of the radical. Under these circumstances the radicals can be removed by squaring both sides of the equation. This process may introduce roots that do not satisfy the original equation. For this reason it is necessary to test each result.

Example 1

Find the root(s) of $1 - x = \sqrt{x + 5}$.

i. $1 - x = \sqrt{x+5}$ (square both sides).
ii. $1 - 2x + x^2 = x + 5$.
iii. $x^2 - 3x - 4 = 0$.
iv. $x = \dfrac{3 \pm \sqrt{9+16}}{2} = \dfrac{3 \pm 5}{2} = \begin{cases} -1 \\ 4 \end{cases}$.
v. Check of $x = -1$:

$$1 - (-1) \stackrel{?}{=} \sqrt{-1+5}$$
$$2 = 2.$$

Check of $x = 4$:

$$1 - 4 \stackrel{?}{=} \sqrt{4+5}$$
$$-3 \neq 3, \qquad \{-1\}.$$

Therefore only -1 is a root.

If there is more than one radical in the equation, the technique of squaring may need to be repeated.

Example 2

Find the root(s) of the equation $\sqrt{x+4} + \sqrt{2x-1} = 6$.

i. $\sqrt{x+4} = 6 - \sqrt{2x-1}$; (square both sides)
ii. $x + 4 = 36 - 12\sqrt{2x-1} + 2x - 1$

$x + 31 = 12\sqrt{2x-1}$; (square both sides)

iii. $x^2 + 62x + 961 = 144(2x - 1)$

$x^2 - 226x + 1105 = 0$;

iv. $x = \dfrac{226 \pm \sqrt{(226)^2 - 4(1105)}}{2}$

$= \dfrac{226 \pm 216}{2} = \begin{cases} 5 \\ 221 \end{cases}.$

v. Check of $x = 5$:

$\sqrt{5+4} + \sqrt{2(5)-1} \stackrel{?}{=} 6$

$3 + 3 = 6.$

Check of $x = 221$:

$\sqrt{221+4} + \sqrt{2(221)-1} \stackrel{?}{=} 6$

$15 + 21 \neq 6.$

Therefore only 5 is a root.

4-4 Exercises

1. Find the root(s) of

$$\sqrt{x-2} = 5.$$

Check your results.

---

2. Find the root(s) of

$$3 - \sqrt{2x-3} = 0$$

and check your results.

■ 1. $\sqrt{x-2} = 5$

$x - 2 = 25$

$x = 27 \qquad \{27\}.$

Check:

$\sqrt{27-2} \stackrel{?}{=} 5$

$5 = 5.$

---

3. Solve $\sqrt{x} + \sqrt{x-16} = 8$.
Check your results.

■ 2. $3 - \sqrt{2x-3} = 0$

$\sqrt{2x-3} = 3$

$2x - 3 = 9$

$x = 6 \qquad \{6\}.$

Check:

$3 - \sqrt{2(6)-3} \stackrel{?}{=} 0$

$0 = 0.$

4. Solve $\sqrt{5x+1} = 3\sqrt{x\ 3}.$
Check your results.

■ 3. $\sqrt{x} + \sqrt{x-16} = 8$
$\sqrt{x-16} = 8 - \sqrt{x}$
$x - 16 = 64 - 16\sqrt{x} + x$
$-80 = -16\sqrt{x}$
$5 = \sqrt{x}$
$25 = x, \qquad \{25\}.$
Check:
$\sqrt{25} + \sqrt{25-16} \stackrel{?}{=} 8$
$5 + 3 = 8.$

---

5. Solve $\sqrt{x} + 1 = \sqrt{x-9}.$
Check your results.

■ 4. $\sqrt{5x+1} = 3\sqrt{x-3}$
$5x + 1 = 9(x-3)$
$5x + 1 = 9x - 27$
$28 = 4x$
$7 = x, \qquad \{7\}.$
Check:
$\sqrt{5(7)+1} \stackrel{?}{=} 3\sqrt{7-3}$
$\sqrt{36} = 3\sqrt{4}$
$6 = 6.$

---

6. Solve $\sqrt{x+11} - 3 = \sqrt{x-4}.$
Check your results.

■ 5. $\sqrt{x} + 1 = \sqrt{x-9}$
$x + 2\sqrt{x} + 1 = x - 9$
$2\sqrt{x} = -10$
$\sqrt{x} = -5$
$x = 25.$
Check:
$\sqrt{25} + 1 \stackrel{?}{=} \sqrt{25-9}$
$5 + 1 \neq 4.$
Therefore there are no roots.

---

7. Solve $\sqrt{x+2} + \sqrt{x-2}$
$= \sqrt{5x-6}.$
Check your results.

■ 6. $\sqrt{x+11} - 3 = \sqrt{x-4}$
$x + 11 - 6\sqrt{x+11} + 9$
$= x - 4$
$6\sqrt{x+11} = 24$
$\sqrt{x+11} = 4$
$x + 11 = 16$
$x = 5, \ \{5\}.$

Check:

$$\sqrt{5+11}-3 \stackrel{?}{=} \sqrt{5-4}$$
$$4-3=1.$$

---

8. Given that

$$T=2\pi\sqrt{\frac{L}{g}},$$

solve for L.

■ 7. $\sqrt{x+2}+\sqrt{x-2}$
$$=\sqrt{5x-6}$$
$$x+2+2\sqrt{(x+2)(x-2)}$$
$$+x-2=5x-6$$
$$2\sqrt{x^2-4}=3x-6$$
$$4x^2-16=9x^2-36x+36$$
$$0=5x^2-36x+52$$
$$x=\frac{36\pm\sqrt{1296-1040}}{10}$$
$$=\frac{36\pm 16}{10}, \qquad \left\{\frac{26}{5}, 2\right\}.$$

Check $x=2$:

$$\sqrt{2+2}+\sqrt{2-2}$$
$$\stackrel{?}{=}\sqrt{5(2)-6}$$
$$2+0=2.$$

Check $x=\frac{26}{5}$:

$$\sqrt{\frac{26}{5}+\frac{10}{5}}+\sqrt{\frac{26}{5}-\frac{10}{5}}$$
$$\stackrel{?}{=}\sqrt{5\left(\frac{26}{5}\right)-6}$$
$$\sqrt{\frac{36}{5}}+\sqrt{\frac{16}{5}}\stackrel{?}{=}20$$
$$\frac{6}{5}\sqrt{5}+\frac{4}{5}\sqrt{5}\stackrel{?}{=}2\sqrt{5}$$
$$2\sqrt{5}=2\sqrt{5}.$$

■ 8. $T=2\pi\sqrt{\frac{L}{g}}$
$$T^2=4\pi^2\left(\frac{L}{g}\right)$$
$$L=\frac{T^2g}{4\pi^2}.$$

4-4 Problems

In Problems 1–6, change the formulas so that the designated quantity is given as a function of the other variables and constants.

1. $V = \pi r^2 h$; r (radius of a circular cylinder).

2. $Z_0 = \sqrt{\frac{L}{C}}$; C (capacitance of transmission lines).

3. $F_0 = \frac{1}{2\pi\sqrt{LC}}$; L (inductance of an LC oscillator).

4. $V = \sqrt{\frac{g\lambda}{2\pi}}$; λ (length of a deep water wave).

5. $k = \sqrt{\frac{6R^2 - a^2}{24}}$; R (radius of gyration).

6. $I = \frac{E}{\sqrt{R^2 + X^2}}$; R (resistance).

7. The discharge over a weir plate may be calculated using the formula

$$Q = CL(H + h_v)^{3/2}.$$

Calculate C, the constant of proportionality, when

$$L = 4.5, \quad H = .30, \quad h_v = 12, \quad Q = 2.5.$$

8. Calculate the length of d.

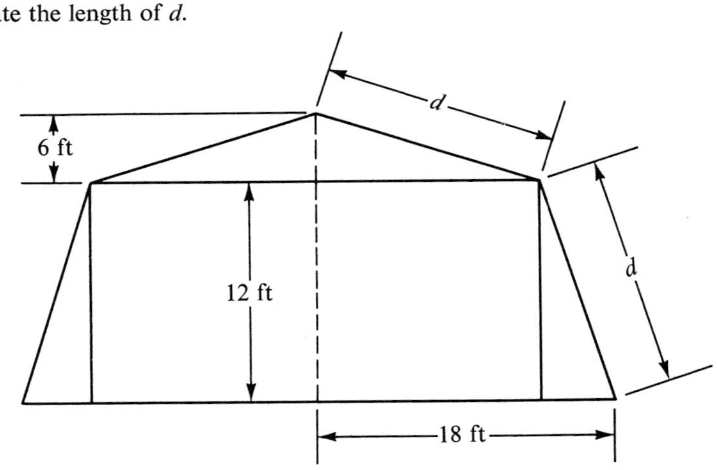

4 Supplementary Exercises and Problems

Write the following expressions in other numerical forms. (4–1)

1. 3^0
2. -5^0
3. $(-16)^0$
4. $\left(\frac{1}{2}\right)^{-1}$
5. -4^{-2}
6. $\frac{-5}{6^{-1}}$
7. $\left(-\frac{2}{3}\right)^{-2}$
8. $\left(\frac{2^{-1}}{5}\right)^{-2}$

Write the following expressions in other literal forms. (4–1)

9. $\left(\frac{x}{y^2}\right)^0$
10. $(w + w^{-1})^2$
11. $\frac{uv^{-3}}{u^{-1}v}$
12. $p^2q^3(p^{-1} + q^{-2})$
13. $(w^0 + w)^{-1}$
14. $\left[\frac{12.3(10^{-6}) \times 4.58(10^3)}{2.14(10^5)}\right]$ (4–1)
15. $\left[\frac{1.416(10^{-3})}{1.68(10^{-2})}\right]^4$ (4–1)
16. $[3.7(10^{-2}) \times 8.1(10^3)]^{-2}$ (4–1)

Write the following expressions in other numerical forms. (4–2)

17. $125^{1/3}$
18. $8^{4/3}$
19. $(-8)^{2/3}$
20. $\left(\frac{1}{49}\right)^{-1/2}$
21. $(.0756)^{3/2}$

22. $(1460)^{2/3}$

Write the following in other literal forms. (4-2)

23. $\dfrac{w^{1/6}}{w^{-2/3}}$

24. $(x^{1/2} + y^{1/2})^2$

25. $\left(\dfrac{1}{t^{1/3}}\right)^{-3/4}$

26. $\sqrt{z^{-1}}\,\sqrt[4]{z^3}$

27. $\dfrac{x^{1/2}y^2}{x^{-1}y^{1/4}}$

28. $\sqrt{\sqrt[3]{x^4}}$

Rewrite the following expressions to eliminate radicals in the denominator. (4-3)

29. $\dfrac{1}{\sqrt{u} - \sqrt{v}}$

30. $\dfrac{\sqrt{8} + 1}{\sqrt{8} - 1}$

31. $\dfrac{\sqrt{3}}{1 + \sqrt{3}}$

32. Solve for T (temperature) in terms of the other variables. (4-4)

$$H_c = \frac{H_0 P_s}{P_0} \sqrt{\frac{T_0}{T_s}}.$$

33. Solve the formula $Q = CL(H + h_v)^{3/2}$ for H in terms of the other variables. (4-4)

34. Solve for R in the formula (4-4)

$$z = \sqrt{R^2 + (x_L - x_C)^2}.$$

35. Solve for v_r in the formula (4-4)

$$f_0 = f_s \sqrt{\frac{c + v_r}{c - v_r}}.$$

36. Solve for x and check the solution(s). (4-4)

$$\sqrt{x - 3} - 6 = 0.$$

37. Solve for w and check the solution(s). (4-4)

$$2\sqrt{w+3} = \sqrt{w+18}.$$

38. Solve for t and check the solution(s). (4-4)

$$7 - \sqrt{3t+7} - \sqrt{3t} = 0.$$

Chapter Five Logarithmic and Exponential Functions

5-1 Exponential and Logarithmic Expressions

In Chapter 4 we gave attention to expressions such as $n^{a/b}$, where a and b are integers. If n is any nonnegative real number, then $n^{a/b}$ can, in theory, be evaluated by using the relationship $n^{a/b} = \sqrt[b]{n^a}$. Practically, a/b has to be a ratio of only certain small integers if the expression is to be evaluated without other techniques. For example, evaluating the right-hand expression of the equation

$$A = 100\left(1 + \frac{.055}{12}\right)^{12(20)}$$

gives the accumulated interest and principal of \$100 invested at an annual rate of 5.5 percent compounded monthly for 20 years. Taking the 240th power of a number by repeated multiplication is not practical. Yet the expression can be evaluated with very little effort by using *logarithms*.

We first turn our attention to discovering what logarithms are. Do not expect something entirely new, because logarithms and exponents are closely related.

If we consider the equation

$$y = b^x, \quad b > 0, \quad b \neq 1, \qquad \text{(Eq. 5-1a)}$$

where x and y are variables and b is a constant, we should notice that for each value of x there will be a corresponding value of y and that y may be considered a function of x. This equation is known as an **exponential equation**, because the independent variable x is an exponent, rather than an algebraic factor or term.

The set of ordered pairs (the function) defined by Eq. 5-1a with a specific *base b greater than zero but unequal to one*, is called a system of logarithms. The elements of the set of x's which constitute the *domain* of the function are called logarithms and are the same as rational exponents. The set of y's of the function, called the *range*, is the set of positive real numbers.

For example, the equation $y = f(x) = 3^x$ defines a system of logarithms. For every rational number x there is only one corresponding value for y. Some pairs of the system are $(-3, 1/27)$, $(-2, \boxed{})$, $(-1/2, \sqrt{3}/3)$, $(0, 1)$, $(1, 3)$, and $(3/2, \boxed{})$.

■ $\frac{1}{9}$; $3\sqrt{3}$.

The applications of systems of logarithms require that we introduce another type of equation that appears to interchange the independent and dependent roles of the variables x and y in the equation $y = b^x$. It permits us to write an expression for x in terms of y and the base b. In a sentence, the new equation says that x *is equal to the exponent to which a fixed number, the base b, must be raised to produce the number y.* When the word exponent is replaced by the word *log*, an abbreviation of logarithm, the sentence is condensed into the mathematical equation

$$x = \log_b y. \qquad \text{(Eq. 5-1b)}$$

Equation 5-1b reads, "x equals log y, base b."

Except for the question of which variable is dependent and which is independent, the relationship between x and y is the same for both Eq. 5-1a and Eq. 5-1b.

Returning to our example, if $y = 3^x$, then $x = \log_3 y$. The first of these two equivalent equations we refer to as the *exponential form*, and the second as the *logarithmic form*.

Exponential form	*Logarithmic form*
$y = f(x) = 3^x$	$\leftrightarrow x = \log_3 y$
$f(0) = 3^0 = 1$	$\leftrightarrow 0 = \log_3 1$
$f(1) = 3^1 = 3$	$\leftrightarrow 1 = \log_3 3$
$f(2) = 3^2 = 9$	$\leftrightarrow 2 = \log_3 9$

$$f(3) = 3^3 = 27 \leftrightarrow \boxed{} \quad \text{(a)}$$

$$\boxed{} \leftrightarrow 4 = \log_3 81 \quad \text{(b)}$$

$$f(-1) = 3^{-1} = \frac{1}{3} \leftrightarrow -1 = \log_3 \boxed{} \quad \text{(c)}$$

$$f(-2) = 3^{-2} = \frac{1}{9} \leftrightarrow \boxed{} \quad \text{(d)}$$

■ a. $3 = \log_3 27$, b. $f(4) = 3^4 = 81$, c. $\frac{1}{3}$, d. $-2 = \log_3 \frac{1}{9}$.

It is essential to be able to change from one form to the other and to understand the relationship that exists in each form.

Example 1

Determine the value of y in the equation $3 = \log_2 y$.

This is equivalent to saying "2 to the third power is equal to y," or $y = 2^3$, so $y = 8$.

Example 2

Determine the value of x in the equation $x = \log_5 125$.

Five must be raised to the power x to equal 125, or $5^x = 125$; but $125 = 5^3$, so $x = 3$.

Example 3

Determine the value of b in the equation $\log_b 1/16 = 4$.

Since $b^4 = 1/16$ is the equivalent exponential equation, and since we know that $(1/2)^4 = 1/16$, we conclude that $b^4 = (1/2)^4$, or $b = 1/2$.

The restrictions that the base b must be greater than zero, but unequal to one, are necessary. If, for example, $b = 1$, then $y = 1^x = 1$ for all values of x. If $b \leq 0$, then the range would have both real and imaginary values. Actually, the most useful systems of logarithms have bases greater than one, as we shall see in subsequent sections.

5-1 Exercises

Convenient numbers are used in the exercises so that the relationships can be easily verified. In Exercises 1–20, write the logarithmic form for the given exponential form or vice versa.

1. $2^3 = 8$

---

2. $4^2 = 16$

■ 1. $3 = \log_2 8$

---

3. $\log_4 16 = 2$

■ 2. $2 = \log_4 16$

---

4. $\log_8 16 = \frac{4}{3}$

■ 3. $4^2 = 16$

---

5. $3^3 = 27$

■ 4. $8^{4/3} = 16$

---

6. $3^{-2} = \frac{1}{9}$

■ 5. $3 = \log_3 27$

---

7. $\log_9 1 = 0$

■ 6. $-2 = \log_3\left(\frac{1}{9}\right)$

---

8. $\log_4\left(\frac{1}{16}\right) = -2$

■ 7. $1 = 9^0$

9. $\left(\frac{1}{2}\right)^3 = \frac{1}{8}$	■ 8. $\frac{1}{16} = 4^{-2}$
10. $8^{2/3} = 4$	■ 9. $3 = \log_{1/2}\left(\frac{1}{8}\right)$
11. $\log_6 216 = 3$	■ 10. $\frac{2}{3} = \log_8 4$
12. $\log_{16} 64 = \frac{3}{2}$	■ 11. $216 = 6^3$
13. $2^{-5} = \frac{1}{32}$	■ 12. $64 = 16^{3/2}$
14. $8^0 = 1$	■ 13. $-5 = \log_2\left(\frac{1}{32}\right)$
15. $\log_{27} 9 = \frac{2}{3}$	■ 14. $0 = \log_8 1$
16. $\log_{10} 100 = 2$	■ 15. $9 = 27^{2/3}$
17. $4^{3/2} = 8$	■ 16. $100 = 10^2$
18. $4^{5/2} = 32$	■ 17. $\frac{3}{2} = \log_4 8$
19. $\log_{10} 1{,}000{,}000 = 6$	■ 18. $\frac{5}{2} = \log_4 32$

---

20. $\log_{10}(.01) = -2$

■ 19. $1{,}000{,}000 = 10^6$

■ 20. $.01 = 10^{-2}$

Determine the numerical value of the literal quantity in Exercises 21–28.

21. $\log_2 8 = x$

---

22. $\log_3 81 = x$

■ 21. $2^x = 8 = 2^3;\ x = 3$

---

23. $\log_2 y = 4$

■ 22. $3^x = 81 = 3^4;\ x = 4$

---

24. $3 = \log_{10} y$

■ 23. $y = 2^4 = 16$

---

25. $\log_b 4 = 2$

■ 24. $y = 10^3 = 1000$

---

26. $\log_b .001 = -3$

■ 25. $4 = b^2;\ b = 2$

---

27. $\log_{10} y = .5$

■ 26.
$$\begin{aligned} b^{-3} &= .001 \\ b^3 &= 1000 \\ b &= 10 \end{aligned}$$

---

28. $\log_{10}\sqrt[3]{100} = x$

■ 27.
$$\begin{aligned} y &= 10^{.5} \\ &= 10^{1/2} \\ &= \sqrt{10} \approx 3.16. \end{aligned}$$

---

29. Make a table of values for the equation $f(x) = 4^x$, when $x = -2, -1, 0, 1, \frac{3}{2}, 2, \frac{5}{2}, 3$.

■ 28.
$$\begin{aligned} 10^x &= 100^{1/3} \\ &= (10^2)^{1/3} = 10^{2/3} \\ x &= \frac{2}{3}. \end{aligned}$$

30. For what values of x will $f(x) = 10^x$ be negative?

■ 29.

x	-2	-1	0	1	$\frac{3}{2}$	2	$\frac{5}{2}$	3
$f(x)$	$\frac{1}{16}$	$\frac{1}{4}$	1	4	8	16	32	64

■ 30. There is no x that makes 10^x less than zero.

5-1 Problems

1. What are the logarithms of 2, 4, 8, 16, 32, 64, 128, and 256 to the base 2? To the base 8? To the base 16?
2. Solve for the numerical value of n when

$$\log_8(n+2) = \frac{5}{3}.$$

3. Write $P = e^{kt}$ in logarithmic form when the base is e.
4. Write

$$\frac{N}{10} = \log_{10}\left(\frac{P_0}{P_i}\right)$$

in exponential form.

5. Write

$$\frac{P_2}{P_1} = 10^{w/4883v}$$

in logarithmic form when the base is 10.

6. Write $-(pH) = \log_{10}(H^+)$ in exponential form.
7. Write

$$\frac{R}{R_0} = e^{b/\tau}$$

in logarithmic form when the base is e.

5-2 The Properties of Logarithms

The importance of logarithms becomes apparent as we investigate their properties. In the following discussion of properties, the base is considered to be greater than one.

Consider the sequence of related exponential and logarithmic statements:

$$2^3 = 8 \quad \leftrightarrow \log_2 8 = 3$$
$$2^2 = 4 \quad \leftrightarrow \log_2 4 = 2$$
$$2^{3+2} = 8 \cdot 4 \leftrightarrow \log_2 (8 \cdot 4) = \boxed{}.$$

■ 3 + 2

This illustration suggests that the logarithm of the *product* of two (or more) factors is equal to the *sum* of the logarithms of the separate factors. The truth of this statement can be demonstrated in a very simple manner.

$$M = b^m \quad \leftrightarrow \log_b M = m$$
$$N = b^n \quad \leftrightarrow \log_b N = n$$
$$M \cdot N = b^{m+n} \leftrightarrow \log_b (M \cdot N) = m + n.$$

The final statement, after substituting equivalent values for m and n, becomes

$$\log_b M \cdot N = \log_b M + \log_b N. \qquad \text{(Prop. 5-2a)}$$

This property is the key to using tables of logarithms discussed in Section 5-3.

The second property we discuss is so nearly like Property 5-2a that a simple example should be sufficiently convincing.

$$3^4 = 81 \leftrightarrow \log_3 81 = 4,$$
$$3^3 = 27 \leftrightarrow \log_3 27 = 3,$$
$$3^{4-3} = \frac{81}{27} \leftrightarrow \log_3 \left(\frac{81}{27}\right) = 4 - 3.$$

The logarithm of a *quotient* equals the *difference* of the logarithm of the dividend minus the logarithm of the divisor; in other words,

$$\log_b\left(\frac{M}{N}\right) = \log_b M - \log_b N. \qquad \text{(Prop. 5-2b)}$$

The final property deals with powers and roots. Examine the details of the next demonstration.

$$2^3 = 8 \leftrightarrow \log_2 8 = 3$$
$$(2^3)^2 = 2^{3 \cdot 2} = 8^2 \leftrightarrow \log_2 8^2 = 3 \cdot 2.$$

The logarithm of a *power* of a number equals the *product* of the power times the logarithm of the number. Examine carefully the following demonstration.

$$N = b^n \quad \leftrightarrow \log_b N = n,$$
$$N^a = b^{an} \leftrightarrow \log_b N^a = an.$$

Substituting $\log_b N$ for n in the final equation gives

$$\log_b N^a = a \log_b N. \qquad \text{(Prop. 5-2c)}$$

This property applies to radicals as well.

Example 1

i. $64 = 2^6 \leftrightarrow \log_2 64 = 6.$

ii. $\sqrt[3]{64} = 64^{1/3}$

$= (2^6)^{1/3}$

$= 2^{6(1/3)} \leftrightarrow \log_2 \sqrt[3]{64} = \left(\frac{1}{3}\right)(6)$

$= \frac{1}{3} \log_2 64.$

5-2 Exercises

1. Write the sequence of statements in equivalent logarithmic form.
   i. $3^2 = 9$,
   ii. $3^3 = 27$,
   iii. $3^{3+2} = 27 \cdot 9$.

2. Write the statements in equivalent logarithmic form.
   i. $5^3 = 125$,
   ii. $5^2 = 25$,
   iii. $\frac{5^3}{5^2} = 5^{3-2} = \frac{125}{25}$.

■ 1. i. $\log_3 9 = 2$,
   ii. $\log_3 27 = 3$,
   iii. $\log_3 (27 \cdot 9) = 3 + 2$.

3. Write in logarithmic form.
   i. $4^{3/2} = 8$,
   ii. $4^{1/2} = 2$,
   iii. $4^{3/2+1/2} = 8 \cdot 2$.

■ 2. i. $\log_5 125 = 3$,
   ii. $\log_5 25 = 2$,
   iii. $\log_5 \frac{125}{25} = 3 - 2$.

4. Write in logarithmic form.
   i. $9^{1/2} = 3$,
   ii. $9^{3/2} = 27$,
   iii. $9^{1/2-3/2} = \frac{3}{27}$.

■ 3. i. $\log_4 8 = \frac{3}{2}$,
   ii. $\log_4 2 = \frac{1}{2}$,
   iii. $\log_4 (8 \cdot 2) = \frac{3}{2} + \frac{1}{2}$.

5. Write in logarithmic form.
   i. $3^2 = 9$,
   ii. $(3^2)^{3/2} = 3^{2(3/2)} = 9^{3/2}$.

■ 4. i. $\log_9 3 = \frac{1}{2}$,
   ii. $\log_9 27 = \frac{3}{2}$,
   iii. $\log_9 \left(\frac{3}{27}\right) = \frac{1}{2} - \frac{3}{2}$.

---

6. Write in logarithmic form.
   i. $2^4 = 16$,
   ii. $(2^4)^{1/2} = 2^{4(1/2)} = 16^{1/2}$.

■ 5. i. $\log_3 9 = 2$,
ii. $\log_3 9^{3/2} = 2 \cdot \frac{3}{2}$.

---

7. $\log_{10} A + \log_{10} B = \log_{10} \boxed{}$.

■ 6. i. $\log_2 16 = 4$,
ii. $\log_2 16^{1/2} = 4 \cdot \frac{1}{2}$.

---

8. $\log_{10} A - \log_{10} B = \log_{10} \boxed{}$.

■ 7. $\log_{10} A + \log_{10} B$
$= \log_{10} (AB)$
(Prop. 5-2a).

---

9. $\log_{10} A + \log_{10} B - \log_{10} C$
$= \log_{10} \boxed{}$.

■ 8. $\log_{10} A - \log_{10} B$
$= \log_{10}\left(\frac{A}{B}\right)$.
(Prop. 5-2b).

---

10. $2 \log_{10} A = \log_{10} \boxed{}$.

■ 9. $\log_{10} A + \log_{10} B - \log_{10} C$
$= \log_{10}(A \cdot B) - \log_{10} C$
$= \log_{10}\left(\frac{A \cdot B}{C}\right)$.

---

11. $\frac{1}{2} \log_{10} B = \log_{10} \boxed{}$.

■ 10. $2 \log_{10} A = \log_{10} A^2$.
(Prop. 5-2c)

---

12. $3(\log_{10} A - \log_{10} B)$
$= \log_{10} \boxed{}$.

■ 11. $\frac{1}{2} \log_{10} B = \log_{10} B^{1/2}$
$= \log_{10} \sqrt{B}$.
(Prop. 5-2c)

---

13. $\frac{2}{3}\log_{10} A + \frac{1}{3}\log_{10} B$

$= \log_{10} \boxed{}$.

■ 12. $3(\log_{10} A - \log_{10} B)$

$= 3\log_{10}\left(\frac{A}{B}\right)$

$= \log_{10}\left(\frac{A}{B}\right)^3$.

---

14. $\log_{10} \sqrt[3]{B^2} = \boxed{} \log_{10} B$.

■ 13. $\frac{2}{3}\log_{10} A + \frac{1}{3}\log_{10} B$

$= \log_{10} A^{2/3} + \log_{10} B^{1/3}$

$= \log_{10}(A^{2/3} \cdot B^{1/3})$

$= \log_{10} \sqrt[3]{A^2 B}$.

---

15. $\log_{10}(B^2\sqrt{B}) = \boxed{} \log_{10} B$.

■ 14. $\log_{10} \sqrt[3]{B^2} = \log_{10} B^{2/3}$

$= \frac{2}{3}\log_{10} B$.

■ 15. $\log_{10}(B^2\sqrt{B})$

$= \log_{10} B^2 B^{1/2}$

$= \log_{10} B^{5/2}$

$= \frac{5}{2}\log_{10} B$.

5-2 Problems

1. What is the numerical value of $\log_{10} 1$? $\log_5 1$? $\log_e 1$?
2. Apply Property 5-2b to $\log_{10}\left(\frac{1}{A}\right)$.
3. $3\log_{10} A + 4\log_{10} B = \log_{10} \boxed{}$.
4. $\log_{10} A - \log_{10} B = \log_{10} \boxed{}$.
5. If $\log_b A = 2$ and $\log_b B = 3$, what is $\log_b(AB)$? $\log_b\left(\frac{A}{B}\right)$? $\log_b\left(\frac{B}{A}\right)$? $\log_b \sqrt{A}$? $\log_b \sqrt{B}$?
6. Given $\log_b \sqrt{y} = 0.24$, find $\log_b y^2$.
7. Given $\log_b\left(\frac{1}{y}\right) = \frac{1}{4}$, find $\log_b y$.

5-3 Common Logarithms—Base 10

Theoretically, any nonnegative number unequal to one can be used as a base for a system of logarithms, but actually only two bases are in general use. The **common system** has a base of ten. The advantages of having a base of ten, which is the same as the base of our numeration system, will become apparent as we use the prepared tables of common logarithms. The **natural system** (Napierian system) has the irrational base called e, which is approximately 2.718. The advantages of using base e do not become apparent until one begins to use the calculus. Many formulas used in technical work involve base-e logarithms because the formulas were developed using the calculus. For this reason we give specific attention to natural logarithms in Section 5-5.

For instructional purposes a very brief table of common logarithms is printed here. The body of Table 5-3a contains, to the nearest ten-thousandths, the logarithms of numbers (N) from 1.0 to 9.9 in steps of tenths. (A larger table of common logarithms appears in the Appendix.)

TABLE 5-3a

$\text{Log}_{10} N = n$

N	.0	.1	.2	.3	.4	.5	.6	.7	.8	.9
1	.0000	.0414	.0792	.1139	.1461	.1761	.2041	.2304	.2553	.2788
2	.3010	.3222	.3424	.3617	.3802	.3979	.4150	.4314	.4472	.4624
3	.4771	.4914	.5051	.5185	.5315	.5441	.5563	.5682	.5798	.5911
4	.6021	.6128	.6232	.6335	.6435	.6532	.6628	.6721	.6812	.6902
5	.6990	.7076	.7160	.7243	.7324	.7404	.7482	.7559	.7634	.7709
6	.7782	.7853	.7924	.7993	.8062	.8129	.8195	.8261	.8325	.8388
7	.8451	.8513	.8573	.8633	.8692	.8751	.8808	.8865	.8921	.8976
8	.9031	.9085	,9138	.9191	.9243	.9294	.9345	.9395	.9445	.9494
9	.9542	.9590	.9638	.9685	.9731	.9777	.9823	.9868	.9912	.9956

From the table we can read directly the approximate logarithm of a number from 1.0 to 9.9.

Example 1

$\log_{10} 2.2 = .3424$

$\log_{10} 8.9 = .9494$

$\log_{10} 6.2 = \boxed{}$

$\log_{10} 3.0 = \square$

$\log_{10} 1.0 = \square$

■ .7924; .4771; .0000.

The table can also serve in the reverse manner. Given the value of a logarithm, the corresponding N can be determined.

Example 2

Find the number whose logarithm (base 10) is .3617.

Notice that the entries in the body of the table increase from left to right and top to bottom. Locate the logarithm nearest to .3617 in the table. It corresponds to $N = 2.3$. Therefore $\log_{10} 2.3 = .3617$.

If $x = \log_b y$, then we may also say, "y is the **antilogarithm** of x, base b." In mathematical notation this is written $y = \text{antilog}_b\, x$. Given this type of equation, it is more meaningful to think "y is equal to the number whose logarithm is x, base b."

Example 3

$\text{antilog}_{10}\, .6021 = 4.0$

$\text{antilog}_{10}\, .8751 = 7.5$

$\text{antilog}_{10}\, .6812 = \square$

$\text{antilog}_{10}\, .0000 = \square$

$\text{antilog}_{10}\, .9085 = \square$

■ 4.8; 1.0; 8.1.

Table 5-3a is not brief because of its limited range from 1.0 to 9.9, but rather because it passes through the range in steps of tenths. Upon investigating most tables of common logarithms, we find the range is from one to ten, but the interval between entries is smaller.

How are the logarithms of numbers outside of the range 1–10 determined if all or most tables have only that range? Examine carefully the following list of logarithmic equations and the question may be partly answered.

$$\begin{aligned}
\log_{10} .054 &= .7324 - 2 \\
\log_{10} .54 &= .7324 - 1 \\
\log_{10} 5.4 &= .7324 \\
\log_{10} 54 &= .7324 + 1 \\
\log_{10} 540 &= .7324 + 2 \\
\log_{10} 5400 &= .7324 + 3 \\
\log_{10} 54000 &= \boxed{\qquad\qquad}
\end{aligned}$$

■ .7324 + 4

The decimal part, called the **mantissa**, of every logarithm in the list is identical and is found as a result of using the table. Only the integer part, called the **characteristic**, changes; it can be determined from the position of the decimal point relative to the significant digits 5 and 4.

Knowing that *integer* values can be *added* to the decimal values given in the table to obtain the logarithms of numbers outside the range of 1–10 is a result of understanding and applying the fundamental properties

1. $\log_{10} 10^n = n$, and
2. $\log_{10} (M\,N) = \log_{10} M + \log_{10} N$.

The first property rewritten in exponential form simply says $10^n = 10^n$. A table of values should not be necessary to determine logarithms of powers of ten.

Example 4

$$\log_{10} 10^{-2} = -2$$

$$\log_{10} 10^{-1} = -1$$

$\log_{10} 10^0 = 0$

$\log_{10} 10 = \square$

$\log_{10} 10^6 = \square$

■ 1; 6.

The second property was introduced and amplified in Section 5-2.

A good technique for determining the logarithm of a number outside the range of the table is to first write the number in *scientific notation*. For example, the number 386,000 in scientific notation is written as 3.86×10^5. One of the factors is in the range of the table of logarithms and the other is a power of ten.

Example 5

Find the logarithm of 3900.

$$\begin{aligned}\log_{10} 3900 &= \log_{10}(3.9 \times 10^3)\\ &= \log_{10} 3.9 + \log_{10} 10^3\\ &= .5911 + 3.\end{aligned}$$

Example 6

Find the logarithm of 0.0045.

$$\begin{aligned}\log_{10} 0.0045 &= \log_{10}(4.5 \times 10^{-3})\\ &= \log_{10} 4.5 + \log_{10} 10^{-3}\\ &= .6532 + (-3)\\ &= .6532 - 3.\end{aligned}$$

It is perfectly correct to combine arithmetically the mantissa (decimal part) and the characteristic (integer part) of a logarithm, but in doing so, we obscure the parts of a logarithm that must be separate when finding antilogarithms from a table.

Example 7

Find the $\text{antilog}_{10}(.7993 + 4)$.

$$\begin{aligned}\text{antilog}_{10}(.7993 + 4) &= (\text{antilog}_{10}\ .7993)(\text{antilog}_{10}\ 4)\\ &= 6.3 \times 10^4.\end{aligned}$$

Example 8

Find the $\text{antilog}_{10}(.9345 - 3)$.

$$\begin{aligned}\text{antilog}_{10}(.9345 - 3) &= (\text{antilog}_{10}\ .9345)(\text{antilog}_{10}(-3))\\ &= 8.6 \times 10^{-3}.\end{aligned}$$

Since most of our computations are in base 10 logarithms, we shall omit the subscript from the notation and assume the base to be 10 unless otherwise stated.

5-3 Exercises

1. $\log 351 =$ ☐

---

2. $\log 38.7 =$ ☐

■ 1. $\log (3.51 \times 10^2) = .5453 + 2.$

---

3. $\log 9720 =$ ☐

■ 2. $\log (3.87 \times 10^1) = .5877 + 1.$

---

4. $\log 2.89 =$ ☐

■ 3. $\log (9.72 \times 10^3) = .9877 + 3.$

---

5. $\log 56.8 =$ ☐

■ 4. $\log 2.89 = .4609.$

---

6. antilog $(.3892 + 1)$ = [ ]

■ 5. $\log(5.68 \times 10^1) = .7543 + 1.$

---

7. antilog $(.5276 + 2)$ = [ ]

■ 6. antilog $(.3892 + 1) = 2.45 \times 10^1.$

---

8. antilog $(.6117 + 5)$ = [ ]

■ 7. antilog $(.5276 + 2) = 3.37 \times 10^2.$

---

9. antilog $(.9304)$ = [ ]

■ 8. antilog $(.6117 + 5) = 4.09 \times 10^5.$

---

10. antilog $(.8651 - 1)$ = [ ]

■ 9. antilog $(.9304) = 8.52.$

---

11. $\log 8700$ = [ ]

■ 10. antilog $(.8651 - 1) = 7.33 \times 10^{-1}.$

---

12. $\log .246$ = [ ]

■ 11. $\log(8.700 \times 10^3) = .9395 + 3.$

---

13. $\log .00238$ = [ ]

■ 12. $\log(2.46 \times 10^{-1}) = .3909 - 1.$

---

14. $\log(8.12 \times 10^5)$ = [ ]

■ 13. $\log(2.38 \times 10^{-3}) = .3766 - 3.$

---

15. $\log(1.97 \times 10^{-6})$ = [ ]

■ 14. $\log(8.12 \times 10^5) = .9096 + 5.$

---

16. antilog $(.9020 - 2)$ = [ ]

■ 15. $\log(1.97 \times 10^{-6}) = .2945 - 6.$

---

17. antilog (.2923 − 3)
 = [ ]

■ 16. antilog (.9020 − 2)
 $= 7.98 \times 10^{-2}$.

---

18. antilog (.9284 − 2)
 = [ ]

■ 17. antilog (.2923 − 3)
 $= 1.96 \times 10^{-3}$.

---

19. antilog (8.2355 − 10)
 = [ ]

■ 18. antilog (.9284 − 2)
 $= 8.48 \times 10^{-2}$.

---

20. antilog (9.4871)
 = [ ]

■ 19. Notice in this problem that the integer part of the logarithm is $8 - 10 = -2$. Some literature writes the integer portion of logarithms in the manner illustrated by this exercise.
antilog (8.2355 − 10)
 = antilog (.2355 − 2)
 $= 1.72 \times 10^{-2}$.

■ 20. antilog (9.4871)
 = antilog (.4871 + 9)
 $= 3.07 \times 10^{9}$.

5-3 Problems

1. On a slide rule, the L scale contains the logarithms of numbers between one and ten appearing on the C or D scale. The correspondence is shown by the slide-rule hair line. Compare slide-rule results with tabular results to make certain you can use the slide rule correctly to obtain logarithms and antilogarithms.
2. Some electronic calculators have the capacity to evaluate logarithms and antilogarithms. If one is available, take the opportunity to use it.
3. Some technical areas use logarithms to a greater extent than others. If the area for which you are training uses logarithms, learn to read the tables that will be required on the job.

5-4 Computations with Logarithms

Products, quotients, roots, and powers can be calculated by using logarithms. Slide rules and modern desk calculators are better for calculating products and quotients. They also help in calculating roots and powers when they are designed with this capacity. But there is one unalterable fact: both the procedures and properties of logarithms must be understood whether powers and roots are calculated by machine or hand.

The examples in this section illustrate the methods for computing products, quotients, roots, and powers with the aid of logarithms. An understanding of all four operations is necessary, but emphasis should be given to the procedures for computing roots and powers.

Example 1

Find the product of 35 and 550 using logarithms.

We use Property 5-2a, ($\log_{10} MN = \log_{10} M + \log_{10} N$).

$$\begin{aligned}\log_{10}(35)(550) &= \log_{10}(3.5 \times 10^1) + \log_{10}(5.5 \times 10^2)\\ &= (.5441 + 1) + (.7404 + 2)\\ &= 1.2845 + 3\\ &= .2845 + 4;\end{aligned}$$

$$\text{antilog } (.2845 + 4) \approx 1.925 \times 10^4;$$

therefore, $35(550) \approx 19250$.

Example 2

Find the product of 0.0084 and 410 using logarithms.

$$\begin{aligned}\log_{10}(.0084)(410, &= \log_{10}(8.4 \times 10^{-3}) + \log_{10}(4.1 \times 10^2)\\ &= (.9243 - 3) + (.6128 + 2)\\ &= 1.5371 - 1\\ &= .5371;\end{aligned}$$

$$\text{antilog } .5371 \approx 3.44;$$

therefore, $(.0084)(410) \approx 3.44$.

Very few complications occur in using logarithms for multiplication. Some difficulties can arise in division, especially when the divisor is greater than the dividend.

Example 3

Find $\dfrac{390}{0.58}$ using logarithms.

We use Property 5-2b, $\left(\log_{10}\dfrac{M}{N}=\log_{10}M-\log_{10}N\right)$.

$$\log_{10}\left(\frac{390}{0.58}\right)=\log_{10}(3.9\times 10^{2})-\log_{10}(5.8\times 10^{-1})$$
$$=(.5911+2)-(.7634-1).$$

The decimal part of the difference must result in a positive number if we intend to use a table to evaluate the antilogarithm. Therefore a 1 is added to the .5911 and subtracted from the 2. So

$$\log_{10}\left(\frac{390}{0.58}\right)=(1.5911+1)-(.7634-1)$$
$$=.8277+2;$$

antilog $(.8277+2)\approx 6.72\times 10^{2}$;

therefore, $\dfrac{390}{0.58}\approx 672.$

Example 4

Find the quotient $\dfrac{1}{87.2}$ using logarithms.

$$\log_{10}\left(\frac{1}{87.2}\right)=\log_{10}1-\log_{10}(8.72\times 10^{1})$$
$$=(.0000)-(.9405+1)$$
$$=(1.0000-1)-(.9405+1)\quad\text{(add and subtract one)}$$
$$=.0595-2;$$

$\text{antilog}_{10}(.0595 - 2) \approx 1.15 \times 10^{-2};$

therefore, $\dfrac{1}{87.2} \approx .0115.$

In Section 5-1 we said that evaluating

$$A = 100\left(1 + \frac{.055}{12}\right)^{12(20)}$$

requires raising

$$\left(1 + \frac{.055}{12}\right) \approx 1.00458$$

to the 240th power. This computation is illustrated in Example 5.

Example 5

Evaluate $(1.00458)^{240}$ using logarithms.

We use Property 5-2c, ($\log_{10} N^a = a \log_{10} N$).

$$\begin{aligned} \log_{10}(1.00458)^{240} &= 240 \log 1.00458 \\ &= 240\,(.001986) \\ &= .47663; \end{aligned}$$

$\text{antilog } .47663 \approx 2.9966;$

therefore, $(1.00458)^{240} \approx 2.9966.$

Concluding the problem started in Section 5-1, the amount A after 20 years is \$100(2.9966), or \$299.66.

Example 5 does not pose some of the arithmetic difficulties that can occur in evaluating radical or exponential expressions. We shall encounter some in Examples 6 and 7.

Example 6

Evaluate $(.0243)^{0.62}$ using logarithms.

$$\begin{aligned}\log_{10}(.0243)^{0.62} &= 0.62\log_{10}(2.43 \times 10^{-2})\\ &= 0.62(.3856 - 2)\\ &= (.2391) - (1.24)\\ &= (2.2391 - 2) - (1.24) \quad \text{(add and subtract two)}\\ &= .9991 - 2;\end{aligned}$$

$$\text{antilog}\,(.9991 - 2) \approx 9.98 \times 10^{-2};$$

therefore, $(.0243)^{0.62} \approx .0998.$

Example 7

Evaluate $(27.4)^{-1.3}$ using logarithms.

$$\begin{aligned}\log_{10}(27.4)^{-1.3} &= (-1.3)\log_{10}(2.74 \times 10)\\ &= (-1.3)(.4378 + 1)\\ &= (-1.3)(1.4378)\\ &= -1.8691\\ &= (2 - 1.8691) - 2 \quad \text{(add and subtract two)}\\ &= .1309 - 2;\end{aligned}$$

$$\text{antilog}\,(.1309 - 2) \approx 1.35 \times 10^{-2};$$

therefore, $(27.4)^{-1.3} \approx .0135.$

There is no "best" method for avoiding negative decimal parts to the logarithms; it is simply an arithmetic problem that may be done in any convenient manner.

5-4 Exercises

Use logarithms for the computations in the exercises.

1. $(.0025)(892) =$ ______________

---

2. $(548)(2.37) = \boxed{}$

■ 1. $\log (.0025)(892)$
$= \log (2.5 \times 10^{-3})$
$+ \log (8.92 \times 10^{2})$
$= (.3979 - 3)$
$+ (.9504 + 2)$
$= 1.3483 - 1$
$= .3483;$
antilog $(.3483) = 2.23.$

---

3. $\dfrac{28.9}{13.2} = \boxed{}$

■ 2. $\log (548)(2.37)$
$= \log (5.48 \times 10^{2})$
$+ \log 2.37$
$= (.7388 + 2) + (.3747)$
$= 1.1135 + 2$
$= .1135 + 3;$
antilog $(.1135 + 3)$
$= 1.30 \times 10^{3}.$

---

4. $\dfrac{28.9}{4850} = \boxed{}$

■ 3. $\log \dfrac{28.9}{13.2}$
$= \log (2.89 \times 10^{1})$
$- \log (1.32 \times 10^{1})$
$= (.4609 + 1)$
$- (.1206 + 1)$
$= .3403;$
antilog $(.3403) = 2.19.$

---

5. $2.00^{100} = \boxed{}$

■ 4. $\log \dfrac{28.9}{4850}$
$= \log (2.89 \times 10^{1})$
$- \log (4.85 \times 10^{3})$
$= (.4609 + 1)$
$- (.6857 + 3)$
$= (1.4609) - (.6857 + 3)$
$= .7752 - 3;$
antilog $(.7752 - 3)$
$= 5.96 \times 10^{-3}.$

---

6. $212^{2.5} = \boxed{}$

■ 5. $\log 2.00^{100} = 100 \log 2.00$
$= 100(.30103)$
$= 30.103$
$= .103 + 30;$
antilog $(.103 + 30)$
$= 1.27 \times 10^{30}.$

---

7. $(\sqrt[5]{31})^2 = \boxed{}$

■ 6. $\log 212^{2.5}$
$= 2.5 \log (2.12 \times 10^2)$
$= 2.5(.3263 + 2)$
$= 2.5(2.3262)$
$= 5.8158;$
antilog $(.8158 + 5)$
$= 6.54 \times 10^5.$

---

8. $\sqrt[7]{.268} = \boxed{}$

■ 7. $\log (31)^{2/5}$

$= \frac{2}{5} \log (3.1 \times 10)$

$= \frac{2}{5} (.4914 + 1)$

$= \frac{2}{5} (1.4914)$

$= .5966;$

antilog $(.5966) = 3.95.$

---

9. $\left(\frac{2.43}{.461}\right)^{1.6} = \boxed{}$

■ 8. $\log (2.68 \times 10^{-1})^{1/7}$

$= \frac{1}{7} \log (2.68 \times 10^{-1})$

$= \frac{1}{7} (.4281 - 1)$

$= \frac{1}{7} (6.4281 - 7)$

$= .9183 - 1;$

antilog $(.9183 - 1)$
$= 8.29 \times 10^{-1}.$

---

10. $7.47(3.45)^{2.4} =$ [ ]

■ 9. $\log\left(\frac{2.43}{.461}\right)^{1.6}$

$= 1.6 \log\left(\frac{2.43}{.461}\right)$

$= 1.6(\log 2.43$
$- \log(4.61 \times 10^{-1}))$
$= 1.6(.3856$
$- (.6637 - 1))$
$= 1.6(.7219)$
$= 1.1550;$

antilog $(.1550 + 1)$
$= 1.43 \times 10^1.$

■ 10. $\log 7.47(3.45)^{2.4}$
$= \log(7.47)$
$+ 2.4 \log(3.45)$
$= (.87332) + 2.4(.53782)$
$= 2.1641;$

antilog $(.1641 + 2)$
$= 1.46 \times 10^2.$

5-4 Problems

1. The quantity of water Q(in cubic feet per second) passing over a V-notched weir is determined by the formula $Q = 2.53H^{2.5}$, where H is the head of water in feet. Calculate the amount of water that passes over the weir in one hour if $H = 1.30$ feet.
2. The response to sound by the human ear is not proportional to the intensity of the sound source. There is, however, a logarithmic relationship. Consequently, the unit of measure for sound amplification is the decibel (N_{db}) defined by the equation

$$N_{db} = \log_{10}\left(\frac{P_2}{P_1}\right)^{10},$$

where P_2 is output and P_1 is input power. Find N_{db} when $P_1 = .005$ watts and $P_2 = 10$ watts.

3. The crushing weight P, in pounds, of an iron column is given by the formula

$$P = 299{,}600 \frac{d^{3.55}}{L^2},$$

where the diameter (d) is in inches and the length (L) is in feet. What weight will crush an iron column 12.25 feet long and 3.5 inches in diameter?

4. The formula

$$P = S\left(\frac{1}{\left(1+\frac{i}{c}\right)^{nc}}\right)$$

is used to calculate the present value, P, of an investment if it is to realize a value S after n years at an annual rate i compounded c times per year. Calculate P when

$$i = 4 \text{ per cent}, \qquad c = 4,$$
$$n = 10 \text{ years}, \qquad s = \$5600.$$

5. The head loss from friction in a pipe line is determined by the formula

$$H = 0.38\,V^{1.86} d^{-1.25}.$$

Calculate H, when $V = 4.5$ and $d = 0.5$.

6. A particular type of bacteria doubles in number every 40 minutes under certain conditions. If a culture of this bacteria is left to multiply for 12 hours, by what factor will its number increase?

5-5 Graphs and Natural Logarithms

The three equations $y = b^x$, $\log_b y = x$, and $y = \text{antilog}_b\, x$ are equivalent. This is to say, if we choose a particular value for $b > 0$, a table of values that satisfies one equation will also satisfy the other two equations, and the graph drawn to represent one of the functions will also represent the other two.

Example 1

If $b = 10$, then $y = 10^x$, $\log_{10} y = x$, and $y = \text{antilog}_{10}\, x$.

x	y
-1.00	0.100
$-.6\overline{6}$	.215
$-.3\overline{3}$	.464
.00	1.00
$.3\overline{3}$	2.15
$.6\overline{6}$	4.64
1.00	10.00

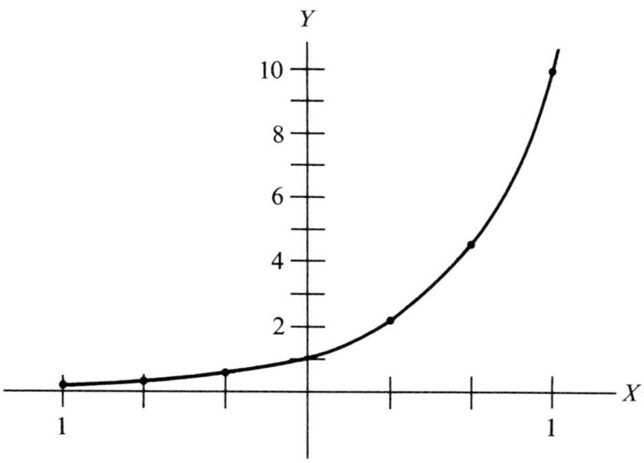

Fig. 5-5a.

Example 2

If $b = 1$, then $y = 1^x$, $\log_1 y = x$, and $y = \text{antilog}_1\, x$.

x	y
-1.00 ⋮ 1.00	(1 for all x)

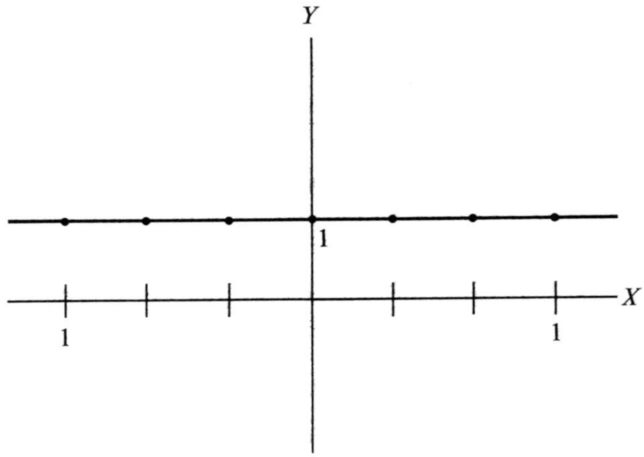

Fig. 5-5b.

Example 3

If $b = \frac{1}{8}$, then $y = \left(\frac{1}{8}\right)^x$, $\log_{1/8} y = x$, and $y = \text{antilog}_{1/8}\, x$.

x	y
-1.00	8
$-.6\overline{6}$	4
$-.3\overline{3}$	☐
$.00$	1
$.3\overline{3}$	☐
$.6\overline{6}$	.25
1.00	.125

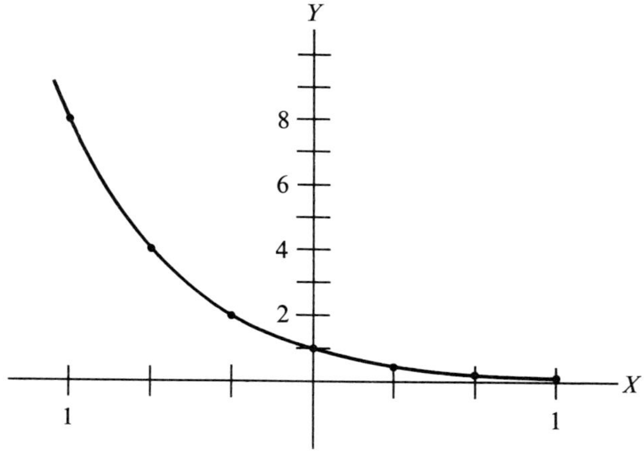

Fig. 5-5c

■ 2; .5.

The three bases used in Examples 1, 2, and 3 were chosen for the purpose of comparison. All three graphs have equal y-intercepts and no x-intercepts. In particular, notice that when $b = 10$ (or simply when $b > 1$), the slope of the graph is everywhere positive. When $b = 1$, the slope is zero over the entire domain of x's. When $b < 1$ (but of course greater than zero), the slope of the graph is everywhere negative.

When systems of logarithms were first produced early in the seventeenth century, the base was not the center of attention. Yet, working independently, John Napier and Jobst Bürgi—who receive principal credit for the invention of logarithms—used bases very closely related to the base of our modern **natural system of logarithms**. The base of natural (or Napierian) logarithms is the irrational number called e, which is approximately equal to 2.7183.

Although such a base may seem very cumbersome, it appears more frequently in scientific and engineering work than do common logarithms because of its special property. The *slope* of the graph of $y = f(x) = e^x$ is *equal* to e^x over the entire domain of the variable x. In other words, the slope function is identical to the function f. If $f(x) = e^x$, then $f'(x) = e^x$.

Example 4

If $b = e \approx 2.7183$, then $y = f(x) = e^x$, $\log_e y = x$, $y = \text{antilog}_e\, x$, and $f'(x) = e^x$.

x	$f(x)$	$f'(x)$
−2.0	.135	.135
−1.5	.223	
−1.0	.368	
−0.5	607	
0.0	1.00	
0.5	1.65	
1.0	2.72	
1.5	4.48	
2.0	7.39	

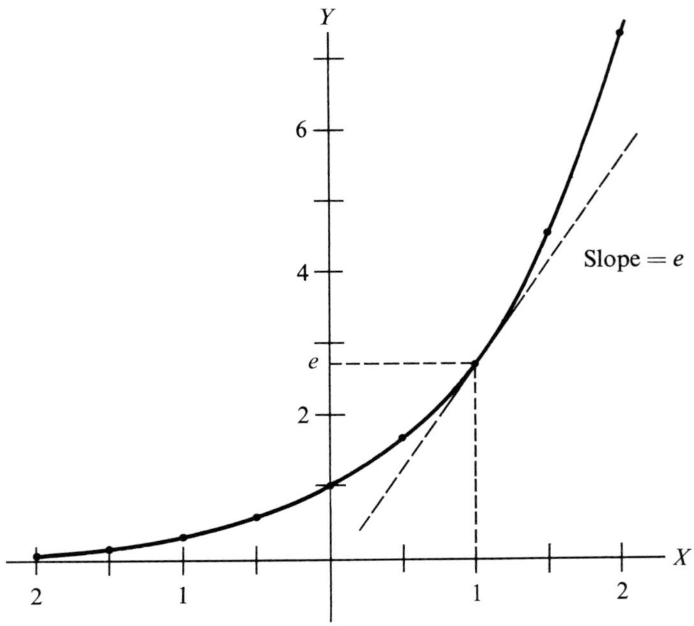

Fig. 5-5d

■ The values for $f'(x)$ are repeated from $f(x)$ for corresponding values of x, for instance, .223, .368, and so forth.

The properties of logarithms discussed in Section 5-2 remain the same for natural logarithms. Recall that the properties were developed in terms of a general base b greater than zero and unequal to one.

The tables of natural logarithms are considerably different from those for common logarithms. Both integer and decimal parts of natural logarithms as well as negative logarithms appear in the tables, because there is no convenient relationship between the decimal numeration system and the number e.

A table of limited range is usually augmented with a partial list of the natural logarithms of the powers of ten:

$$\log_e 10 = 2.302585$$
$$\log_e 10^2 = \boxed{}$$
$$\log_e 10^3 = 6.907755$$
$$\log_e 10^4 = 9.210340$$
$$\log_e 10^5 = 11.512926$$
$$\log_e 10^6 = 13.815511$$
$$\log_e 10^7 = 16.118096$$
$$\log_e 10^8 = 18.430681$$
$$\log_e 10^9 = 20.723266$$

■ $\log_e 10^2 = 2 \log_e 10 = 2(2.302585)$
$= 4.605170.$

Since only two bases, ten and e, are in common use, $\log_{10}$ is abbreviated to log and the base ten is understood, and $\log_e$ becomes ln where the base e is understood. All other bases are designated in the usual way with subscript notation.

Example 5

$$\log_{10} 7.24 = .8597 \leftrightarrow \log 7.24 = .8597$$
$$\log_e 7.24 = 1.9796 \leftrightarrow \ln 7.24 = 1.9796$$

5-5 Exercises

Use tables of natural logarithms.

1. $\ln 3.25 =$ [ ]

---

2. $\ln 9.47 =$ [ ]

■ 1. $\ln 3.25 = 1.17865$.

---

3. $\ln 23.8 =$ [ ]

■ 2. $\ln 9.47 = 2.24813$.

---

4. $\ln 5300 =$ [ ]

■ 3. $\ln (2.38 \times 10^1)$
$= .86710 + 2.30259$
$= 3.16969$.

---

5. antiln 1.74047
= [ ]

■ 4. $\ln (5.3 \times 10^3)$
$= 1.66771 + 6.90775$
$= 8.57546$.

---

6. antiln 2.2126
= [ ]

■ 5. antiln 1.74047 = 5.700.

---

7. antiln 4.3167
= [ ]

■ 6. antiln 2.2126 = 9.14.

---

8. $\ln e =$ [ ]

■ 7. antiln 4.3167
= antiln (2.0141
+ 2.3026)
$= 7.49 \times 10^1$.

---

9. $\ln 1 =$ [ ]

■ 8. $\ln e = 1.0000$.

---

10. $\ln .0246 =$ [ ]

■ 9. $\ln 1 = 0.0000$.

---

11. $\ln .000832 =$ [ ]

■ 10. $\ln (2.46 \times 10^{-2})$
$= .90016 - 4.60517$
$= -3.70501.$

---

12. antiln 1.0000
= [ ]

■ 11. $\ln (8.32 \times 10^{-4})$
$= 2.11866 - 9.21034$
$= -7.09168.$

---

13. antiln (− 1.2214)
= [ ]

■ 12. antiln 1.0000 = e
= 2.71828.

---

14. antiln (− 8.6784)
= [ ]

■ 13. antiln (−1.2214)
= antiln (2.3026
− 1.2214 − 2.3026)
= antiln (1.0812
− 2.3026)
$= 2.95 \times 10^{-1}.$

---

15. antiln (−16.3871)
= [ ]

■ 14. antiln (−8.6784)
= antiln (9.2103
− 8.6784 − 9.2103)
= antiln (.5319
− 9.2103)
$= 1.70 \times 10^{-4}.$

---

16. $\ln e^{4.5} =$ [ ]

■ 15. antiln (−16.3871)
= antiln (18.4207
− 16.3871 − 18.4207)
= antiln (2.0336
− 18.4207)
$= 7.64 \times 10^{-8}.$

---

17. $\ln e^{-.002} =$ [ ]

■ 16. $\ln e^{4.5} = 4.5 \ln e$
$= 4.5.$

---

18. $\ln \left(\frac{32.4}{2.5}\right)^{.2} =$ [ ]

■ 17. $\ln e^{-.002} = -.002 \ln e$
$= -.002.$

■18. $\ln\left(\frac{32.4}{2.5}\right)^{.2}$

$= .2(\ln 32.4 - \ln 2.5)$
$= .2(3.47816 - .91629)$
$= .51237.$

5-5 Problems

1. The increasing energy (E) of a certain quantity of gas expanding at a constant temperature from volume V_0 to volume V_1 is given by the formula

$$E = P_0 V_0 \ln\left(\frac{V_1}{V_0}\right).$$

Find E, if $P_0 = 87.5$, $V_0 = 246$, and $V_1 = 472$.

2. One curve on a universal time-constant chart is the graph of the function

$$i = \frac{E}{R} e^{-t/RC}.$$

Graph the function $i = f(t)$ from $t = 0$ to $t = 3$ in steps of 0.2, if $E = 10$, $R = 1$, and $C = 1$.

3. A power line hanging between utility poles forms a curve called a catenary. The length of the line is determined by the equation

$$L = a(e^{x/a} - e^{-x/a}),$$

where a is a constant determined by the weight per unit length and tension on the cable and x is one-half the span. Find the length of a power line hanging between two poles 350 feet apart if $a = 2150$.

4.

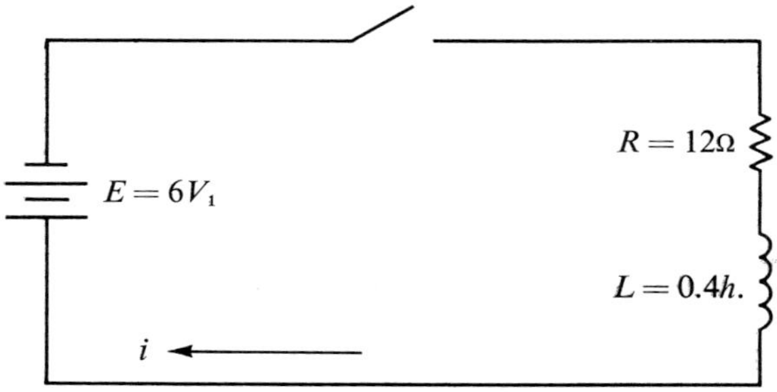

Solve for i at $t = .03$ seconds, using the equation

$$i = \frac{E}{R}(1 - e^{-Rt/L}).$$

5. After X rays penetrate to a depth of x centimeters, the intensity $I = I_0 e^{-\mu x}$, where I_0 is the intensity at the surface. Graph (I/I_0) as a function of x from 0.0 to 0.2 centimeters when $\mu = 12.9$ (the absorption coefficient for lead).

5-6 Change of Base and Interpolation

A change in the base of logarithms can be made using the formula

$$\log_b N = \frac{\log_a N}{\log_a b}. \qquad \text{(Eq. 5-6a)}$$

Assume for example that we want to calculate the value of ln 3.42 but have only tables of common logarithms. Dividing log 3.42 by log e will give the desired quantity.

$$\ln 3.42 = \frac{\log 3.42}{\log 2.718} = \frac{.53403}{.43425} = 1.23.$$

The important point to remember is that *one* table of logarithms can serve to evaluate logarithms of any base b, $b > 0$.

As a review and exercise, we shall show how to arrive at Eq. 5-6a. Let us start with the equation $N = b^n$ and its equivalent logarithmic form ____________.

■ $\log_b N = n$.

If $N = b^n$, then the logarithms of both N and b^n are equal; in other words,

$$\log_a N = \log_a b^n, \quad a > 0, a \neq 1,$$
$$= n \log_a b.$$

Since $n = \log_b N$, by substitution,

$$\log_a N = \log_b N \log_a b,$$

which may be rearranged to read

$$\log_b N = \frac{\log_a N}{\log_a b}.$$

Example 1

Using common logarithms, find $\log_{16} 86$.

$$\log_{16} 86 = \frac{\log 86}{\log 16} = \frac{1.93450}{1.20412} = 1.607.$$

Computations with logarithms result in approximations because most of the numbers given in the body of a logarithm table are rounded-off irrational numbers.

The accuracy of a logarithmic calculation is limited by the number of significant figures given in the table. We may improve the accuracy of the numbers given in a table by *one* significant digit using a method called **linear interpolation**. Let us illustrate the method with an example.

Suppose we want to approximate the value of log 3.147 using a table where both 3.14 and 3.15, but nothing between, appear in the number column. Since 3.147 is 7/10 of the distance between 3.14 and 3.15, we adjust the logarithm for 3.147 by 7/10 of the difference between the logarithm of 3.14 and the logarithm of 3.15. This is equivalent to forming a proportion of the differences (illustrated in Fig. 5-6a) and solving for x when $y = .007$.

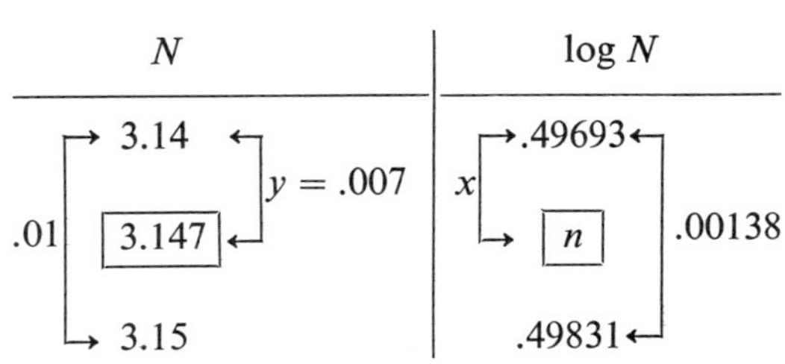

Fig. 5-6a

$$\frac{y}{.01} = \frac{x}{.00138} \qquad \text{(Eq. 5-6b)}$$

$$\frac{.007}{.01} = \frac{x}{.00138}$$

$$\begin{aligned} x &= .7(.00138) \\ &= .000966. \end{aligned}$$

The value x is added to the logarithm of 3.14 to obtain n, the approximate logarithm of 3.147.

$$\begin{aligned} \log 3.147 &= n \\ &\approx .49693 + x \\ &\approx .49693 + .000966 \\ &\approx .497896 \text{ or } .49790. \end{aligned}$$

We know that Eq. 5-6b is a linear equation. The method of *linear* interpolation is so named because we are approximating a nonlinear logarithmic function with a linear function. Figure 5-6b is a graphical exaggeration of the numerical example just completed.

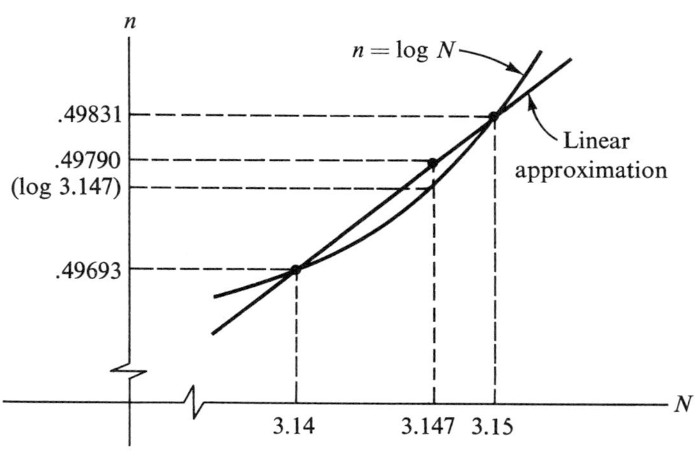

Fig. 5-6b

The actual difference between .49790 and log 3.147 is less than one hundred-thousandth.

The method of linear interpolation also applies to finding an N, given the logarithm of N.

Example 2

Find, by linear interpolation, antilog .6321 using four-place logarithmic tables.

	N			$\log N$	
.01	4.28	y	.0007	.6314	.0011
	$\boxed{N}$			$\boxed{.6321}$	
	4.29			.6325	

■ i. $\dfrac{y}{.01} = \dfrac{.0007}{.0011}$

$y = .00636\ldots,$

ii. $\text{antilog } .6321 = N$

$= 4.28 + y$

$\approx 4.286.$

5-6 Exercises

1. Use common logarithms to calculate $\log_2 32$.

---

2. Use common logarithms to calculate $\log_{12} 100$.

■ 1. $\log_2 32.0 = \dfrac{\log 32}{\log 2}$

$= \dfrac{1.5051}{.3010} = 5.0003.$

---

3. Evaluate ln 4.32 using common logarithms.

■ 2. $\log_{12} 100 = \dfrac{\log 100}{\log 12}$

$= \dfrac{2.0000}{1.0792} = 1.8532.$

---

4. Evaluate ln .0582 using common logarithms.

■ 3. $\ln 4.32 = \dfrac{\log 4.32}{\log e} = \dfrac{.63548}{.43425} = 1.4634.$

---

5. If $\log 9.11 = .95952$ and $\log 9.12 = .95999$, use the method of linear interpolation to find log 9.113.

■ 4. $\ln .0582 = \dfrac{\log (5.82 \times 10^{-2})}{\log e} = \dfrac{.76492 - 2}{.43425} = \dfrac{-1.23508}{.43425} = -2.8447.$

---

6. If $\ln 3.21 = 1.16627$ and $\ln 3.22 = 1.16938$, use the method of linear interpolation to find ln 3.218.

■ 5.

		N		$\log N$	
.01		9.11	.003	.95952	.00047
		[9.113]	x	[n]	
		9.12		.95999	

$$\frac{x}{.00047} = \frac{.003}{.01}$$

$$\begin{aligned} n &= x + .95952 \\ &= .3(.00047) + .95952 \\ &= .95966. \end{aligned}$$

---

7. If $\log 6.28 = .79796$ and $\log 6.29 = .79865$, use the method of linear interpolation to find antilog .79854 to four significant figures. (*Note*: The method is very similar to Exercise 5.)

■ 6.

		N		$\log N$	
.01		3.21	.008	1.16627	.00311
		[3.218]	x	[n]	
		3.22		1.16938	

$$\frac{x}{.00311} = \frac{.008}{.01}$$

$$\begin{aligned} n &= x + 1.16627 \\ &= .8(.00311) + 1.16627 \\ &= 1.16876. \end{aligned}$$

8. If ln 75.2 = 6.62274 and ln 75.3 = 6.62407, find antiln 6.2355 to four significant figures by the method of linear interpolation.

■ 7.

	N			$\log N$	
.01	6.28	y	.00058	.79796	.00069
	N			.79854	
	6.29			.79865	

$$\frac{y}{.01}=\frac{.00058}{.00069}$$

$$\begin{aligned} N &= y + 6.28 \\ &= .01\left(\frac{58}{69}\right) + 6.28 \\ &= 6.288. \end{aligned}$$

■ 8.

	N			$\log N$	
.1	75.2	y	.00081	6.62274	.00133
	N			6.62355	
	75.3			6.62407	

$$\frac{y}{.1}=\frac{.00081}{.00133}$$

$$\begin{aligned} N &= y + 75.2 \\ &= .1\left(\frac{81}{133}\right) + 75.2 \\ &= 75.26. \end{aligned}$$

5-6 Problems

1. In the formula $q = q_0 e^{kt}$,

$$k = -\frac{\ln 2}{1800}.$$

Evaluate k using common logarithms.

2. Evaluate the expression $\frac{1}{20}(\ln 10 - \ln 9)$ using common logarithms.

3. Evaluate $\frac{1}{2}(e^x - e^{-x})$ when $e = 2.718$ and $x = 1.36$, using common logarithms.

4. Evaluate $\dfrac{\ln 5.24}{\ln .375}$, using common logarithms.
5. Some tables print what are called *proportional parts* to aid in the interpolation of logarithms. The value that is called x in the computation of Exercise 5 of this section is the value that is provided in the tables. Examine tables of this kind and practice using the proportional parts.

5-7 Logarithmic and Exponential Equations

Many interesting and important problems have exponential relationships that make it important for you to know how to solve exponential and logarithmic equations. The following list gives examples of quantities that have this relationship.

Time and compounded interest.

Altitude and atmospheric pressure.

Temperature and rate of chemical reaction.

Time and growth of bacteria.

Temperature and thermionic emission.

Time and radioactivity.

Temperature difference and heat transfer.

Sound intensity and audibility.

Thickness of medium and light intensity.

Current and inductance.

Spacing of transmission lines and impedance.

Time and charge of a capacitor.

pH and hydrogen ion concentration.

Vapor pressure and temperature.

In Section 5-1 we introduced simple logarithmic and exponential equations similar to $\log_{10} y = 4$ and $9^x = 27$. It was easy for you to find the values that satisfied these types of equations, because only simple numbers with perfect powers and roots were used. We are now ready to go beyond this type of problem—to those that cannot be solved by inspection alone. We must depend upon the properties of logarithms and tables of logarithms.

Example 1

Solve $\log(x + 12) - \log x = 1.32$, for x.

i. $\log\left(\dfrac{x+12}{x}\right) = 1.32$ (by Prop. 5-2b),

ii. $\dfrac{x+12}{x} = \text{antilog } 1.32$

$= \boxed{}$,

iii. $12 = 20.9x - x$,

iv. $x = \dfrac{12}{19.9} = .603.$

■ 2.09×10^1

There is no general solution to logarithmic equations; however the technique of solving them often involves first combining the logarithmic terms using the properties of logarithms, and then taking the antilogarithm of both sides of the equation; this transforms the problem into an equation of the type studied in earlier chapters.

Example 2

Solve $x \log 3.21 - \log 24.1 = 0$ for x.

i. $x \log 3.21 = \log 24.1$,

ii. $x = \dfrac{\log 24.1}{\log 3.21}$

$\approx \boxed{} \approx \boxed{}.$

■ $x \approx \dfrac{1.38202}{.50651} \approx 2.73.$

In line (ii) of Example 2, it is very important to recognize that

$$\frac{\log 24.1}{\log 3.21} \neq \log (24.1 - 3.21).$$

We point this out because it is a common error in calculation.

Usually the first step in solving an exponential equation involves transforming it into a logarithmic equation by writing the logarithm of both sides of the equation. This step permits the unknown exponents to become simple factors in the terms of the equation.

Example 3

Solve $3^{2x-1} = 5$ for x.

i. $3^{2x-1} = 5,$

ii. $(2x - 1) \log 3 = \log 5,$

iii. $2x - 1 = \dfrac{\log 5}{\log 3}$

$$= \frac{.69897}{.47712} = 1.465,$$

iv. $x = \dfrac{1.465 + 1}{2} = 1.231.$

Example 4

Radioactive substances decay at a rate that depends upon the number of unstable particles present. If we start with 100 milligrams of active thorium (Th^{234}), in 24.1 days it will decay to 50 milligrams of active thorium according to the formula $N = N_0 e^{-t/\tau}$, where N_0 is the initial amount, N the amount after t days, and τ is the constant mean life of the unstable particles. Calculate the value of τ and the time for 100 milligrams to decay to 10 milligrams.

Part 1:

i. $\dfrac{50}{100} = e^{-24.1/\tau}$

ii. $\ln .5 = -\frac{24.1}{\tau} \ln e,$ $(\ln e = 1)$

iii. $\tau = -\frac{24.1}{\ln .5} = \frac{-24.1}{-.693} = 34.8.$

Part 2:

iv. $\frac{10}{100} = e^{-t/34.8},$ (use the value of τ calculated in Part 1.)

v. $\ln .1 = -\frac{t}{34.8} \ln e,$

vi. $t = -34.8(\ln .1)$
$= -34.8(-2.3026)$
$= 80.0$ days.

5-7 Exercises

1. Solve for x in the equation

$$\log (3x)^2 = 1.2344.$$

---

2. Solve for w when

$$\log (w^2 - 4) - \log (w + 2) = 0.$$

■ 1. $2 \log 3x = 1.2344$
$\log 3x = .6172$
$3x = \text{antilog } .6172$
$x = \frac{4.142}{3} = 1.381.$

---

3. Solve for x when

$$2^x = 25.$$

■ 2. $\log (w^2 - 4) - \log (w + 2) = 0$
$\log \left(\frac{w^2 - 4}{w + 2}\right) = 0$
$\log (w - 2) = 0$
$w - 2 = \text{antilog } 0 = 1$
$w = 3.$

---

4. Solve for n when

$$125^n = 48(5)^{n+1}.$$

■ 3.
$$2^x = 25$$
$$x \log 2 = \log 25$$
$$x = \frac{\log 25}{\log 2}$$
$$= \frac{1.39794}{.30103}$$
$$= 4.64.$$

---

5. Solve for z when

$$4^{z+2} = 8^{2z-1}.$$

■ 4.
$$125^n = 48(5)^{n+1}$$
$$n \log 125 = \log 48 + (n+1) \log 5$$
$$n(\log 125 - \log 5) = \log 48 + \log 5$$
$$n = \frac{\log 48 + \log 5}{\log 125 - \log 5}$$
$$= \frac{1.68124 + .69897}{2.09691 - .69897}$$
$$= \frac{2.38021}{1.39794} = 1.70.$$

---

6. Solve for x when

$$(3)^{x+2}(2)^{x+1} = 7^{3x}.$$

■ 5. Since the bases of both exponential expressions are powers of 2, we can write the logarithm of both sides using *base 2*.

$$4^{z+2} = 8^{2z-1}$$
$$(z+2) \log_2 4 = (2z-1) \log_2 8$$
$$(z+2)2 = (2z-1)3$$
$$2z + 4 = 6z - 3$$
$$4z = 7$$
$$z = 7/4.$$

(Other bases can be used as well.)

---

7. Solve for x when

$$\frac{(1.06)^x - 1}{.06} = 12.$$

■ 6.
$$(3)^{x+2}(2)^{x+1} = 7^{3x}$$
$$(x+2) \log 3 + (x+1) \log 2 = 3x \log 7$$

$$(x+2)(.477) + (x+1)(.301) = 3(.845)x$$
$$1.757x = 1.255$$
$$x = .714.$$

---

8. Transform the equation

$$M = 2.5 \log A - \log a^{2.5}$$

so that a is given in terms of the other variables. (*Note*: There are several correct forms.)

■ 7. $\dfrac{(1.06)^x - 1}{.06} = 12$

$$1.06^x = 1.72$$
$$x \log 1.06 = \log 1.72$$
$$x = \frac{\log 1.72}{\log 1.06} = \frac{.23553}{.02531}$$
$$= 9.12.$$

---

9. Solve for p when

$$2 \log p = 1 + \log\left(p + \frac{11}{10}\right).$$

■ 8. $M = 2.5 \log A - \log a^{2.5}$

$$M = 2.5(\log A - \log a)$$
$$\frac{M}{2.5} = \log \frac{A}{a}$$
$$\text{antilog}\left(\frac{M}{2.5}\right) = \frac{A}{a}$$
$$a = \frac{A}{\text{antilog}\left(\frac{M}{2.5}\right)}.$$

Other correct forms are:

1. $a = 10^{\log A - M/2.5}$;
2. $a = A(10)^{-M/2.5}$.

■ 9. $2 \log p = 1 + \log\left(p + \frac{11}{10}\right)$

$$\log p^2 - \log\left(p + \frac{11}{10}\right) = 1$$
$$\log\left(\frac{p^2}{p + \frac{11}{10}}\right) = 1$$
$$\frac{p^2}{p + \frac{11}{10}} = 10$$

$$p^2 - 10p - 11 = 0$$
$$(p + 1)(p - 11) = 0$$
$$p = 11$$
only.

(Logarithms of negative values are undefined.)

5-7 Problems

1. In the formula $A = P(1 + r)^t$,

$$P = \text{(principal)},$$
$$r = \text{(annual rate of interest)},$$
$$t = \text{(years)},$$
$$A = \text{(accumulated interest and principal)}.$$

How many years are required for a principal investment of \$500 to grow to \$1000 at an annual rate of 6.5 percent compounded annually?

2. The formula in Problem 1 is modified slightly when the interest is compounded quarterly rather than annually. It becomes

$$A = P\left(1 + \frac{r}{4}\right)^{4t}.$$

Repeat Problem 1 and find t when the interest is compounded quarterly.

3. The number of bacteria in a certain culture increases according to the equation

$$N = N_0\, e^{.008t},$$

where N_0 is the number at the beginning and N is the number after time t. Of course e is the base of the natural logarithms. Compute the time for the bacteria to double.

4. The pressure of the atmosphere in pounds per square inch at a height of x feet above sea level is given by the formula

$$P = 14.72e^{-kx},$$

where e is the base of the natural logarithms and k is constant. Observations at 1122 feet give $P = 14.11$. What is the value of k?

5. Open parallel transmission lines have a characteristic impedance, Z_0, which is determined by the formula

$$Z_0 = 276 \log \frac{a}{b}.$$

If Z_0 is to equal 300 ohms, find the distance (a) that must separate the lines when $b = .040$ inches.

6. A certain hydrocarbon in solution decomposes into other substances at a rate proportional to the amount still unchanged, according to the formula

$$x = x_0 e^{kt},$$

where x_0 is the amount present at $t = 0$. If 30 pounds are reduced to 10 pounds in 4 hours, when will 95 percent of the hydrocarbon be decomposed?

7. The current (i) emitted from a heated tungsten filament is given by the equation

$$i = 60T^2 e^{B/T}.$$

When $T = 2000$, $i = .093$. Find i when $T = 3000$. (*Hint: B* is constant and must be determined first from the initial data.)

8. In chemistry, pH is a measure of hydrogen ion concentration (H^+). Determine the concentration (H^+) when the pH $= 7$, if the relationship is

$$\text{pH} = \log \frac{1}{(\text{H}^+)}.$$

9. Show that if $\ln i = -\frac{Rt}{L} + \ln \frac{E}{R}$, then $i = \frac{E}{R} e^{-Rt/L}$.

10. Solve the equation

$$i = \frac{E}{R}(1 - e^{-Rt/L})$$

for t.

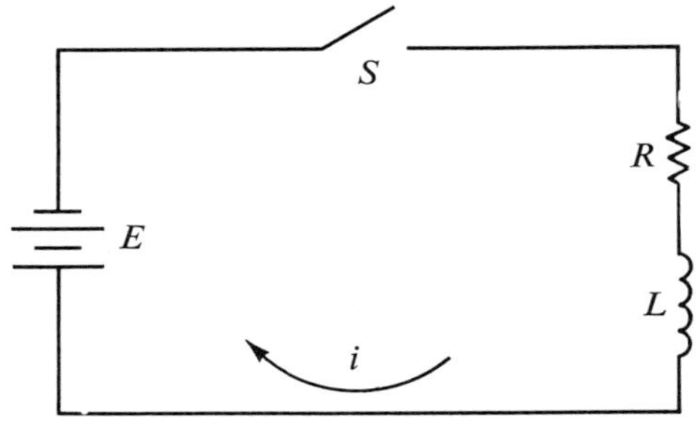

5 Supplementary Exercises and Problems

1. Why isn't $b = 1$ or $b \leq 0$ a suitable base for a system of logarithms? (5-1)

Write the following in either exponential or logarithmic form and determine the value of the variable by inspection. (5-1)

2. $x = \log_2 8$.
3. $3 = \log_b 10^3$.
4. $10^0 = y$.
5. $9^x = 27$.
6. $\log_5 25 = x$.
7. $0.5 = \log_{16} y$.
8. $\log_2\left(\frac{1}{8}\right) = x$.
9. $\log_b\left(\frac{2}{3}\right) = \frac{1}{2}$.
10. $b^{.5} = 1.2$.

Given $\log_b 5 = .2330$ and $\log_b 7 = .2617$, find the following numbers. (5-2)

11. $\log_b 35$
12. $\log_b\left(\frac{7}{5}\right)$
13. $\log_b\left(\frac{1}{5}\right)$
14. $\log_b\left(\frac{5}{7}\right)$
15. $\log_b\left(\frac{1}{7}\right)$
16. $\log_b 49$
17. $\log_b 125$
18. $\log_b \sqrt{1.4}$
19. $\dfrac{\log_b 5}{\log_b 7}$
20. $\log_b (25b)$

21. $\log_b y^3 + \log_b y^2 = \boxed{} \log_b y.$ (5-2)
22. $\log_b y^2 - \log_b \sqrt{y^3} = \boxed{} \log_b y.$ (5-2)
23. $\boxed{}(\log_b \sqrt{y} - \log_b \sqrt[3]{y}) = \log_b y.$ (5-2)
24. Express P as a function of Q when (5-2)

$$\frac{1}{3}\log_3 P + 2\log_3 Q = 1.$$

If $f(x) = \log_{10} x$, evaluate the following using a table of common logarithms. (5-3)

25. $f(2.35)$
26. $f(.00438)$
27. $f(5/9)$
28. $f(87{,}400)$
29. $f(\pi)$
30. $f(.001)$

If $h(x) = \text{antilog}_{10}\, x$, evaluate the following using a table of common logarithms. (5-3)

31. $h(.1248)$
32. $h(2.3145)$
33. $h(.2341 - 2)$
34. $h(.7643 - 4)$
35. $h(-.2654)$
36. $h(-1.3654)$

37. Compute $10^{1.6541}$. (5-3)
38. Compute $10^{.3842-2}$. (5-3)

Use logarithms to make the following computations. (5-4)

39. $\dfrac{4810}{34.6}$
40. $\dfrac{1.46}{.00287}$
41. $(2.46)^{1.2}$
42. $(.0655)^{0.3}$
43. $(25.5)^{-.6}$
44. $(1.08)^{20}$
45. $\left(\dfrac{1}{1.125}\right)^{15}$
46. $\sqrt[5]{10.21}$
47. $6.28(1.50)^{.4}$
48. $(2.6)^{.2} - (2.6)^{-.2}$

49. Air pressure loss in a pipe due to friction is calculated using the formula

$$f = \frac{CLQ^2}{rd^5}.$$

Calculate f with the aid of logarithms when

$$L = 25 \text{ (feet)}, \qquad r = \frac{114.7}{14.7},$$

$$Q = .07 \text{ (cubic feet per second)}, \quad d = \frac{5}{8} \text{ (inches)},$$

$$C = \frac{.1025}{d^{.3}}.$$ (5-4)

50. The formula for fixed percentage depreciation is

$$L_k = P\left(\frac{L}{P}\right)^{k/n}.$$

Find the salvage value L_k at $k = 8$ years, given the initial cost $P = \$8{,}525$ and scrap value $L = \$500$ after $n = 10$ years. (5-4)

Use a table of natural logarithms for the following exercises. (5-5)

51. ln 8.63
52. ln 28.4
53. antiln .3540
54. antiln 2.743
55. ln .00243
56. ln 49800
57. antiln (−2.412)
58. antiln 9.876
59. ln $(\pi^{4.55})$
60. ln $\left[\dfrac{1}{(3.12)^{1.8}}\right]$
61. $\ln\sqrt{x} - \ln x^2 = \boxed{}\ \ln x.$ (5-5)
62. $\dfrac{\ln 3.21}{\ln(.5)^{.3}}$ (5-5)
63. Compute $e^{-4.21}$. (5-5)
64. Compute $e^{.1325}$. (5-5)
65. Make a table of logarithms with a base of 12 from 1 to 12 in steps of one. (Use the change-of-base formulas.) (5-6)
66. Use linear interpolation to find log 8.643. (5-6)
67. Use linear interpolation to find ln 4.386. (5-6)
68. $\text{Log}_{12} 2.97 = .43807$ and $\log_{12} 3.46 = .49952$. Find by linear interpolation $\log_{12} 3.20$. Compare the result to the value obtained by using the change-of-base formula. (5-6)
69. Find $\log_\pi 10$. (5-6)
70. Find ln π using common logarithms. (5-6)
71. Find log π using natural logarithms. (5-6)
72. Compute $\dfrac{1}{\ln 10}$. (5-6)
73. Compute $\dfrac{1}{\log e}$. (5-6)
74. Solve $2^t = 7$ for t. (5-7)
75. Solve $3^{w+1} = 2^w$ for w. (5-7)

76. Solve $2 \log y - \log 10y = 0$ for y. (5-7)

77. Solve $i = \frac{E}{R} e^{-t/RC}$ for t. (5-7)

78. Solve $\log (x + 1) - \log x = 1.4$ for x. (5-7)

79. Solve $6^{\log x} = 6x$ for x. (5-7)

80. If $\frac{I}{I_0} = 80\%$ when $x = 0.5$ centimeters, find x when

$$I = I_0 e^{-kx} \text{ and } \frac{I}{I_0} = 50\%.$$ (5-7)

Chapter Six Angles, Triangles, and Trigonometric Ratios

6-1 Angles and the Tangent of Angles

Trigonometry, like many branches of mathematics, has a very long history. Significant elements of trigonometry existed more than three thousand years ago, and aspects of analytical trigonometry appeared in the sixteenth century very much as they do today.

In the beginning the central purpose of trigonometry was to determine unknown parts of triangles through calculation rather than direct measurement. Today this is still one of the chief applications of trigonometry in technical fields. The sides and angles of triangles play an important role in technical measurements.

Let us first devote some attention to the concept of angles. An angle can be thought of as the rotation of a half-line about its endpoint, from its initial position to its terminal position. The point about which it rotates is called the

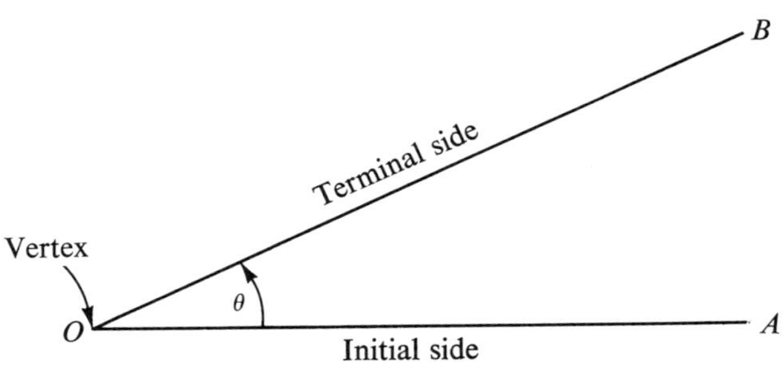

Fig. 6-1a

vertex of the angle. Greek letters are frequently used to name angles. In Fig. 6-1a, half-line OA is the initial side, half-line OB the terminal side, and the name of the angle is θ (theta).

If the angle is less than one-fourth of a complete revolution, it is an **acute angle**. One-fourth of a complete revolution forms a **right angle**, and a revolution that describes more than a right angle but less than two right angles is an **obtuse angle**.

Angles can be measured in *degrees*, *minutes*, and *seconds*, or in *degrees* and *fractions of degrees*. A right angle has a measure of ninety degrees. There are sixty minutes in a degree and sixty seconds in a minute.

Example 1

Change 84 degrees, 30 minutes, and 15 seconds (84° 30′ 15″) to degrees.

$$\begin{aligned} 84^\circ\, 30'\, 15'' &= 84 + \frac{30}{60} + \frac{15}{3600} \\ &= 84 + .5 + .004166\ldots \\ &= 84.504166\ldots^\circ \end{aligned}$$

Example 2

Change 112.258° to degrees, minutes, and seconds.

$$\begin{aligned} 112.258^\circ &= 112^\circ + 60(.258)' \\ &= 112^\circ + 15.48' \\ &= 112^\circ + 15' + 60(.48)'' \\ &= 112^\circ\, 15'\, 28.8''. \end{aligned}$$

If the rotation is counterclockwise, the angle is *positive*. A clockwise rotation produces a *negative* angle. A counterclockwise rotation of 250 degrees produces an angle whose terminal side is in the same position as a -110-degree angle.

Let us now superimpose an angle on the rectangular coordinate system. If the vertex of the angle is at the origin and the initial side is along the positive x-axis, the angle is said to be in **standard position**. The distance from any point P on the terminal side of the angle to the origin is called the **radius vector** and is designated by r. (See Fig. 6-1b.)

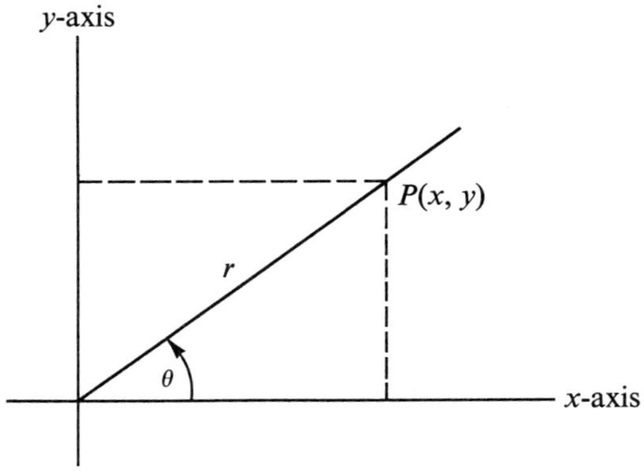

Fig. 6-1b

We may consider the terminal side of the angle to be the graph of an equation. The equation of this line is $y = mx$, where m is the slope of the line.

$$mx = y \qquad \text{(equation of terminal side)}$$

$$m = \frac{y}{x} \qquad \text{(dividing each member by } x \neq 0\text{)}$$

Notice that the coordinates of every point on the line give the same value when the ordinate is divided by the abscissa.

Example 3

Let $y = .5x$ be the equation of the terminal side of angle θ (Fig. 6-1c). $P_1(2,1)$, $P_2(4,2)$, and $P_3(6.5,3.25)$ are points on the graph of $y = .5x$.

$$m = \frac{y}{x} = \frac{1}{2} = \frac{2}{4} = \frac{3.25}{6.5} = .5.$$

No matter what point on the line is selected, the value of y/x is always .5.

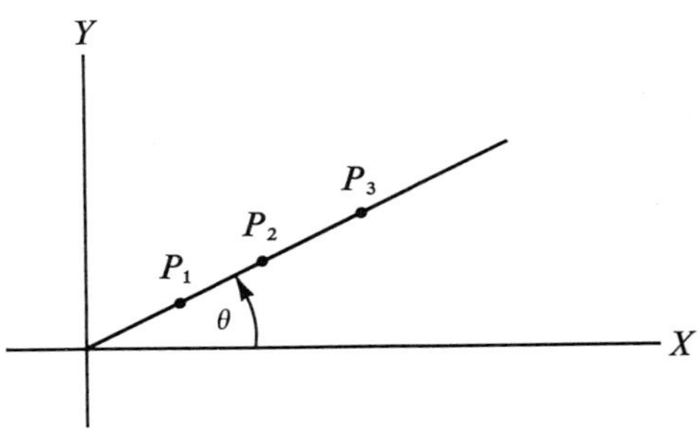

Fig. 6-1c

Now instead of being termed the slope of the line, the value of m is named in relation to the standard-position angle θ. It is called the **tangent** of the angle θ and is written

$$\tan \theta = \frac{y}{x}. \qquad \text{(Eq. 6-1a)}$$

The slope of the terminal side of a standard-position angle is numerically the same as the tangent of the angle.

Example 4

Suppose the terminal side of a standard-position angle passes through the point (2,2). Find θ and $\tan \theta$.

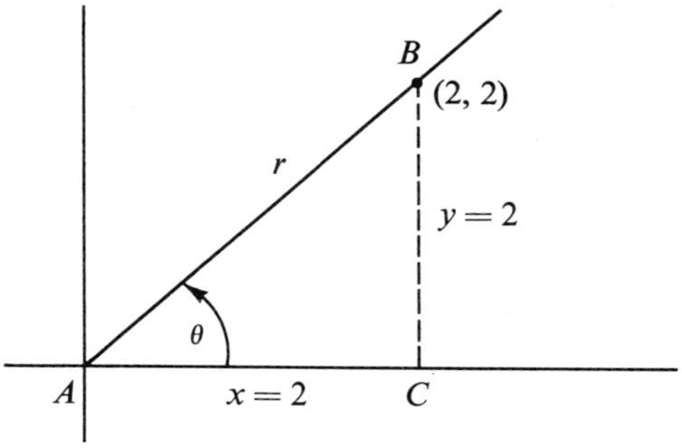

Fig. 6-1d

i. If $x = 2$ and $y = 2$, then $\tan \theta = y/x = 2/2 = 1$.
ii. To find θ, we must examine the geometry of the right triangle ABC formed by including, in Fig. 6-1d, the line segment $\overline{BC}$. Since angle ACB is 90 degrees and the triangle is isosceles, θ must be ☐ degrees, and tan ☐ = 1.

■ 45; 45°.

If for the next example we should choose the terminal side of a standard-position angle to pass through (3,8), we would have no difficulty calculating $\tan\theta$, but we would have considerable difficulty determining the value of θ. Very few angles can be determined by the methods used in Example 4. Fair approximations of θ may result from accurate drawings and direct measurement. Analytic methods, found in more advanced texts, are actually used to evaluate θ and $\tan\theta$ to any degree of accuracy desired. The results appear in such tables as Table 4 in the Appendix. With each angle θ, there is only one value for $\tan\theta$.

Example 5

If $\theta = 65°$, then from a table of tangents we read that $\tan 65° = 2.1445$. This is to say, for every point $P(x, y)$ on the terminal side of a 65° angle, $y/x = 2.1445$.

Example 6

If $x = 2$ and $y = 1.5$, then $\tan\theta = 1.5/2 = .75$. From a table of tangents, we find that θ is about 36.8°.

A table of tangents is important in solving problems that reduce to the general equation $\tan\theta = y/x$, where any one of the variables θ, y, or x is unknown.

Example 7

If $\theta = 30°$ and $x = 4.00$, find y.

$$\tan\theta = \frac{y}{x} \quad \text{or} \quad y = x\tan\theta$$

$$= 4\tan 30°$$

$$= 4(.5774)$$

$$= 2.31.$$

Example 8

A Coast Guard station determines how far from shore small craft go by measuring the angle of depression while sighting a ship from a known coastal elevation. If the observation sight is 121.3 feet above sea level, how far out to sea is a craft when the angle of depression is 3°?

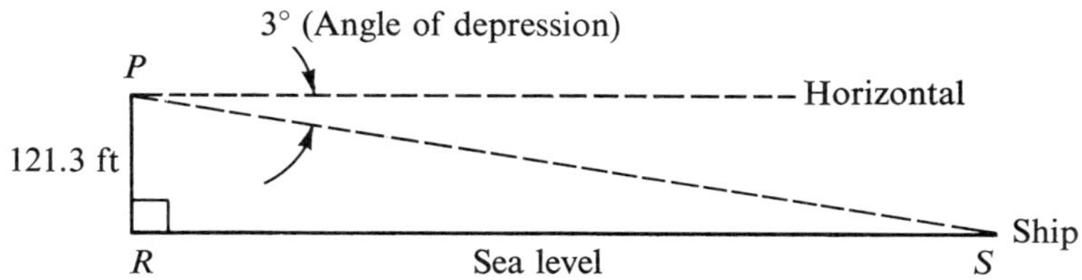

i. Let the complement of 3° be the standard-position angle and let $\overline{PR}$ be the initial side and $\overline{PS}$ the terminal side.

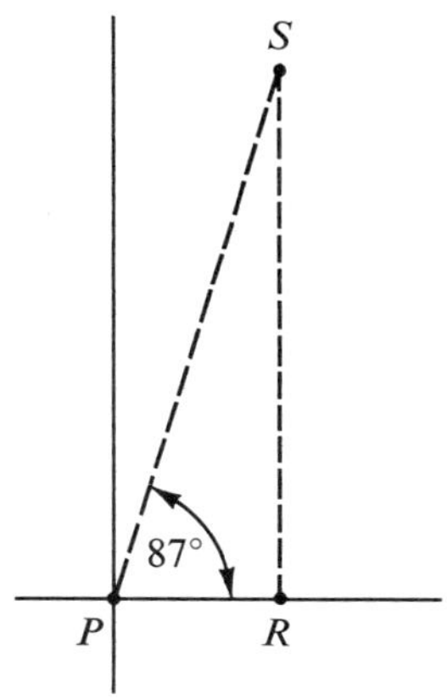

ii. $\tan 87° = \dfrac{\overline{SR}}{121.3}$

$$\begin{aligned}\overline{SR} &= 121.3 \tan 87° \\ &= 121.3(19.081) \\ &= 2314 \text{ feet} \approx 0.44 \text{ miles.}\end{aligned}$$

6-1 Exercises

1. Change 27.5° to degrees, minutes, and seconds.

---

2. Change 113.22° to degrees, minutes, and seconds.

■ 1. $27.5° = 27° + .5(60')$
$= 27°\ 30'.$

---

3. Change 318.828° to degrees, minutes, and seconds.

■ 2. $$\begin{aligned} 113.22^\circ &= 113^\circ + .22(60') \\ &= 113^\circ + 13.2' \\ &= 113^\circ + 13' \\ &\quad + .2(60'') \\ &= 113^\circ\, 13'\, 12''. \end{aligned}$$

---

4. Change 37° 47′ 20″ to degrees.

■ 3. $$\begin{aligned} 318.828^\circ &= 318^\circ + .828(60') \\ &= 318^\circ + 49.68' \\ &= 318^\circ + 49' \\ &\quad + .68(60'') \\ &= 318^\circ\, 49'\, 40.8''. \end{aligned}$$

---

5. Change 111° 15′ 35″ to degrees.

■ 4. $37^\circ\, 47'\, 20''$
$$\begin{aligned} &= \left(37 + \frac{47}{60} + \frac{20}{3600}\right)^\circ \\ &= (37 + .7833 + .0055)^\circ \\ &= 37.7888^\circ. \end{aligned}$$

---

6. Change 245° 37′ 18″ to degrees.

■ 5. $111^\circ\, 15'\, 25''$
$$\begin{aligned} &= \left(111 + \frac{15}{60} + \frac{35}{3600}\right)^\circ \\ &= (111 + .25 + .0097)^\circ \\ &= 111.2597^\circ. \end{aligned}$$

---

7. Use a table to find
   a. tan 15°,
   b. tan 32°,
   c. tan 0°.

■ 6. $245^\circ\, 37'\, 18''$
$$\begin{aligned} &= \left(245 + \frac{37}{60} + \frac{18}{3600}\right)^\circ \\ &= (245 + .6166 + .0050)^\circ \\ &= 245.6216^\circ. \end{aligned}$$

---

8. Use a table to find
   a. tan 85°,
   b. tan 20.8°,
   c. tan 68.3°.

■ 7. a. tan 15° = .2679;
   b. tan 32° = .6249;
   c. tan 0° = 0.

---

9. Use a table to find θ, if

a. $\tan \theta = .0349$,
b. $\tan \theta = .2773$,
c. $\tan \theta = .9195$.

■ 8. a. $\tan 85° = 11.430$;
b. $\tan 20.8° = .3799$;
c. $\tan 68.3° = 2.513$.

---

10. Use a table to find θ, if

a. $\tan \theta = 1.0875$,
b. $\tan \theta = 31.82$,
c. $\tan \theta = .7054$.

■ 9. a. $\tan 2.0° \doteq .0349$;
b. $\tan 15.5° = .2773$;
c. $\tan 42.6° = .9195$.

---

11. Find θ.

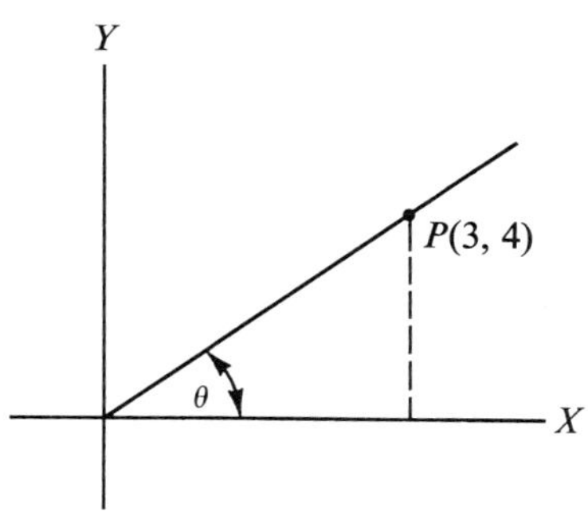

■ 10. a. $\tan 47.4° = 1.0875$;
b. $\tan 88.2° = 31.82$;
c. $\tan 35.2° = .7054$.

---

12. Find y.

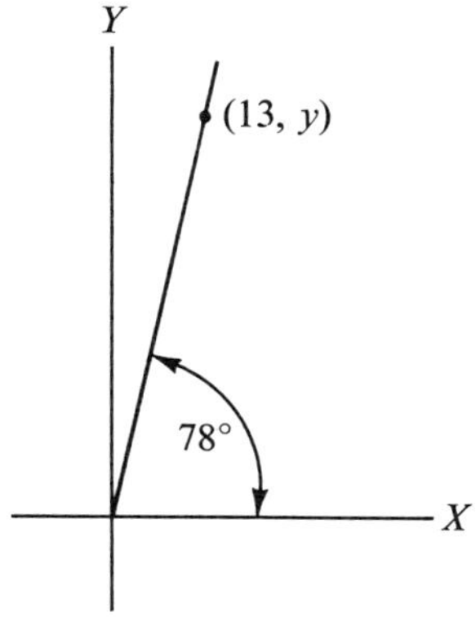

■ 11. $$\tan \theta = \frac{4}{3} = 1.3333$$
$$\theta = 53.1°.$$

13. Find x.

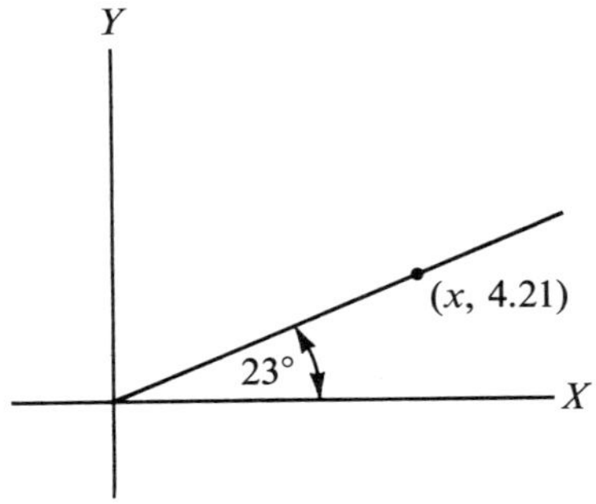

■12. $\tan 78° = \frac{y}{13}$.

$$y = 13 \tan 78° = 13(4.705) = 61.2.$$

■13. $\tan 23° = \frac{4.21}{x}$.

$$x = \frac{4.21}{\tan 23°} = \frac{4.21}{.4245} = 9.92.$$

6-1 Problems

1. What angle does the line $y = 3x$ make with the x-axis?
2. If from a point 120 feet from the base of a tree, the angle of elevation of the top of the tree is 37°, how tall is the tree?
3. The side of an equilateral triangle is 18 feet. How long is its altitude?
4. The span of a roof is 28 feet, and its rise at the center of the span is 10 feet. What angle does the roof make with the horizontal?
5. A roadway rises 28 feet in a horizontal distance of 1000 feet. What angle does the road make with the horizontal?
6. When an airplane flying at 30,000 feet is directly over town C, the angle of depression of lake B is 11° 20′. How far is B from C?

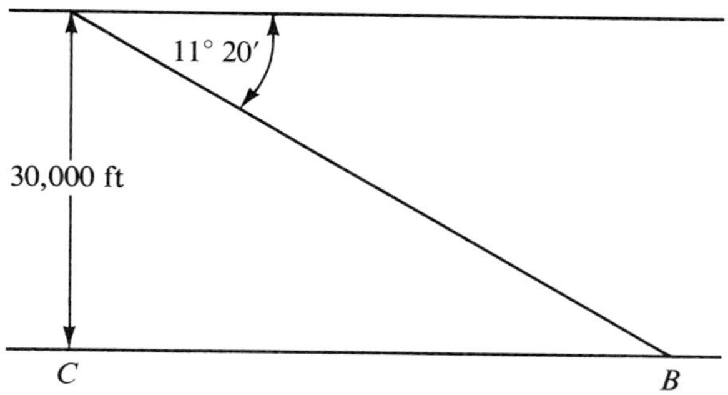

7. Concrete steps are to be designed for an embankment with a 22° slope. What is the ratio of the riser to tread lengths for every step? If the riser is 8 inches, what is the tread length?

6-2 Sines, Cosines, and Right Triangles

We saw in the last section that, if the coordinates of a point on the terminal side of a standard-position angle are known, then the angle can be determined by using the relationship $\tan \theta = y/x$.

Consider the information given in Fig. 6-2a. If the terminal side of the angle is a line segment of specific length and at a specific angle, then P is a

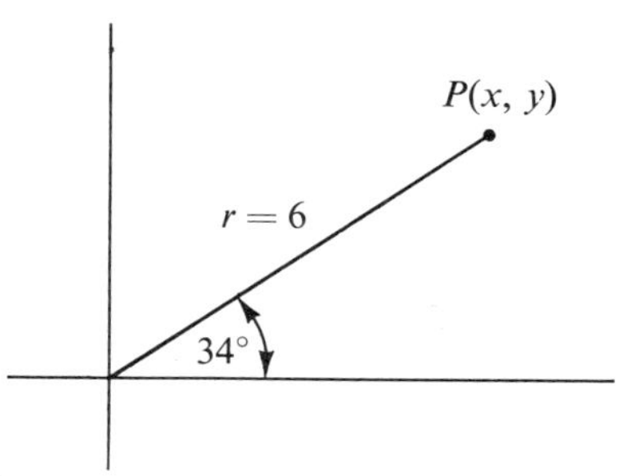

Fig. 6-2a

point with specific coordinates. Can they be determined by calculation? The answer is yes, but the calculations must be based on the use of a trigonometric ratio other than the tangent.

Just as the values of y/x are named and tabulated in terms of θ, so are the values of x/r and y/r. (See Table 4, Appendix.)

$$\text{sine } \theta = \frac{y}{r} = \frac{\text{ordinate}}{\text{radius vector}}. \qquad \text{(Eq. 6-2a)}$$

$$\text{cosine } \theta = \frac{x}{r} = \frac{\text{abscissa}}{\text{radius vector}}. \qquad \text{(Eq. 6-2b)}$$

The common abbreviations of sine θ and cosine θ are sin θ and cos θ.

To calculate the ordinate of P in Fig. 6-2a, we write

$$\sin 34° = \frac{y}{6}$$
$$y = 6 \sin 34°$$
$$= 6(.5592) = 3.36.$$

The abscissa can now be calculated by the relationship $r^2 = x^2 + y^2$ (Pythagorean theorem), $\tan \theta = y/x$, or by $\cos \theta = x/r$. Using the last, we find

$$\cos 34° = \frac{x}{6}$$
$$x = 6 \cos 34°$$
$$= 6(\boxed{}) = 4.97.$$

■ .8290

Since only three values x, y, and r make up the three ratios sine θ, cosine θ, and tangent θ, we should look for a simple arithmetic relationship between them. Notice that

$$\frac{y}{x} = \frac{y/r}{x/r}, \quad \text{or} \quad \tan \theta = \boxed{\phantom{\frac{\sin\theta}{\cos\theta}}}.$$

■ $\tan \theta = \dfrac{\sin \theta}{\cos \theta}$. (Eq. 6-2c)

Another useful but perhaps less obvious relationship exists to account for the peculiar arrangement of the trigonometric tables. The tables show, for example, that $\sin 45° = \cos 45°$, $\sin 40° = \cos 50°$, $\sin 31° = \cos 59°$, or $\sin 22.8° = \cos 67.2°$. This relationship exists so long as the two angles have a sum of 90°. Figure 6-2b may help provide an explanation. The standard-position angles θ and $90° - \theta$ are complementary.

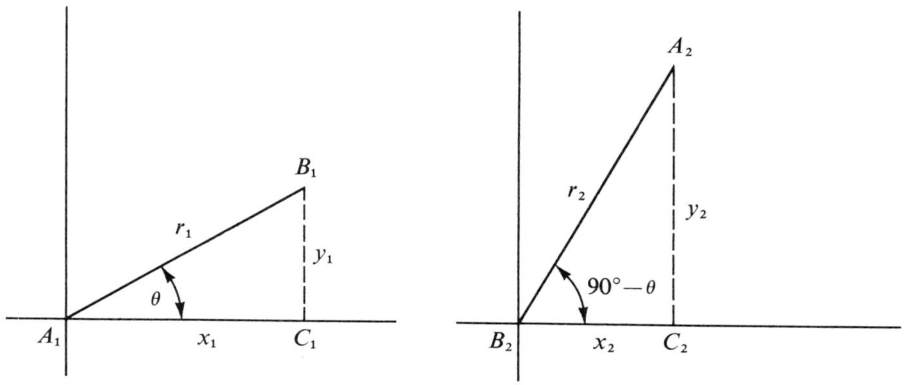

Fig. 6-2b

Drawing perpendiculars $\overline{B_1C_1}$ and $\overline{A_2C_2}$ produces two similar triangles, $A_1B_1C_1$ and $A_2B_2C_2$. As a consequence, we may write

$$\frac{y_1}{r_1} = \frac{x_2}{r_2}$$

or

$$\sin \theta = \cos(90° - \theta). \qquad \text{(Eq. 6-2d)}$$

The words sine and cosine may have a little more significance when we realize that the prefix *co* comes from the word *co*mplementary, and the sine and cosine are *complementary* trigonometric functions of angles. We may therefore say that the sine and cosine of complementary angles are equal. Knowing this, we should begin to understand why tables of sines and cosines can be abbreviated and also why slide rules need only contain sine scales rather than both sine and cosine scales.

Equations 6-2c and 6-2d are true for all values of θ and therefore are identities. They are used in trigonometric problems to such an extent that they are worth memorizing.

Our study of trigonometry so far has brought us to the point that, given any two of the values $\theta(0° < \theta < 90°)$, r, x, or y, we can calculate the others by using the trigonometric ratios $\sin\theta$, $\cos\theta$, or $\tan\theta$. In the context of *right* triangles, if we are given either the acute angle A or B and one side, or two sides, then the other values may be calculated. (See Fig. 6-2c.)

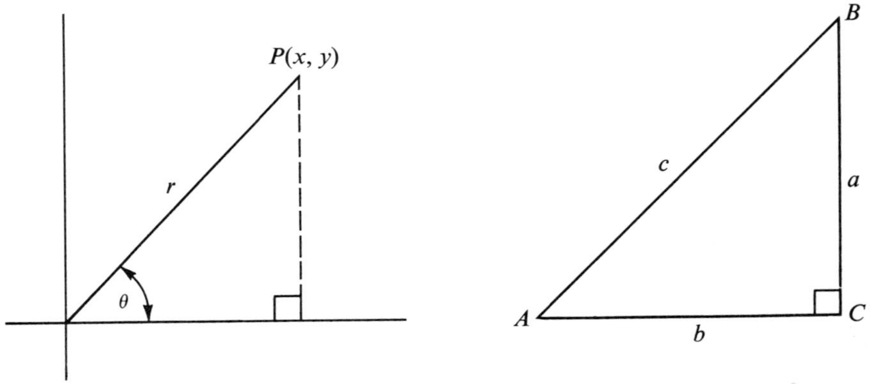

Fig. 6-2c

Example 1

If $A = 20°$ and $b = 12.0$, find B, a, and c.

i. Since $A + B + C = 180°$,

$B = \boxed{\qquad}$.

ii. $\tan 20° = \dfrac{a}{12}$

$a = 12 \tan 20°$

$= \boxed{\qquad}$.

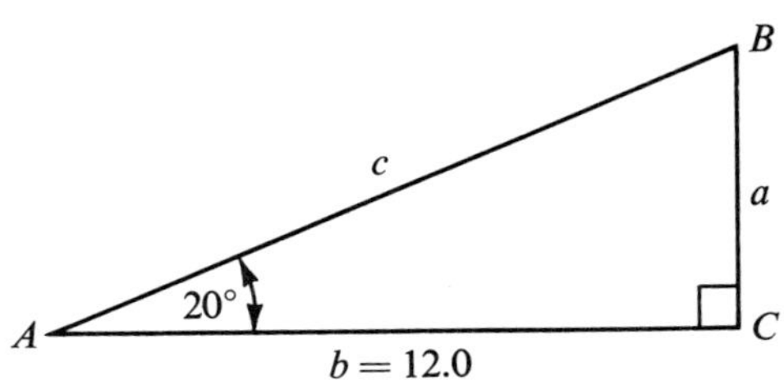

iii. $\cos 20° = \dfrac{12}{c}$

$c = \dfrac{12}{\cos 20°}$

$= \boxed{\qquad}$.

■ $B = 70°$; $a = 4.37$; $c = 12.8$.

Example 2

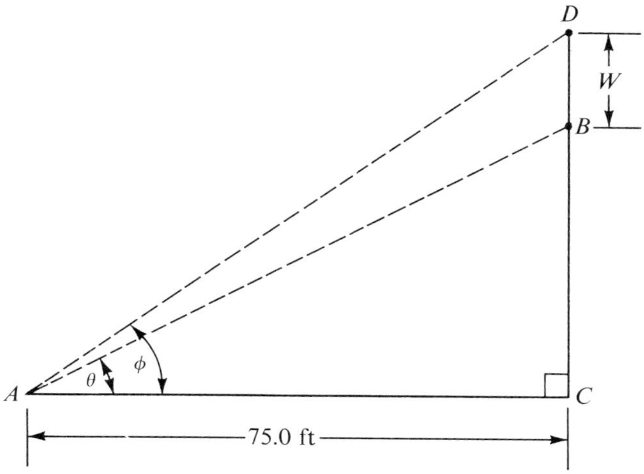

If $\angle\theta = 22.0°$ and $\angle\phi = 28.2°$, find the length W.

i. $\tan\theta = \dfrac{\overline{BC}}{\overline{AC}}$

$$\overline{BC} = \overline{AC}\tan\theta$$
$$= 75.0\,(\tan 22.0°).$$

ii. $\overline{DC} = 75.0\,(\tan 28.2°).$

iii. $W = \overline{DC} - \overline{BC} = 75.0(\tan 28.2° - \tan 22.0°)$

$$= 75.0(.5362 - .4040)$$
$$= 75.0(.1322)$$
$$= 9.92.$$

Example 3

Neither dimension B nor H of the symmetrical dovetail can be measured directly with calipers. However, calipers can be used to make the measurements of 2.20 inches and 1.50 inches shown in the figure when the two plugs of known diameter are inserted. Calculate B and H.

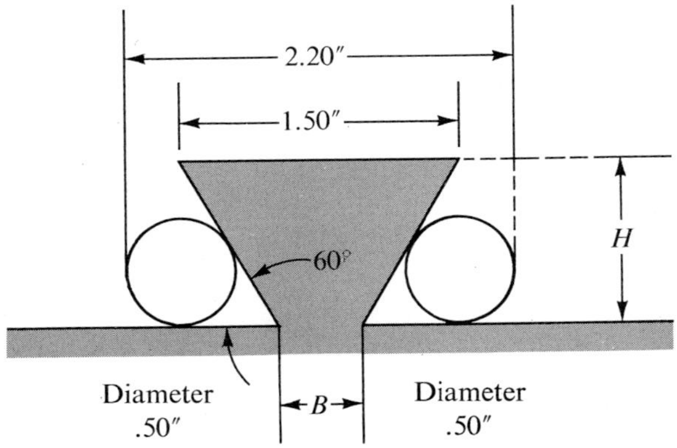

Part 1 (Solving for B)
The figure is redrawn in a general manner and auxiliary lines added to aid in the solution.

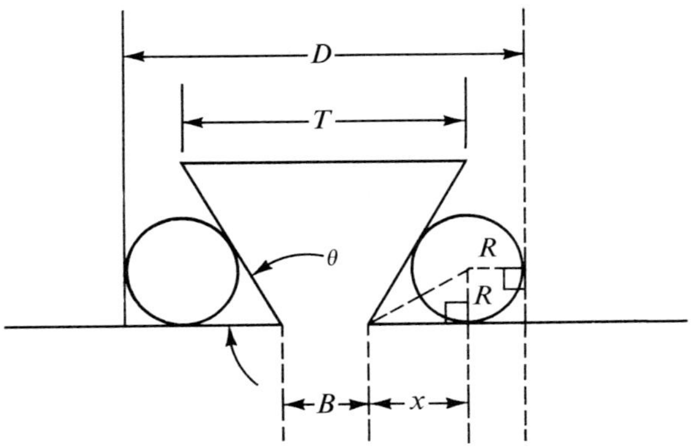

i. $\tan\left(\frac{\theta}{2}\right)=\frac{R}{x}$ or $x=\frac{R}{\tan\left(\frac{\theta}{2}\right)}$;

ii. $\frac{D}{2}-\frac{B}{2}=x+R$

$$\begin{aligned}B&=D-2x-2R\\&=D-\frac{2R}{\tan\left(\frac{\theta}{2}\right)}-2R\\&=2.20-\frac{.50}{.5774}-.50\\&=.834.\end{aligned}$$

Part 2 (Solving for H)
We use the value of B found in *Part* 1.

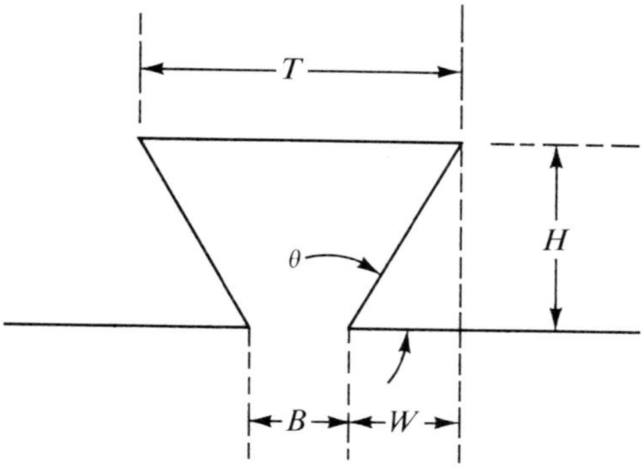

i. $W=\frac{T}{2}-\frac{B}{2}$;

ii. $\tan\theta=\frac{H}{W}$ or $H=W\tan\theta$

$$=\left(\frac{T-B}{2}\right)\tan\theta$$

$$=\left(\frac{1.50-.834}{2}\right)1.7321$$

$$=.577.$$

6-2 Exercises

1. Use tables to find
   a. sin 10°,
   b. sin 28.6°,
   c. sin 62.4°.

---

2. Use tables to find
   a. cos 27.5°,
   b. cos 85.8°,
   c. cos 0°.

■ 1. a. sin 10° = .1736;
b. sin 28.6° = .4787;
c. sin 62.4° = .8862.

---

3. Use tables to find θ, when
   a. $\sin\theta = .2924$,
   b. $\sin\theta = .5678$,
   c. $\sin\theta = .9860$.

■ 2. a. $\cos 27.5° = .8870$;
b. $\cos 85.8° = .07324$;
c. $\cos 0° = 1.0000$.

---

4. Use tables to find θ, when
   a. $\cos\theta = .3190$,
   b. $\cos\theta = .6561$,
   c. $\cos\theta = .9986$.

■ 3. a. $\theta = 17.0°$;
b. $\theta = 34.6°$;
c. $\theta = 80.4°$.

---

5. What angle has a sine equal to 2 ?

■ 4. a. $\theta = 71.4°$;
b. $\theta = 49.0°$;
c. $\theta = 3.0°$.

---

6. Given $\theta = 24°$ and $x = 12$, find y and r. (See Fig. 6-2d.)

■ 5. For all angles θ, $\sin\theta \neq 2$. The ordinate is never greater than the radius vector.

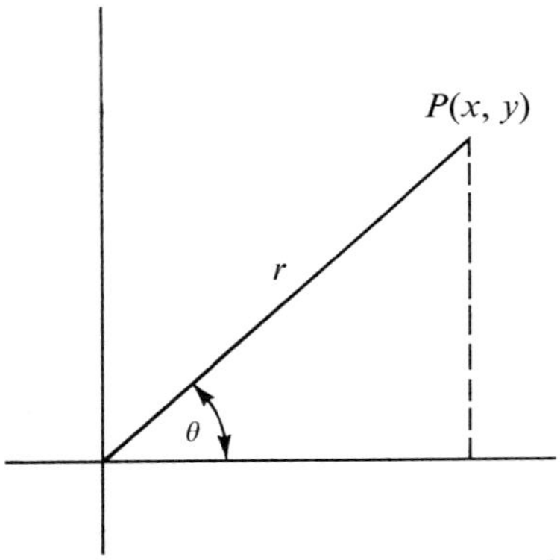

Fig. 6-2d

---

7. Given $\theta = 81°$ and $r = 7$, find x and y. (See Fig. 6-2d.)

■ 6. i. $\tan 24° = y/12$
$$y = 12(\tan 24°) = 5.34;$$
ii. $\cos 24° = 12/r$
$$r = 12/\cos 24° = 13.13.$$

---

8. Given $x = 12$ and $y = 14$, find θ and r. (See Fig. 6-2d.)

■ 7. i. $\sin 81° = y/7$
$y = 7 \sin 81°$
$= 6.91$;
ii. $\cos 81° = x/7$
$x = 7 \cos 81°$
$= 1.095$.

---

9. Given $A = 42°$ and $a = 15$, find b, c, and B. (See Fig. 6-2e.)

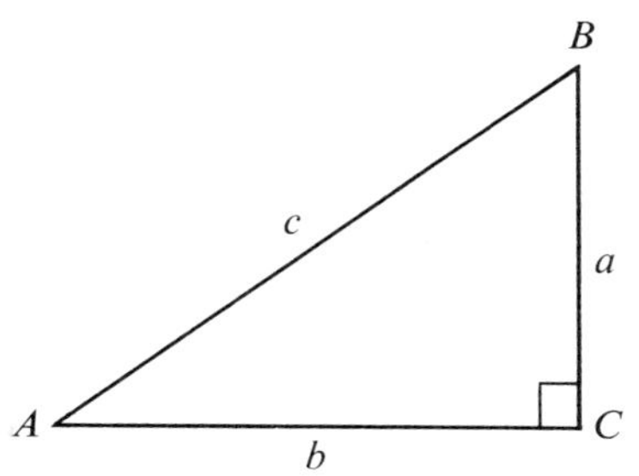

Fig. 6-2e

■ 8. i. $\tan \theta = 14/12 = 1.166$
$\theta = 49.4°$;
ii. $\sin 49.4° = 14/r$
$r = 14/\sin 49.4°$
$= 18.45$.

---

10. Given $A = 32.2°$ and $c = 14.26$, find a, b, and B. (See Fig. 6-2e.)

■ 9. i. $\tan 42° = 15/b$
$b = 15/\tan 42°$
$= 16.66$;
ii. $\sin 42° = 15/c$
$c = 15/\sin 42°$
$= 22.42$;
iii. $B = 90° - 42° = 48°$.

---

11. Given $b = 8.4$ and $c = 11.4$, find A, B, and a. (See Fig. 6-2e.)

■ 10. i. $\sin 32.2°$
$= a/14.26$
$a = 14.26 \sin 32.2°$
$= 7.599$;
ii. $\cos 32.2°$
$= b/14.26$
$b = 14.26 \cos 32.2°$
$= 12.07$;
iii. $B = 90° - 32.2°$
$= 57.8°$.

12. Find $\overline{BC}$, if $\overline{AB} = 500$ feet,
$\angle B = 90°$,
$\angle A = 51.6°$.

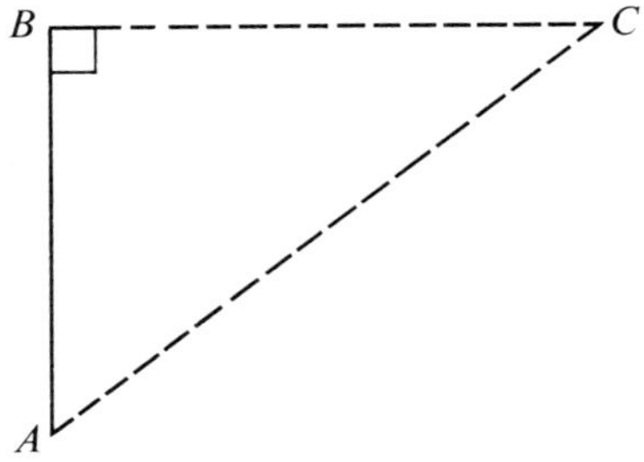

■ 11. i. $\cos A = 8.4/11.4$
$= .737$
$A = 42.5°$;
ii. $B = 90° - 42.5°$
$= 47.5°$;
iii. $\sin 42.5°$
$= a/11.4$
$a = 11.4 \sin 42.5$
$= 7.70$.

■ 12. $\tan 51.6° = \overline{BC}/500$
$\overline{BC} = 500 \tan 51.6°$
$= 630.9$ ft.

6-2 Problems

1. In an AC circuit, the resistance (R) is a quantity plotted on the x-axis and the reactance (X) is plotted on the y-axis. The impedance (Z) is numerically equal to the radius vector.
   a. Find Z and the angle θ when $R = 25$ and $X = 10$.
   b. Find X when $Z = 35$ and $R = 21$.

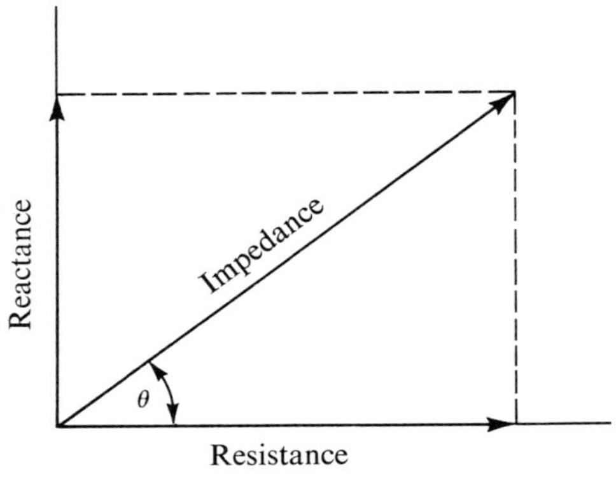

2. Find θ.

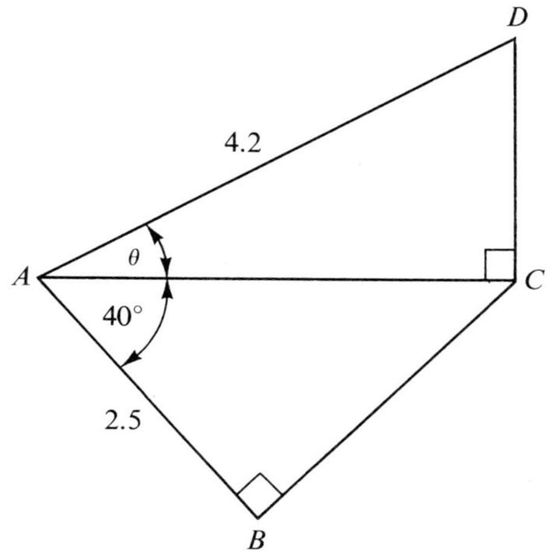

3. Find x.

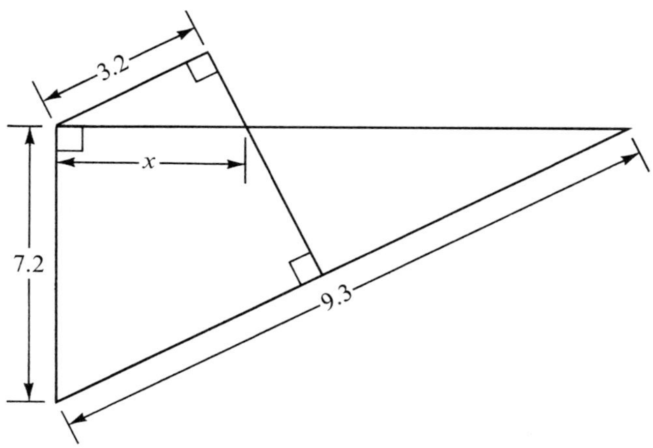

4. A sphere 11 inches in diameter is placed in a right circular cone 17 inches in diameter and 15 inches deep. Determine how far the top of the sphere is above or below the rim of the cone.

5. Find R so that the minimum distance separating the curbs is 40 feet.

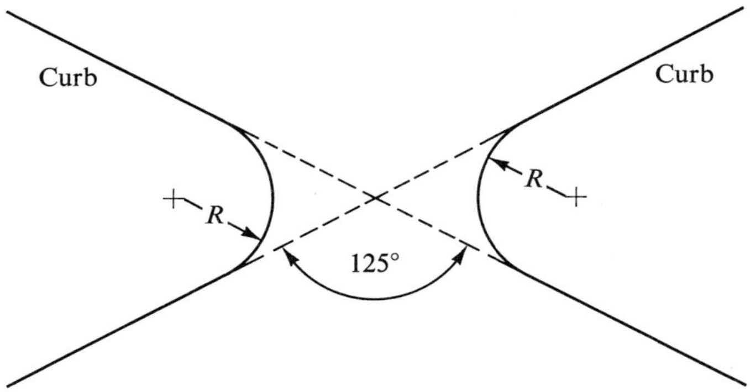

6. M can be calculated when plugs of known diameter (D) are placed as illustrated and T is measured. Angle A is also known. Write a formula for M in terms of the other variables.

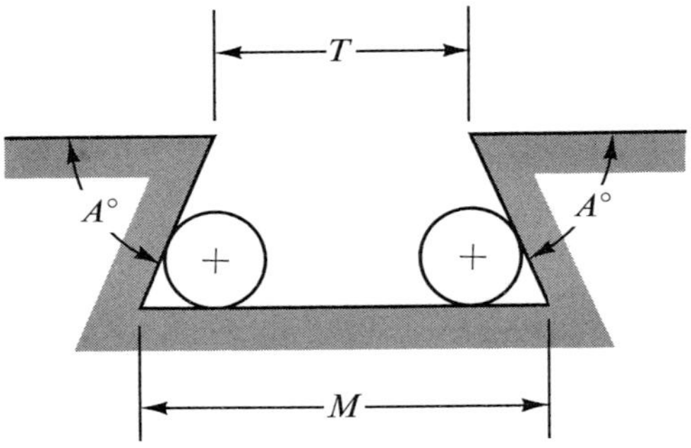

7. The diameter of the larger circle is 5 inches. Find the diameter of the smaller circle.

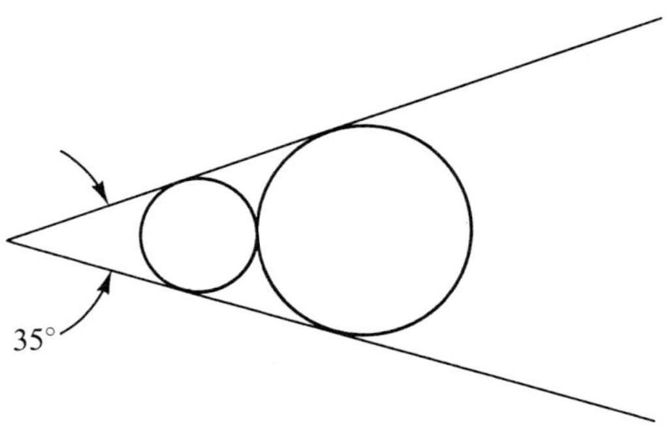

8. Find θ.

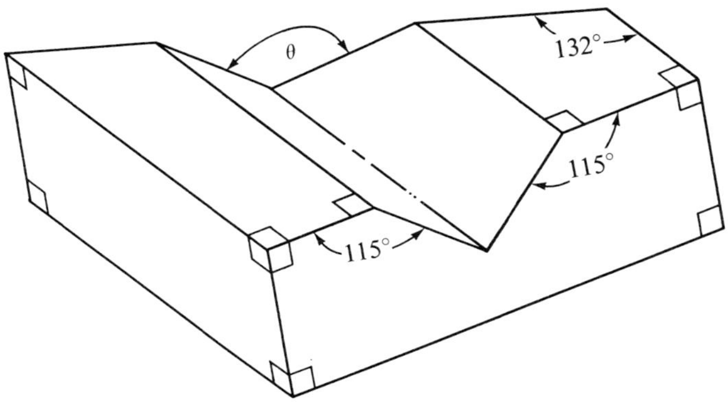

9. Given $\angle AOB = 20°$ and $\angle BOC = 30°$, find $\angle POC$.

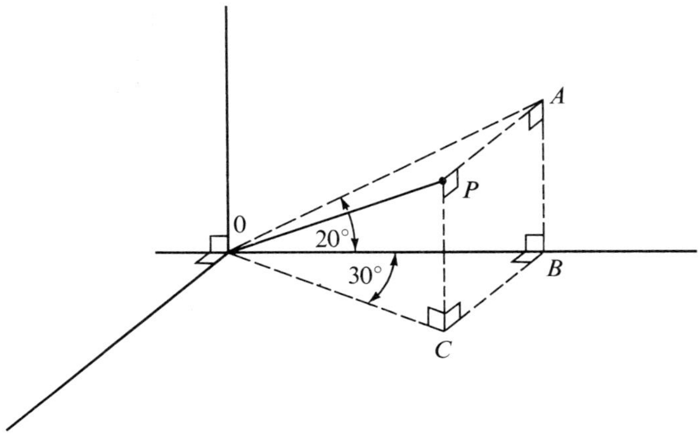

10. Find θ.

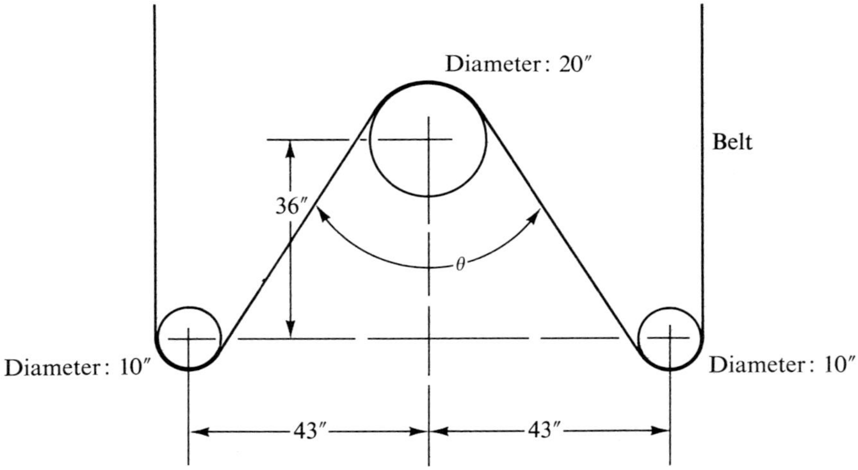

6-3 The Law of Sines—Oblique Triangles

We have used trigonometry to determine the complete solution of right triangles. Not all triangles are right triangles, of course, but the concepts of right triangles are still useful in dealing with **oblique triangles** (triangles without right angles).

First we consider an oblique triangle whose angles are all acute. Soon we shall show that this restriction on oblique triangles is unnecessary. But by approaching the problem this way, we do not encounter too many new ideas immediately.

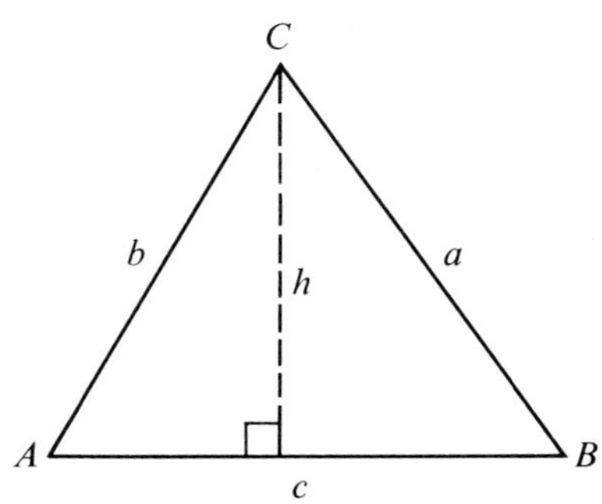

Fig. 6-3a

In Fig. 6-3a, if ABC is any acute triangle and h is the altitude from C to c, then $\sin A = h/b$ and $\sin B = h/a$. Solving each equation for h, we obtain $h = b \sin A$ and $h = a \sin B$.

Therefore, $b \sin A = a \sin B$, or rearranged algebraically,

$$\frac{b}{\sin B} = \boxed{}.$$

■ $$\frac{b}{\sin B} = \frac{a}{\sin A}. \qquad \text{(Eq. 6-3a)}$$

If instead the altitude is drawn from A to a, then $c \sin B = b \sin C$ or

$$\frac{b}{\sin B} = \frac{c}{\sin C}. \qquad \text{(Eq. 6-3b)}$$

Combining into a single expression the results of Equations 6-3a and 6-3b, we have what is called the *law of sines*,

$$\frac{a}{\sin A} = \frac{b}{\sin B} = \frac{c}{\sin C}. \qquad \text{(Eq. 6-3c)}$$

The sine law can be used in the solution of triangles when at least one of the ratios of Equation 6-3c is known in addition to one other measurement.

Example 1

Given: $a = 10$,
$b = 9$,
$A = 37°$.

Find: B, C, and c.

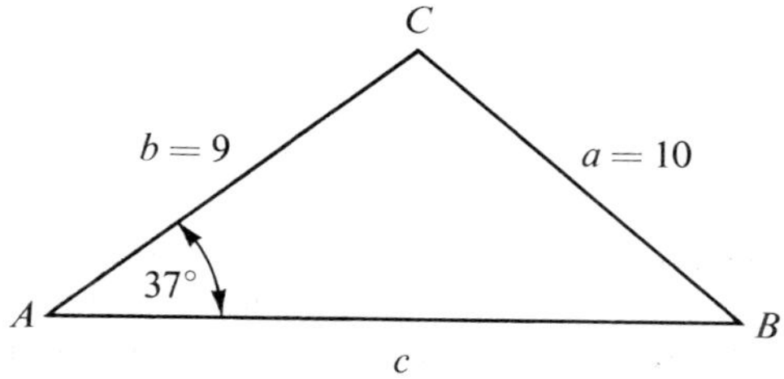

i. $\dfrac{a}{\sin A} = \dfrac{b}{\sin B}$

$$\sin B = \frac{b \sin A}{a}$$

$$= \frac{9 \sin 37°}{10} = .541$$

$$B = 32.8°;$$

ii. $C = 180° - (A + B) = 180° - 69.8° = 110.2°;$

iii. $\dfrac{a}{\sin A} = \dfrac{c}{\sin C}$

$$c = \frac{a \sin C}{\sin A} = \frac{10 \sin 110.2°}{\sin 37°}.$$

Now we have a problem. Angle C is obtuse. We shall delay the computation of Example 1 until we have considered the value of trigonometric ratios corresponding to angles greater than 90°.

Recall the definitions of $\sin\theta$, $\cos\theta$, and $\tan\theta$ in terms of x, y, and r.

$\sin\theta = \square$, $\cos\theta = \square$, $\tan\theta = \square$.

■ $\dfrac{y}{r}$; $\dfrac{x}{r}$; $\dfrac{y}{x}$.

The definitions do *not* change as θ increases or as the terminal side of the angle appears in other quadrants of the coordinate system. The reason that we give special attention to angles greater than 90 degrees is that most tables do not contain entries beyond 90 degrees. To do so would be redundant. Let us see why.

Figure 6-3b is drawn so that $P_1P_2P_3P_4$ is a rectangle whose sides are parallel to the coordinate axes and whose diagonals intersect at the origin.

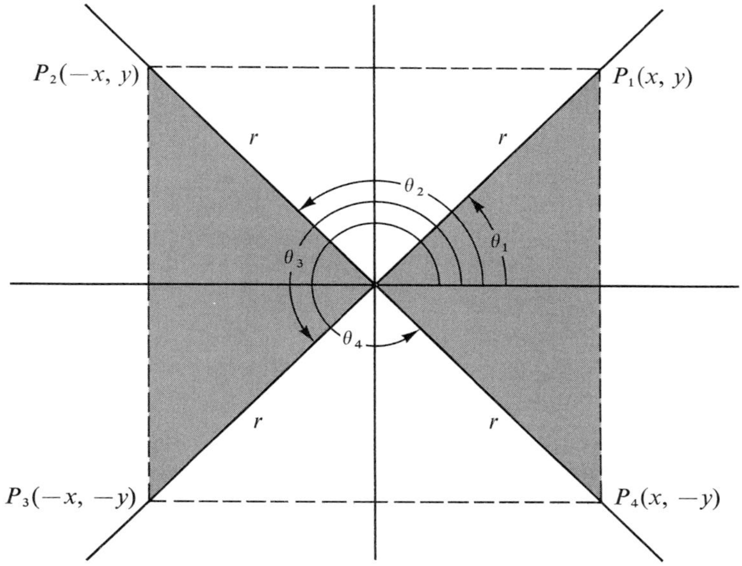

Fig. 6-3b

The geometry of the figure requires that the four shaded triangles be congruent and that $\theta_1 = 180° - \theta_2$, $\theta_1 = \theta_3 - 180°$, and $\theta_1 = 360° - \theta_4$. Consequently if

$$\sin\theta_1 = \frac{y}{r},$$

then

$$\sin\theta_2 = \frac{y}{r} = \sin\theta_1 = \sin(180° - \theta_2), \qquad \text{(Eq. 6-3d)}$$

$$\sin\theta_3 = \frac{-y}{r} = -\sin\theta_1 = -\sin(\theta_3 - 180°),$$

$$\sin\theta_4 = \frac{-y}{r} = -\sin\theta_1 = -\sin(360° - \theta_4).$$

The sines of the four angles are equal in absolute value and differ only in sign. For example,

$$\sin 30° = .5000,$$

$$\sin 150° = \sin(180° - 150°) = \boxed{},$$

$$\sin 210° = -\sin(210° - 180°) = \boxed{},$$

$$\sin 330° = -\sin(360° - 330°) = \boxed{}.$$

■ .5000; —.5000; —.5000.

Figure 6-3b can also be used in interpreting the cosine and tangent of angles greater than 90 degrees.

If

$$\cos \theta_1 = \frac{x}{r},$$

then

$$\cos \theta_2 = \boxed{} = -\cos \theta_1 = -\cos(180° - \theta_2),$$

$$\cos \theta_3 = \boxed{} = -\cos \theta_1 = -\cos(\theta_3 - 180°),$$

$$\cos \theta_4 = \boxed{} = \cos \theta_1 = \cos(360° - \theta_4).$$

■ $\frac{-x}{r}$; $\frac{-x}{r}$; $\frac{x}{r}$.

If

$$\tan \theta_1 = \frac{y}{x},$$

then

$$\tan\theta_2 = \boxed{} = -\tan\theta_1 = -\tan(180° - \theta_2),$$

$$\tan\theta_3 = \boxed{} = \tan\theta_1 = \tan(\theta_3 - 180°),$$

$$\tan\theta_4 = \boxed{} = -\tan\theta_1 = -\tan(360° - \theta_4).$$

■ $\dfrac{y}{-x}$; $\dfrac{-y}{-x}$; $\dfrac{-y}{x}$.

Now we can evaluate c in step (iii) of Example 1.

$$c = \frac{10 \sin 110.2°}{\sin 37°} = \frac{10(\sin(180° - 110.2°))}{\sin 37°}$$

$$= \frac{10 \sin 69.8°}{\sin 37°} = \frac{10(.9385)}{.6018} = 15.6.$$

To show that we are justified in using the sine law in connection with obtuse angles of triangles, consider Figure 6-3c.

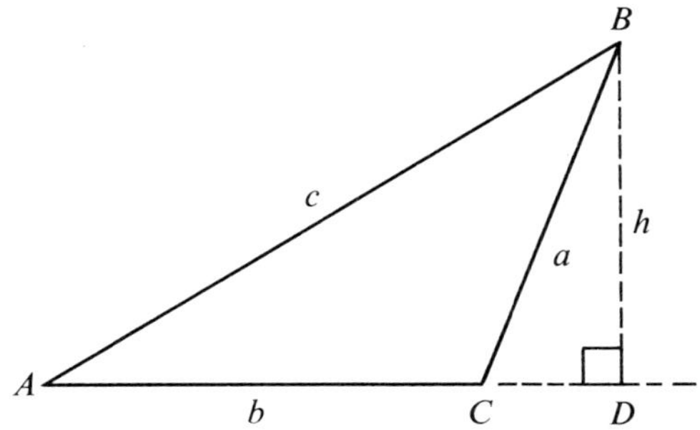

Fig. 6-3c

The altitude from B is external to the triangle, because angle C is obtuse. Now $\sin A = h/c$ and $\sin(180° - C) = h/a$. Solving both for h and equating them, we find that

$$c \sin A = a \sin(180° - C),$$

or

$$\frac{c}{\sin(180° - C)} = \frac{a}{\sin A}.$$

Since $\sin(180° - C) = \sin C$ by Equation 6-3d,

$$\frac{c}{\sin C} = \frac{a}{\sin A}.$$

One further problem occurs in connection with the use of the law of sines.

Example 2

Given: a = 7, b = 9, $A = 37°$. Find B.

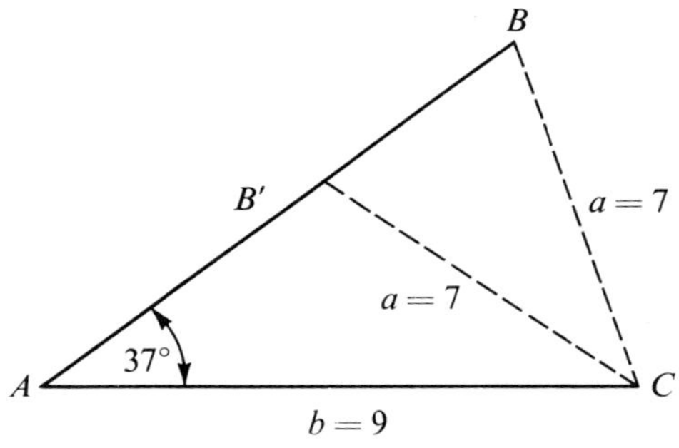

When we draw this triangle, we make a disconcerting discovery. It is not one triangle, but two! That is, two distinctly different triangles may have the given measurements. A complete solution to the problem involves finding the parts of both triangles.

Part 1 (B) $\quad \frac{a}{\sin A} = \frac{b}{\sin B}$,

$$\sin B = \frac{b \sin A}{a} = \frac{9 \sin 37°}{7} = .7737.$$

$$B = 50.7°.$$

Part 2 (B') $\quad B' = 180° - B = 129.3°.$

This problem points up the importance of making careful drawings when working with triangles. Many serious errors may be avoided simply by taking time to make a scale drawing.

6-3 Exercises

1. With the aid of tables (or slide rule) find
   a. $\sin 125°$,
   b. $\sin 337°$,
   c. $\sin 240.3°$.

---

2. With the aid of tables (or slide rule) find
   a. $\cos 321°$,
   b. $\cos 110°$,
   c. $\cos 261.5°$.

■ 1. a. $\sin 125° = \sin(180° - 125°) = \sin 55° = .8192$;
b. $\sin 337° = -\sin(360° - 337°) = -\sin 23° = -.3907$;
c. $\sin 240.3° = -\sin(240.3° - 180°) = -\sin 60.3° = -.8686$.

---

3. Find:
   a. $\tan 99°$,
   b. $\tan 192.4°$,
   c. $\tan 285.1°$.

■ 2. a. $\cos 321° = \cos(360° - 321°) = \cos 39° = .7771$;
b. $\cos 110° = -\cos(180° - 110°) = -\cos 70° = -.3420$;
c. $\cos 261.5° = -\cos(261.5° - 180°) = -\cos 81.5° = -.1478$.

---

4. Solve the triangle when $c = 8$, $a = 10$, $A = 42°$.

■ 3. a. $\tan 99° = -\tan(180° - 99°) = -\tan 81° = -6.314$;
b. $\tan 192.4° = \tan(192.4° - 180°) = \tan 12.4° = .2199$;

c. $\tan 285.1°$

$= -\tan(360° - 285.1°)$

$= -\tan 74.9°$

$= -3.706.$

---

5. Solve for b, B, and C when $c = 11.2$, $a = 8.4$, and $A = 37.3°$.

■ 4.

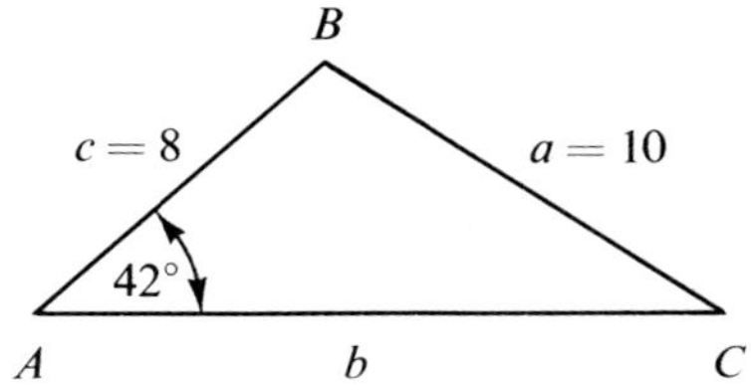

i. $$\frac{a}{\sin A} = \frac{c}{\sin C} = \frac{b}{\sin B}$$

$$\frac{10}{\sin 42°} = \frac{8}{\sin C}$$

$$\sin C = \frac{8 \sin 42°}{10}$$

$$C = 32.4°;$$

ii. $B = 180°$

$- (42° + 32.4°)$

$= 105.6°;$

iii. $$\frac{10}{\sin 42°} = \frac{b}{\sin 74.4°}$$

$$b = \frac{10 \sin 74.4°}{\sin 42°}$$

$$= 14.4.$$

6. Solve the triangle, given $a = 31.7$, $c = 27.8$, $A = 61.6°$.

■ 5.

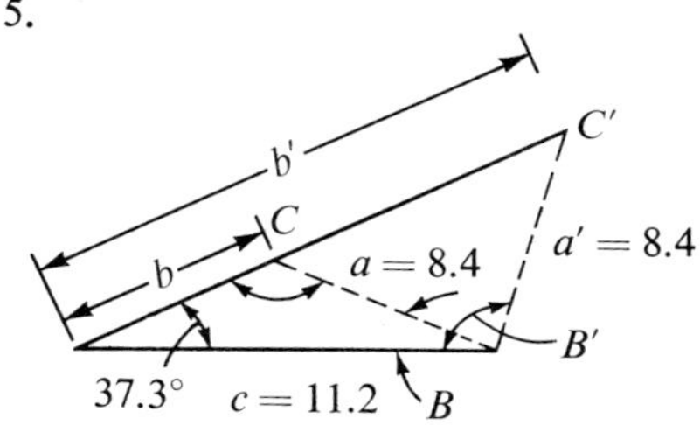

Part 1 (C)

i. $\dfrac{8.4}{\sin 37.3°} = \dfrac{11.2}{\sin C}$

$\sin C = \dfrac{11.2 \sin 37.3°}{8.4}$

$= .8080$

$C = 126.1°;$

ii. $B = 180° - (37.3° + 126.1°)$

$= 16.6°;$

iii. $\dfrac{8.4}{\sin 37.3°} = \dfrac{b}{\sin 16.6°}$

$b = \dfrac{8.4 \sin 16.6°}{\sin 37.3°}$

$= 3.96.$

Part 2 (C')

i. $C' = 180° - C$

$= 53.9°;$

ii. $B' = 180° - (37.3° + 53.9°)$

$= 88.8°;$

iii. $\dfrac{8.4}{\sin 37.3°} = \dfrac{b'}{\sin 88.8°}$

$b' = \dfrac{8.4 \sin 88.8°}{\sin 37.3°}$

$= 13.9.$

7. Solve the triangle, given $a = 6.5$, $c = 14.2$, $A = 47°$.

■ 6.

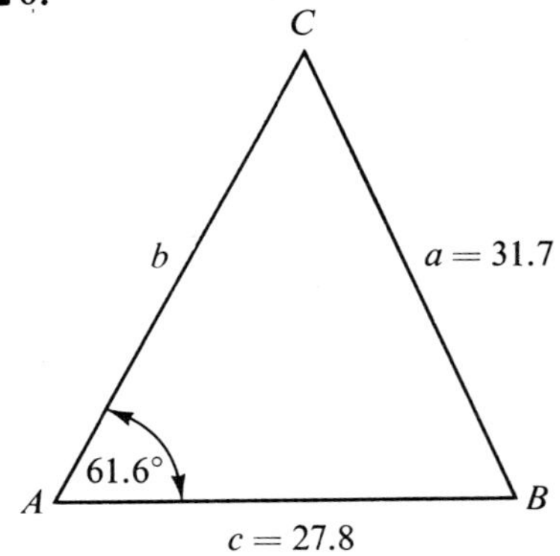

i. $\dfrac{31.7}{\sin 61.6°} = \dfrac{27.8}{\sin C}$

$\sin C = \dfrac{27.8 \sin 61.6°}{31.7}$

$C = 50.5°;$

ii. $B = 180° - (61.6° + 50.5°) = 67.9°;$

iii. $\dfrac{31.7}{\sin 61.6} = \dfrac{b}{\sin 67.9°}$

$b = 33.4.$

■ 7.

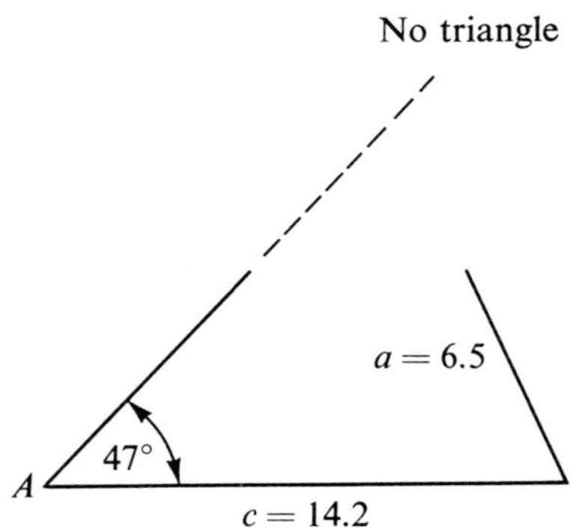

6-3 Problems

1. Find x.

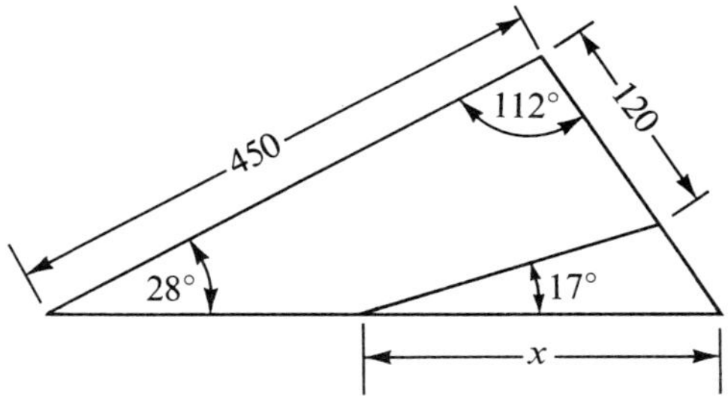

2. Find F_1, when $F_2 = 248$, $\theta = 28°$, and $R = 470$.

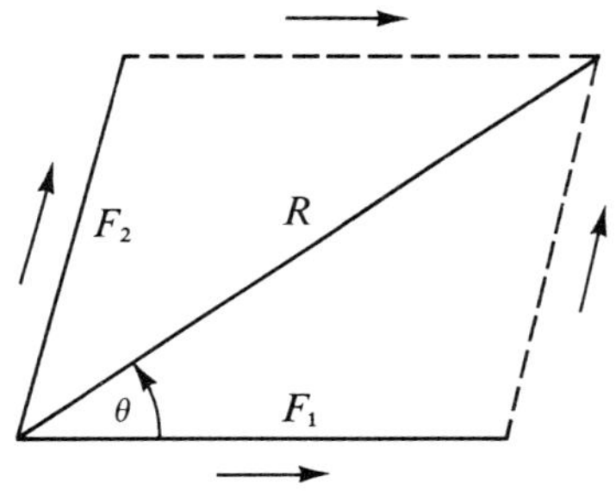

3. What is the diameter of the largest circle that can be drawn in the interior of of triangle ABC?

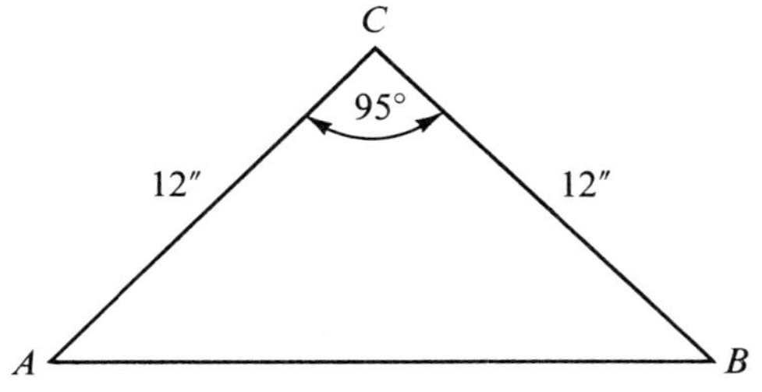

4. $\angle ABC = 110°$, $\overline{BC} = 12$ feet,
$\angle BCD = 40°$, $\overline{CD} = 25$ feet,
$\angle CDE = 45°$.

Find lengths $\overline{DE}$ and angle θ so that points A, B, E, and F will be collinear.

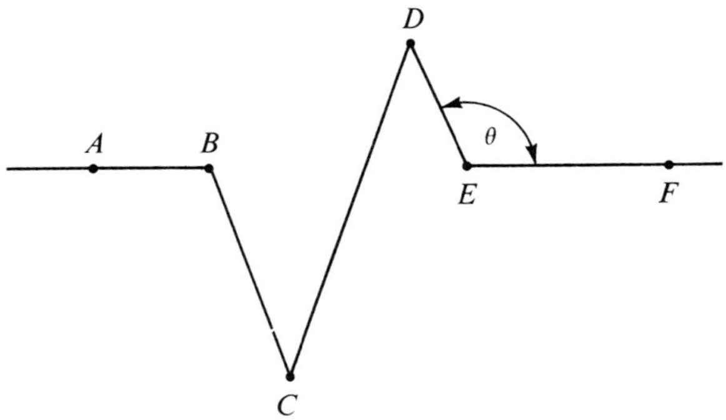

5. Determine the length of the side of the largest regular hexagon which can be cut from a square which is 10 inches on a side.

6-4 Law of Cosines—Oblique Triangles

For every right triangle the square of the hypotenuse is equal to the sum of the squares of the legs, symbolically $c^2 = a^2 + b^2$. If we subtract from the right member of this expression the quantity $2ab \cos C$, the resulting equation is true not only for right triangles, but for *all* triangles:

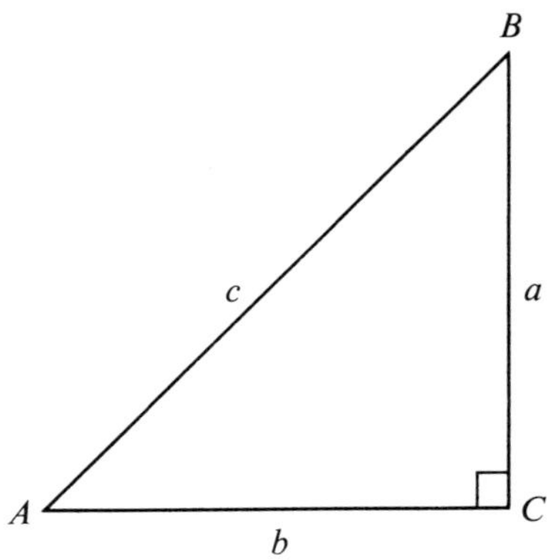

$$c^2 = a^2 + b^2 - 2ab \cos C. \qquad \text{(Eq. 6-4a)}$$

This equation is known as the **law of cosines** and sometimes is referred to as the generalized Pythagorean theorem. (The derivation of Equation 6-4a appears in the exercises of this section.)

The law of cosines is an important relationship because it provides a means of solving all specific triangles that cannot be solved by using the law of sines (Eq. 6-3c).

Example 1

If $b = 7$, $a = 9$, and $C = 45°$, find c.

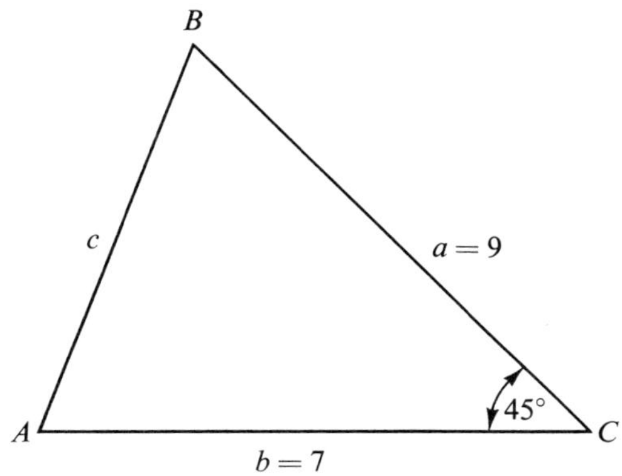

$$\begin{aligned} c^2 &= a^2 + b^2 - 2ab \cos C \\ &= 9^2 + 7^2 - 2(9)(7)(.707) \\ &= 130 - 89 \\ &= 41; \end{aligned}$$

$$c = \sqrt{41} \approx 6.4.$$

If three sides of a triangle are known, the law of cosines may be used to find the angles.

Example 2

Given the sides $c = 6$, $b = 9$, and $a = 10$ of a triangle, find the angles.

i. For convenience, we solve the law of cosines for cos C before substituting values.

$$c^2 = a^2 + b^2 - 2ab \cos C$$

$$\begin{aligned} \cos C &= \frac{a^2 + b^2 - c^2}{2ab} \\ &= \frac{10^2 + 9^2 - 6^2}{2(10)(9)} \\ &= \frac{145}{180} = .8055 \end{aligned}$$

$$C = 36.3°.$$

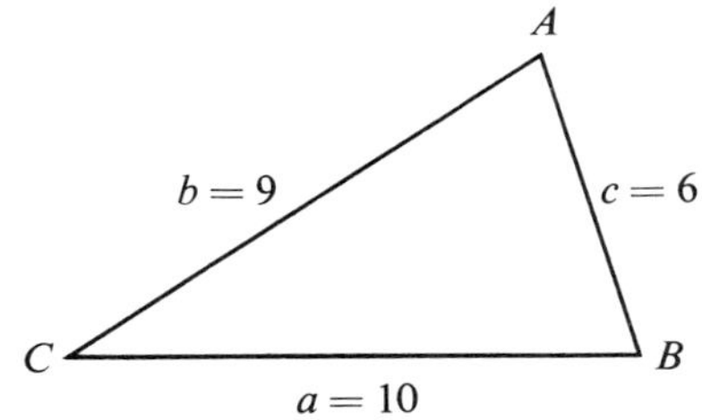

ii. To determine the value of another angle we may elect to use either the law of cosines again or the law of sines. Since the amount of computation is considerably less with the law of sines, we proceed with it.

$$\frac{b}{\sin B} = \frac{c}{\sin C}$$

$$\sin B = \frac{b \sin C}{c} = \frac{9(.5920)}{6} = .8880$$

$$B = 62.6°.$$

iii. $A = 180° - (36.3° + 62.6°) = 81.1°.$

Several comments are now in order regarding special problems that arise in applying the law of cosines.

1. The law is valid even for obtuse triangles. Remember that if $90° \leq A < 180°$, then $\cos A = -\cos(180° - A)$.
2. A, B, and C may be assigned to any of the vertices of a triangle. This is in effect saying that $a^2 = b^2 + c^2 - 2bc \cos A$ and $b^2 = a^2 + c^2 - 2ac \cos B$ are equally suitable forms for the law of cosines.
3. When only the three sides of a triangle are given, it is difficult to determine immediately whether or not one of the angles is obtuse. Therefore, it is advisable to solve for the angles opposite the smaller sides first, because if there is an obtuse angle it *must* be opposite the largest side.

Example 3

Find angles A, B, and C when $a = 7$, $b = 8$, and $c = 12$.

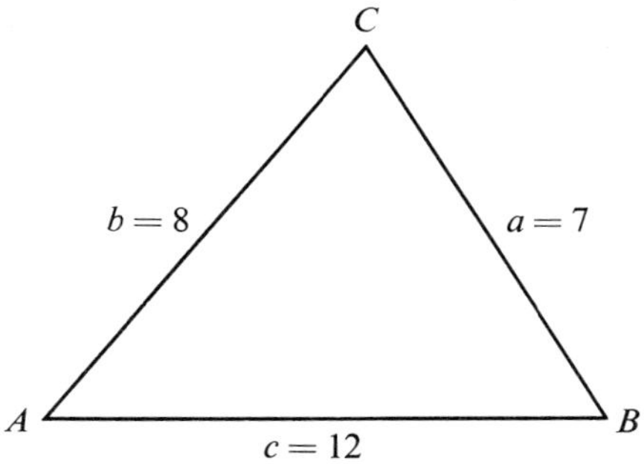

i. $\cos A = \dfrac{b^2 + c^2 - a^2}{2bc}$

$= \dfrac{64 + 144 - 49}{192}$

$= \dfrac{159}{192} = .8281$

$A = 34.1°;$

ii. $\dfrac{a}{\sin A} = \dfrac{b}{\sin B},$

$\sin B = \dfrac{b \sin A}{a} = \dfrac{8(.5606)}{7} = .6406$

$B = 39.8°;$

iii. $C = 180° - (34.1° + 39.8°) = 106.1°.$

6-4 Exercises

1. Complete the frames in the following sequence of statements in terms of the acute triangle ABC.

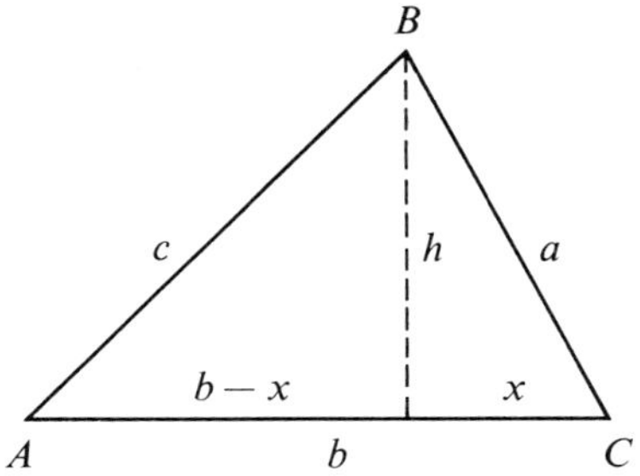

i. $h^2 = c^2 - \boxed{}$.
ii. $h^2 = a^2 - \boxed{}$.
iii. Equating (i) and (ii),
$c^2 - (b - x)^2 = a^2 - x^2$ or
$c^2 = a^2 + b^2 - \boxed{}$.

iv.

$$\cos C = \frac{x}{a}, \text{ or}$$

$x = \boxed{}$.

v. Substituting the value of x into the final statement of step (iii), $c^2 =$ $\boxed{}$.

---

2. Complete the frames in the following sequence of statements in terms of the obtuse triangle ABC.

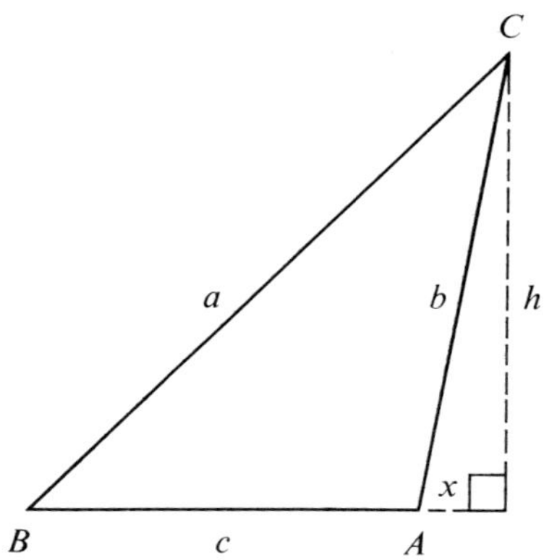

i. $h^2 = b^2 - \boxed{}$.
ii. $h^2 = a^2 -$ $\boxed{}$.
iii. Equating (i) and (ii),
$a^2 - \boxed{}$
$= b^2 - \boxed{}$,
or $a^2 = b^2 + c^2 + \boxed{}$.
iv. $\cos (180° - A) = x/b$, or
$x = \boxed{}$.
v. But $\cos (180° - A)$
$= -\cos \boxed{}$.
vi. Therefore x
$= -b \cos \boxed{}$.

■ 1. i. $h^2 = c^2 - (b - x)^2$;
ii. $h^2 = a^2 - x^2$;
iii. $c^2 - (b - x)^2$
$= a^2 - x^2$ or
$c^2 = a^2 + b^2 - 2bx$;
iv. $\cos C = \frac{x}{a}$ or
$x = a \cos C$;
v. $c^2 = a^2 + b^2$
$- 2ab \cos C$.

This exercise establishes the justification for the law of cosines for acute triangles. Since drawing the altitude h from B to b is completely arbitrary, we could have drawn it from A to a and established that $b^2 = a^2 + c^2 - 2ac \cos B$, or from C to c and established that $a^2 = b^2 + c^2 - 2bc \cos A$.

vii. Substituting the value of x into the final statement of step (iii), we obtain $a^2 =$ [ ].

---

3. Complete the solution of the triangle when $a = 9$, $b = 10$, and $C = 47°$.

■ 2. i. $h^2 = b^2 - x^2$;
ii. $h^2 = a^2 - (c + x)^2$;
iii. $a^2 - (c + x)^2 = b^2 - x^2$, or $a^2 = b^2 + c^2 + 2cx$;
iv. $\cos(180° - A) = \frac{x}{b}$, or $x = b \cos(180° - A)$;
v. $\cos(180° - A) = -\cos A$;
vi. therefore $x = -b \cos A$;
vii. $a^2 = b^2 + c^2 - 2bc \cos A$.

This exercise establishes the justification for the law of cosines for obtuse triangles.

---

4. If $a = 11$, $b = 14$, and $C = 112°$, find c, A, and B.

■ 3.

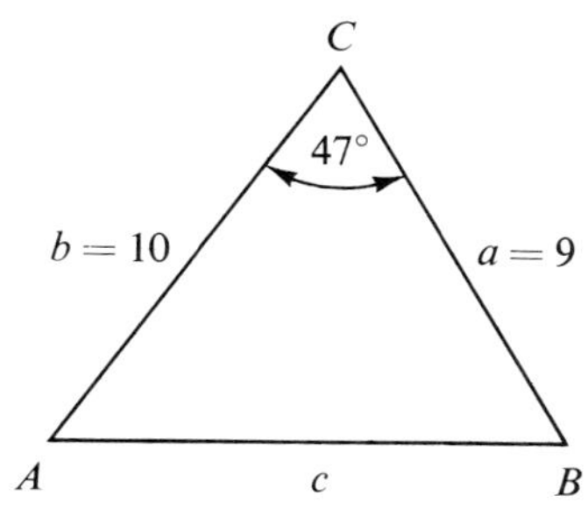

i. $c^2 = a^2 + b^2 - 2ab \cos C$
$= 81 + 100 - 2(9)(10)(.682)$
$= 58.24$
$c = 7.63$;

ii. $\dfrac{7.63}{\sin 47^\circ} = \dfrac{9}{\sin A}$

$A = 59.6^\circ$;

iii. $B = 180^\circ - (47^\circ + 59.6^\circ) = 73.4^\circ$.

---

5. Find the angles of a triangle if the sides are $a = 5$, $b = 6$, and $c = 7$.

■ 4.

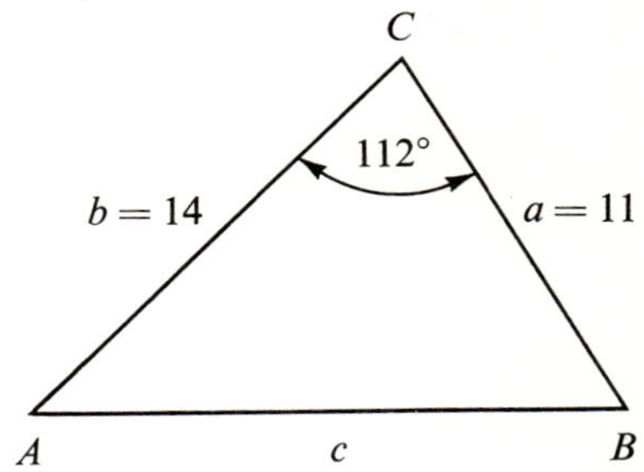

i. $c^2 = a^2 + b^2 - 2ab \cos C$
$= 121 + 196 - 308 \cos 112^\circ$
$= 317 + 308 \cos 68^\circ$
$= 432.4$

$c = 20.79$;

ii. $\dfrac{20.79}{\sin 112^\circ} = \dfrac{20.79}{\sin 68^\circ} = \dfrac{11}{\sin A}$

$A = 29.4^\circ$;

iii. $B = 180^\circ - (112^\circ + 29.4^\circ) = 38.6^\circ$.

---

6. Find the angles of the triangle with sides $a = 0.8$, $b = 1.2$, and $c = 1.8$.

■ 5.

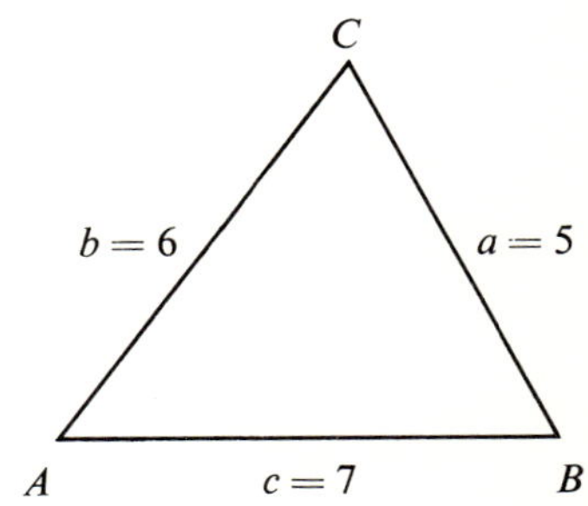

i. $\cos A = \dfrac{b^2 + c^2 - a^2}{2bc}$

$= \dfrac{36 + 49 - 25}{2(6)(7)}$

$= .7143$

$A = 44.4°;$

ii. $\dfrac{5}{\sin 44.4°} = \dfrac{6}{\sin B}$

$B = 57.1°;$

iii. $C = 180° - (44.4° + 57.1°) = 78.5°.$

■ 6.

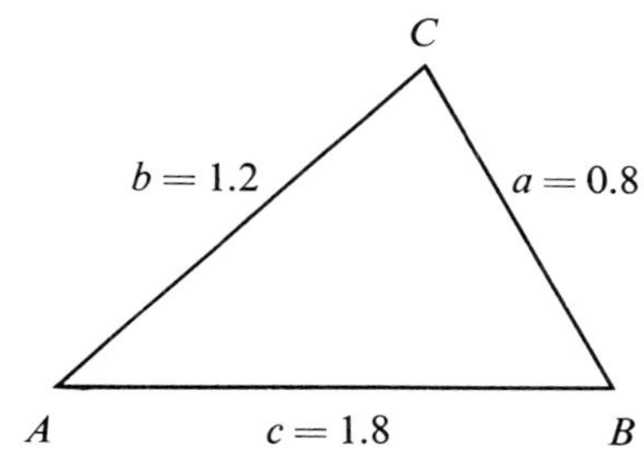

i. $\cos A$

$= \dfrac{b^2 + c^2 - a^2}{2bc}$

$= \dfrac{1.44 + 3.24 - .64}{2(1.2)(1.8)}$

$= .9351$

$A = 20.8°;$

ii. $\dfrac{0.8}{\sin 20.8°} = \dfrac{1.2}{\sin B}$

$B = 32.2°;$

iii. $C = 180° - (20.8° + 32.2°)$

$= 127.0°.$

6-4 Problems

1. Survey notes give the following data:

$$\overline{DC} = 500 \text{ feet},$$
$$\angle ACD = 32^\circ,$$
$$\angle ADC = 129^\circ,$$
$$\angle BDC = 43^\circ,$$
$$\angle BCD = 113^\circ.$$

Find the length $\overline{AB}$.

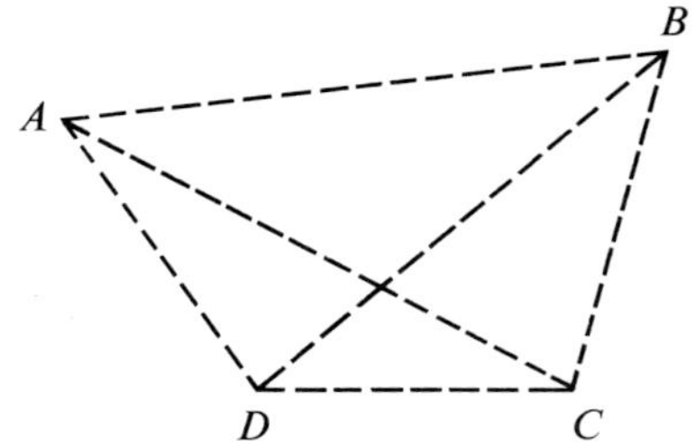

2. Nine evenly-spaced holes are to be drilled in the circumference of a circle 7¼ inches in diameter. What should be the distance from center to center of the holes?

3.

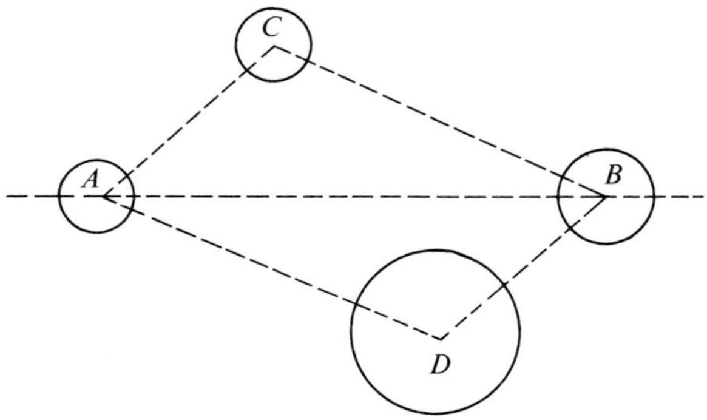

$$\overline{AB} = 3.000 \text{ inches}$$
$$\overline{AD} = 2.125 \text{ inches}$$
$$\overline{AC} = 1.250 \text{ inches}$$
$$\overline{BD} = 1.312 \text{ inches}$$
$$\overline{CB} = 2.200 \text{ inches}$$

In order to locate holes on a boring machine, holes A and B are placed in line with the motion of the table; C and D must be located in terms of parallel and perpendicular distances from A along $\overline{AB}$. Determine these values.

4. A highway relocation survey gives the following measurements:

$$\angle ADB = 105^\circ,$$
$$\angle ACB = 33^\circ,$$
$$\overline{DB} = 118 \text{ feet},$$
$$\overline{BC} = 254 \text{ feet}.$$

Find $\overline{AB}$.

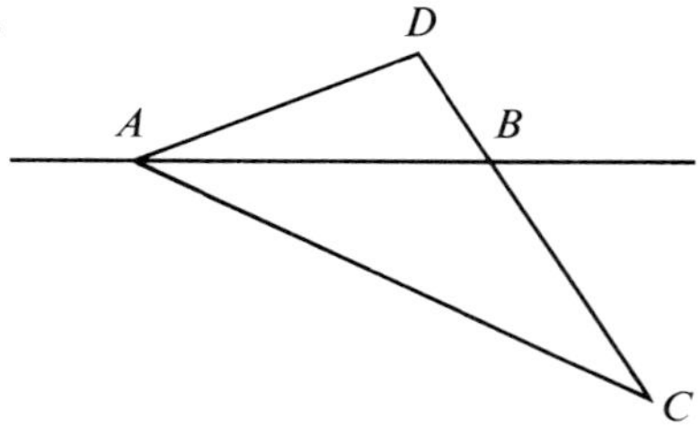

5. All three pulleys have an eight-inch diameter. The measurements from center to center are:

$$\overline{AB} = 12.0 \text{ inches},$$
$$\overline{BC} = 16.0 \text{ inches},$$
$$\overline{AC} = 18.5 \text{ inches}.$$

Find the length of the belt to the nearest tenth of an inch.

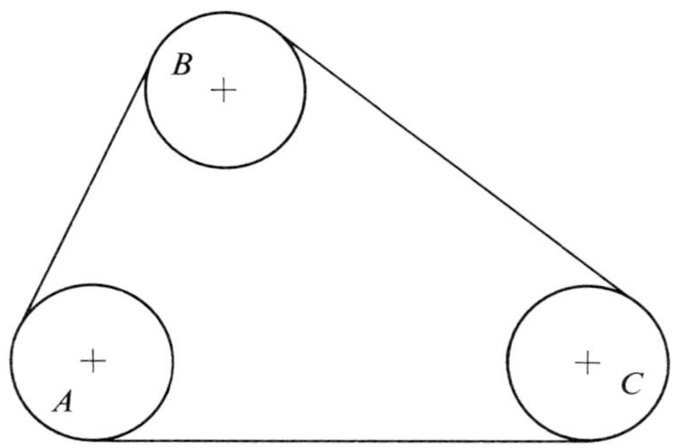

6.

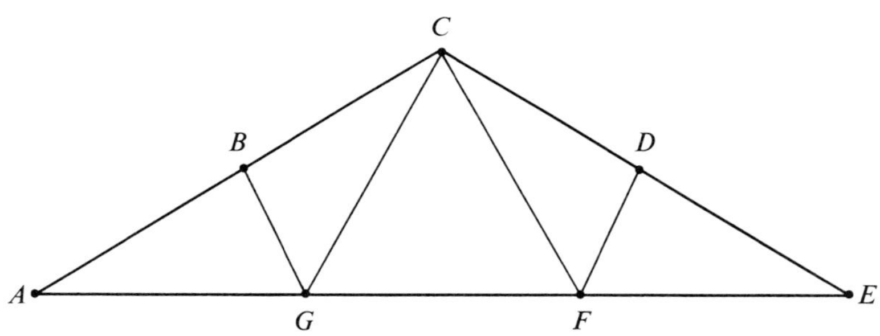

Points G and F divide the span $\overline{AE}$ into thirds. B is the midpoint of line segment $\overline{AC}$, and D is the midpoint of line segment $\overline{CE}$. Determine all the dimensions of this truss design when the span is 34.0 feet and the altitude is 8.5 feet.

7. A pole is standing vertically on a 30% slope. If guy wires are attached 25 feet up the pole and make a 35° angle with the pole, how long are the guy wires that are used uphill and downhill from the pole?

6-5 Area Formulas

Several formulas for computing area require that the height (h), or an equivalent dimension, be known. Frequently the height is an auxiliary dimension that must be determined by calculation.

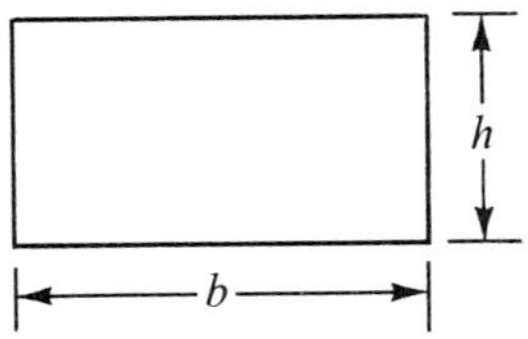

Rectangle
$A = bh$

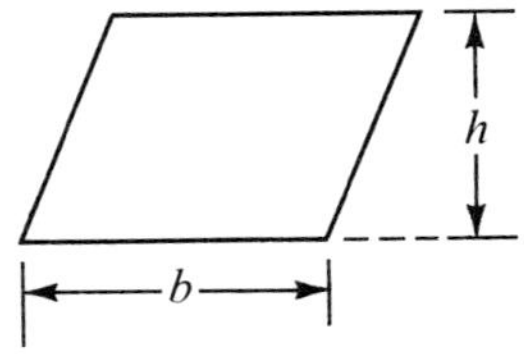

Parallelogram
$A = bh$

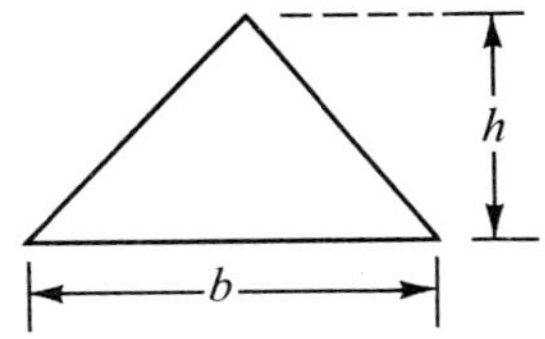

Triangle
$A = (1/2)bh$

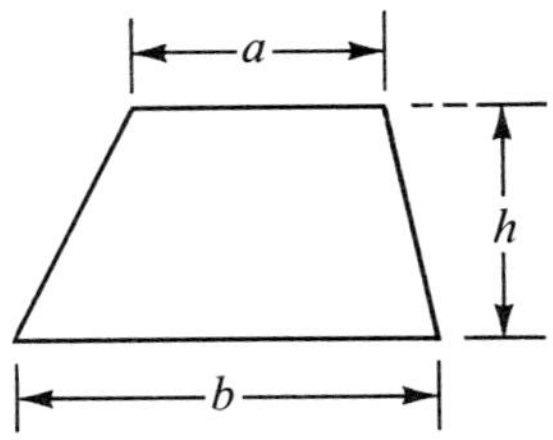

Trapezoid
$A = (1/2)h(a + b)$

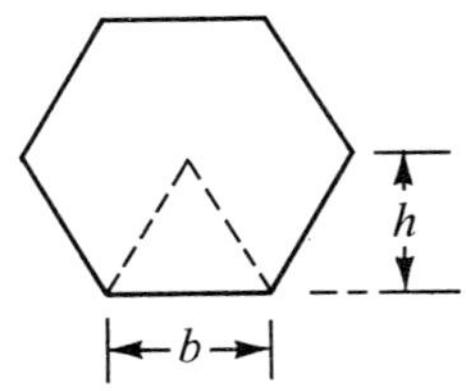

Regular hexagon
$A = 3bh$

Example 1

Assume that in Fig. 6-5a only a, b, and C are known. Write a formula for the area in terms of the known values.

i. $\sin C = \dfrac{h}{a}$, or

$h = a \sin C$;

ii. $(\text{Area}) = \dfrac{1}{2} bh$, or

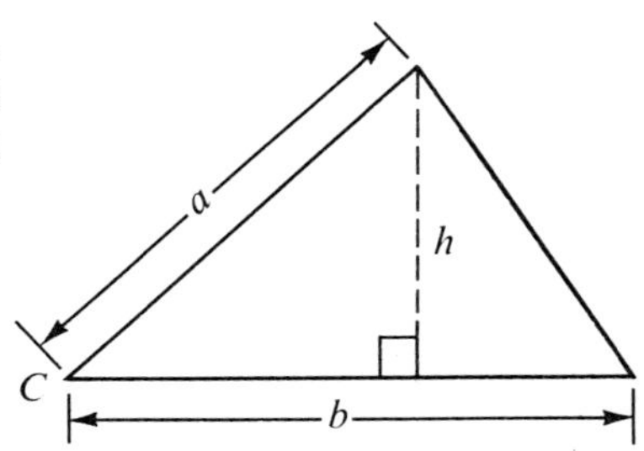

Fig. 6-5a

$$(\text{Area}) = \frac{1}{2} ab \sin C. \qquad \text{(Eq. 6-5a)}$$

As a general rule, if the dimensions given for a polygon are sufficient to make it a specific figure, then the area can be calculated.

Example 2

Write a formula for the area of a regular polygon with n sides, each of length b.

i. Isosceles triangle AOB is one of the n congruent triangles of the regular polygon.

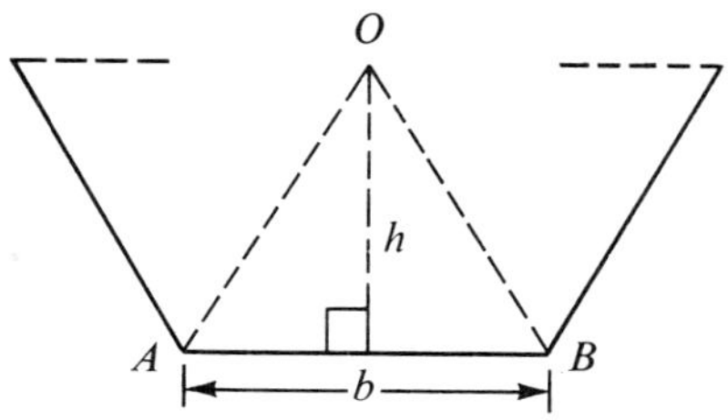

Fig. 6-5b

ii. $\frac{1}{2}\angle AOB = \frac{360°}{2n} = \frac{180°}{n}$;

iii. $\tan\frac{180°}{n} = \frac{b/2}{h}$, or

$h = \boxed{}$;

iv. $(\text{Area})_n = \left(\frac{1}{2}bh\right)n$

$= \frac{1}{2}b\left(\frac{b}{2\tan(180°/n)}\right)n$, or

$$(\text{Area})_n = \frac{b^2n}{4\tan(180°/n)}. \qquad \text{(Eq. 6-5b)}$$

■ $h = \frac{b}{2\tan(180°/n)}$.

When only three sides of a triangle are given, writing a concise formula that involves only the three sides with or without trigonometric expressions is a bit more challenging than the problems in Examples 1 and 2. Our objective, however, is simply to solve for area, which in the case just suggested may be done by first using the law of cosines to find an unknown angle and then using Equation 6-5a.

There is another approach using what is called Hero's formula. Its justification depends on trigonometric relationships that are introduced later in the text.

If the three sides of the triangle are a, b, and c, Hero's formula is

$$(\text{Area}) = \sqrt{s(s-a)(s-b)(s-c)}, \qquad \text{(Eq. 6-5c)}$$

where

$$s = \frac{a+b+c}{2}. \qquad \text{(Eq. 6-5d)}$$

6-5 Exercises

1. Calculate the area of triangle ABC when $a = 8$, $b = 9$, and $C = 47°$.

---

2. Calculate the area of triangle ABC when $b = 12.3$, $a = 14.7$, and $C = 112° 15'$.

■ 1.

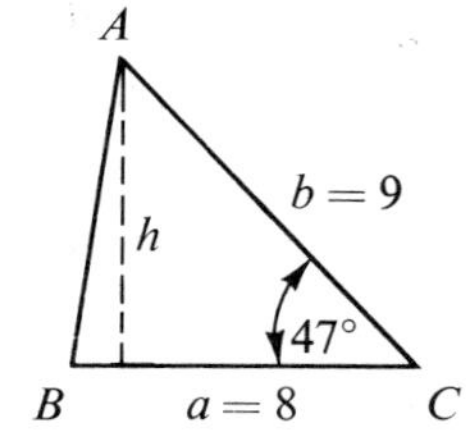

$$(\text{Area}) = \frac{1}{2} ab \sin C$$

$$= \frac{8(9) \sin 47°}{2}$$

$$= 26.3.$$

3. Find the area of triangle ABC when $a = 11.5$, $b = 14.5$, and $c = 22$.

■ 2.

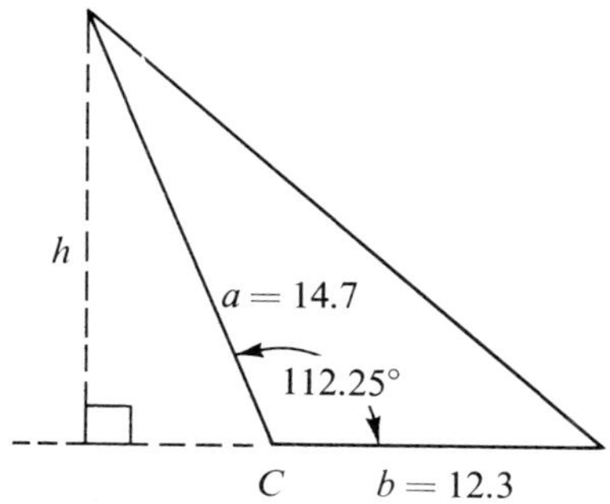

Equation 6-5a was derived using a figure that may have suggested $C < 90°$. Actually, the equation is perfectly general.

i. $\sin C = \sin (180° - C)$

$$= \frac{h}{a} \quad \text{or} \quad h = a \sin C$$

ii. $(\text{Area}) = \frac{1}{2} bh$

$$= \frac{1}{2} ab \sin C$$

$$= \frac{14.7(12.3) \sin 112.25°}{2}$$

$$= 83.67.$$

4. Find the area of triangle ABC.

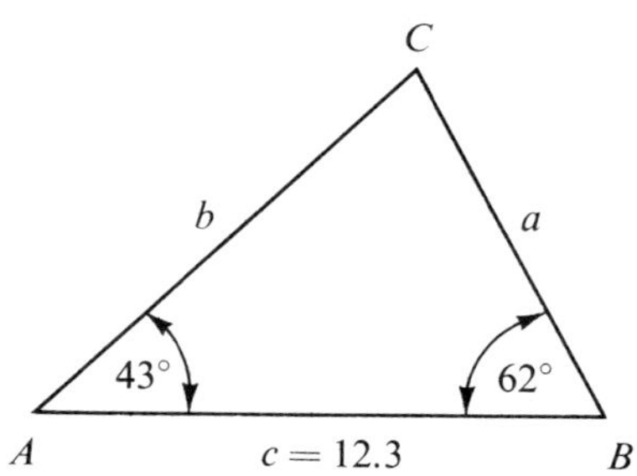

■ 3.

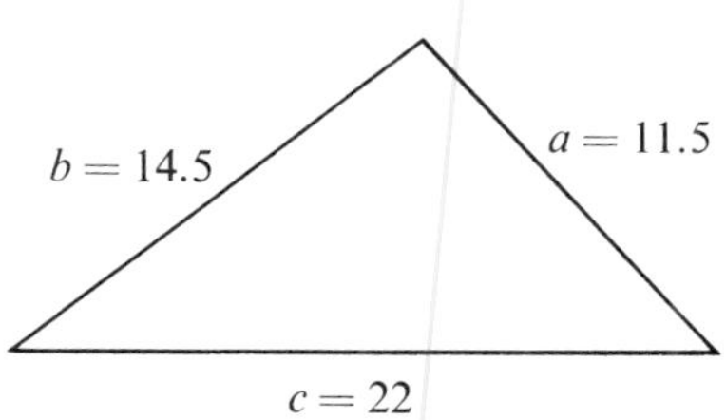

Using Hero's formula:

$$s = \frac{a+b+c}{2}$$

$$= \frac{11.5 + 14.5 + 22}{2} = 24.0$$

(Area)

$$= \sqrt{24(24-11.5)(24-14.5)(24-22)}$$

$$= 75.50.$$

---

5. Find the area of a regular octagon with sides of 60 feet.

■ 4. i. $C = 180° - (43° + 62°)$

$= 75°;$

ii. $\dfrac{c}{\sin C} = \dfrac{b}{\sin B}$

$$b = \frac{c \sin B}{\sin C};$$

iii. Let h be the height from C to c.

$\sin A = h/b,$ or

$$h = b \sin A$$

$$= \left(\frac{c \sin B}{\sin C}\right) \sin A;$$

iv. $(\text{Area}) = \dfrac{1}{2}(\text{base})(\text{height})$

$$= \frac{1}{2} c \left[\frac{c \sin B}{\sin C}\right] \sin A, \quad \text{or}$$

$$(\text{Area}) = \frac{c^2 \sin B \sin A}{2 \sin C} \qquad \text{(Eq. 6-5e)}$$

$$= \frac{(12.3)^2(\sin 62°)(\sin 43°)}{2 \sin 75°}$$

$$= 47.1.$$

■ 5. $(\text{Area})_n = \dfrac{b^2 n}{4 \tan (180°/n)},$

$$(\text{Area})_8 = \frac{(60)^2(8)}{4 \tan 22.5°}$$

$$= 17{,}380 \text{ square feet.}$$

6-5 Problems

1. Find the area of trapezoid $AA'B'B$.

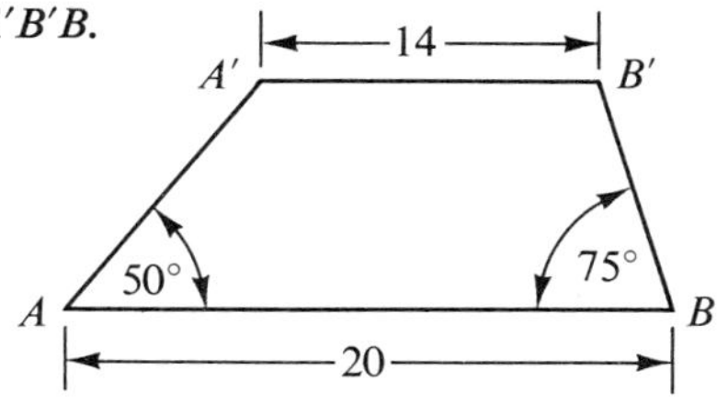

2. Assuming the gables are symmetrical and at right angles to each other, find the area of the roof surface.

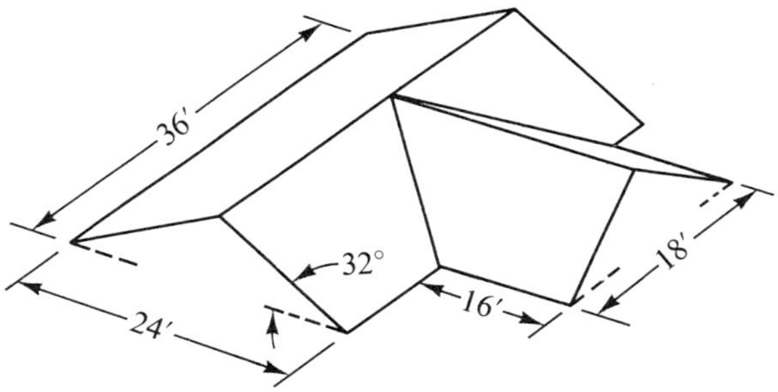

3. Find the capacity of this horizontal cylinder for each foot of depth. (Length: 10 feet; Diameter: 6 feet.)

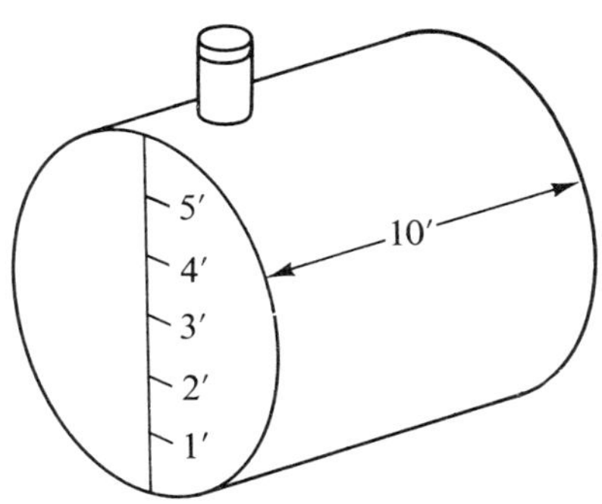

4. A counterweight on a 12-inch flywheel is in the form of a segment subtending a central angle of 40°. Find its weight if it is 3.00 inches thick and made of a metal with a density of 0.28 pounds per cubic inch.

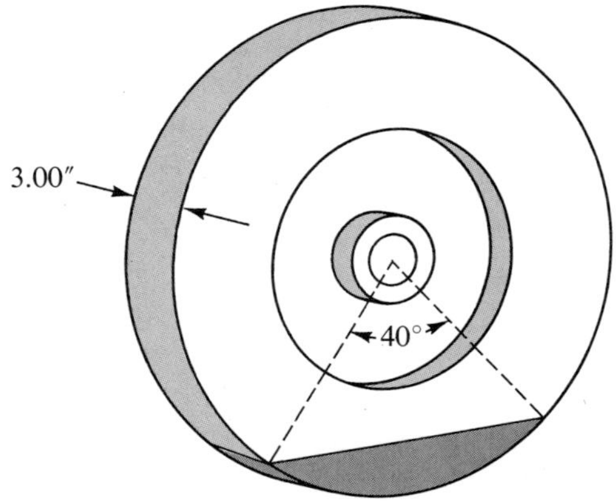

Diameter: 12 inches

5. If the area of $A'B'C'D'$ is 30 square inches, find the area of $ABCD$ (see Fig 6-5c).
6. If the area of $ABCD$ is 120 square feet, find the area of $A'B'C'D'$ (see Fig 6-5c).

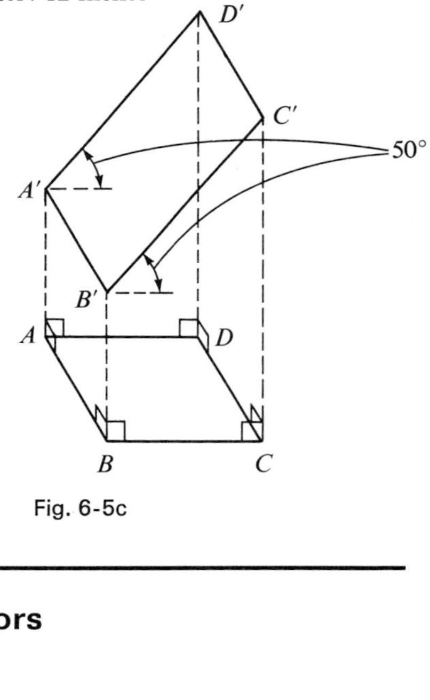

Fig. 6-5c

6-6 Introduction to Vectors

Two cables, each having a tensile strength of 1000 pounds, can suspend 2000 pounds in the manner illustrated in Fig. 6-6a; but they can support only 1414 pounds suspended in the manner illustrated in Fig. 6-6b and only 1000 pounds in the manner shown in Fig. 6-6c. In the first illustration, ordinary arithmetic appears to be an adequate model for describing the net effect of the two cables; but it is not satisfactory in the second and third illustrations, where we are dealing with a quantitative situation that is not adequately described by arithmetic and numbers common to arithmetic.

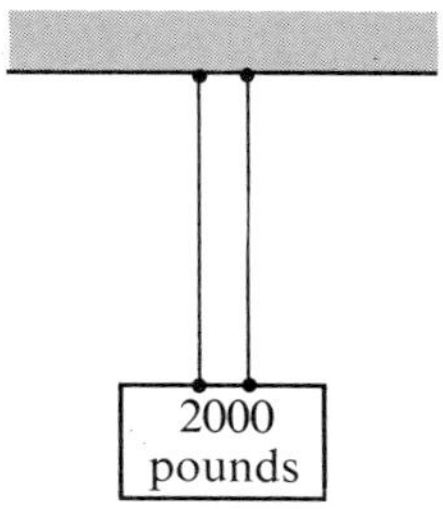

Fig. 6-6a

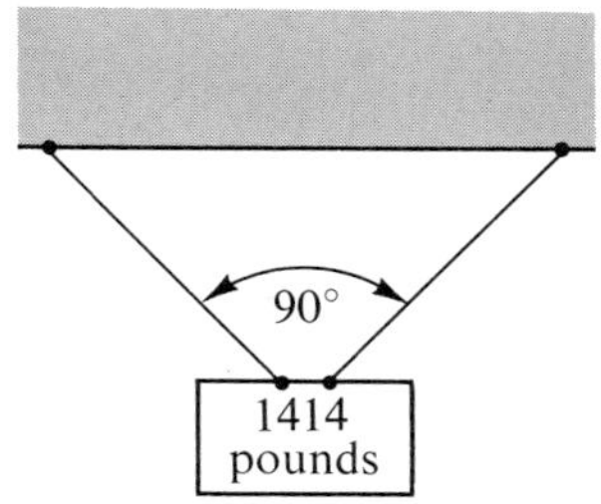

Fig. 6-6b

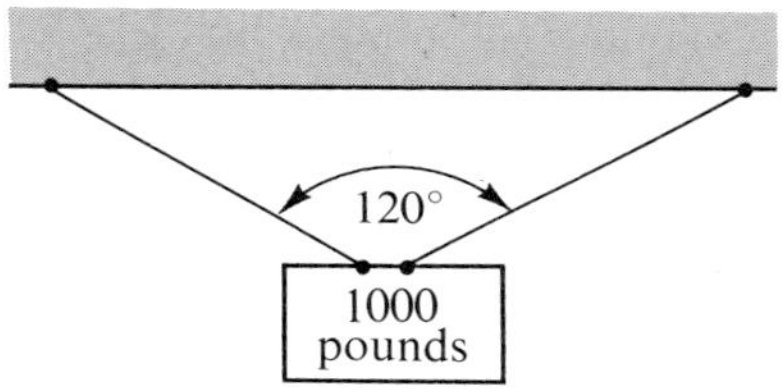

Fig. 6-6c

Examine the following two lists.

List 1	*List* 2
Social Security number	Gravitational force
Day of the month	Acceleration
Sheets of paper in a ream	Stress
Charge on a capacitor	Centrifugal force
Density	Torque
Wage	Velocity
Inventory of paper clips	Electric field strength

Both lists identify quantitative matters; the distinction is that the first list contains items that can each be appropriately described by a single real number, whereas the second lists ideas that need two or more numbers to adequately describe them, one for **magnitude** and the other(s) for **direction**. List 1 contains **scalar** quantities, and List 2 contains **vector** quantities.

In scientific and technical fields the treatment of vectors is often limited to special cases of the more general mathematical treatment of vectors. Our discussion is limited to **two-dimensional vectors**. The general analytic format of a vector is an array of numbers in a single row or single column. The number of elements in the row or column determines the dimension of the vector. The vector

$$(x, y, z) \quad \text{or} \quad \begin{bmatrix} x \\ y \\ z \end{bmatrix}$$

is an example of a three-dimensional vector.

The vector

$$(x_1, x_2, x_3, \ldots x_n) \quad \text{or} \quad \begin{bmatrix} x_1 \\ x_2 \\ x_3 \\ \vdots \\ x_n \end{bmatrix}$$

is n-dimensional. Two-dimensional row vectors are our concern here. They are the most elementary but yet have extensive application to science and technology.

In many technical applications the two numbers of a two-dimensional vector represent *magnitude* and *direction*. For example, the vector expression of force tells the magnitude of the force and the direction in which it is being exerted. The vector for velocity tells the speed and direction of motion.

The geometric representation of two-dimensional vectors is a very great aid in understanding their purpose and properties. The magnitude is represented by a line drawn to scale. The direction is represented by the line's orientation and sense in a plane. In order for direction to be meaningful, points of reference must be clearly understood. For this reason, geometric representations of vectors are placed on the coordinate plane.

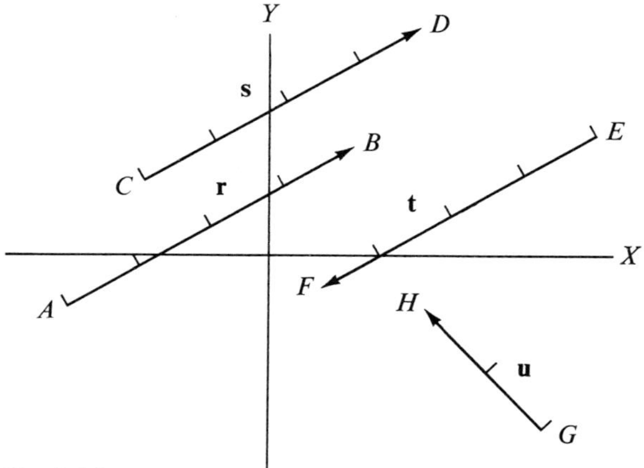

Fig. 6-6d

Four vectors are illustrated in Fig. 6-6d. An arrow head is placed on the **terminal point** of a vector. The opposite end is the **initial point.** The methods of identifying vectors vary: $\overrightarrow{AB}$, $\vec{r}$, **r**, $\underset{\cdot}{r}$ all identify the same vector in Fig. 6-6d. We shall use the boldface letter.

If we assume that line segments $\overline{AB}$, $\overline{CD}$, and $\overline{EF}$ are parallel and equal in length, we can point out some of the properties of two-dimensional vectors.

1. Vectors **r** and **s** are *equal* because they have equal magnitudes and directions. We write **r** = **s**. Graphically, the **translation** of a vector does not change its value. This is to say, if lines are drawn connecting the terminal points and initial points of two vectors and the resulting figure is a parallelogram, then the vectors are equal (Fig. 6-6e).

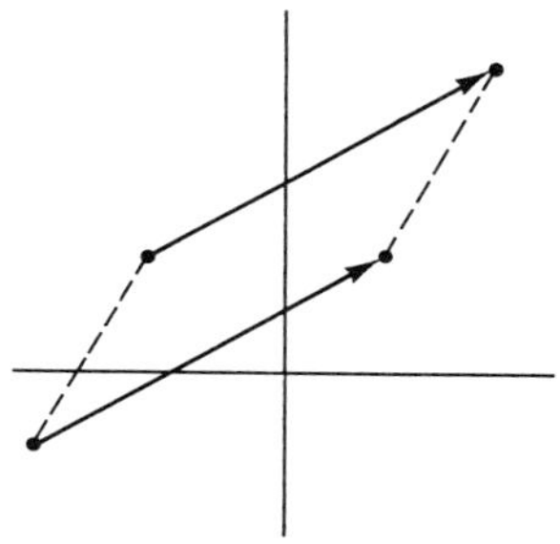

Fig. 6-6e

2. Vectors **r** and **t** are equal in magnitude *only*. This may be expressed writing $|\mathbf{r}| = |\mathbf{t}|$ using the notation of absolute value, or $r = t$, where only the scalar values of the vectors are indicated.
3. Since **r** and **t** are oriented in the same manner but opposite in sense, we may write $\mathbf{r} = -\mathbf{t}$, where the negative sign has the effect of reversing the sense.
4. Since **r** has twice the magnitude of **u**, we may write $|\mathbf{r}| = 2|\mathbf{u}|$.

Numerical computation is significantly simplified by translating all vectors into **standard position**. A vector in standard position is drawn with the *initial point* on the origin of the coordinate system (Figure 6-6f).

In standard position the magnitude and direction of every vector can be clearly identified with either the radius vector r and angle θ, or the coordinates of the terminal point, x and y. For example, from Figure 6-6f,

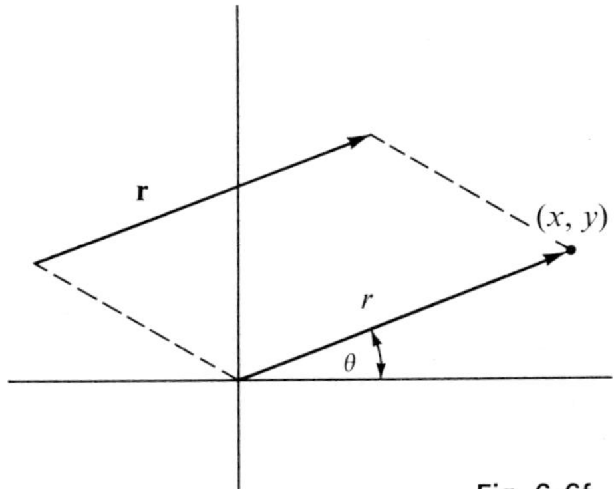

Fig. 6-6f

$\mathbf{r} = (r, \theta)$,	(Polar form)	(Eq. 6-6a)

or

$\mathbf{r} = (x, y)$.	(Rectangular form)	(Eq. 6-6b)

The two forms, **polar** and **rectangular**, are equivalent. Show how x and y can be expressed in terms of r and θ.

■ $x = r \cos \theta$,

■ $y = r \sin \theta$.

Show how r and θ can be expressed in terms of x and y.

■ $r = \sqrt{x^2 + y^2}$ (Pythagorean theorem)

■ $\tan \theta = \dfrac{y}{x}$.

Example 1

If $\mathbf{r} = (16.2,\ 114°)$, what is $\mathbf{r}$ in rectangular form?

i. $x = r \cos \theta$
$= 16.2 \cos 114° = -16.2 \cos 66° = -6.59;$

ii. $y = r \sin \theta$
$= 16.2 \sin 114° = 16.2 \sin 66° = 14.8.$

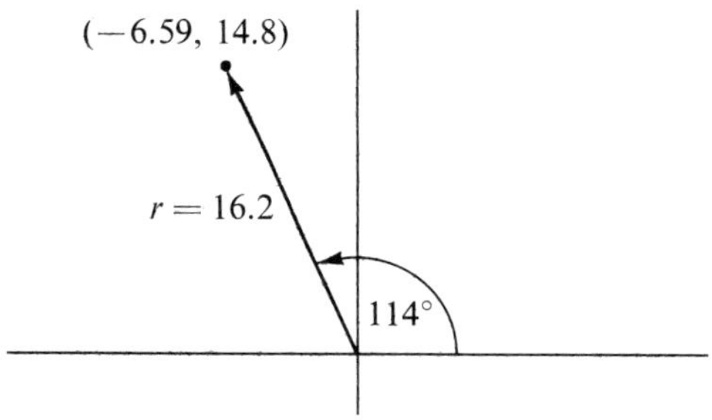

Example 2

If $\mathbf{r} = (3.28,\ -7.12)$, what is $\mathbf{r}$ in polar form?

i. $r = \sqrt{x^2 + y^2}$
$= \sqrt{3.28^2 + (-7.12)^2}$
$= 7.84;$

ii. $\tan \theta = \dfrac{y}{x} = -\dfrac{7.12}{3.28} = -2.170$

$\theta = (360° - 65.3°) = 294.7°.$

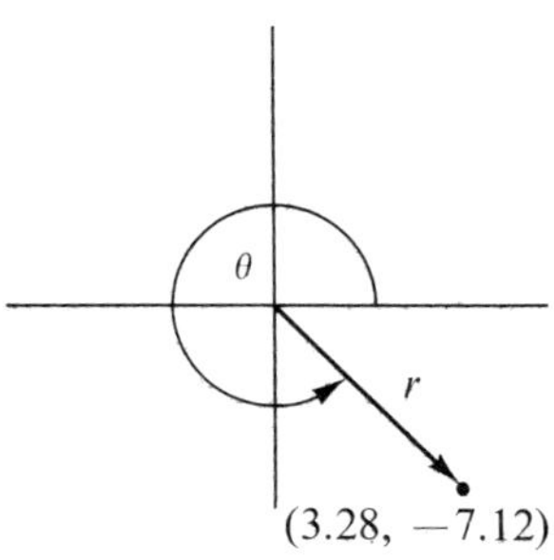

6-6 Exercises

1. Find x and y.

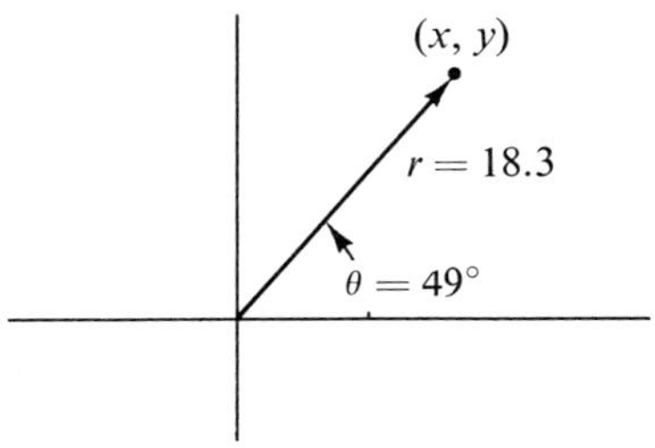

2. Find x and y.

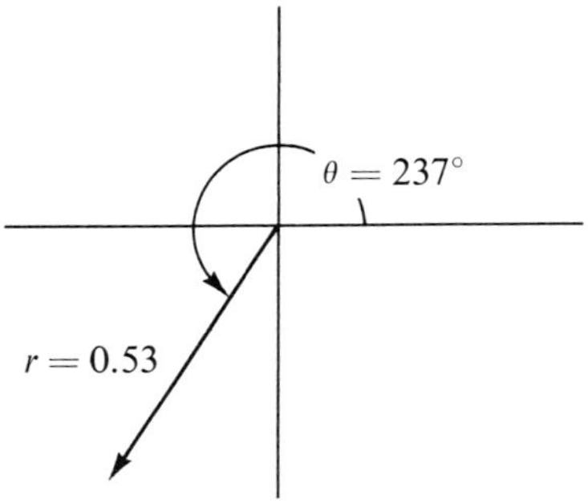

■ 1. $x = r \cos \theta$
$= 18.3 \cos 49° = 12.0;$
$y = r \sin \theta$
$= 18.3 \sin 49° = 13.8.$

3. Find r and θ.

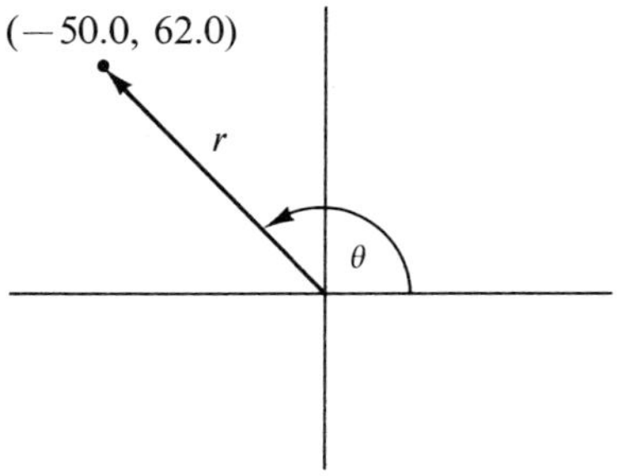

■ 2. $x = r \cos \theta$
$= 0.53 \cos 237°$
$= -0.53 \cos 57°$
$= -.289;$
$y = r \sin \theta$
$= 0.53 \sin 237°$
$= -0.53 \sin 57° = -.445.$

---

4. Find θ and y.

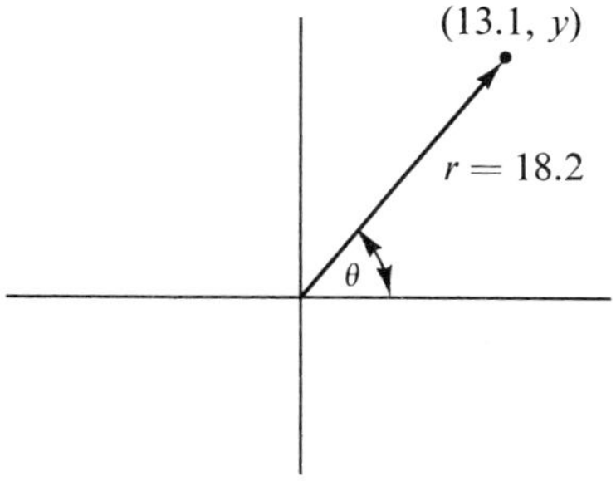

■ 3. $r = \sqrt{x^2 + y^2}$
$= \sqrt{(-50)^2 + (62)^2}$
$= 79.6;$

$\tan\theta = -\frac{62}{50} = -1.2400$
$\theta = 180° - 51.1°$
$= 128.9°.$

---

5. Find x and r.

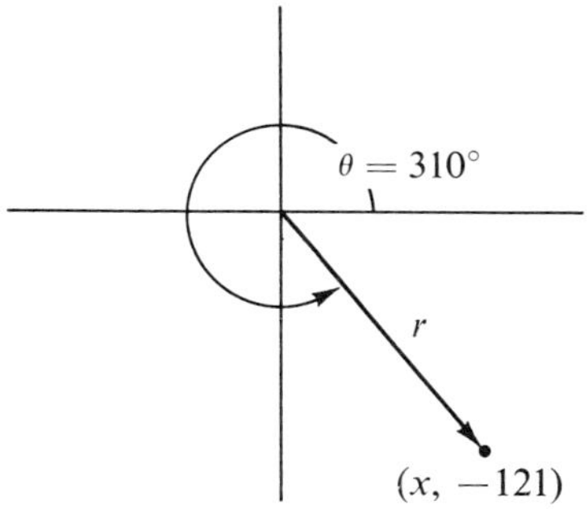

■ 4. $\cos\theta = \frac{x}{r} = \frac{13.1}{18.2} = .720$
$\theta = 43.9°;$
$y = r\sin\theta$
$= 18.2 \sin 43.9°$
$= 12.6.$

---

6. Change $\mathbf{r} = (-21.2, 7.2)$ to polar form.

■ 5. $\tan\theta = \frac{y}{x}$

$x = \frac{y}{\tan\theta} = \frac{-121}{\tan 310°}$

$= \frac{-121}{-\tan 50°}$

$= \frac{121}{1.1918} = 101.5;$

$\sin\theta = \frac{y}{r}$

$r = \frac{y}{\sin\theta} = \frac{-121}{\sin 310°}$

$$= \frac{-121}{-.7660}$$
$$= 158.$$

---

7. If $\mathbf{r} = (8.66, 213°)$, find $-\mathbf{r}$.

■ 6. $\tan\theta = -\dfrac{7.2}{21.2} = -.3396$

$$\theta = 180° - 18.76°$$
$$= 161.24°;$$
$$r = \sqrt{(-21.2)^2 + (7.2)^2}$$
$$= 22.39;$$
$$\mathbf{r} = (22.4, 161.2°).$$

---

8. If $\mathbf{s} = (-.06, .05)$, find $-\mathbf{s}$.

■ 7. $-\mathbf{r} = (8.66, 213° - 180°)$
$= (8.66, 33°)$.

---

9. Represent $\mathbf{r}$ in standard rectangular form and position.

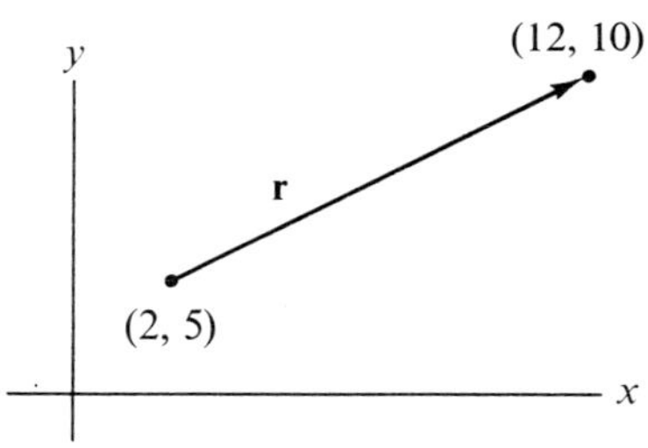

■ 8. $-\mathbf{s} = (.06, -.05)$.

---

10. Represent vector $\mathbf{s}$ in standard polar form and position.

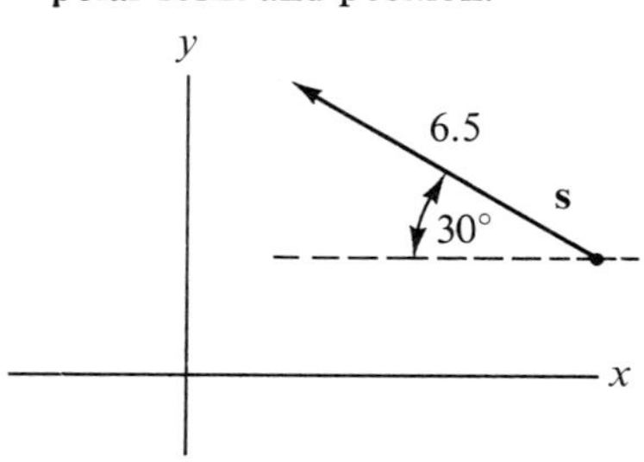

■ 9. $x = 12 - 2 = 10$,
$y = 10 - 5 = 5$,
$\mathbf{r} = (10, 5)$.

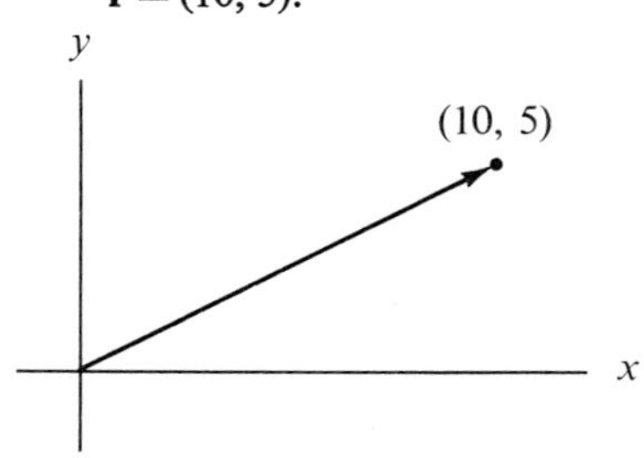

■ 10. $\mathbf{s} = (6.5, 180° - 30°)$
$= (6.5, 150°)$.

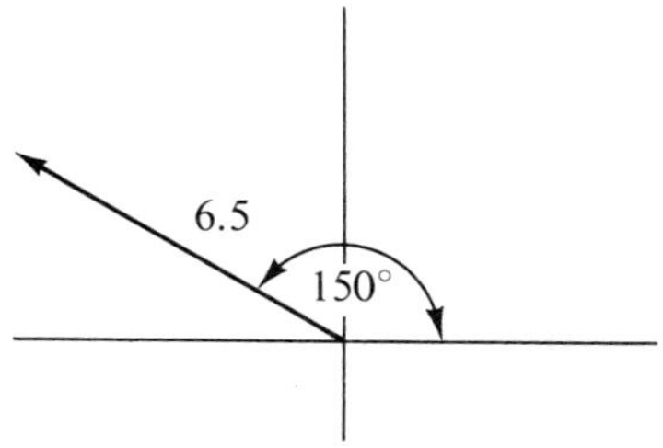

6-6 Problems

1. If $\mathbf{r} = (-4.7, 8.1)$, find $|\mathbf{r}|$.
2. If $\mathbf{r} = (5.2, 8.7)$, find $-\mathbf{r}$.
3. If a vector not in standard position has its initial point at $(-3, -4)$ and terminal point at $(5, 2)$, find the polar form and rectangular form in standard position.
4. If $\mathbf{r} = (-10.4, 8.6)$, find a vector of equal absolute value that is perpendicular to $\mathbf{r}$.
5. A plane is flying 350 miles per hour in a northwesterly direction; express its velocity as a vector in polar form and rectangular form.
6. This vector diagram represents the cable tensions and gravitational force on the weight of Figure 6-6b. Using the information given about the figure, write the polar and rectangular forms of vectors $\mathbf{t}_1$, $\mathbf{t}_2$, and $\mathbf{w}$.

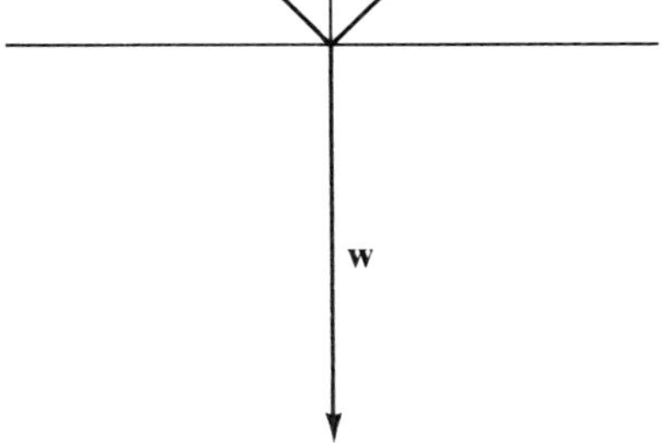

6-7 Operations with Two-Dimensional Vectors

Many words in the English language have multiple meanings. The context in which a word is used is the key to interpreting correctly those words of multiple meaning. Now we shall discover the meaning of *addition* and *subtraction* in terms of vectors. Fortunately there are similarities between the definitions you have known for some time and the definitions that these two words have for vectors.

If $\mathbf{r}_1 = (x_1, y_1)$ and $\mathbf{r}_2 = (x_2, y_2)$, then their *sum* is

$$\mathbf{r}_1 + \mathbf{r}_2 = (x_1 + x_2, y_1 + y_2). \qquad \text{(Def. 6-7a)}$$

Example 1

Find $\mathbf{r}_1 + \mathbf{r}_2$ when $\mathbf{r}_1 = (6, 3)$ and $\mathbf{r}_2 = (2, 4)$.

■ $\mathbf{r}_1 + \mathbf{r}_2 = (6 + 2, 3 + 4) = (8, 7)$.

The graphical interpretation of Example 1 is very interesting. The understanding we gain from graphs will greatly affect our insight into problem solving.

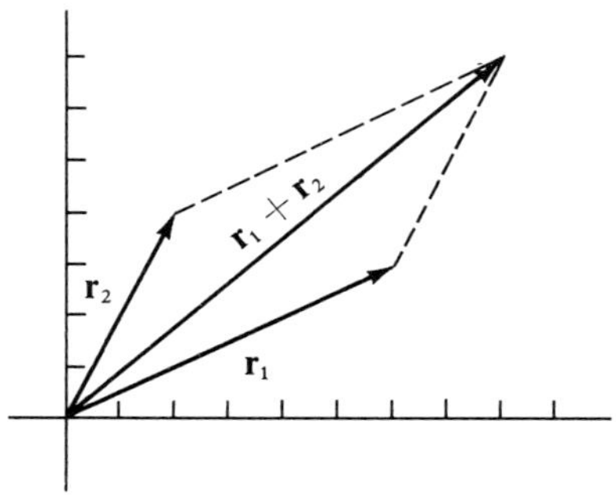

Fig. 6-7a

The definition has the effect of translating $\mathbf{r}_2$ so that the initial point of $\mathbf{r}_2$ coincides with the terminal point of $\mathbf{r}_1$ and forms a parallelogram; $\mathbf{r}_1 + \mathbf{r}_2$ is the diagonal.

Example 2

Find $\mathbf{r}_1 + \mathbf{r}_2$ when $\mathbf{r}_1 = (4, 62°)$ and $\mathbf{r}_2 = (5, 205°)$.

Both vectors must first be changed to rectangular form.

i. $x_1 = 4 \cos 62° = 1.88,$
$y_1 = 4 \sin 62° = 3.53.$

ii. $x_2 = 5 \cos 205°$
$= -5 \cos 25° = -4.53;$
$y_2 = 5 \sin 205°$
$= -5 \sin 25° = -2.11.$

iii. $\mathbf{r}_1 + \mathbf{r}_2 = (x_1 + x_2, y_1 + y_2)$
$= \boxed{}.$

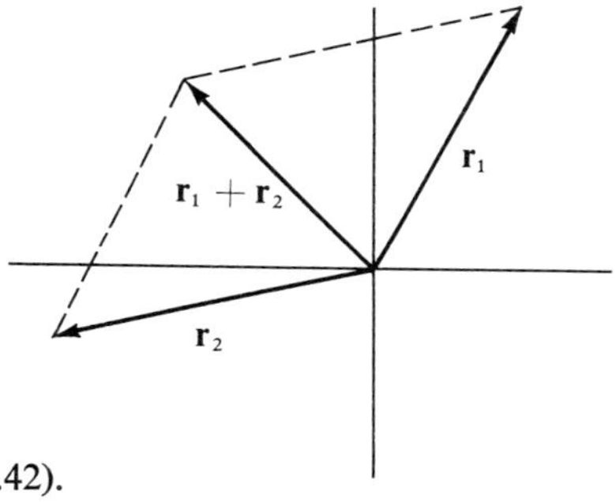

■ $(1.88 - 4.53, 3.53 - 2.11) = (-2.65, 1.42).$

iv. In polar form, $(-2.65, 1.42)$ is equivalent to $\boxed{}$.

■ $\tan \theta = \dfrac{1.42}{-2.65} = -.536$

$\theta = 180° - 28.2° = 151.8°;$

$|\mathbf{r}_1 + \mathbf{r}_2| = \sqrt{(-2.65)^2 + (1.42)^2}$
$= 3.02.$

Therefore

$\mathbf{r}_1 + \mathbf{r}_2 = (3.02, 151.8°).$

Now, read again the first paragraph of Section 6-6 and Problem 6 of that section. The vector diagram (Fig. 6-7b) that corresponds to the conditions represented in Figure 6-6b is drawn so that the point where the cables are attached to the load corresponds to the origin of the coordinate system.

	Polar form	Rectangular form
t_1	(1000, 135°)	(−707, 707)
t_2	(1000, 45°)	(707, 707)
w	(1414, 270°)	(0, −1414)

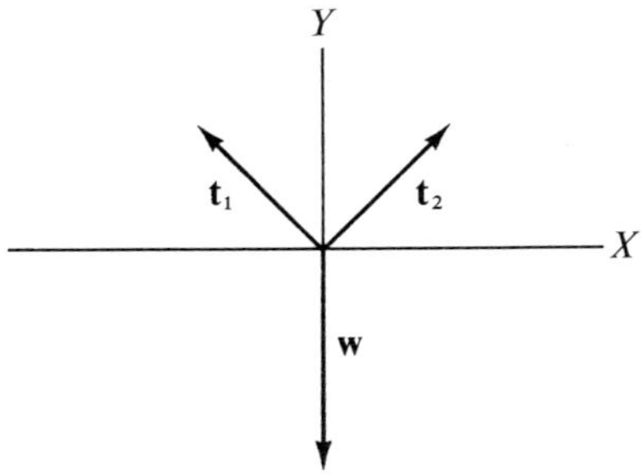

Fig 6-7b

Notice that $\mathbf{t}_1 + \mathbf{t}_2 = (-707 + 707, 707 + 707) = (0, 1414)$. It should be apparent that the vector sum (0, 1414) is the negative of $\mathbf{w} = (0, -1414)$.

In the mathematics of real numbers, recall that subtraction is defined in terms of addition; that is, $a - b = a + (-b)$. An analogous definition applies to the subtraction of vectors; $\mathbf{r} - \mathbf{s} = \mathbf{r} + (-\mathbf{s})$.

Example 3

Find $\mathbf{r} - \mathbf{s}$, when $\mathbf{r} = (8, 2)$ and $\mathbf{s} = (3, -5)$.

i. If $\mathbf{s} = (3, -5)$, then $-\mathbf{s} = (-3, 5)$.
ii. $\mathbf{r} - \mathbf{s} = \mathbf{r} + (-\mathbf{s}) = (8 - 3, 2 + 5) = (5, 7)$.

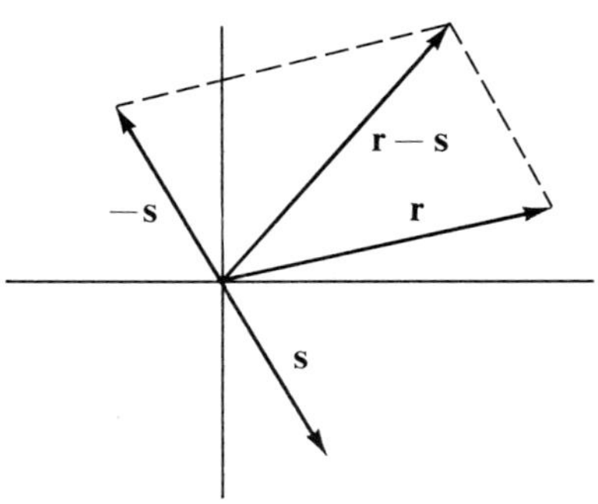

If c is a scalar and $\mathbf{r} = (x, y)$, then

$$c\mathbf{r} = (cx, cy) \qquad \text{(Def. 6-7b)}$$

defines the *product of a vector and a scalar*. The effect of multiplying by c is quite varied, depending upon the value given c. When $c < 0$, the sense of the vector is changed. If $|c| < 1$, then the magnitude is decreased, and if $|c| > 1$, then the magnitude is increased.

Example 4

If $\mathbf{r} = (-3.2, 5.1)$, find $(-.5)\mathbf{r}$.

$(-.5)\mathbf{r} = (-.5\,(-3.2), -.5\,(5.1)) = (1.6, -2.55)$.

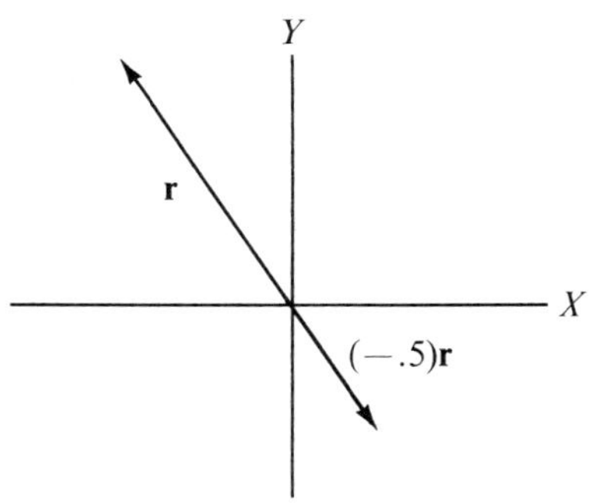

6-7 Exercises

1. If $\mathbf{r} = (4, 9)$ and $\mathbf{s} = (-6, -2)$, find and graph $\mathbf{r} + \mathbf{s}$.

2. For the same values given in Exercise 1, find and graph $\mathbf{r} - \mathbf{s}$.

■ 1. $\mathbf{r} + \mathbf{s} = (4 - 6, 9 - 2) = (-2, 7)$.

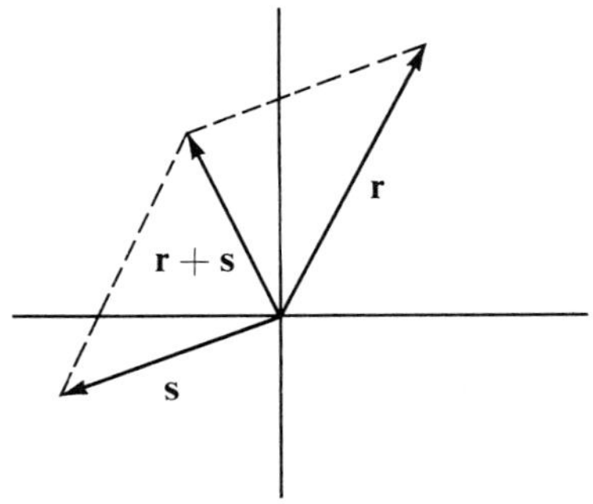

---

3. For the same values given in Exercise 1, find $\mathbf{s} - \mathbf{r}$, graph, and compare the results with Exercise 2.

■ 2. $\mathbf{r} - \mathbf{s} = \mathbf{r} + (-\mathbf{s})$
$= (4 + 6, 9 + 2)$
$= (10, 11)$.

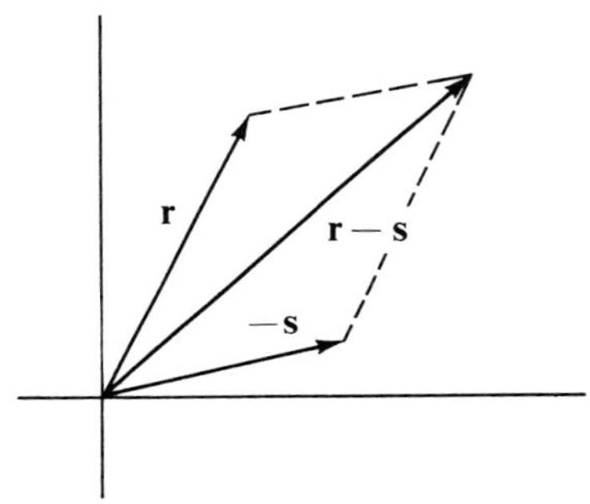

---

4. If $\mathbf{u} = (8, 40°)$ and $\mathbf{v} = (5, -60°)$, find $\mathbf{u} + \mathbf{v}$.

■ 3. $\mathbf{s} - \mathbf{r} = \mathbf{s} + (-\mathbf{r})$
$= (-6 - 4, -2 - 9)$
$= (-10, -11)$.

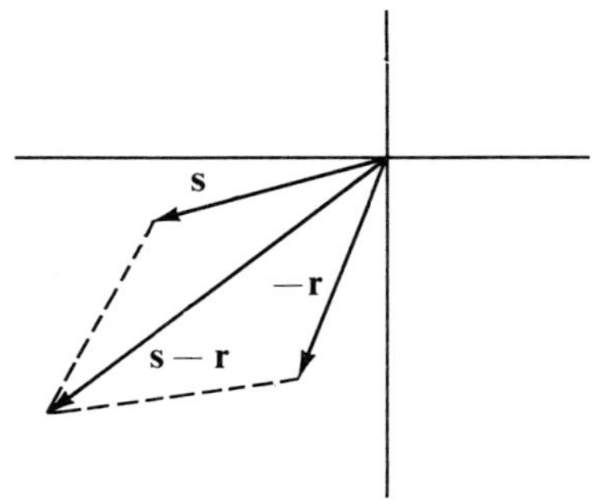

5. If $\mathbf{z} = (-.5, 3.5)$ and $\mathbf{w} = (-4.2, -6.7)$, find $2(\mathbf{z} + \mathbf{w})$.

■ 4.
i. $x_u = 8 \cos 40° = 6.12$
$y_u = 8 \sin 40° = 5.14$;
ii. $x_v = 5 \cos 60° = 2.5$
$y_v = -5 \sin 60°$
$= -4.33$;
iii. $\mathbf{u} + \mathbf{v} = (8.62, 0.81)$
$= (8.62, 4.4°)$.

6. Solve for the value of $\mathbf{t}$ when $\mathbf{r} + 3\mathbf{t} = \mathbf{s}$, $\mathbf{r} = (5, 3)$, and $\mathbf{s} = (-2, 9)$.

■ 5. $2(\mathbf{z} + \mathbf{w}) = 2(-.5 - 4.2, 3.5 - 6.7)$
$= 2(-4.7, -3.2)$
$= (-9.4, -6.4)$.

7. Find s in polar form.

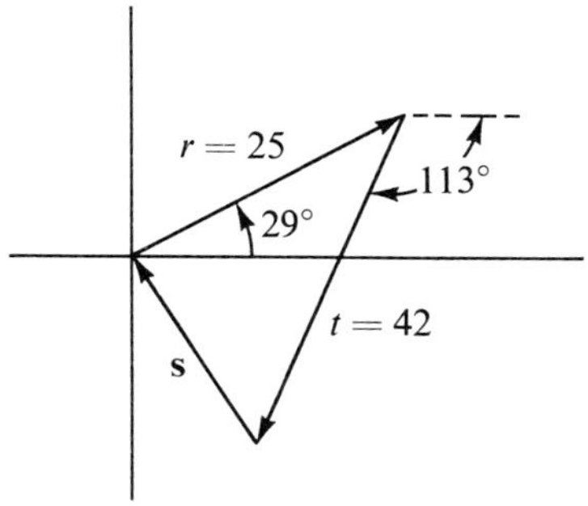

■ 6. $\mathbf{t} = \frac{1}{3}(\mathbf{s} - \mathbf{r})$
$= \frac{1}{3}(-7, 6)$
$= \left(-\frac{7}{3}, 2\right)$.

■ 7.
i. $\mathbf{r}$ and $\mathbf{t}$ represented in standard rectangular form are:
$\mathbf{r} = (25, 29°)$
$= (25 \cos 29°, 25 \sin 29°)$
$= (21.9, 12.1)$;
$\mathbf{t} = (42, -113°)$
$= (-42 \cos 67°, -42 \sin 67°)$
$= (-16.4, -38.7)$.
ii. $\mathbf{r} + \mathbf{t} + \mathbf{s} = (0, 0)$
$\mathbf{s} = -\mathbf{r} - \mathbf{t}$,
$= (-5.5, 26.6)$.

6-7 Problems

1. When two or more forces (forces are vector quantities) act on the same point, the sum of the forces is the single force that has the same effect as the others combined. The *sum* of forces is often called the **resultant**. Find the resultant of two forces A and B acting on a point, when A has a magnitude of 35 pounds at 20° and B has a magnitude of 40 pounds at 60°.
2. If a boat with a speed of 8 miles per hour in still water heads north in a river whose water is flowing eastward at 3 miles per hour, what speed and direction does the boat actually travel?
3. Find the resultant of the following forces:

A	38 pounds	121°,
B	76 pounds	261°,
C	57 pounds	317°.

4. Find the resultant of the following forces:

A	20 pounds	47°,
B	30 pounds	121°,
C	50 pounds	180°,
D	70 pounds	300°.

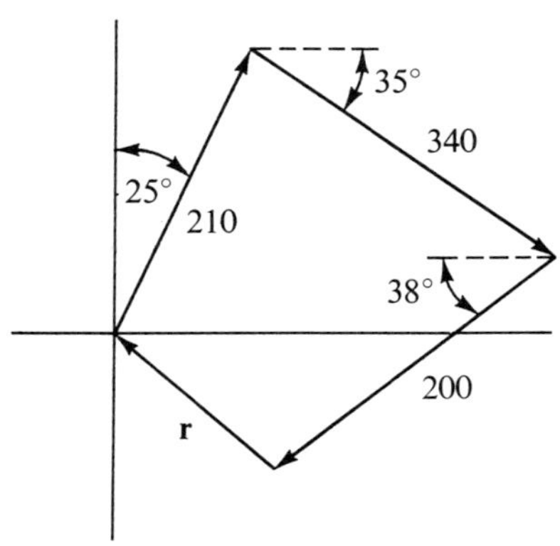

Fig. 6-7c

5. The sum of the four vectors in Fig. 6-7c is equal to (0, 0). Find the polar form of **r** (the dotted lines are parallel to the x-axis).
6. The surface tension forces (s) and adhesive force (A) are in equilibrium; in other words, their sum is zero.
   a. Determine $\mathbf{s}_{LV}$ when

$$\mathbf{A} = (10, 180°),$$

$$\mathbf{s}_V = (25, 90°), \quad \mathbf{s}_L = (8, 270°).$$

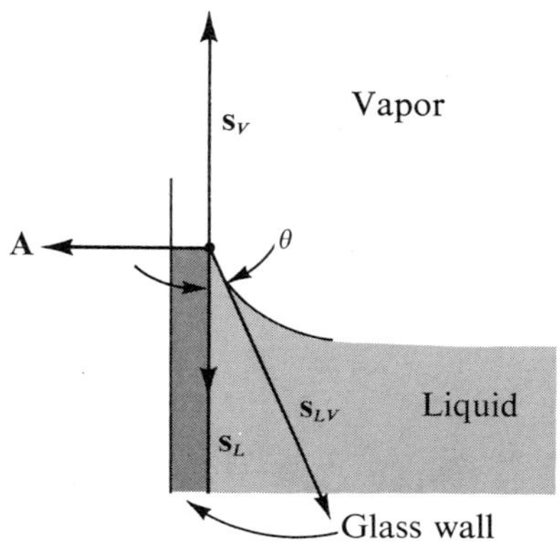

b. What is the value of the contact angle θ?

7. The wrecking ball is at the end of its upward swing. If the weight $\mathbf{w} = (2200, 270°)$ and $\theta = 35°$, find the tension (**T**) and the force that causes the wrecking ball to swing back.

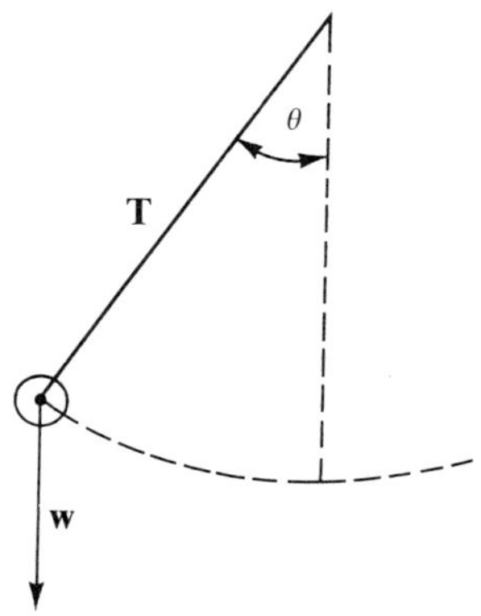

8. The guy wire has a maximum safe tensile strength of 5.3 tons, and the boom has a maximum safe crushing strength of 7.4 tons. If the safe load is exceeded, which will fail first? What is the maximum safe load?

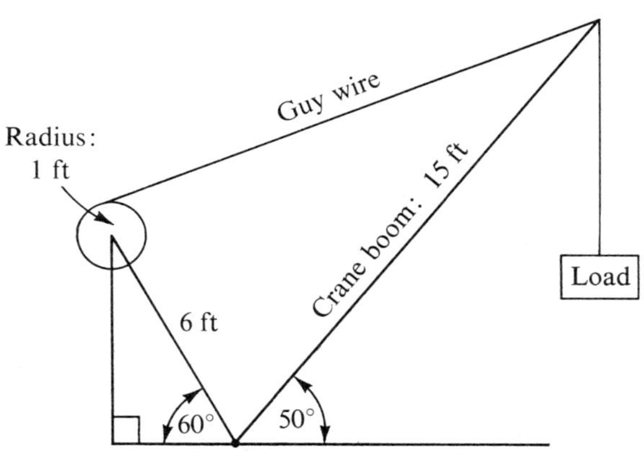

6 Supplementary Exercises and Problems

1. Change $28° 37' 2''$ to degrees and decimal parts of a degree. (6-1)
2. Change $48.369°$ to degrees, minutes, and seconds. (6-1)
3. An angular velocity is $12° 10' 5''$ per 3.62 seconds. What is the angular velocity in degrees per minute? (6-1)
4. Use a table or slide rule to find
   a. tan 37.5°,
   b. tan 78.7°,
   c. tan 12.3°. (6-1)
5. Use a table or slide rule to find θ, when
   a. $\tan \theta = 7.821$,
   b. $\tan \theta = .3462$,
   c. $\tan \theta = .0487$. (6-1)
6. At what angle must an airplane climb to clear 120 feet 375 feet from takeoff? (6-1)
7. The maximum safe angle of descent is 10° when flying below an altitude of 1000 feet. How far from the touch-down point must the plane start its descent from 1000 feet? (6-1)
8. Find y when $\tan 62.3° = \dfrac{y}{14}$. (6-1)
9. Find x when $\tan 13.7° = \dfrac{12{,}500}{x}$. (6-1)
10. Find θ. (6-1)

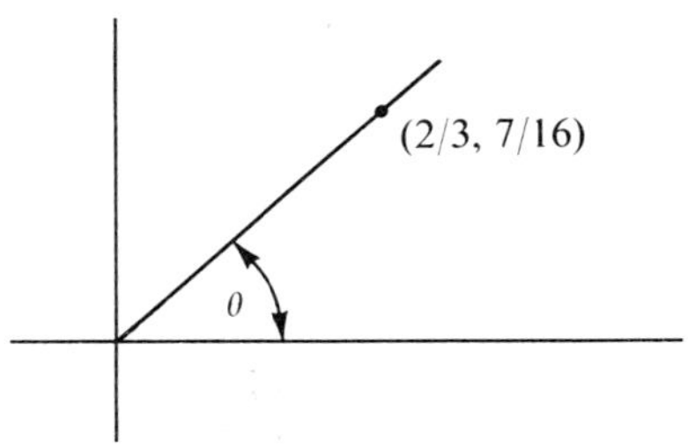

11. Find θ. (6-1)

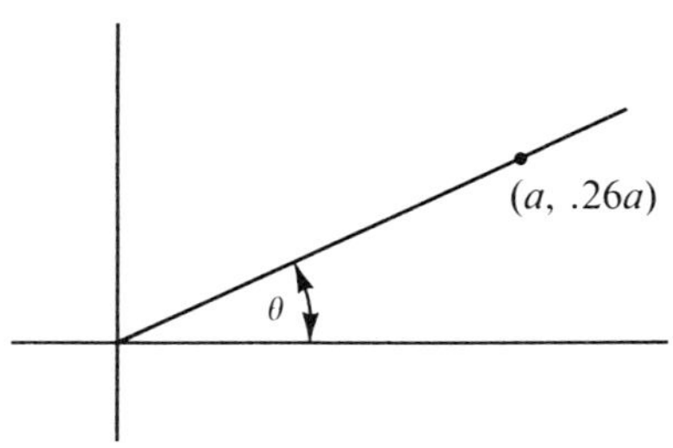

12. Using a table or slide rule, find the following values:
    a. $\sin 15.2°$,
    b. $\cos 15.2°$,
    c. $\sin 72.6°$,
    d. $\cos 17.4°$. (6-2)
13. If $\sin \theta = 3m$ and $\cos \theta = 7m$, find θ. (6-2)
14. If $\sin \theta = .3821$, find $\cos(90° - \theta)$. (6-2)
15. If $\sin \alpha = \cos \beta$, and $\alpha - \beta = 20°$, find α and β. (6-2)
16. Using a table or slide rule, find θ when
    a. $\sin \theta = .2222$,
    b. $\cos \theta = .7888$,
    c. $\cos \theta = .7070$,
    d. $\sin \theta = .0000$. (6-2)
17. Find r and y. (6-2)

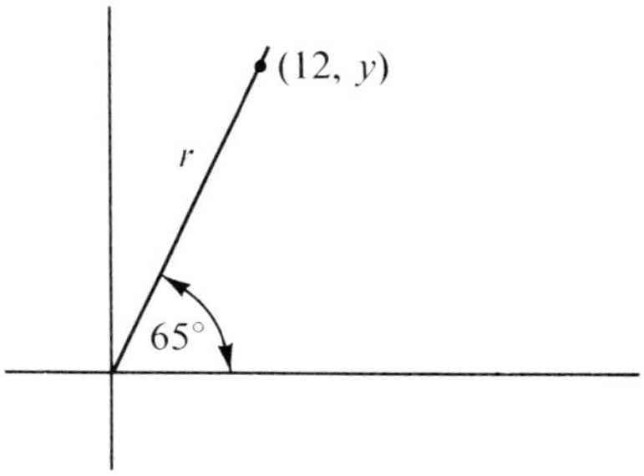

18. Describe a method of measuring the angle RQP using only a measuring tape and information from trigonometric tables. (6-2)

19. The slide rule contains an ST scale that is used to evaluate both $\sin \theta$ and $\tan \theta$ in a limited interval of values for θ. Examine a slide rule and trigonometric tables and determine where $\sin \theta \approx \tan \theta$ to within .001. Prepare an informal justification for this occurrence. (6-2)
20. What angle (to the nearest one-tenth degree) on the slide rule corresponds to the minimum angle on the S scale and maximum angle on the ST scale? (6-2)

21. The safe range for operating a swinging crane is between 32 and 78 degrees. What is the maximum distance over which it can operate from a stationary position if the crane is 32 feet long? What is the maximum height it can elevate a load from level ground? (6-2)
22. A weather balloon is anchored with a 1200-foot cord. What is its altitude if it makes a 65° angle with level ground? (6-2)
23. Find the length of a side of a regular octagon inscribed in a circle with a diameter of 4 feet. (6-2)
24. Calculate the length of the diagonal $\overline{AB}$ for this rectangular prism. (6-2)

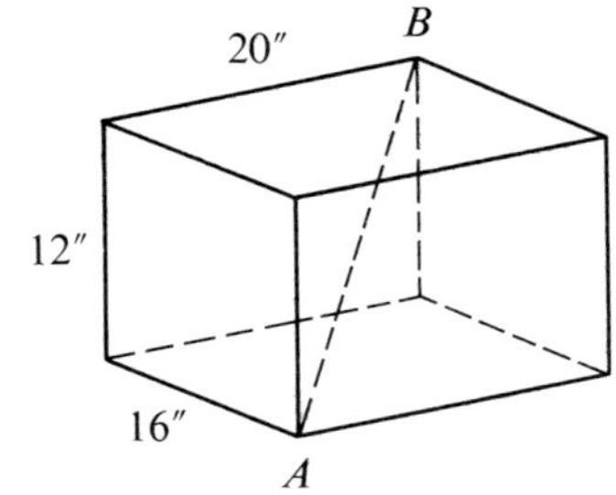

25. Find the angle that $\overline{BC}$ makes with $\overline{AD}$ when $\overline{AB} = 3.214$ inches, $\overline{BC} = 4.300$ inches, and $\overline{CD} = 1.831$ inches. (6-2)

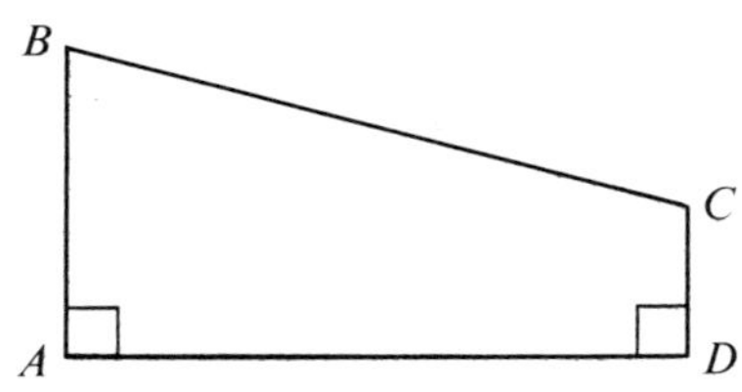

26. Find the radius of the circle when the distance $\overline{T_1T_2} = 1000$ feet. T_1 and T_2 are points of tangency. (6-2)

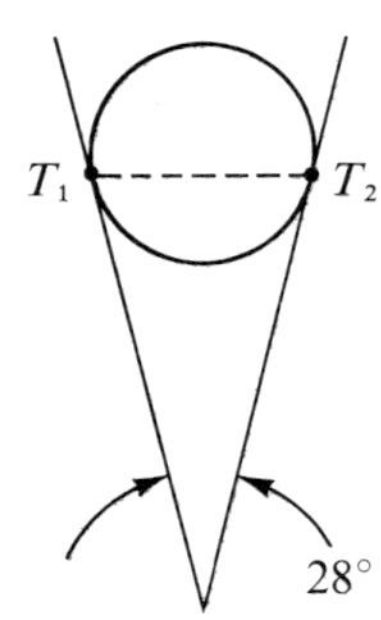

27. The 0, 10, 20, . . . , 60 values are evenly spaced on the arc dial. Where should 10, 20, 40, and 50 be placed on the three-inch linear scale relative to the zero? (6-2)

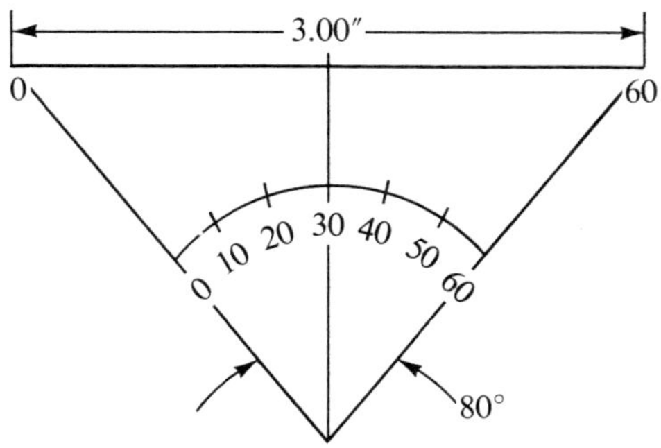

28. Find the radius of the circle. (6-2)

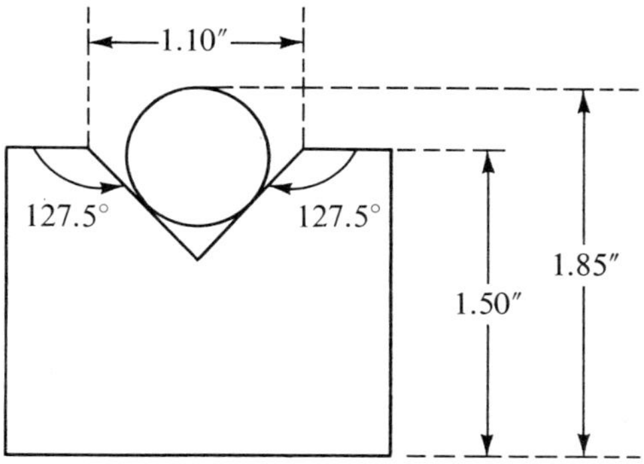

29. Find the length w on the threads of a worm gear. (6-2)

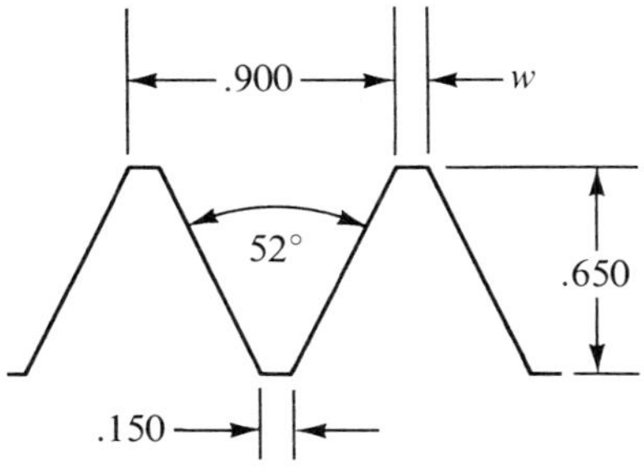

30. $\angle ABC = 132°$,
$\overline{BC} = 35.5$ ft,
$\angle BCD = 93°$.
Find $\overline{CD}$ and $\angle CDE$. (6-2)

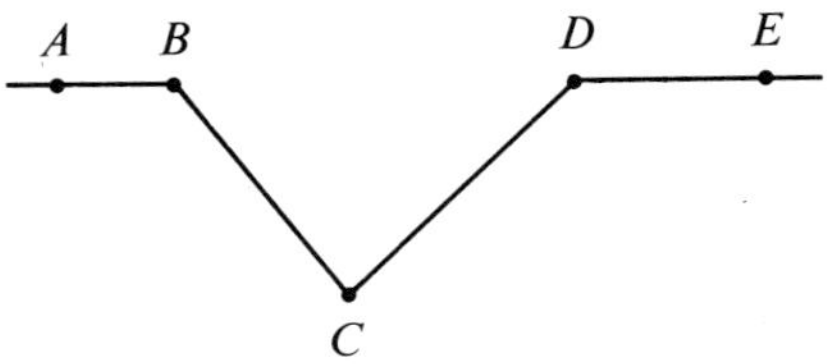

31. With the aid of tables or slide rule, find
 a. sin 340°,
 b. sin 98°,
 c. sin(−75°). (6-3)
32. With the aid of tables or slide rule, find
 a. cos 237°,
 b. cos(−110°),
 c. cos 720°. (6-3)
33. Using tables or slide rule, find
 a. tan 180°,
 b. tan(−20°),
 c. tan 116°. (6-3)
34. Find all the values for θ such that $0° \leq \theta < 360°$ when
 a. $\sin \theta = -.6060$,
 b. $\cos \theta = -.7841$,
 c. $\tan \theta = -2.3021$. (6-3)
35. Find all the values for θ such that $0° \leq \theta < 360°$ when
 a. $\sin \theta = .3412$,
 b. $\cos \theta = .2315$,
 c. $\tan \theta = .4681$. (6-3)
36. Find sides b and a. (6-3)

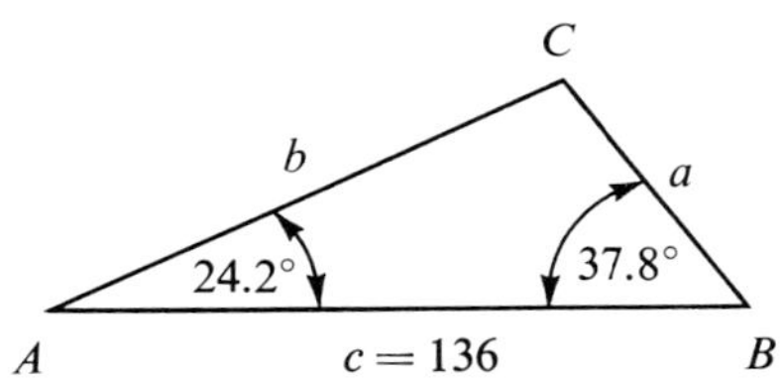

37. Find C and b. (6-3)

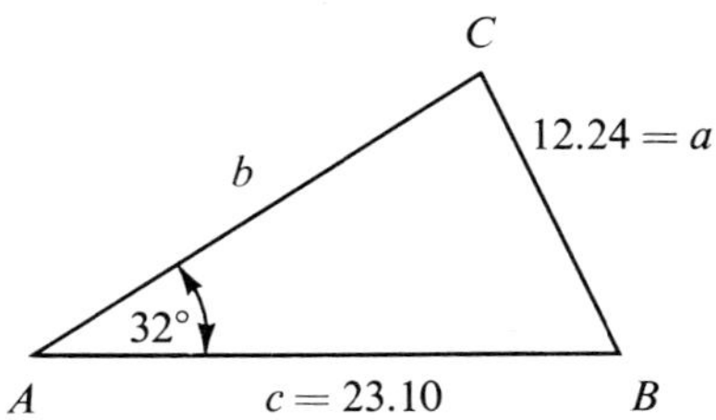

38. Find the height of triangle MNP from vertex P. (6-3)

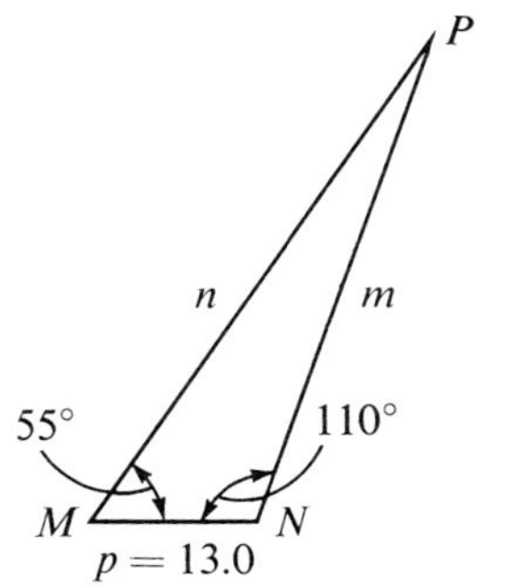

39. Write the law of cosines for a^2 and b^2. (6-4)

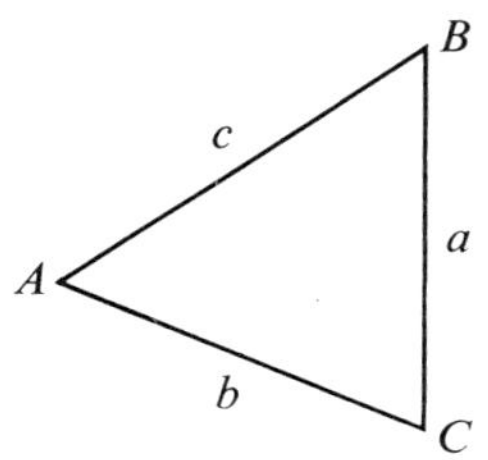

40. Solve the law of cosines for cos A. (6-4)
41. Find the length of the unknown side of the triangle. (6-4)

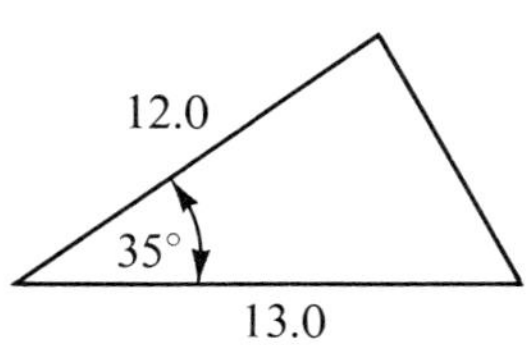

42. Find length x. (6-4)

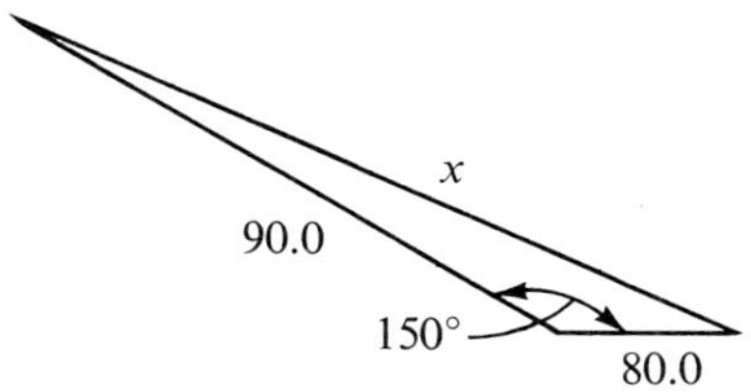

43. Find angle C (6-4)

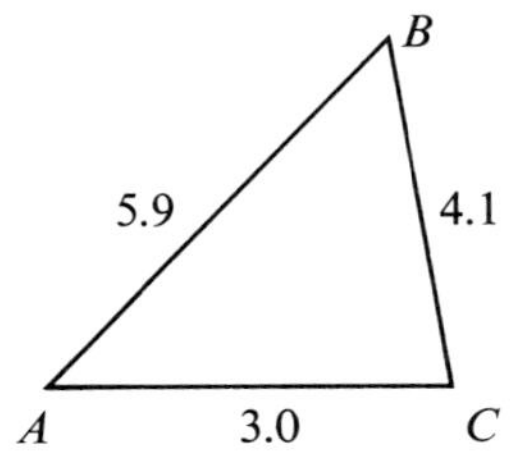

44. If the quadrilateral is a parallelogram and $F_1 = 200$, $F_2 = 175$, and $\theta = 78°$, then find R. (6-4)

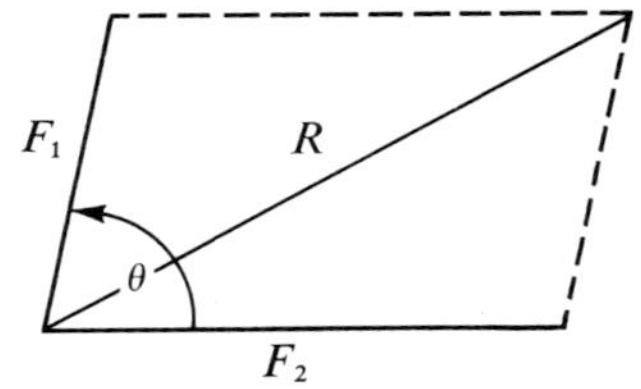

45. The jib of a crane is the rigid member that does the lifting. The jib makes a 75° angle with the horizontal. If the jib swings through an angle of 80° about its vertical axis, what is the angle between the initial and final positions of the jib? (6-4)

46. Find the area. (6-5)

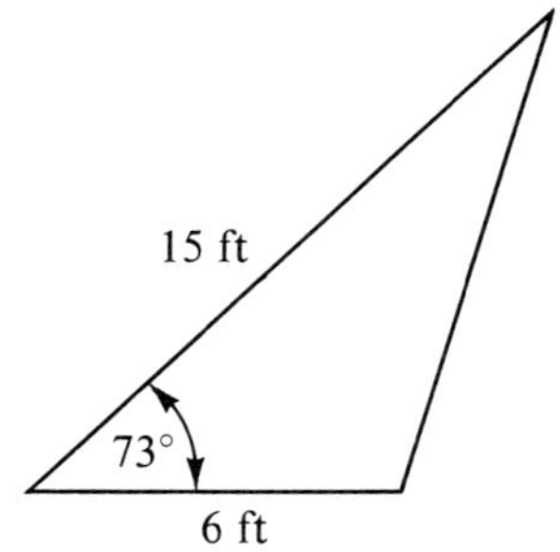

47. Gravel is piled in the shape of a cone. The diameter of the base is 42 feet and the angle the sloping sides form with the horizontal is 40°. Find the volume in cubic yards. (6-5)

48. The area of triangle ABC is 90 square inches. Find angle C.

(6-5)

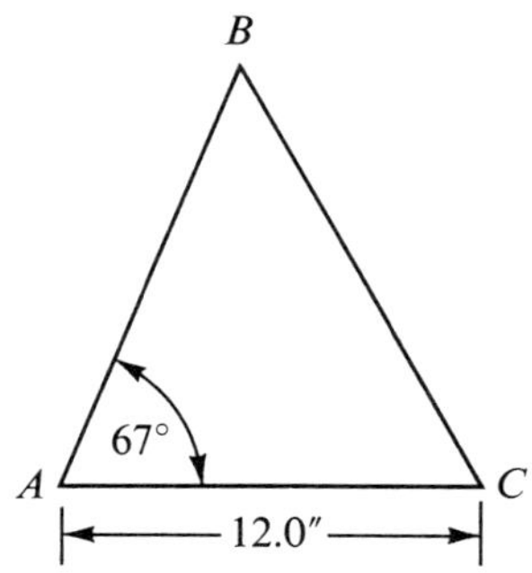

49. Inorganic fertilizer is piled into the square corner of a warehouse. Find the volume in cubic feet.

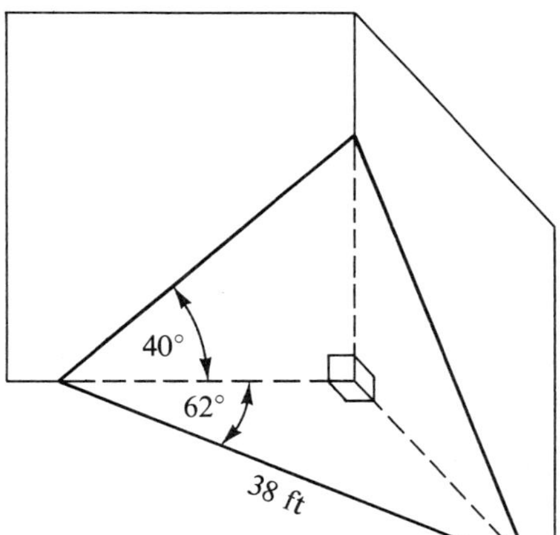

50. $\overline{AB} = 22$, $\angle CAB = 38°$,
$\overline{CD} = 16$, $\angle CBA = 48°$.
Find the altitude of triangle ABD from the base $\overline{AB}$. (6-5)

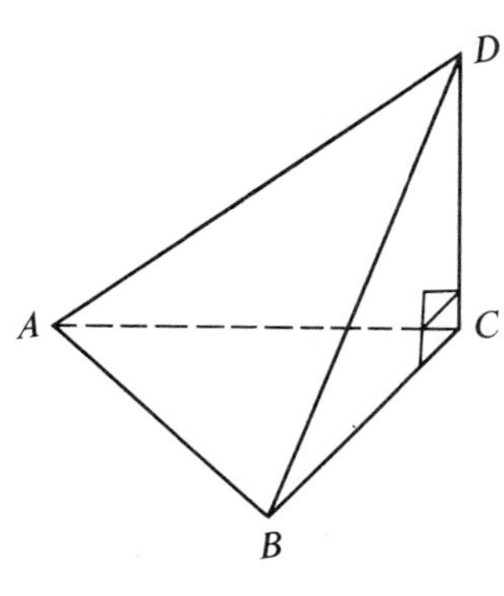

51. Write $\mathbf{r} = (-3.5, 6.2)$ in polar form. (6-6)
52. Write $\mathbf{s} = (12.3, -4.7)$ in polar form. (6-6)
53. Change $\mathbf{t} = (7.81, 139°)$ to rectangular form. (6-6)
54. Change $\mathbf{v} = (187, -38°)$ to rectangular form. (6-6)
55. If $\mathbf{r} = (1000, 212°)$, find $-\mathbf{r}$. (6-6)
56. If $\mathbf{u} = (-13, 25)$, find $-\mathbf{u}$. (6-6)
57. If $\mathbf{u} = (3700, 1200)$ and $\mathbf{v} = (-2100, 700)$, find $\mathbf{u} + \mathbf{v}$ and illustrate geometrically. (6-7)
58. The forces $\mathbf{F}_1 = (24, 6)$ and $\mathbf{F}_2 = (16, -4)$ act simultaneously on the same object. Determine $\mathbf{F}_1 + \mathbf{F}_2$ and illustrate graphically. (6-7)
59. If $\mathbf{r} = (13.0, 130°)$ and $\mathbf{s} = (15.0, 25°)$, find $\mathbf{r} + \mathbf{s}$; represent the results in polar form. (6-7)
60. Using the values of $\mathbf{r}$ and $\mathbf{s}$ from the preceding exercise, calculate $\mathbf{r} - \mathbf{s}$; represent the results in polar form. (6-7)
61. If $\frac{1}{4}\mathbf{s} = \mathbf{t} - \mathbf{s}$, find $\mathbf{t}$ when $\mathbf{s} = (3.1, -7.5)$. (6-7)
62. Find $\mathbf{r}$. (6-7)

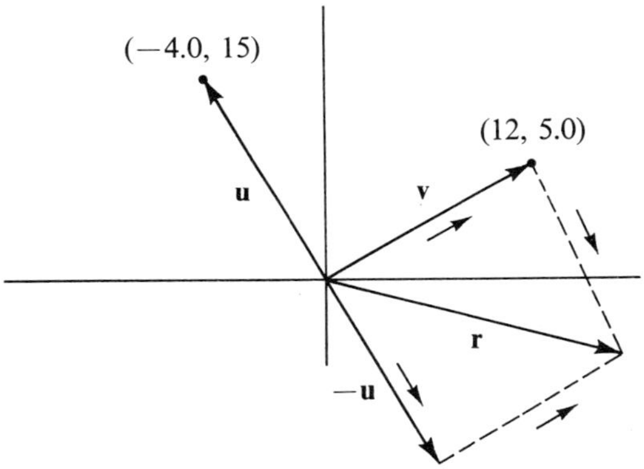

63. The "rimpull" on a loaded truck of weight $\mathbf{W}$ due to the grade of the road is represented by $\mathbf{R}$. Find $|\mathbf{R}|$ when $|\mathbf{W}| = 40{,}000$ pounds. (6-7)

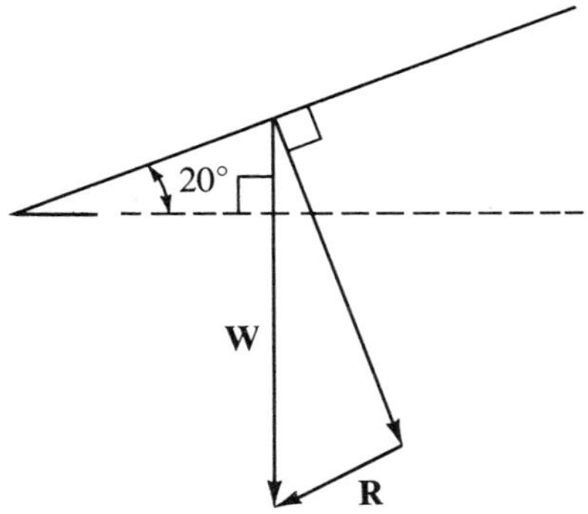

Chapter Seven Trigonometric Functions of Real Numbers

7-1 Radians

In Chapter 6 we used the degree as a unit of angular measurement. There is another unit which is sometimes more convenient. This unit, called a **radian**, relates the central angle of a circle to the radius of a circle. Since

$$C = 2\pi r, \qquad \frac{C}{r} = 2\pi. \qquad \text{(Eq. 7-1a)}$$

That is, the ratio of the circumference of a circle to its radius is 2π.

A central angle whose arc is equal in length to the radius is defined as one radian. Therefore,

$$2\pi \text{ radians} = 360°,$$

$$\pi \text{ radians} = 180°,$$

and

$$1 \text{ radian} = \frac{180°}{\pi} \approx 57.3°.$$

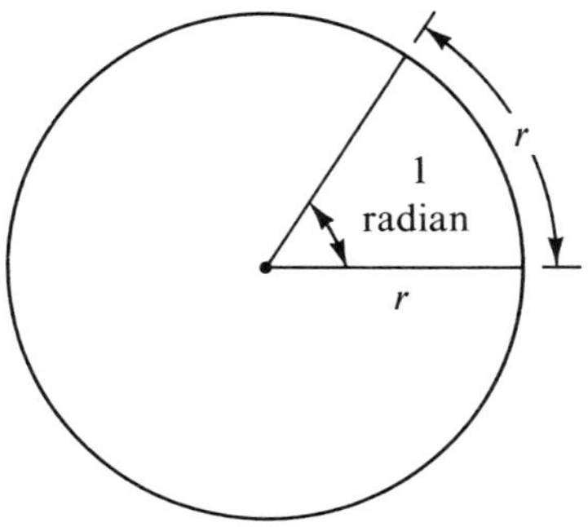

Fig. 7-1a

Example 1

Find the number of degrees in 3.2 radians.

$$3.2 = \frac{3.2\pi}{\pi} = \frac{3.2(180^\circ)}{3.14} = 183.4^\circ.$$

It is sometimes desirable to express radians in terms of π.

Example 2

Express 5.32 radians in terms of π.

$$5.32 = \frac{5.32\pi}{\pi} = \frac{5.32\pi}{3.14} = 1.69\pi.$$

Example 3

Find the number of radians in 62 degrees.

$$62^\circ = \frac{62^\circ\pi}{\pi} = \frac{62^\circ\pi}{180^\circ} = .344\pi, \quad \text{or} \quad 1.08 \text{ radians.}$$

The trigonometric functions of angles given in radians can be found directly from tables giving the angles in radians. They can also be found by converting radians into degrees and using the degree table.

7-1 Exercises

1. Change from degree to radian measure:
   a. 30°,
   b. 135°.

2. Change from radian measure to degrees:
   a. $\frac{\pi}{2}$,
   b. $\frac{\pi}{6}$.

■ 1. a. $\frac{30^\circ\pi}{180^\circ} = .167\pi$, or .523;
b. $\frac{135^\circ\pi}{180^\circ} = .75\pi$, or 2.36.

3. Change from degree to radian measure:
a. 270°,
b. −110°.

■ 2. a. $\frac{\pi}{2}\left(\frac{180°}{\pi}\right) = 90°$;
b. $\frac{\pi}{6}\left(\frac{180°}{\pi}\right) = 30°$.

---

4. Change from radian measure to degrees:
a. $\frac{3\pi}{2}$,
b. .48.

■ 3. a. $\frac{270°\pi}{180°} = 1.5\pi$,
or 4.72;
b. $\frac{-110°\pi}{180°} = -.612\pi$,
or −1.92.

---

5. Find the following trigonometric values:
a. $\sin\frac{\pi}{4}$,
b. $\sin\frac{3\pi}{2}$.

■ 4. a. $\frac{3\pi}{2}\left(\frac{180°}{\pi}\right) = 270°$;
b. $\frac{(.48)(180°)}{\pi} = 27.5°$.

---

6. Find:
a. $\cos\frac{\pi}{3}$,
b. $\cos\frac{5\pi}{6}$.

■ 5. a. $\sin\frac{\pi}{4} = .707$;
b. $\sin\frac{3\pi}{2}$
$= -\sin\left(\frac{3\pi}{2} - \pi\right)$
$= -\sin\frac{\pi}{2}$
$= -1.000$.

---

7. Find:
a. $\tan\frac{\pi}{6}$,
b. $\tan\frac{3\pi}{4}$.

■ 6. a. $\cos\frac{\pi}{3} = .500$;
b. $\cos\frac{5\pi}{6}$
$= -\cos\left(\pi - \frac{5\pi}{6}\right)$

$$= -\cos\frac{\pi}{6}$$

$$= -.866.$$

---

8. Find:
   a. sin .63,
   b. cos 1.93,
   c. tan .79.

■ 7. a. $\tan\frac{\pi}{6} = .577;$

b. $\tan\frac{3\pi}{4}$

$$= -\tan\left(\pi - \frac{3\pi}{4}\right)$$

$$= -\tan\frac{\pi}{4}$$

$$= -1.000.$$

---

9. Find θ in radians:
   a. $\sin\theta = .6060$,
   b. $\cos\theta = .4818$,
   c. $\tan\theta = 1.4605$.

■ 8. a. sin .63 = .589;
   b. cos 1.93 = −.352;
   c. tan .79 = 1.009.

---

10. Find:
   a. sin 4.62,
   b. cos 5.3,
   c. tan 2.

■ 9. a. $\theta = .6510$ radians;
   b. $\theta = 1.0681$ radians;
   c. $\theta = .9704$ radians.

■ 10. a. sin 4.62

$$= -\sin(4.62 - \pi)$$

$$= -\sin 1.478$$

$$= -.9957;$$

b. cos 5.3

$$= \cos(2\pi - 5.3)$$

$$= \cos .983$$

$$= .5548.$$

c. $\tan 2$

$$= -\tan(\pi - 2)$$
$$= -\tan 1.14$$
$$= -2.184.$$

7-1 Problems

The first three problems are an extension of the instruction of this section. The ideas they present are used extensively in the sections that follow.

1. One of the advantages of radian measure is the convenient relationship that exists between the central angle and corresponding arc length. When θ is given in radian measure, (arc length) $= r\theta$.

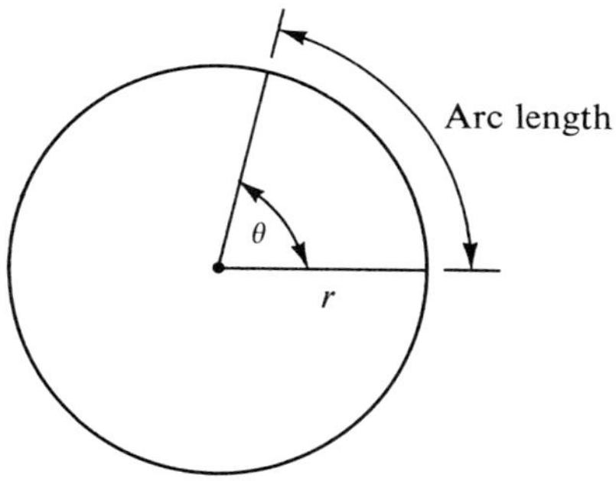

Calculate the missing quantities in the following table.

Central angle	Radius	Arc length
π radians	4.00 in.	______
1.55 radians	______	16.0 in.
______	5.52 in.	12.5 in.

2. If the terminal side of an angle is changing position at a constant rate, the relationship between time, rate of change of angle, and the angle at any time is analogous to the relationship between time, velocity, and distance. If $t =$ (time), $\omega =$ (angular velocity), and $\theta =$ (the angle), then

$$\omega t = \theta.$$

Calculate the missing quantities in the following table.

Angular velocity (ω)	Angle (θ)	Time (t)
0.4 radians per second	___ radians	12 seconds
$\pi/2$ radians per second	6.4 radians	___ seconds
___ radians per second	60π radians	.5 seconds

3. If $\omega =$ (radians per second) and $r =$ (radius in inches), what would ωr represent?
4. Change the angular velocity of 6.28π radians per second to degrees per second; to revolutions per second.
5. Find the length of the belt.

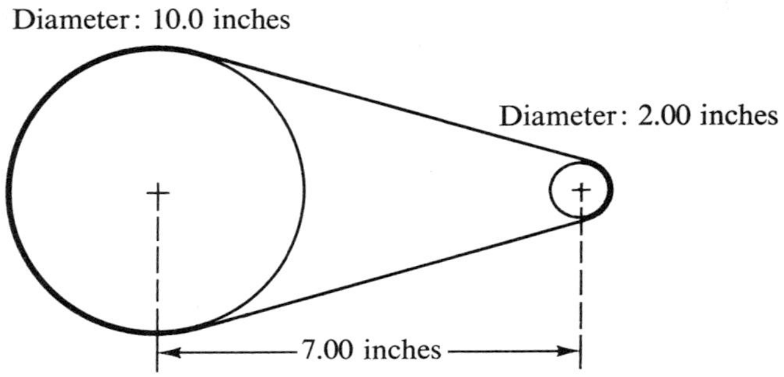

6. When Q makes one revolution because of its contact with wheel P, P turns $4\pi/5$ radians. Find the radius of wheel P when the radius of Q is one inch.

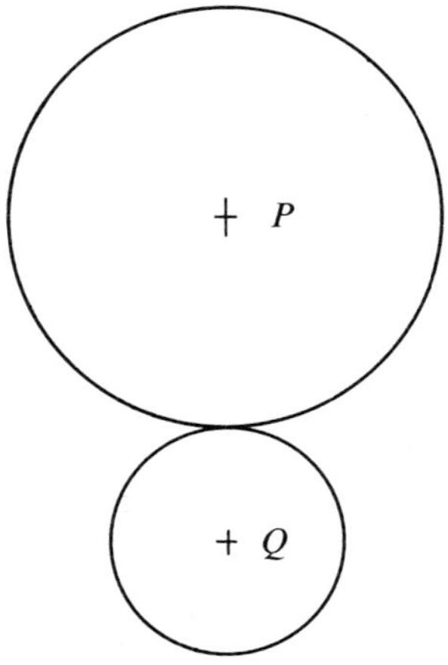

7. If the angular velocity of a rolling wheel 34 inches in diameter is 15π radians per second, what is the linear velocity of the wheel?
8. The radius of the earth's equator is 6371 kilometers. What is the length of one arc degree on the equator?
9. Calculate the circumference of the earth's 40th parallel.

7-2 Trigonometric Functions: $f(x) = \sin x$ and $g(x) = \cos x$

Up to this point we have considered trigonometry as an aid to the solution of triangles. In this section we introduce the concept of trigonometric functions. Recall from Chapters 1 and 3 that a *function* is a special relationship between two sets of numbers, called a domain and a range, such that for each element of the domain there is only one corresponding element from the range. The relationship is often defined by an equation and is easily graphed on a coordinate system. The functions of this section are defined by the equations $f(x) = \sin x$ and $g(x) = \cos x$.

Let us first turn our attention to $f(x) = \sin x$. To produce any part of the set of ordered pairs $(x, f(x))$ that comprise the function f, we rely on the use of trigonometric tables.

After completing the following table, graph the results.

$f(x) = \sin x$

$f(x)$	x (radians)
.000	.0
.479	.5
.841	1.0
.997	1.5
.909	2.0
.598	2.5
.141	3.0
−.351	3.5
	4.0
−.978	4.5
	5.0
−.706	5.5
	6.0
.215	6.5

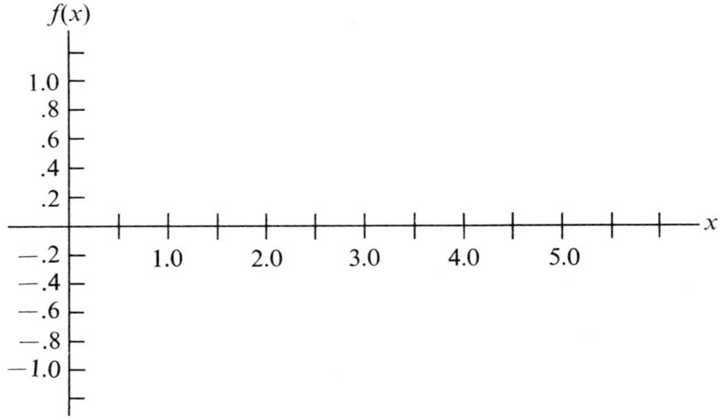

$f(x)$	x (radians)
−.757	4.0
−.959	5.0
−.279	6.0

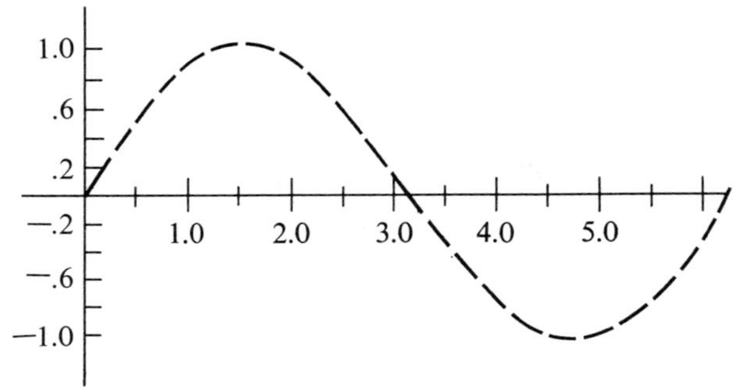

Fig. 7-2a

The graph of $f(x) = \sin x$ is called a **sine wave**. Anyone who is at all familiar with alternating current produced by an alternator should recognize this as a typical graph of voltage plotted against time.

If the graph of Figure 7-2a is extended to approximately $x = 12.6$, we discover that what already appears is repeated (Figure 7-2b). Functions that have the characteristic of repeating all of the values of the range in a definite order over a specific interval of the domain are called **periodic functions**. The smallest interval of domain over which this repetition occurs is called the **period**.

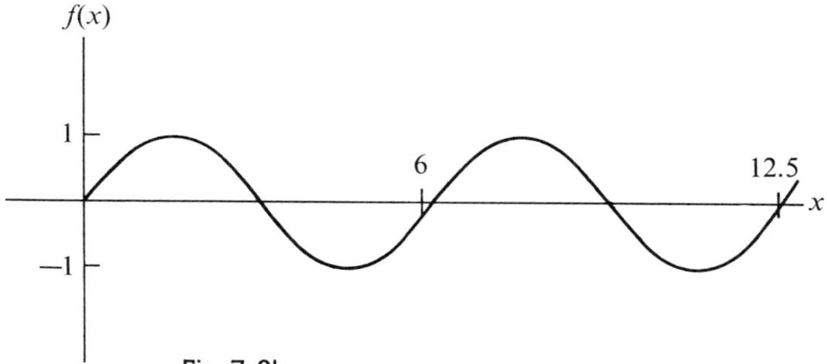

Fig. 7-2b

For the function $f(x) = \sin x$ the period is $\boxed{}$.

■ 2π, or approximately 6.28

Stated more precisely, if $f(x) = f(x + P)$ over the domain of the function, then the function is periodic with period P. In terms of the sine function, for all x,

$$f(x) = \sin x = \sin(x + 2\pi).$$

All of the basic trigonometric functions are periodic.

The sine wave can be produced graphically in a very simple and interesting manner. Let the variable x be measured in radians and the radius vector $r = 1$. Then the line segment (Fig. 7-2c) $\overline{OP} = \sin x$. Each corresponding value x and $\overline{OP}$ are translated to the accompanying coordinate system.

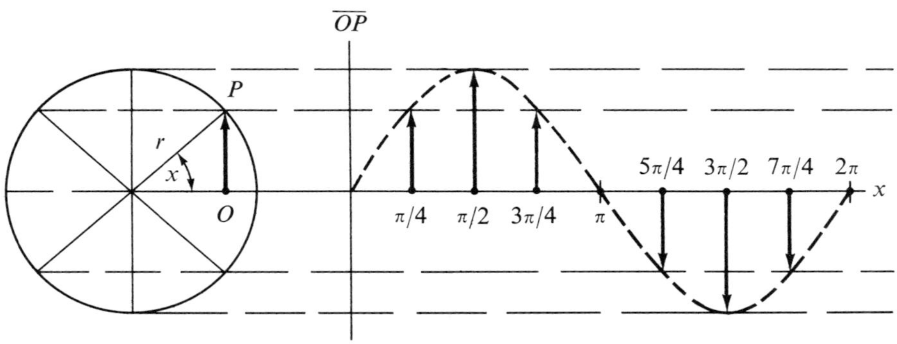

Fig. 7-2c

We would expect the function defined by $g(x) = \cos x$ to have similarities to the sine function. The graph can be produced quickly using a slight variation of the technique used in producing Fig. 7-2c. Again we shall let x be in radians and the radius vector $r = 1$. Then the line segment $\overline{OQ} = \cos x$. (Fig. 7-2d.)

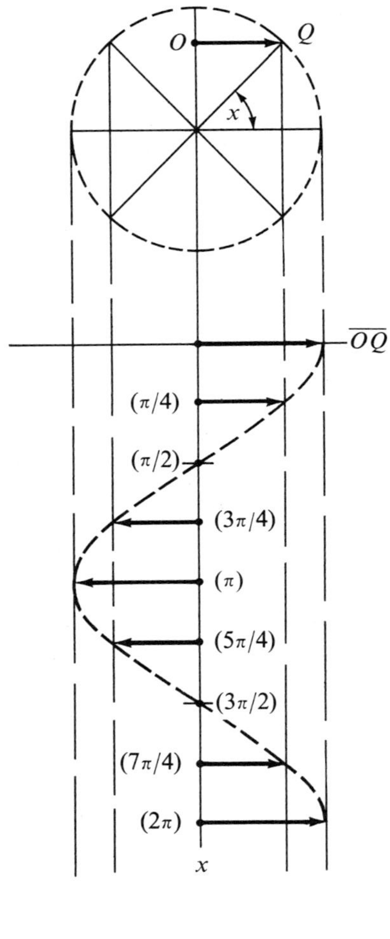

Fig. 7-2d

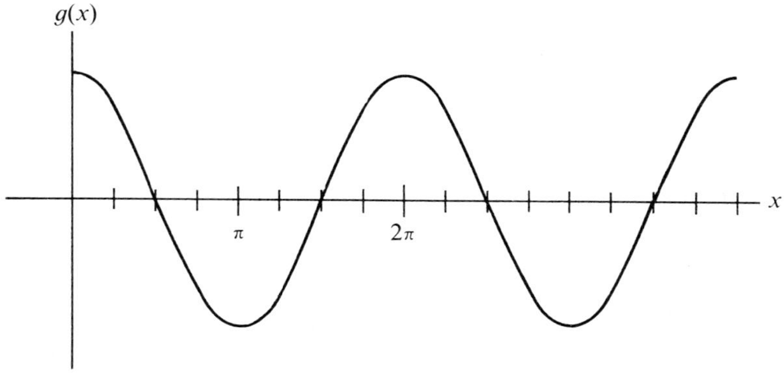

Fig. 7-2e

Extending the graph illustrated in Fig. 7-2d beyond $x = 2\pi$ will produce the graph in Fig. 7-2e. As we would expect, $g(x) = \cos x$ is periodic with a period equal to $\square$.

■ 2π

Having looked only graphically at the cosine function, let us check some of the values, using trigonometric tables.

$$g\left(\frac{\pi}{2}\right) = \cos\frac{\pi}{2} = \square,$$

$$g(\pi) = \cos\pi = \square,$$

$$g\left(\frac{\pi}{4}\right) = \cos\frac{\pi}{4} = \square.$$

■ 0; -1; .707.

7-2 Exercises

1. If $f(x) = \sin x$, find
   a. $f(0)$,
   b. $f(\pi/3)$,
   c. $f(4.21)$,
   d. $f(-1)$.

---

2. If $g(x) = \cos x$, find
   a. $g(0)$,
   b $g(\pi/3)$,
   c. $g(4.21)$,
   d. $g(-1)$.

■ 1. a. $f(0) = \sin 0 = 0$;
b. $f(\pi/3) = \sin \pi/3 = .866$;
c. $f(4.21) = \sin 4.21 = -.875$;
d. $f(-1) = \sin(-1) = -.841$.

---

3. Complete the table and graph.

$f(x) = \cos x.$

$f(x)$	x (radians)
	0.00
	0.50
	1.00
	1.50
	2.00
	2.50
	3.00
	3.50
	4.00
	4.50
	5.00
	5.50
	6.00
	6.50

■ 2. a. $g(0) = \cos 0 = 1.000$;
b. $g(\pi/3) = .5000$;
c. $g(4.21) = -.4848$;
d. $g(-1) = .540$.

4. $\overline{OP}$ in Fig. 7-2c is called the ***y-projection*** of the radius vector (**r**). What is the y-projection of $\mathbf{r} = 8$ when $x = \pi/3$? $x = 7\pi/6$?

■ 3. $f(x) = \cos x$

$f(x)$	x
1.00	0.00
.878	0.50
.540	1.00
.071	1.50
—.416	2.00
—.801	2.50
—.990	3.00
—.936	3.50
—.654	4.00
—.211	4.50
.284	5.00
.709	5.50
.960	6.00
.977	6.50

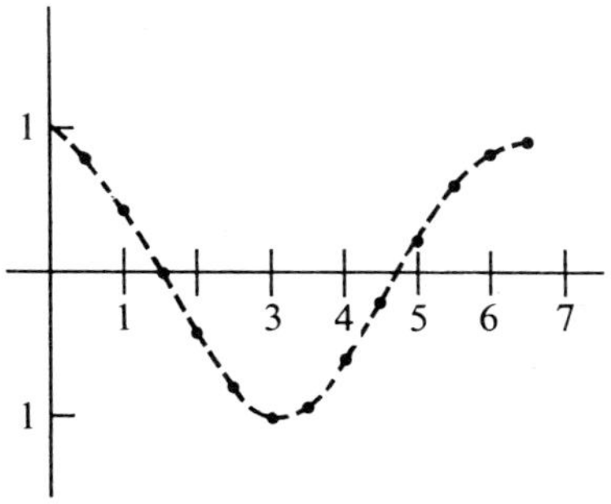

5. $\overline{OQ}$ in Fig. 7-2d is called the ***x-projection*** of the radius vector (**r**). What is the x-projection of $\mathbf{r} = 5.2$ when $x = \pi/8$? $x = 5\pi/8$?

■ 4. a. $\overline{OP} = 8 \sin \pi/3$
$= 8(.866) = 6.92;$

b. $\overline{OP} = 8 \sin 7\pi/6$
$= 8(-\sin \pi/6)$
$= 8(-.500)$
$= -4.00.$

6. If $f(x) = \sin x = .707$, what is x ?

■ 5. a. $\overline{OQ} = 5.2 \cos \pi/8$
$= 5.2(.924) = 4.81;$

b. $\overline{OQ} = 5.2 \cos 5\pi/8$
$= 5.2(-.383)$
$= -1.99.$

---

7. If $g(x) = \cos x = -.866$, what is x?

■ 6. x is $\pi/4 + n(2\pi)$ or $3\pi/4 + n(2\pi)$, n any integer value. (These radian measures correspond to $45° + n\ 360°$ and $135° + n\ 360°$.)

---

8. If $h(x) = \sin x$, what is the value of $h(120)$?

■ 7. x is $5\pi/6 + n(2\pi)$ or $7\pi/6 + n(2\pi)$, n any integer value. (These radian measures are equivalent to $150° + n\ 360°$ and $210° + n\ 360°$.)

■ 8. Since the period of $\sin x$ is $2\pi \approx 6.2832$,

$$\begin{aligned} \sin 120 &= \sin(.61948 \\ &\quad + (19)(6.2832)) \\ &= \sin .6195 \\ &= .5861. \end{aligned}$$

7-2 Problems

1. Review the definition of periodic functions given in Section 7-2. Make a list of some common periodic functions. (Example: Pendulum displacement is a periodic function of time.)
2. Use a physical science reference to discover the definition of *simple harmonic motion*. How is it related to the subject matter of this section?
3. Plot the area of the shadow on the screen as a function of θ.

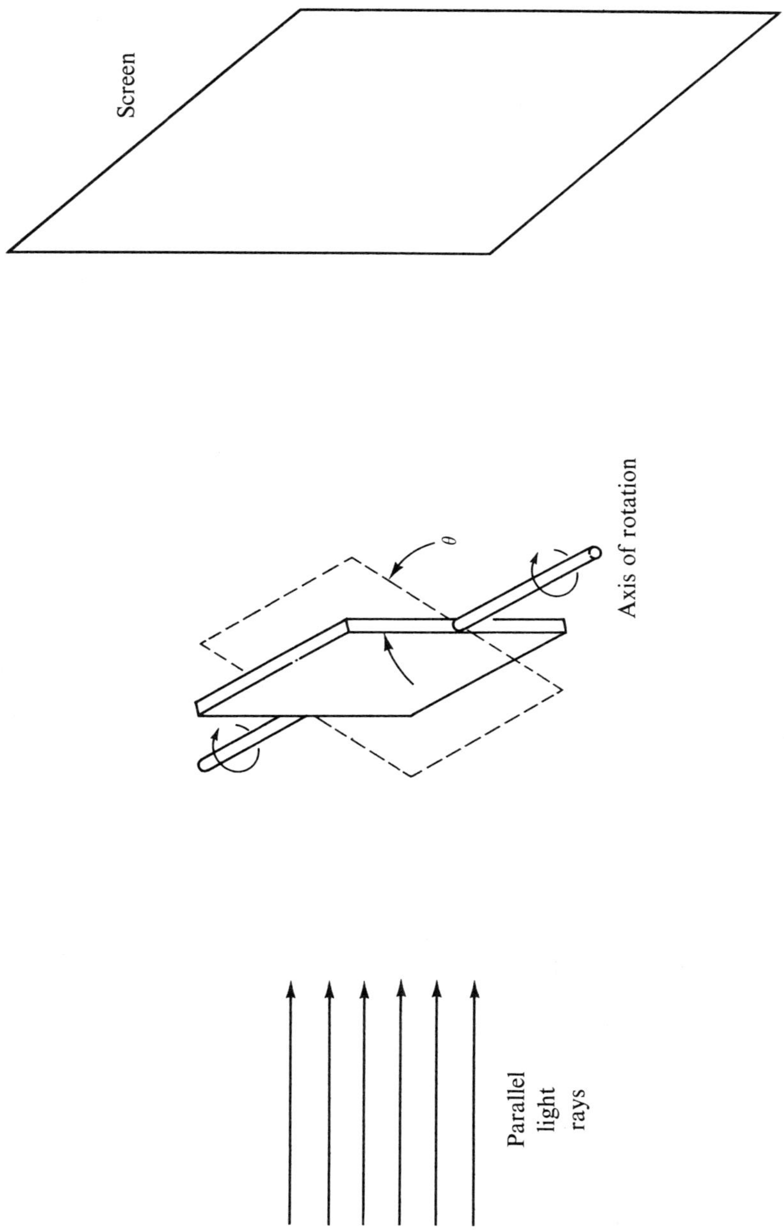
Screen
θ
Axis of rotation
Parallel light rays

4. Plot the position of the top of the piston as a function of θ. (*Caution:* It is not a sine wave.)

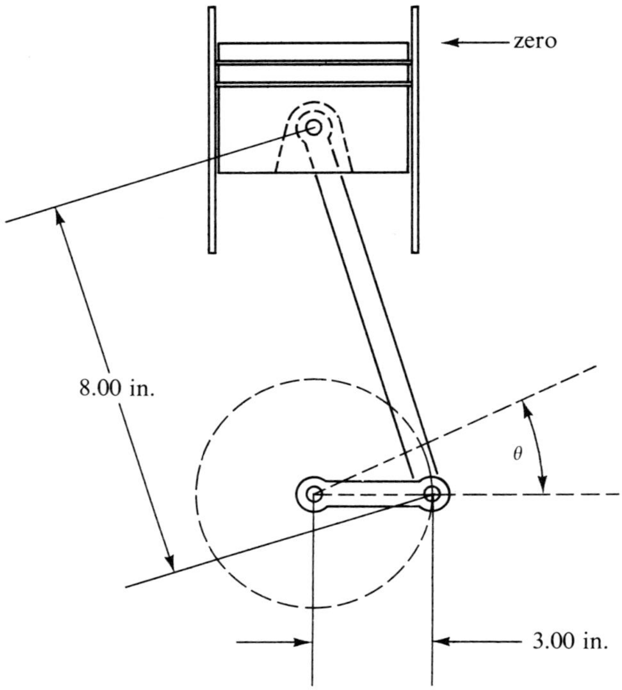

7-3 Amplitude and Period

A slightly more general sine function than that which was studied in Section 7-2 is defined by the equation

$$f(x) = a \sin bx. \qquad \text{(Eq. 7-3a)}$$

The effect of constant a and constant b can be discovered by graphing several examples where only one of the constants is changed while the other remains unchanged.

To determine the effect of the coefficient a, for convenience let $b = 1$ and a have the values 1, 3, and $-1/2$. Therefore, the functions we shall graph for comparison can be represented by the equations

$$f(x) = \sin x, \quad g(x) = 3 \sin x, \quad \text{and} \quad h(x) = -\frac{1}{2} \sin x.$$

x (radians)	$f(x) = \sin x$	$g(x) = 3 \sin x$	$h(x) = -\frac{1}{2} \sin x$
0	.000	.00	.000
$\pi/4$	.707	2.12	−.353
$\pi/2$	1.000	3.00	−.500
$3\pi/4$	.707	2.12	−.353
π	.000	.00	.000
$5\pi/4$	−.707	−2.12	.353
$3\pi/2$			
$7\pi/4$			
2π	.000	.00	.000

■ −1.000; −3.00; .500;
− .707; −2.12; .353.

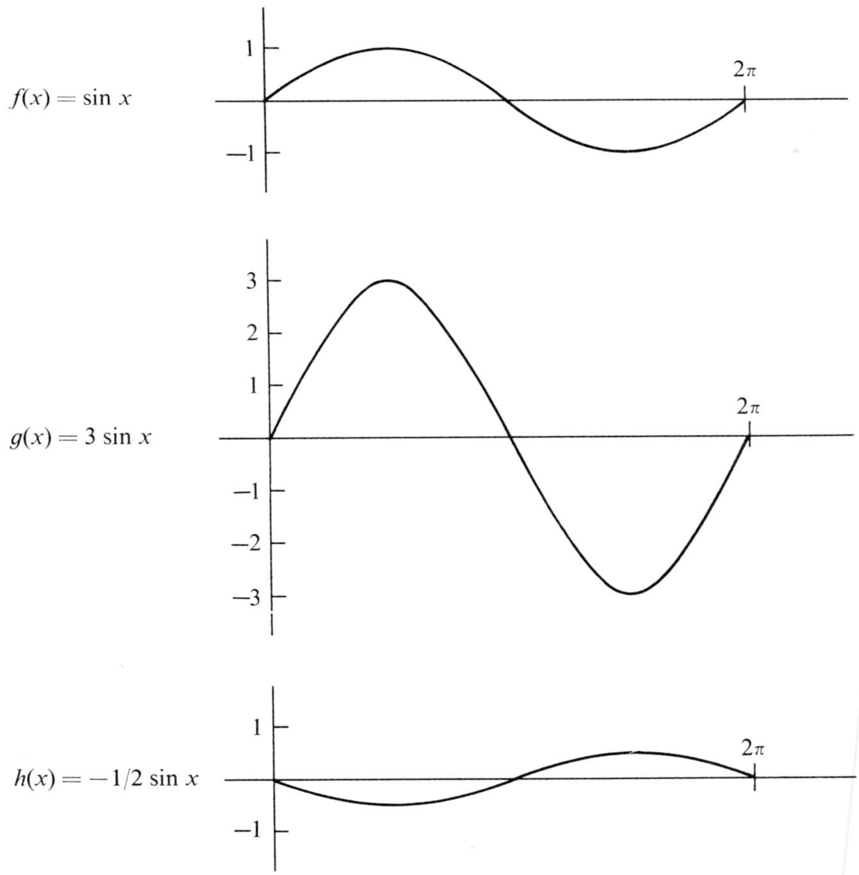

Fig. 7-3a

The coefficient a changes the range of the sine function. For f, the range is from -1 to 1. For g, the range is from [ ], and for h, it is from [ ].

■ -3 to 3; $-\frac{1}{2}$ to $\frac{1}{2}$.

These values are the *minimum* (m) and *maximum* (M) values of the periodic function.

The word **amplitude** may be defined in terms of minimum (m) and maximum (M). For a periodic function,

$$(\text{Amplitude}) = \frac{1}{2}|M - m|. \qquad \text{(Eq. 7-3b)}$$

The amplitudes of the three functions in Fig. 7-3a are one, [ ], and [ ].

■ 3; $\frac{1}{2}$.

In terms of Eq. 7-3a, $|a|$ is equal to the amplitude.

General usage of the word amplitude also includes $|M|$ and $|M - m|$ as accepted definitions. We shall encounter some periodic functions in Section 7-5 where these latter definitions are used.

To discover the effect of b on Eq. 7-3a, we shall graph the three functions that result from letting $a = 1$ and b equal 1, 2, and 2/3.

x (radians)	$f(x) = \sin x$	$g(x) = \sin 2x$	$h(x) = \sin \frac{2x}{3}$
0	.000	.000	.000
$\pi/6$	.500	.866	.342
$\pi/3$	.866	.866	.643
$\pi/2$	1.000	.000	.866
$2\pi/3$	.866	−.866	.985
$5\pi/6$	.500	−.866	.985
π	.000	.000	.866
$7\pi/6$	−.500	.866	.643
$4\pi/3$	−.866	.866	.342
$3\pi/2$	−1.000	.000	.000
$5\pi/3$	[ ]	[ ]	[ ]
$11\pi/6$	[ ]	[ ]	[ ]
2π	[ ]	[ ]	[ ]

■ $-.866$; $-.866$; $-.342$;
$-.500$; $-.866$; $-.643$;
$.000$; $.000$; $-.866$.

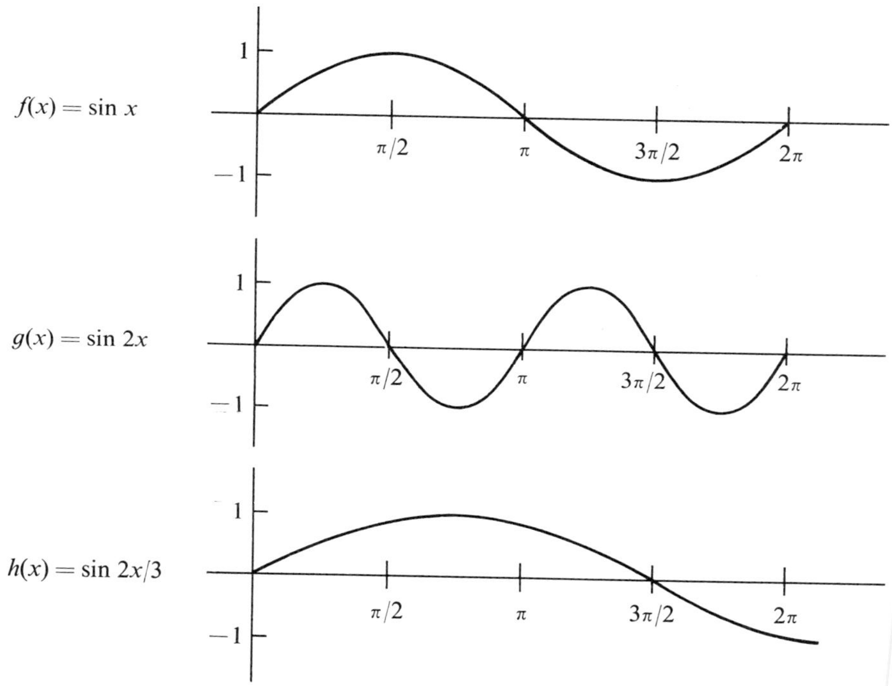

Fig. 7-3b

The constant b changes the period of the sine function. The period of $f(x) = \sin x$ was discussed in Section 7-2. It is ☐. The period of $g(x) = \sin 2x$ is ☐. One complete period of h is not drawn in Fig. 7-3b; however, $3\pi/2$ is one-half the period and therefore, the period of $h(x) = \sin 2x/3$ is ☐.

■ 2π; π; 3π.

If we designate the period with the letter p, then for Eq. 7-3a, when:

$$b = 1, \quad p = 2\pi;$$

$$b = 2, \quad p = \frac{2\pi}{2} = \pi;$$

$$b = \frac{2}{3}, \quad p = \frac{2\pi}{2/3} = 3\pi.$$

In general, when $f(x) = a \sin bx$, then $p = 2\pi/b$ because

$$\begin{aligned} f(x) &= f\left(x + \frac{2\pi}{b}\right) \\ &= a \sin b\left(x + \frac{2\pi}{b}\right) \\ &= a \sin(bx + 2\pi) \\ &= a \sin bx. \end{aligned}$$

Example 1

What is the period of $f(x) = \sin 4x/3$?
Since the period of $\sin x$ is 2π, the period of $\sin 4x/3$ is

$$\frac{2\pi}{4/3} = \frac{3\pi}{2}.$$

Demonstrate that the result is correct.

■ $$\begin{aligned} f(x) &\stackrel{?}{=} f\left(x + \frac{3\pi}{2}\right) \\ &\stackrel{?}{=} \sin\frac{4}{3}\left(x + \frac{3\pi}{2}\right) \\ &\stackrel{?}{=} \sin\left(\frac{4x}{3} + 2\pi\right) \\ &\stackrel{?}{=} \sin\frac{4x}{3} \\ &= f(x). \end{aligned}$$

An examination of $f(x) = a \cos bx$ gives results analogous to those just demonstrated for the sine function.

Example 2

Write an equation for a cosine function with the amplitude 3 and the period 4.

We know $f(x) = \cos x$ has an amplitude equal to one and a period of 2π. We want an equation in the form $g(x) = a \cos bx$. Therefore, $|a| = 3$ and $2\pi/b = 4$, or $b = \pi/2$; so $g(x) = 3 \cos(\pi/2)x$.

7-3 Exercises

1. Find the amplitude and period of $f(x) = 2 \sin 2x$.

---

2. Find the amplitude and period of $f(x) = \frac{1}{3} \cos \frac{x}{4}$.

■ 1. i. $|a| = 2$;
ii. $p = \frac{2\pi}{b} = \frac{2\pi}{2} = \pi$.

---

3. Find the period and amplitude of $f(x) = \pi \sin \pi x$.

■ 2. i. $|a| = \frac{1}{3}$;
ii. $p = \frac{2\pi}{b} = \frac{2\pi}{1/4} = 8\pi$.

---

4. Find the period and amplitude of $f(t) = Y \sin 2\pi ft$.

■ 3. i. $p = \frac{2\pi}{b} = \frac{2\pi}{\pi} = 2$;
ii. $|a| = \pi$.

---

5. Find the period and amplitude of $f(t) = A \cos \omega t$.

■ 4. i. $p = \frac{2\pi}{b} = \frac{2\pi}{2\pi f} = \frac{1}{f}$;
ii. $|a| = Y$.

---

6. Write an equation for a sine function when $a = 2$ and $p = \pi$.

■ 5. i. $p = \frac{2\pi}{b} = \frac{2\pi}{\omega}$;
ii. $|a| = A$.

---

7. Write an equation for a sine function when $a = 3$ and $p = 2$.

■ 6. i. $p = \frac{2\pi}{b}$, or $b = \frac{2\pi}{p}$

$= \frac{2\pi}{\pi} = 2;$

ii. $f(x) = 2 \sin 2x.$

---

8. Write an equation for a cosine function when $a = 1/2$ and $p = 1/2$.

■ 7. i. $b = \frac{2\pi}{p} = \frac{2\pi}{2} = \pi;$

ii. $f(x) = 3 \sin \pi x.$

---

9. Write an equation for a cosine function when $a = .24$ and $p = 1.35$.

■ 8. i. $b = \frac{2\pi}{p} = \frac{2\pi}{1/2} = 4\pi;$

ii. $f(x) = \frac{1}{2} \cos 4\pi x.$

---

10. Write an equation for a sine function when $a = A$ and $p = 1/f$.

■ 9. i. $b = \frac{2\pi}{p} = \frac{2\pi}{1.35} = 4.65;$

ii. $f(x) = .24 \cos 4.65x.$

---

11. The ordinate of a sine curve is 12 when 5/6 of its period is complete. Find the amplitude of the curve.

■ 10. i. $b = \frac{2\pi}{p} = \frac{2\pi}{1/f} = 2\pi f;$

ii. $f(t) = A \sin 2\pi ft.$

---

12. At what value of x between 0 and 2π will the sine curve be 75 percent of its amplitude ?

■ 11. i. $x = \frac{5}{6}(2\pi) = \frac{5\pi}{3};$

ii. $|a| = \left|\frac{12}{\sin 5\pi/3}\right|$

$= \left|\frac{12}{-.8660}\right|$

$= 13.87.$

13. The amplitude of a cosine curve is 20. Find the ordinate $\pi/6$ radians past its minimum.

■ 12. i. $.75 = |\sin x|$;
ii. $x = .27\pi, .73\pi, 1.27\pi,$ and 1.73π.
(Note that $.27\pi$ corresponds to 48.6°.)

■ 13.
$$f(x) = 20\cos\left(\pi + \frac{\pi}{6}\right) = -20\cos\frac{\pi}{6} = -17.3.$$

7-3 Problems

1. A rotating vector is sometimes referred to as a *phasor*. One complete revolution of a vector through 2π radians constitutes one cycle, and through 4π radians constitutes two cycles. The rate at which cycles of the vector occur is called *frequency*.

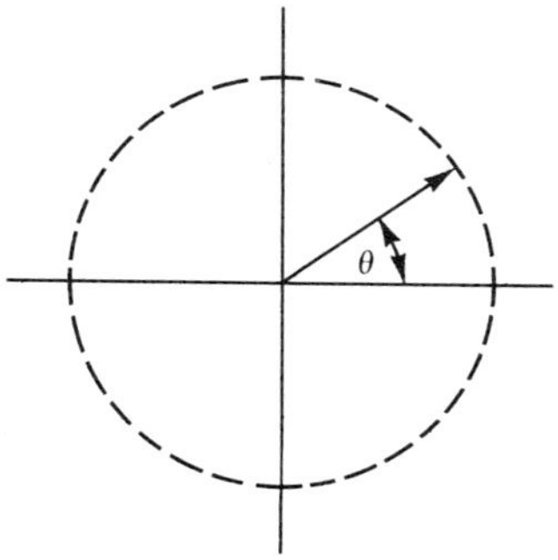

How long does one cycle take if 60 cycles occur in one second? If one cycle takes .01 second, what is the frequency in cycles per second? (Cycles per second are called Hertz.)

2. If $f =$ (frequency in cycles per second) and $t =$ (time in seconds), what does the product ft represent? What does the product $2\pi f$ represent? What does the product $2\pi ft$ represent?
3. Consider the function $f(t) = \sin 2\pi ft$, where $f = 10$ cycles per second. Plot $f(t)$ as a function of t, when t varies from zero to 0.2 seconds.

4. Recall the definition of period given in terms of real numbers. *Period* can also be defined in terms of time. What is the period of the function given in Problem 3?
5. If the period of a sine function is .01 seconds, what is its frequency? What mathematical relationship exists between period and frequency?
6. Aircraft alternators have a frequency of 400 cycles per second. Find the angular velocity in radians per second. What is the period of each cycle?
7. Electric motors designed to operate at constant angular velocities using commercial power sources are used to turn special purpose alternators at other specific angular velocities. Design a belt drive to meet the specifications illustrated.

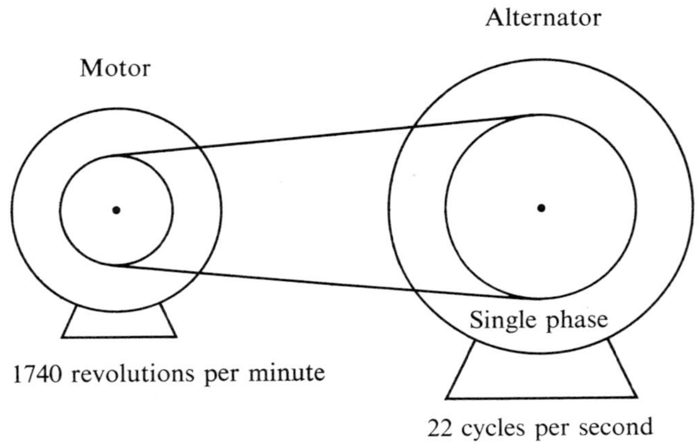

8. Write the equation of the alternator voltage in Problem 7 if it has the form $v = V \sin bt$, where $V = 18$ and v is a function of t.

7-4 Phase

Up to this point we have always seen the graph of a sine function pass through the origin of the coordinate system when $x = 0$.

$f(x) = a \sin bx$

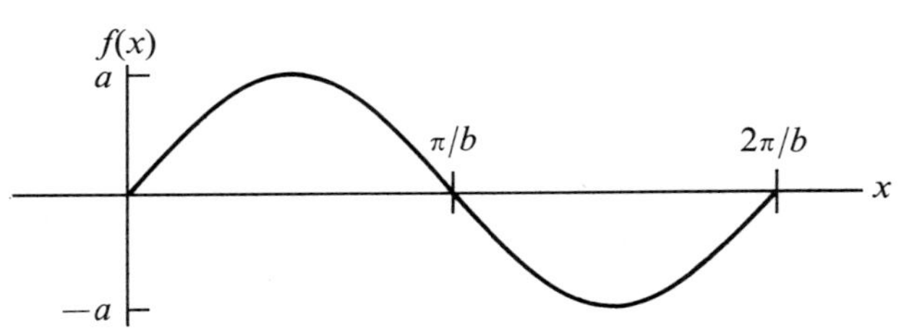

Fig. 7-4a

This need not always be the case. In this section we study a function that has the characteristics shown in Fig. 7-4a, but does not pass through (0, 0).

Consider the equation $g(x) = a \sin(bx + \pi/6)$. When $x = 0$,

$$g(0) = a \sin\left(b(0) + \frac{\pi}{6}\right) = a \sin \frac{\pi}{6},$$

which clearly is not zero. If we continue to calculate values for $g(x)$ over an interval $P = 2\pi/b$, and if we graph the results, we obtain the familiar sine wave, but slightly *shifted* from the usual position illustrated in Fig. 7-4a.

$g(x) = a \sin(bx + \pi/6)$

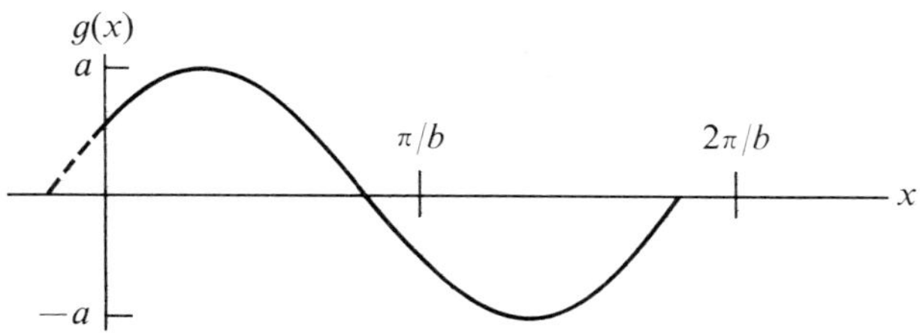

Fig. 7-4b

If $b = 1$, every point of the graph of the g function is shifted $\pi/6$ to the left of the f function.

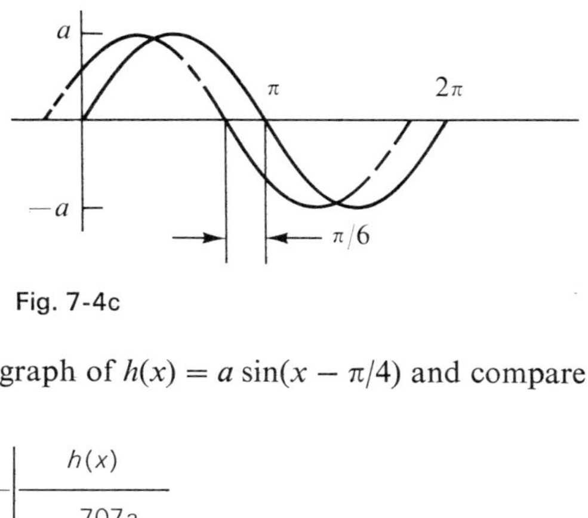

Fig. 7-4c

Prepare a graph of $h(x) = a \sin(x - \pi/4)$ and compare with $f(x) = a \sin x$.

■

x	$h(x)$
0	$-.707a$
$\pi/4$	0
$\pi/2$	$.707a$
$3\pi/4$	a
π	$.707a$
$5\pi/4$	0
$3\pi/2$	$-.707a$
$7\pi/4$	$-a$
2π	$-.707a$

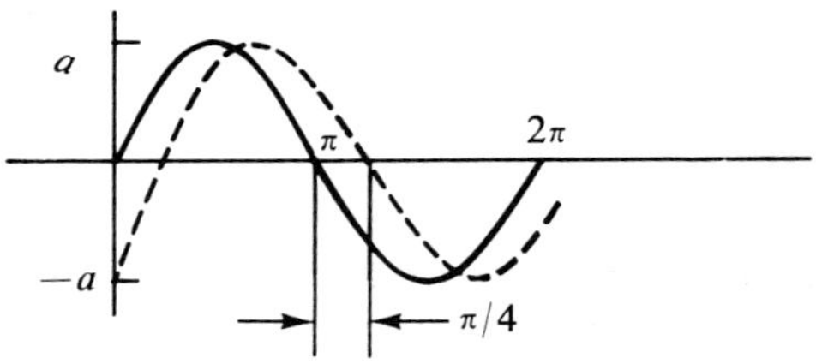

Fig. 7-4d

Pay particular attention to the fact that subtracting $\pi/4$ has shifted the graph of h to the right of the graph of f.

The general form of the equation of the sine function we are examining is

$$f(x) = a \sin(bx + c). \qquad \text{(Eq. 7-4a)}$$

We have seen in Section 7-3 that $|a|$ is the amplitude and b affects the period. The constant c, called **initial phase**, causes what is known as **phase shift**.

Our examples so far have shown that when b is *equal* to one, c is the value of the phase shift. If b is *unequal* to one, we must calculate the phase shift. Observe that by factoring b from the parentheses in Eq. 7-4a, we have

$$f(x) = a \sin b\left(x + \frac{c}{b}\right).$$

In general

$$\frac{c}{b} = \text{(phase shift)}. \qquad \text{(Eq. 7-4b)}$$

Example 1

Graph $g(t) = 12 \sin 20\pi t$ and $f(t) = 12 \sin(20\pi t + \pi/3)$ on the same coordinate system.

Function	Amplitude $\lvert a \rvert$	Period $2\pi/b$	Phase shift c/b
g	12	1/10	0
f	12	1/10	1/60

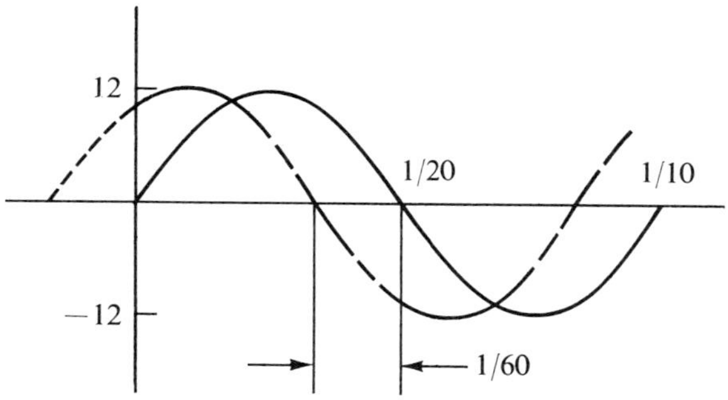

To help us understand the need for introducing c into the sine equation, we shall investigate a problem of electricity. Let us assume that we have an alternating current generator. This could be as simple as a loop of wire that rotates in a magnetic field. When the loop moves so that it cuts across lines of force of the magnetic field, an electromotive force is produced in the wire.

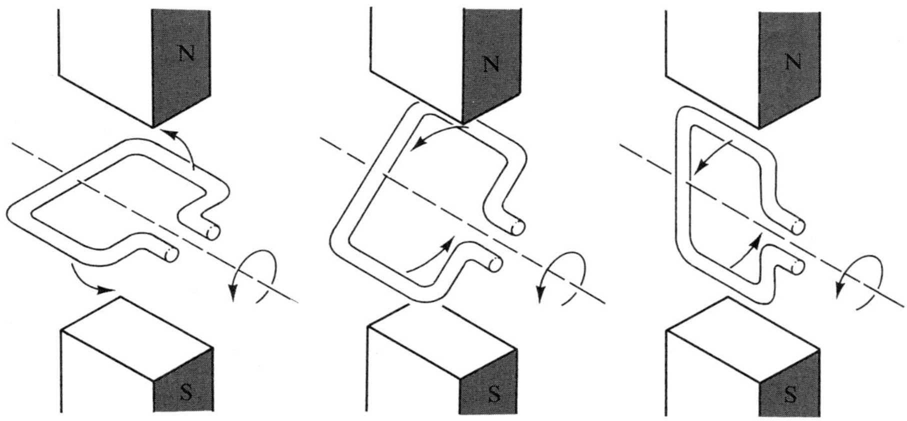

Fig. 7-4e Fig. 7-4f Fig. 7-4g

The amplitude of the electromotive force is proportional to the rate at which the loop of wire passes across the magnetic field that exists between the north and south magnets. When the loop is in the position illustrated in Fig. 7-4e, the wire is moving parallel to the magnetic field and the force is zero. In the position illustrated in Fig. 7-4g, the force is maximum because the wire is moving at right angles to the magnetic field. If the loop of wire is rotated at a constant angular velocity (ω), we may represent the electromotive force (e) as a function of time by the equation

$$f(t) = e = E \sin \omega t. \qquad \text{(Eq. 7-4c)}$$

If an electromotive force varying in the manner of Eq. 7-4c is applied to a pure resistive (R) circuit, the current (i) in the circuit varies in a manner similar to e.

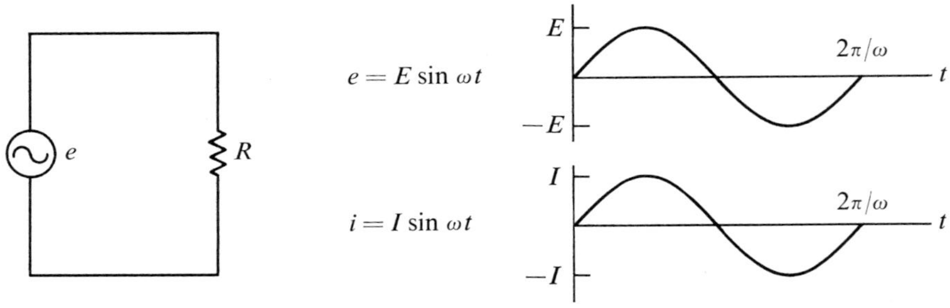

Fig. 7-4h

Quite a different situation occurs when an electromotive force like that defined in Eq. 7-4c is applied to an RL or RC circuit.

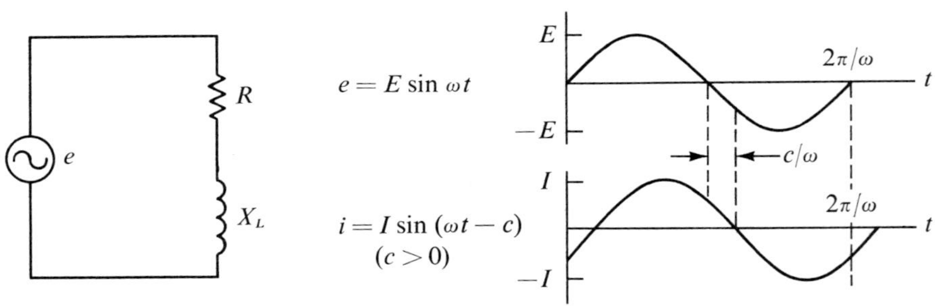

Fig. 7-4i

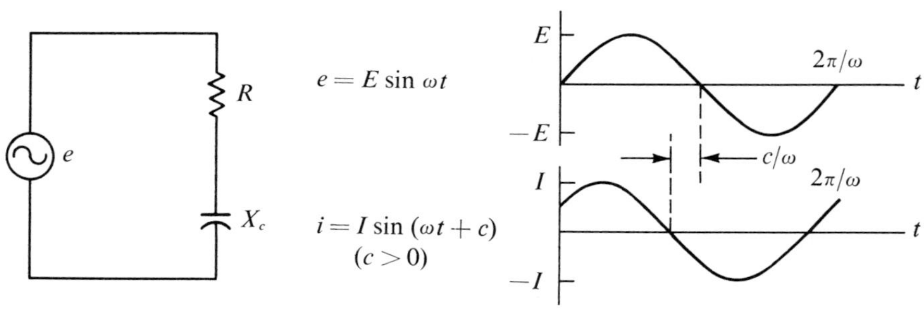

Fig. 7-4j

Figure 7-4i shows the current (i) **lagging** (e) by the **phase shift** c/ω. The word lagging is used because in terms of t (time), similar parts of the curve of i occur after the curve of e. The value of c is determined by the equation

$$c = \arctan \frac{X_L}{R}, \qquad \text{(Eq. 7-4d)}$$

where the symbol *arctan* is to be interpreted as "the number whose tangent is."

Figure 7-4j shows the current (i) **leading** (e) by the **phase shift** of c/ω. Here points of the curve of i occur before similar points of e. In this case

$$c = \arctan \frac{X_c}{R}. \qquad \text{(Eq. 7-4e)}$$

Example 2

If $R = 12$ ohms and $X_L = 8$ ohms, write the equation for the current (i) when $f(t) = e = E \sin 120\pi t$.

i. $c = \arctan \dfrac{X_L}{R}$

$\quad = \arctan \dfrac{8}{12} = \arctan .667 = .187\pi;$

ii. $i = I \sin(120\pi t - .187\pi)$.

Phase shift is therefore a result of a nonzero value for c in the equation $f(x) = a \sin(bx + c)$. The function f *leads* or *lags* $g(x) = a \sin bx$ according to whether c is *greater than* or *less than* zero, respectively.

7-4 Exercises

1. Sketch the graph of

$$f(t) = 2 \sin\left(t + \frac{\pi}{5}\right).$$

2. Sketch the graph of

$$f(t) = 3\sin\left(t - \frac{\pi}{5}\right).$$

■ 1.

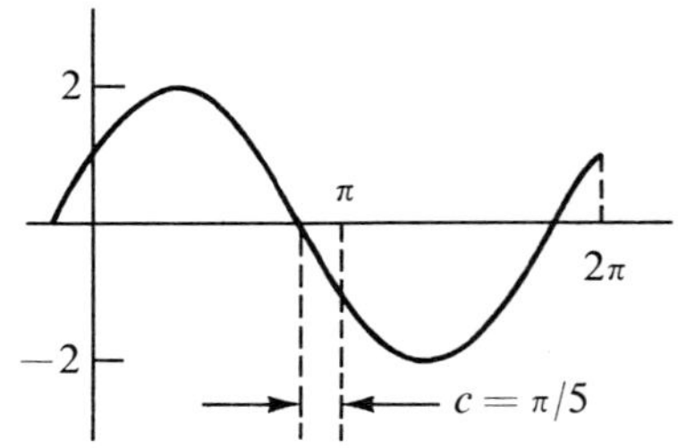

3. The discussion of phase shift of the sine function is analogous for cosine functions. Sketch the graph of

$$f(t) = 2\cos\left(t - \frac{\pi}{6}\right).$$

■ 2.

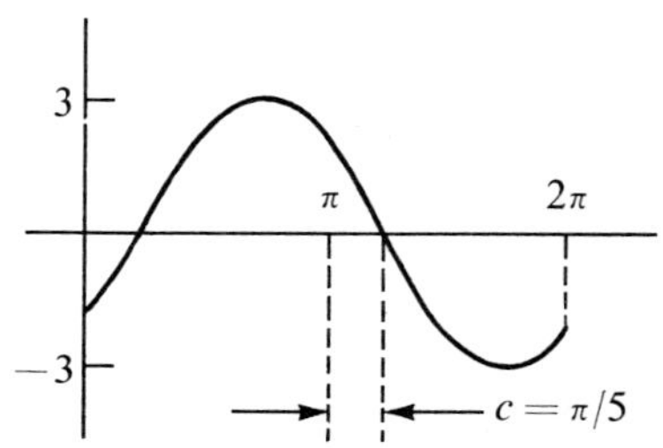

4. Sketch the graph of

$$f(t) = 4\cos\left(t + \frac{\pi}{4}\right).$$

■ 3.

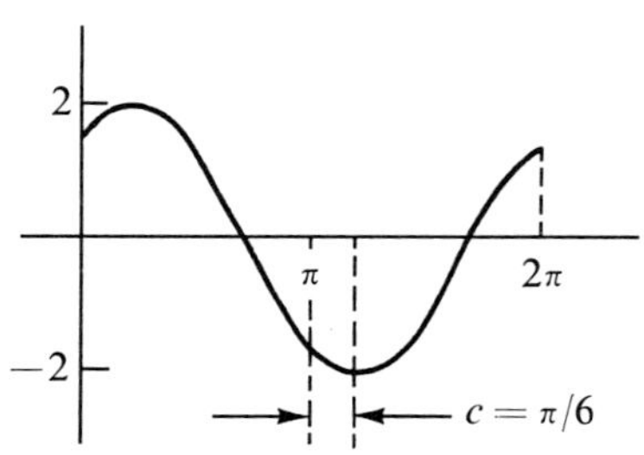

5. Determine the phase shift for
$$f(x) = 2 \sin\left(5x + \frac{\pi}{9}\right).$$

■ 4.

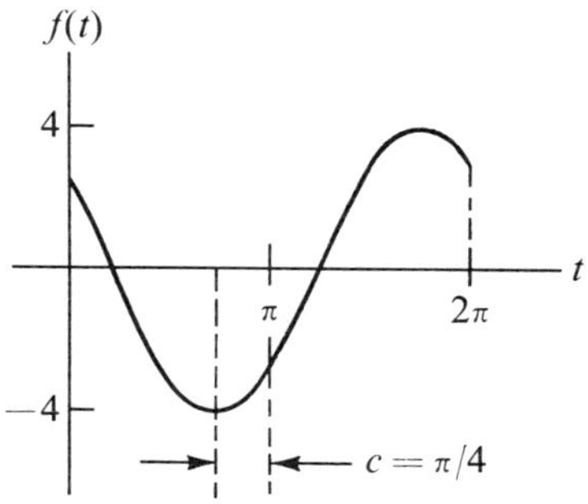

6. Sketch the graph of
$$g(t) = 5 \sin\left(2\pi t + \frac{\pi}{12}\right);$$
indicate the period and phase shift.

■ 5. $\frac{c}{b} = \frac{\pi/9}{5} = \frac{\pi}{45}.$

7. Sketch the graph of
$$h(t) = 16 \sin\left(12\pi t - \frac{\pi}{5}\right);$$
indicate the period and phase shift.

■ 6.

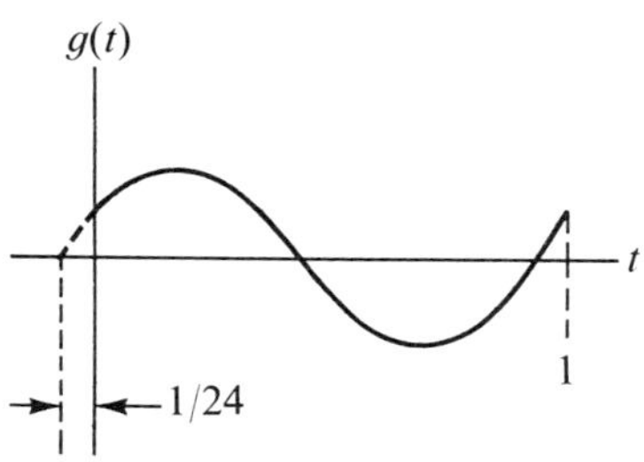

8. Sketch for comparison the graphs of
$$f(x) = \cos x$$
and
$$g(x) = \sin\left(x + \frac{\pi}{2}\right).$$

■ 7.

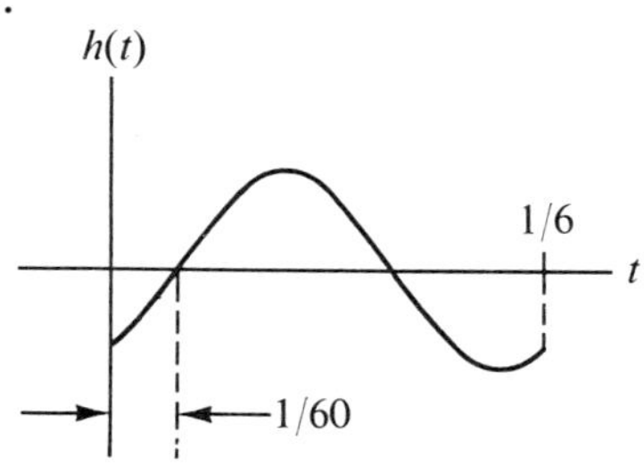

9. Evaluate $f\left(\frac{1}{4}\right)$ when

$$f(t) = 3\sin\left(2\pi t + \frac{\pi}{3}\right).$$

■ 8.

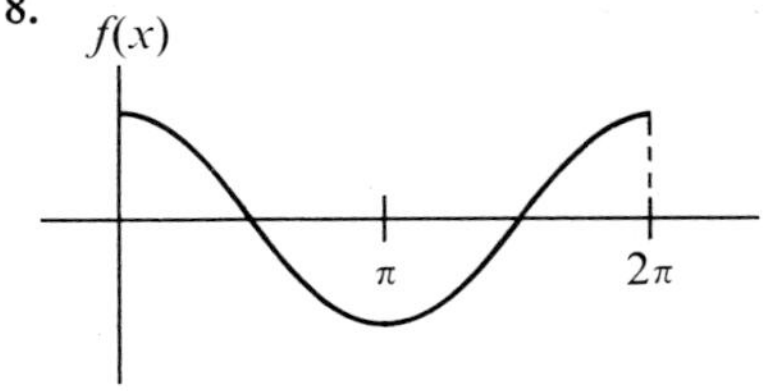

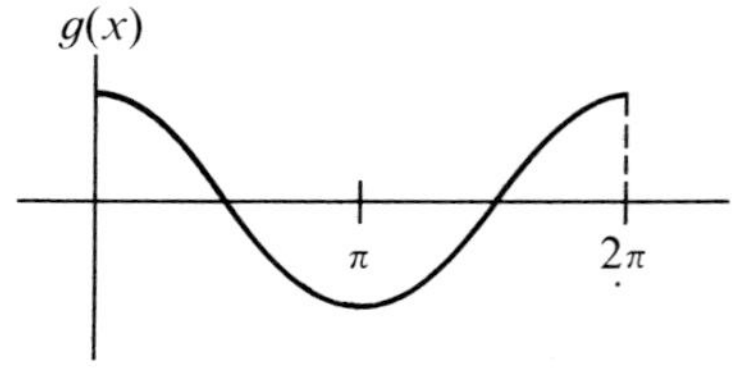

(The functions f and g are identical.)

---

10. If $f\left(\frac{\pi}{9}\right) = 4$ when

$$f(t) = 4\sin(t + c),$$

find the constant c.

■ 9. $f\left(\frac{1}{4}\right) = 3\sin\left(2\pi\left(\frac{1}{4}\right) + \frac{\pi}{3}\right)$

$= 3\sin\frac{5\pi}{6}$

$= 3(.500) = 1.5.$

---

11. If $f(x) = 3\cos\left(x + \frac{\pi}{8}\right) = 3$, find x, such that $0 \leq x \leq 2\pi$.

■ 10. i. $4 = 4\sin\left(\frac{\pi}{9} + c\right)$;

ii. $1 = \sin\left(\frac{\pi}{9} + c\right)$.

Therefore,

iii. $\frac{\pi}{9} + c = \frac{\pi}{2}$ or

$c = \frac{\pi}{2} - \frac{\pi}{9} = \frac{7\pi}{18}$.

(Actually $c = \frac{7\pi}{18} + n2\pi$, $n = 1, 2, \ldots$)

---

12. If $g\left(\frac{\pi}{6}\right) = 1.5$ when

$$g(x) = 2 \sin(x + c),$$

find c such that $0 \leq c \leq \pi$.

■ 11. i. $3 = 3 \cos\left(x + \frac{\pi}{8}\right)$;

ii. $1 = \cos\left(x + \frac{\pi}{8}\right)$;

iii. $x + \frac{\pi}{8} = 0,\ 2\pi$

$$x_1 = -\frac{\pi}{8}$$

$$x_2 = 2\pi - \frac{\pi}{8} = \frac{15\pi}{8}.$$

---

13. If $g(x) = -2$ when

$$g(x) = 4 \cos\left(x - \frac{\pi}{9}\right),$$

find x such that $0 \leq x \leq 2\pi$.

■ 12. i. $1.5 = 2 \sin\left(\frac{\pi}{6} + c\right)$;

ii. $.75 = \sin\left(\frac{\pi}{6} + c\right)$;

iii. $\frac{\pi}{6} + c = .27\pi$

$$c = .27\pi - .17\pi = .10\pi.$$

---

14. Write an equation that lags $f(x) = 2 \sin 6t$ by a phase shift of $\frac{\pi}{30}$.

■ 13. i. $-2 = 4 \cos\left(x - \frac{\pi}{9}\right)$;

ii. $-.5 = \cos\left(x - \frac{\pi}{9}\right)$;

iii. $x - \frac{\pi}{9} = \frac{2\pi}{3}$ and $\frac{4\pi}{3}$

$$x_1 = \frac{2\pi}{3} + \frac{\pi}{9} = \frac{7\pi}{9}$$

$$x_2 = \frac{4\pi}{3} + \frac{\pi}{9} = \frac{13\pi}{9}.$$

15. Write an equation that leads

$$h(x) = 3 \cos \omega t$$

by a phase shift of $\frac{\pi}{4}$.

■ 14. $\frac{c}{b} = \frac{\pi}{30}$, $c = \frac{\pi}{5}$,

$$g(x) = 2 \sin\left(6t - \frac{\pi}{5}\right).$$

■ 15. $\frac{c}{b} = \frac{\pi}{4}$, $c = \frac{\omega\pi}{4}$,

$$k(x) = 3 \cos\left(\omega t + \frac{\omega\pi}{4}\right).$$

7-4 Problems

1. If in an RL circuit, $R = 15$ ohms and $X_L = 11$ ohms, what is the *initial phase constant*? Determine the *phase shift* of the current (i) if the period of the electromotive force (e) is $\pi/60$.
2. If in an RC circuit, $R = 5000$ ohms, $X_C = 3500$ ohms, and the frequency of the electromotive force is 60 cycles per second, by what constant phase shift does the current lead or lag the electromotive force?
3. Find i at $t = .01$ seconds, if $I = 15$.

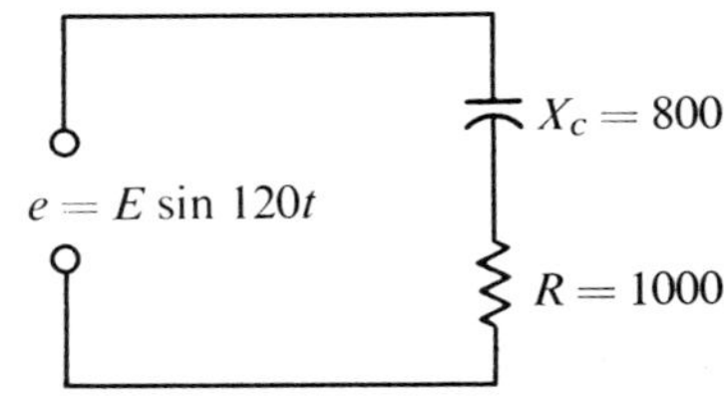

4. What is t when the current (i) reaches its first maximum?

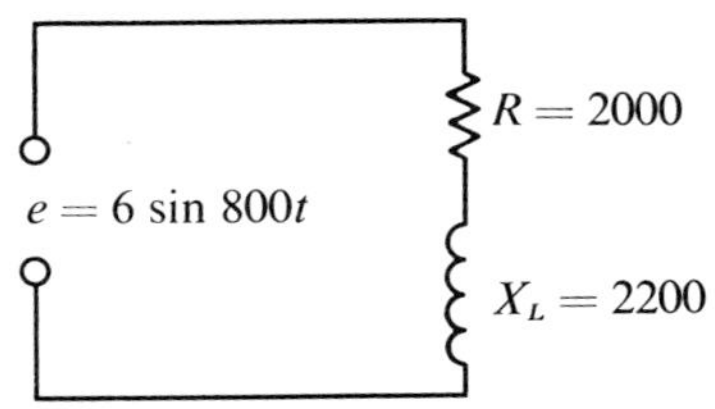

5. Single phase alternators produce a voltage $(v) = V \sin(\omega t + \alpha)$.
   a. If the frequency of the voltage is 60 cps, find ω.
   b. If $V = 10$ volts and $\alpha = \pi/2$, find v at .01 second.

7-5 Composite and Other Trigonometric Functions

An endless variety of functions may be developed by combining two or more functions.

Example 1

Graph $f(x) = \sin x + \cos x$.

We can compute $f(0)$, $f(\pi/6)$, $f(\pi/3)$, and so forth up to 2π, and graph these values. Another, possibly easier, method is to add the corresponding ordinates of the sine and cosine functions geometrically. Let $f(x) = g(x) + h(x)$, where $g(x) = \sin x$ and $h(x) = \cos x$.

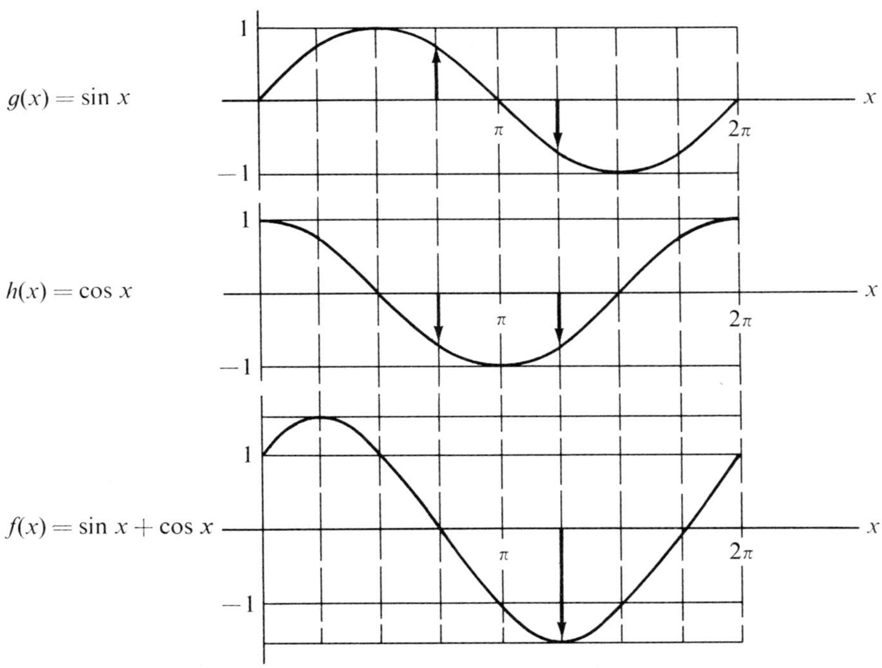

Fig. 7-5a

After we examine Fig. 7-5a, our intuition tells us that f is periodic, and the period is [ ].

■ 2π

The maximum of the function f appears to occur when $x =$ [ ], and the minimum at $x =$ [ ].

■ $\frac{\pi}{4}$; $\frac{5\pi}{4}$.

Therefore the amplitude of the composite function f is [ ].

■ i. $f\left(\frac{\pi}{4}\right) = \sin\frac{\pi}{4} + \cos\frac{\pi}{4}$
$= .707 + .707 = 1.414 = M;$

ii. $f\left(\frac{5\pi}{4}\right) = \sin\frac{5\pi}{4} + \cos\frac{5\pi}{4}$
$= -.707 - .707 = -1.414 = m;$

iii. $(\text{amplitude}) = \frac{1}{2}|M - m|$
$= \frac{1}{2}|1.414 + 1.414| = 1.414.$

A more general form of the composite function illustrated in Fig. 7-5a will be analyzed further in Chapter 8.

Example 2

Graph $f(t) = 2 + \sin t$.

Again, the graph may be prepared with either an algebraic or geometric method. Let $g(t) = 2$ and $h(t) = \sin t$.

$f(t) = 2 + \sin t$

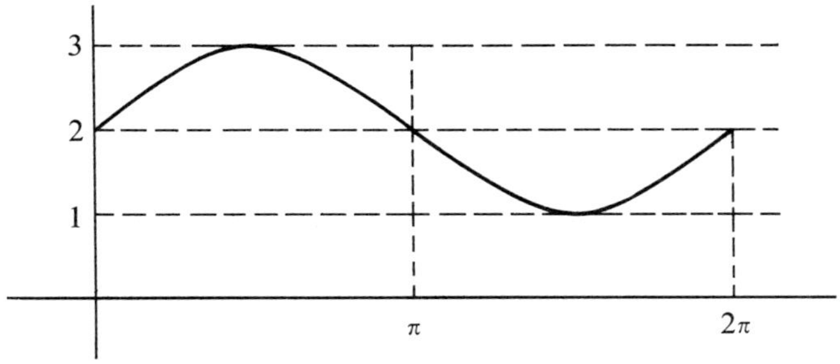

Fig. 7-5b

It should not be difficult to see that the equation $f(t) = 2 + \sin t$ is periodic with $p = 2\pi$. The amplitude, according to Eq. 7-3b, is one. It is for functions of this type that $|M|$ and $|M - m|$ are often used for defining amplitude.

$$\frac{1}{2}|M - m| = \square$$

$$|M| = \square$$

$$|M - m| = \square$$

■ 1; 3; 2.

The next two examples illustrate composite functions that are defined in part by trigonometric expressions, but the functions are *not* periodic.

Example 3

Graph $f(x) = x + \sin x$.

Either an algebraic or geometric method may be used to graph this function. Let $g(x) = x$ and $h(x) = \sin x$.

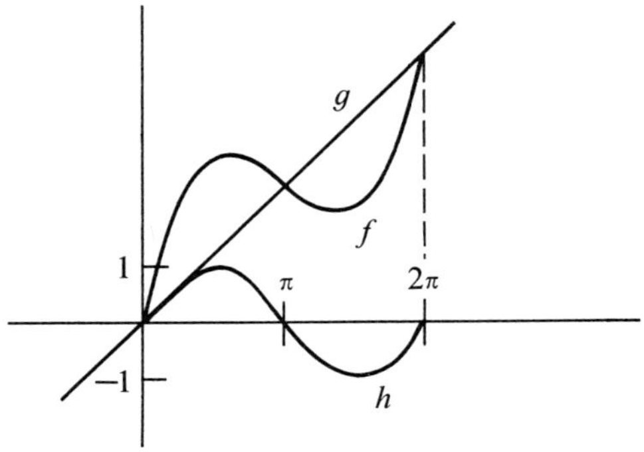

Fig. 7-5c

Example 4

Graph $f(x) = x \sin x$.

x	$\sin x$	$x \sin x$
0	0	0
$\pi/4$	.707	.56
$\pi/2$	1.000	1.57
$3\pi/4$	.707	1.67
π	0	0
$5\pi/4$	−.707	−2.78
$3\pi/2$	−1.000	−4.71
$7\pi/4$	−.707	−3.89
2π	0	0
$9\pi/4$	.707	5.00
$5\pi/2$	1.000	7.85
$11\pi/4$	.707	6.11
3π	0	0
$13\pi/4$	−.707	−7.22
$7\pi/2$	−1.000	−10.99
$15\pi/4$	−.707	−8.33
4π	0	0

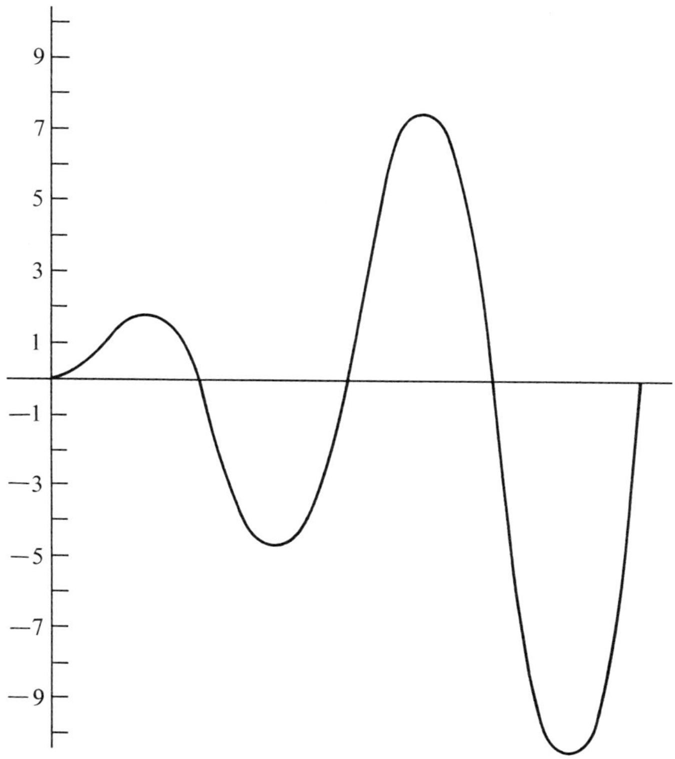

Fig. 7-5d

Rectifiers convert alternating current to pulsating direct current. Rectifiers have many applications. For example, they are a necessary part of broadcast receivers that use commercial power sources. If the input voltage of a full-wave

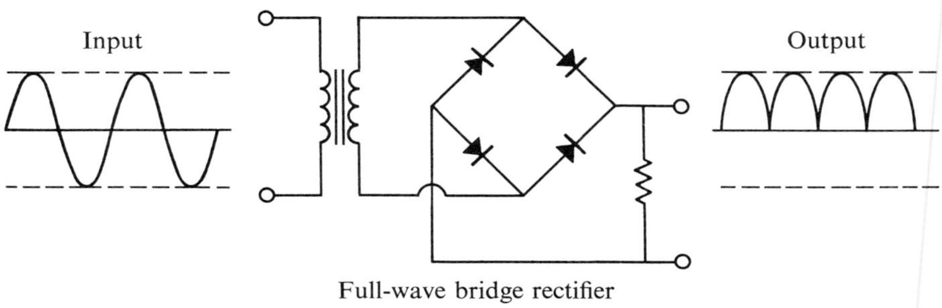

Fig. 7-5e

rectifier is

$$e_i = E_i \sin \omega t, \qquad \text{(Eq. 7-5a)}$$

then the output voltage is

$$e_0 = E_0 |\sin(\omega t + c)|. \qquad \text{(Eq. 7-5b)}$$

The amplitude, $|E_0|$, and the phase shift, (c/ω), are determined by the circuit components. The relationship between the period of the input and the period of the output is always the same. If, for example, the period of the input is $2\pi/\omega$, the period of the output is ☐.

■ π/ω

7-5 Exercises

1. Sketch the graph of

$$f(x) = \frac{x}{2} + \sin x.$$

Is f periodic?

2. Graph $f(x) = \frac{x^2}{10} + \cos 2x$.

Is f periodic?

■ 1.

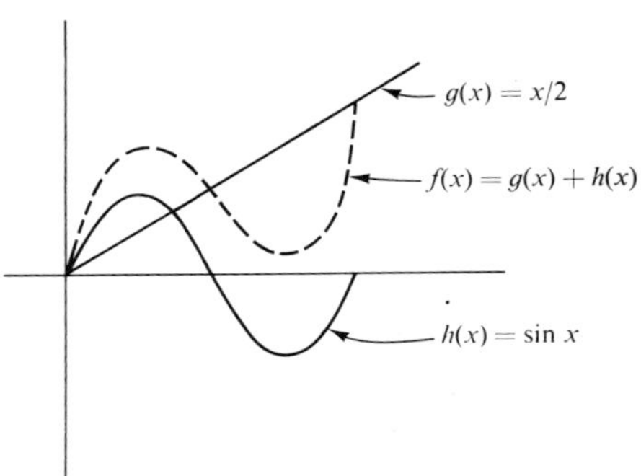

f is not periodic.

---

3. Sketch the graph of

$$f(x) = \sin 2x + \sin x.$$

Is f periodic ?

■ 2.

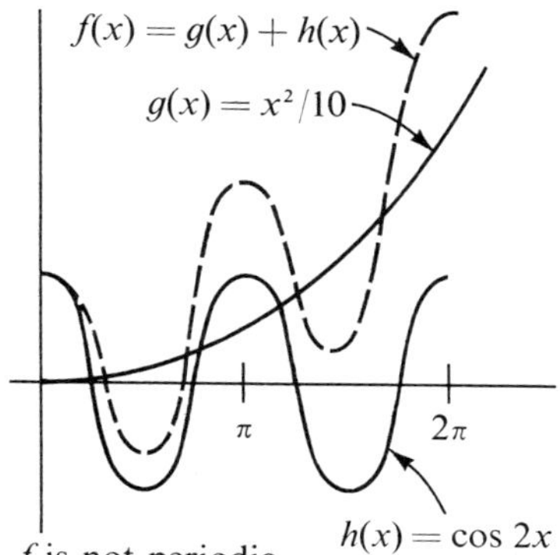

f is not periodic.

---

4. Sketch the graph of

$$f(x) = \cos x - \sin x.$$

Is f periodic ?

■ 3.

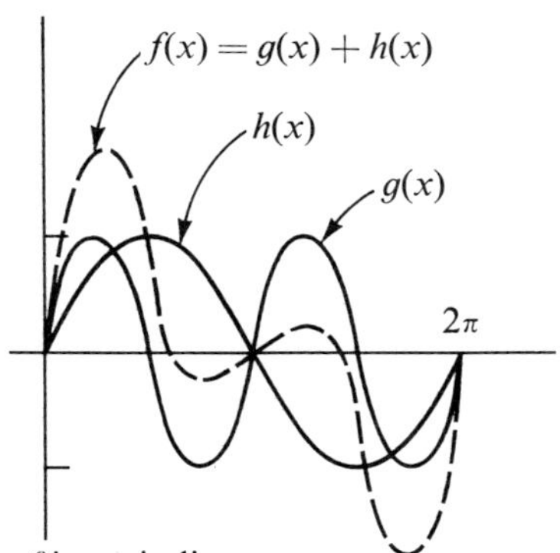

f is periodic.

5. Sketch the graph of

$$f(x) = \sin x + \cos\left(x + \frac{\pi}{2}\right).$$

■ 4.

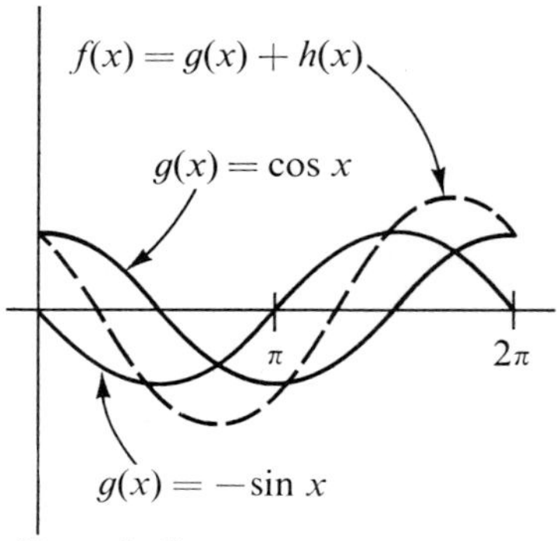

f is periodic.

---

6. Sketch the graph of

$$f(x) = \sin^2 x$$

and the graph of

$$g(x) = \sin x$$

on the same coordinate system. What are the period and amplitude of f?

■ 5.

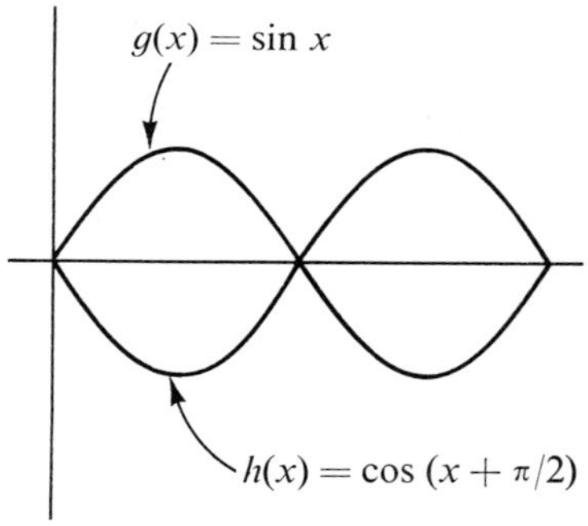

$f(x) = 0$

7. Sketch the graph of $f(x) = \tan x$. What are the period and amplitude ?

■ 6.

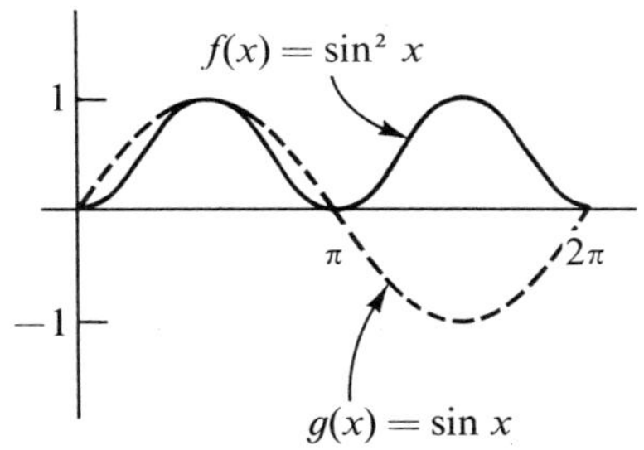

(Period) $= \pi$
(Amplitude) $= 1/2|M - n| = 1/2$, or
$|M| = 1$ or
$|M - n| = 1$

■ 7.

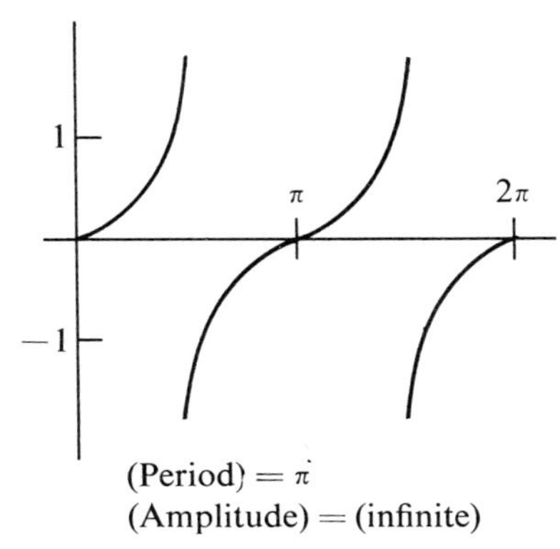

(Period) $= \pi$
(Amplitude) = (infinite)

7-5 Problems

1. Problem 4 of Section 7-2 involving the displacement of a piston is an example of a composite trigonometric function in which displacement (d) is a function of the angle of rotation of the crank shaft (θ). Find the equation for this function. (*Hint*: Generalize the specific solution for d in terms of θ.)
2. Harmonics in music can often be represented mathematically as composite trigonometric functions. If $y = Y_1 \sin \omega t$ is the fundamental tone (first harmonic),

then $y = Y_1 \sin \omega t + Y_2 \sin 2\omega t$ is the fundamental tone sounded with its second harmonic. What is the relationship between the frequencies of the first and second harmonics?

3. Harmonics in electronics are integer multiples of the fundamental wave frequency. If an oscilloscope shows the presence of a third harmonic with a period of .03 microseconds, what is the first harmonic frequency?
4. In electronics, the phase and period relationship between two periodic functions can be viewed on an oscilloscope by having one function produce vertical deflection and the other produce horizontal deflection. Identical results are obtained by graphing the set of ordered pairs $(g(t), h(t))$, where g and h are periodic functions of the same variable t. Prepare a graph for $g(t) = \sin t$ and $h(t) = \sin (t + \pi/2)$. (*Note*: The graph will be complete when t goes from 0 to 2π.)
5. The result in Problem 4 is called a Lissajous figure. Repeat Problem 4 when $g(t) = \sin t$ and $h(t) = \sin 2t$.

7-6 Related Trigonometric Functions

In Chapters 1 and 3 we discovered that for a given linear or quadratic function f, there is a related function f' that gives the slope or rate of change of the dependent variable with respect to the independent variable. In this section we shall look for similar relationships with simple trigonometric functions.

Consider $f(x) = \sin x$. If a function f' exists, it most certainly must be periodic, because the values of the slope of $\sin x$ repeat each period of f. The graph of f' must also pass through zero when the graph of f is at its maximum and minimum.

Let us look at a carefully prepared graph of $f(x) = \sin x$, Figure 7-6a, where the scale on the vertical and horizontal axes are the same. Often a geometric estimate of slopes plotted on a coordinate system suggests the approximate nature of f'.

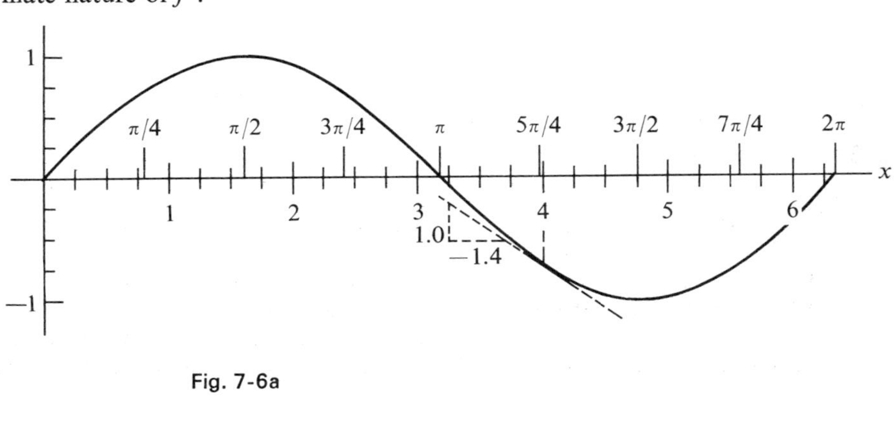

Fig. 7-6a

TABLE 7-6a

x	Approximate slope of sin x
0	[ ]
$\pi/4$	[ ]
$\pi/2$	0.0
$3\pi/4$	[ ]
π	[ ]
$5\pi/4$	$-.07$ $(-1.0/1.4)$
$3\pi/2$	[ ]
$7\pi/4$	[ ]
2π	[ ]

■ (0, 1.0); ($\pi/4$, .7); ($3\pi/4$, $-.7$); (π, -1.0); ($3\pi/2$, 0); ($7\pi/4$, .7); (2π, 1.0).

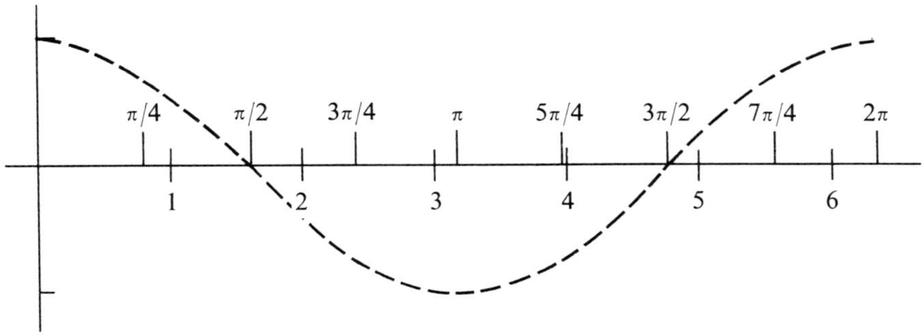

Fig. 7-6b

The graph, Figure 7-6b, of the values in Table 7-6a certainly resembles a simple cosine curve. This is in fact true. If

$$f(x) = \sin x,$$

then

$$f'(x) = \cos x. \qquad \text{(Eq. 7-6a)}$$

With a graphical approach we could discover also that if

$$f(x) = \cos x,$$

then

$$f'(x) = -\sin x. \qquad \text{(Eq. 7-6b)}$$

Applications of Eqs. 7-6a and 7-6b exist to a very great extent in electricity and electronics. For example, when a voltage (v) varying in the manner of a sine wave (f) is applied to a capacitor, then the current (i) varies in the manner of a cosine wave (f'). The equations are also useful in situations involving circular motion and harmonic motion. For example, if f describes the displacement of the end of a vibrating reed in terms of time, then f' describes velocity in terms of time.

One word of caution is necessary at this point. If $f(t) = \sin \omega t$, we may not assume that f' has the value $\cos \omega t$. The correct equation is $f'(t) = \omega \cos \omega t$. Further development of these results is given in Chapter 11.

7-6 Exercises

1. For what value of x does $f(x) = \sin x$ have a slope of .5 ?

---

2. For what value of x does $f(x) = \cos x$ have a slope of $-.8$?

■ 1. $f'(x) = \cos x = .5$

$x = \arccos .5$

$= \dfrac{\pi}{3}$ and $\dfrac{5\pi}{3}$.

(Arccos means "the angle whose cosine is.")

---

3. For what value of x does $f(x) = \sin x$ have a slope of 2 ?

■ 2. $f'(x) = -\sin x = -.8$

$x = \arcsin .8$

$= .29\pi$ and $.70\pi$.

(Arcsin means "the angle whose sine is.")

---

4. Recall that a relative maximum or minimum of a function occurs when its slope is zero. Use this fact to find a relative maximum or a relative minimum of $f(x) = \sin x$.

■ 3. $f'(x) = \cos x = 2$.

(no solution)

---

5. If a pendulum moves so that its distance from the lowest position is given by $f(x) = \sin x$, when is its speed zero ?

■ 4. $f'(x) = \cos x$

$$x = \arccos 0$$
$$= \frac{\pi}{2} \text{ and } \frac{3\pi}{2}.$$

---

6. If the voltage (v) applied to a capacitor is

$$v = f(t) = \cos \omega t,$$

then

$$i = Cf'(t)$$
$$= -C\omega \sin \omega t.$$

Find i when $t = .001$, $\omega = 200\pi$, and $C = .0001$.

■ 5. (speed) $= f'(x)$

$$= \cos x = 0$$
$$x = \arccos 0$$
$$= \frac{\pi}{2} \text{ and } \frac{3\pi}{2}.$$

■ 6. $i = (-.0001)200\pi$

$$\sin 200\pi(.001)$$
$$= .0628(.5878)$$
$$= .0369.$$

7-6 Problems

1. If ω is angular velocity, then the vertical displacement $f(t)$ is given by the equation $f(t) = r \sin \omega t$. The equation that expresses the rate of change of the f function is the equation that gives the slope; in other words, $f'(t) = \omega r \cos \omega t$.

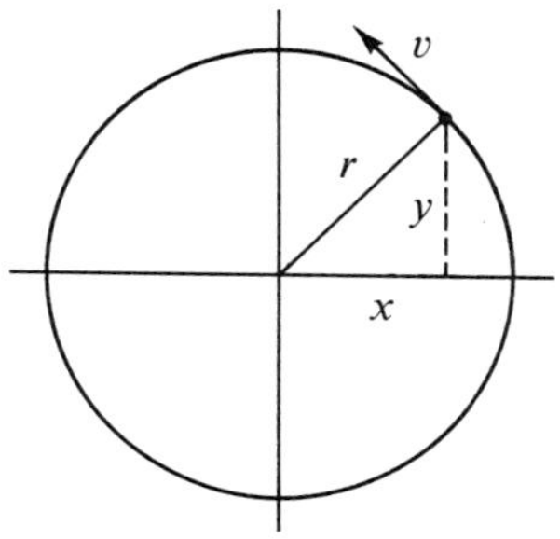

Complete the chart for $\omega = 6.28$ radians and $r = 2$ inches.

t (time, seconds)	$f(t)$ (displacement, inches)	$f'(t)$ (velocity, in./sec)
.0 seconds	______	______
.2 seconds	______	______
.4 seconds	______	______
.6 seconds	______	______
.8 seconds	______	______
1.0 seconds	______	______

2. Assume that the piston on a motor moves with simple harmonic motion. The displacement function and velocity function are given by the equations

$$f(t) = d = r \sin \omega t$$

and

$$f'(t) = v = r\omega \cos \omega t.$$

Find the velocity of the piston at the end of its stroke if the engine turns at 3000 rpm and $r = 6.25$ centimeters.

3. From Problem 2, find the velocity of the piston when $f(t) = 0$.

7 Supplementary Exercises and Problems

1. Change from degrees to radians:
   a. 45°,
   b. 76.5°,
   c. 160°. (7-1)
2. Change from radians to degrees:
   a. .643,
   b. 1.56,
   c. $.78\pi$. (7-1)
3. Convert to radian measure:
   a. 1 degree,
   b. 1 minute,
   c. 1 second. (7-1)
4. Convert to degree measure:
   a. $\pi/6$ radians,
   b. $\pi/8$ radians,
   c. .01 radians. (7-1)
5. Find:
   a. $\sin \pi/16$,
   b. $\cos 1.500$,
   c. $\tan 2.5\pi$. (7-1)

6. If a central angle with radius 4 inches intercepts an arc of 5 inches, find the central angle in radians and degrees. (7-1)
7. What is the angular velocity of the second hand of a clock in radians per second? (7-1)
8. What is the number of revolutions per second made by a 26-inch wheel at 75 miles per hour? (7-1)
9. Plan a pulley system that causes the driven pulley to turn at 1.5 radians per second when the driver pulley turns at 20 radians per second. (7-1)
10. Calculate the circumference of the earth's 50th parallel.
(Earth's equatorial radius: 6371 kilometers.) (7-1)
11. What is the arc length of one degree along the 45th parallel on the earth's surface? (7-1)
12. What is the angle between two sightings of geographic north from two points on the 45th parallel on the earth's surface separated by a distance of 1000 kilometers? (7-1)
13. Fill in the chart for the proper sign of $f(x)$ and $g(x)$. (7-2)

	$0 < x < \pi/2$	$\pi/2 < x < \pi$	$\pi < x < 3\pi/2$	$3\pi/2 < x < 2\pi$
$f(x) = \sin x$ $g(x) = \cos x$				

14. If $f(x) = \sin x$, find
a. $f(\pi)$, b. $f(.5321)$, c. $f(\pi/3)$, d. $f(1.2431)$. (7-2)
15. If $f(x) = \sin x$, find
a. $f(1)$, b. $f(10)$, c. $f(15\pi)$, d. $f(12)$. (7-2)
16. If $h(x) = \cos x$, find
a. $h(\pi)$, b. $h(.3761)$, c. $h(\pi/4)$, d. $h(1.500)$. (7-2)
17. If $h(x) = \cos x$, find
a. $h(1)$, b. $h(0)$, c. $h(7\pi)$, d. $h(16)$. (7-2)
18. If $f(x) = \sin x$, find $f(60)$ and $f(-30)$. (7-2)
19. If $h(x) = \cos x$, find $h(80)$ and $h(-42)$. (7-2)
20. Graph $g(x) = \sin(-x)$ from 0 to 2π. (7-3)
21. Graph $k(x) = \cos(-x)$ from 0 to 2π. (7-3)
22. Sketch $f(x) = \sin x$ and $g(x) = -3 \sin x$ on the same coordinate system. (7-3)
23. Sketch $f(x) = \cos x$ and $g(x) = \cos 3x$ on the same coordinate system. (7-3)
24. Determine the amplitude and period of

$$f(x) = -\frac{1}{2} \cos \frac{\pi x}{5}. \qquad (7\text{-}3)$$

25. Determine the amplitude and period of $g(x) = 7.3 \sin 2.8\pi x$. (7-3)
26. Determine the period of $h(t) = 110 \sin 2\pi ft$. (7-3)
27. Determine the period of $f(t) = 48 \sin 240\pi t$. (7-3)
28. Write the equation of a sine function with amplitude 90 and period $3\pi/5$. (7-3)

29. If $f(t) = a \sin 80\pi t$ and $f(.001) = 18.2$, find $|a|$. (7-3)
30. If $g(t) = 25 \cos 36t$ and $g(t_1) = 18$, find t_1. (7-3)
31. If $h(t) = 72 \sin bt$ and $h(0.2) = -60$, find the period of h. (7-3)
32. A phasor has a period of .082 seconds. What is the frequency? (7-3)
33. Graph one cycle of $g(t) = -\cos 12.5t$. What is the period? (7-3)
34. Graph one cycle of $f(t) = 6 \sin 1256t$. What is the period? (7-3)
35. Sketch the graph of $f(x) = 2 \sin(x - \pi)$ and the graph of $g(x) = 2 \sin x$ on the same coordinate system. (7-4)
36. Sketch the graph of $f(x) = -3 \sin(x + \pi/2)$ and the graph of $g(x) = \cos x$ on the same coordinate system. (7-4)

Determine the phase shift, period, and amplitude of the following periodic functions.

37. $f(x) = 12 \cos(300x - 0.2\pi)$. (7-4)
38. $g(x) = \dfrac{\sin(12\pi x - \pi/6)}{4}$. (7-4)
39. $h(t) = -16 \sin\left(400t + \dfrac{\pi}{3}\right)$. (7-4)
40. $f(t) = 1.6 \cos\left(\dfrac{5\pi t}{3} + \dfrac{2\pi}{9}\right)$. (7-4)
41. Write an equation for a sine function with the appropriate amplitude, period, and phase shift that is equivalent to $f(t) = 16 \cos 2\pi t$. (7-4)
42. Write an equation for a sine function equivalent to

$$g(t) = -4 \cos\left(2\pi \mathrm{f}t + \frac{\pi}{3}\right). \quad (7\text{-}4)$$

43. If $f(1/5) = 12$ when $f(t) = 15 \sin(\pi t + c)$, find the constant c. (7-4)
44. If $f(.005) = 8$ when $f(t) = 12 \sin(44\pi t + c)$, find the constant c. (7-4)
45. If $-4 = 12 \sin\left(x + \dfrac{\pi}{9}\right)$, find x. (7-4)
46. If $8.75 = 20 \cos\left(x - \dfrac{3\pi}{8}\right)$, find x. (7-4)
47. Sketch the graph of $f(x) = -|\cos x|$. What are the period and amplitude of f? (7-5)
48. Sketch the graph of $g(x) = (\cos x)^2$ and $h(x) = \cos x$ on the same coordinate system. What are the period and amplitude of g? (7-5)
49. Sketch the graph of $f(t) = \left|\dfrac{\sin t}{\cos t}\right|$. What are the period and amplitude of f? (7-5)
50. Sketch the graph of $h(t) = \sin t \cos t$. What are the period and amplitude of h? (7-5)
51. The graph of $f(t) = (1/2)^t \sin 2t$ is a *damped sine wave*. Sketch the graph from $t = -\pi$ to 2π. Is f periodic? (7-5)
52. Carefully plot $f(x) = \cos x$ over one cycle using equal vertical and horizontal scales. Estimate the slope of the curve at intervals at $\pi/4$. Graph the results. Determine the probable equation of the function f'. (7-6)

Chapter Eight Trigonometric Identities and Conditional Equations

8-1 Fundamental Trigonometric Identities

What we think of as fundamental properties of real numbers with the operations of addition and multiplication are actually identities. That is, they are equations that are true for all replacements of the variables. For example,

$$x + (-x) = 0, \qquad \text{(Eq. 8-1a)}$$

$$x \cdot \frac{1}{x} = 1, \quad x \neq 0, \qquad \text{(Eq. 8-1b)}$$

and

$$ax + bx = (a + b)x \qquad \text{(Eq. 8-1c)}$$

are fundamental postulates, and they are identities. These and other identities are used in the solution of conditional equations. They enable us to write equivalent conditional equations that eventually are so basic that the solutions are obvious.

Example 1

Notice how Eqs. 8-1a, 8-1b, and 8-1c are used to solve $4x + 3 = x - 1$.

i. $4x + (3 + (-3)) + (-x) = (x + (-x)) + (-3) - 1,$
$4x - x = -4.$ (by Eq. 8-1a)

ii. $(4 - 1)x = -4,$
$3x = -4.$ (by Eq. 8-1c)

iii. $\frac{1}{3}(3)x = \frac{1}{3}(-4),$

$$x = -\frac{4}{3}. \qquad \text{(by Eq. 8-1b)}$$

Fundamental trigonometric identities play an essential role in the simplification and solution of conditional trigonometric equations.

We introduced in Chapter 6 the important identity

$$\tan\theta = \frac{\sin\theta}{\cos\theta}. \qquad \text{(Eq. 6-2d)}$$

To illustrate to a small degree how this identity alone is essential to the solution of a conditional trigonometric equation, consider the next example.

Example 2

Solve the equation

$3\tan\theta\cos\theta - .5 = 2\sin\theta$

for θ in the interval from zero to 2π.

i. $3\tan\theta\cos\theta - .5 = 2\sin\theta;$

ii. $3\left(\frac{\sin\theta}{\cos\theta}\right)\cos\theta - .5 = 2\sin\theta;$

iii. $3\sin\theta - 2\sin\theta = .5$

$$\sin\theta = .5$$

$$\theta = \arcsin .5 = \frac{\pi}{6} \text{ and } \frac{5\pi}{6}.$$

There are three trigonometric ratios in addition to sine, cosine, and tangent that have not been needed in earlier problems but are essential in order for us to proceed. In terms of Fig. 8-1a, they are

$$\text{cotangent } \theta = \frac{x}{y}, \qquad \text{(Eq. 8-1d)}$$

$$\text{secant } \theta = \frac{r}{x}, \qquad \text{(Eq. 8-1e)}$$

$$\text{cosecant } \theta = \frac{r}{y}. \qquad \text{(Eq. 8-1f)}$$

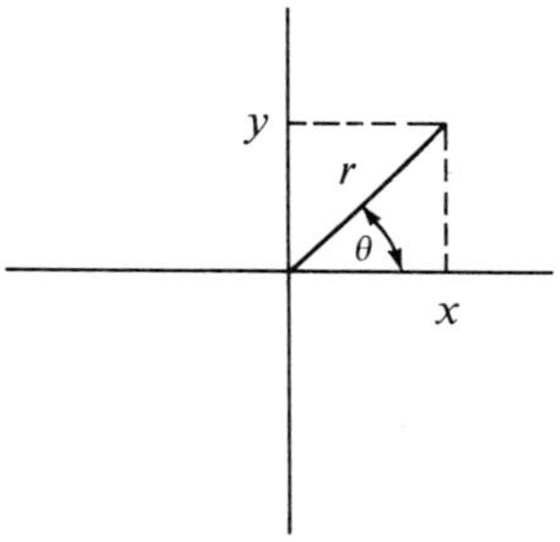

Fig. 8-1a

Notice that cotangent θ (abbreviated cot θ) is the reciprocal of tan θ, secant θ (abbreviated sec θ) is the reciprocal of [ ], and cosecant θ (abbreviated csc θ) is the reciprocal of [ ].

■ cos θ; sin θ.

The following identities should be immediately apparent.

$$\sin \theta = 1/\csc \theta, \qquad \text{(Eq. 8-1g)}$$

$$\cos \theta = 1/\sec \theta, \qquad \text{(Eq. 8-1h)}$$

$$\tan \theta = \frac{1}{\cot \theta}, \qquad \text{(Eq. 8-1i)}$$

$$\cot \theta = \frac{\cos \theta}{\sin \theta}. \qquad \text{(Eq. 8-1j)}$$

Three other fundamental identities result from variations of the Pythagorean theorem, $x^2 + y^2 = r^2$. Dividing the right and left sides of the equation by r^2 gives

$$\frac{x^2}{r^2} + \frac{y^2}{r^2} = 1,$$

or

$$\cos^2 \theta + \sin^2 \theta = 1. \qquad \text{(Eq. 8-1k)}$$

Dividing instead by x^2 gives

$$1 + \frac{y^2}{x^2} = \frac{r^2}{x^2},$$

or

$$1 + \tan^2 \theta = \sec^2 \theta. \qquad \text{(Eq. 8-1m)}$$

Dividing by y^2 gives

$$\frac{x^2}{y^2} + 1 = \frac{r^2}{y^2},$$

or

$$\cot^2 \theta + 1 = \csc^2 \theta. \qquad \text{(Eq. 8-1n)}$$

The trigonometric identities in this section are true for all values of the variable for which the function is defined, that is, where the ratios are finite numbers. Also, the variables may be either the measurement of angles or real numbers.

Example 3

Verify that Eq. 8-1k is true for $\theta = 3\pi/4$.

The verification can be done exactly, because the sine and cosine of $3\pi/4$ are exactly $1/\sqrt{2}$ and $-1/\sqrt{2}$.

$$\cos^2\left(\frac{3\pi}{4}\right) + \sin^2\left(\frac{3\pi}{4}\right) \stackrel{?}{=} 1,$$

$$\left(-\frac{1}{\sqrt{2}}\right)^2 + \left(\frac{1}{\sqrt{2}}\right)^2 \stackrel{?}{=} 1,$$

$$\frac{1}{2} + \frac{1}{2} \stackrel{?}{=} 1,$$

$$1 = 1.$$

The successful solution of trigonometric equations is dependent upon how well basic identities can be recalled. Some effort should be made to form good mental images of the fundamental identities.

8-1 Exercises

The main purpose of these exercises is to help you become familiar with the fundamental identities.

1. Verify Eq. 8-1n for $\theta = \dfrac{3\pi}{4}$.

---

2. Verify Eq. 8-1j for $\theta = \dfrac{3\pi}{4}$.

■ 1. $$\cot^2\left(\frac{3\pi}{4}\right) + 1 \stackrel{?}{=} \csc^2\left(\frac{3\pi}{4}\right)$$

$$(-1)^2 + 1 \stackrel{?}{=} (\sqrt{2})^2$$

$$2 = 2.$$

Show that each expression in Exercises 3–11 has a value of one for all values of x for which the expressions are defined.

3. $\dfrac{\tan x \csc x}{\sec x}$

■ 2. $\cot\left(\dfrac{3\pi}{4}\right) \stackrel{?}{=} \dfrac{\cos(3\pi/4)}{\sin(3\pi/4)}$

$-1 \stackrel{?}{=} \dfrac{-1/\sqrt{2}}{1/\sqrt{2}}$

$-1 = -1.$

---

4. $\csc^2 x(1 - \cos^2 x)$

■ 3. $\dfrac{\tan x \csc x}{\sec x}$

$= \dfrac{\sin x}{\cos x} \cdot \dfrac{\csc x}{1} \cdot \dfrac{1}{\sec x}$

$= \dfrac{\sin x}{\cos x} \cdot \dfrac{1}{\sin x} \cdot \dfrac{\cos x}{1}$

$= \dfrac{\sin x}{\sin x} \cdot \dfrac{\cos x}{\cos x}$

$= 1.$

---

5. $\dfrac{\sin x(\csc x - \sin x)}{\cos^2 x}$

■ 4. $\csc^2 x(1 - \cos^2 x)$

$= \csc^2 x(\sin^2 x)$

$= \left(\dfrac{1}{\sin^2 x}\right) \cdot (\sin^2 x)$

$= 1.$

---

6. $\cos^2 x(1 + \tan^2 x)$

■ 5. $\dfrac{\sin x(\csc x - \sin x)}{\cos^2 x}$

$= \dfrac{\sin x\left(\dfrac{1}{\sin x} - \sin x\right)}{\cos^2 x}$

$= \dfrac{1 - \sin^2 x}{\cos^2 x}$

$= \dfrac{\cos^2 x}{\cos^2 x}$

$= 1.$

---

7. $\dfrac{\sin x \cos x \tan x}{1 - \cos^2 x}$

■ 6. $\cos^2 x(1 + \tan^2 x)$

$$= \cos^2 x \sec^2 x$$

$$= \cos^2 x \left(\frac{1}{\cos^2 x}\right)$$

$$= 1.$$

---

8. $\dfrac{\sin x \tan x + \cos x}{\sec x}$

■ 7. $\dfrac{\sin x \cos x \tan x}{1 - \cos^2 x}$

$$= \frac{\sin x \cos x \left(\dfrac{\sin x}{\cos x}\right)}{1 - \cos^2 x}$$

$$= \frac{\sin^2 x}{1 - \cos^2 x}$$

$$= \frac{\sin^2 x}{\sin^2 x}$$

$$= 1.$$

---

9. $\dfrac{\cos^2 x - \sin^2 x}{1 - 2\sin^2 x}$

■ 8. $\dfrac{\sin x \tan x + \cos x}{\sec x}$

$$= \frac{\sin x \left(\dfrac{\sin x}{\cos x}\right) + \cos x}{\dfrac{1}{\cos x}}$$

$$= \left(\frac{\sin^2 x}{\cos x} + \cos x\right) \cos x$$

$$= \sin^2 x + \cos^2 x$$

$$= 1.$$

---

10. $\dfrac{\tan^2 x \sec^2 x - \tan^4 x}{\tan^2 x}$

■ 9. $\dfrac{\cos^2 x - \sin^2 x}{1 - 2\sin^2 x}$

$$= \frac{(1 - \sin^2 x) - \sin^2 x}{1 - 2\sin^2 x}$$

$$= \frac{1 - 2\sin^2 x}{1 - 2\sin^2 x}$$

$$= 1.$$

11. $\tan^2 x \cos^2 x + \cot^2 x \sin^2 x$

■ 10. $$\frac{\tan^2 x \sec^2 x - \tan^4 x}{\tan^2 x} = \frac{\tan^2 x(\sec^2 x - \tan^2 x)}{\tan^2 x} = \sec^2 x - \tan^2 x = (1 + \tan^2 x) - \tan^2 x = 1.$$

■ 11. $$\tan^2 x \cos^2 x + \cot^2 x \sin^2 x = \left(\frac{\sin^2 x}{\cos^2 x}\right)\cos^2 x + \left(\frac{\cos^2 x}{\sin^2 x}\right)\sin^2 x = \sin^2 x + \cos^2 x = 1.$$

8-1 Problems

1. Evaluate each of the following expressions by using first fundamental identities, then tables.

   a. $\sqrt{1 - \cos^2 .38}$,

   b. $\sqrt{\sec^2 .74 - 1}$,

   c. $\dfrac{\sec .42}{\tan .42 + \cot .42}$.

2. Given $\sin \theta = b$, write the value of the other five trigonometric ratios in terms of b.
3. Find the value of the six trigonometric ratios for $\theta = .6283$ radians.
4. Find the secant, cosecant, and cotangent of $\theta = 152°$.
5. Write a formula for the area of the circle in terms of θ and m.

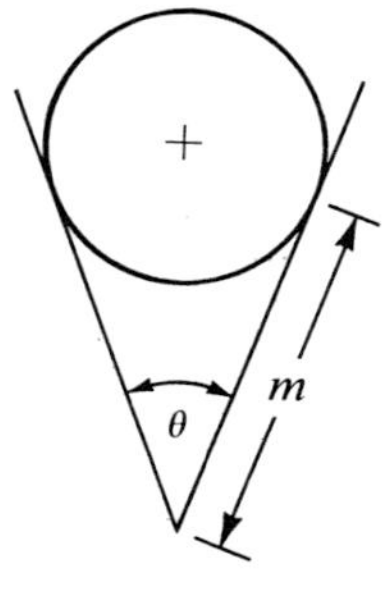

6. The slide rule has only the S, T, and ST trigonometric scales, but it can be used to find secants, cosecants, and cotangents of angles by using appropriate trigonometric identities. Give attention to using the slide rule for all six ratios.
7. Sketch the graph of $f(x) = \csc x$.

8-2 Sin $(\theta \pm \phi)$, Cos$(\theta \pm \phi)$, and Tan $(\theta \pm \phi)$

In this section we develop identities that enable us to consider again certain composite trigonometric equations, but this time we shall consider them mathematically rather than graphically as was done in Section 7-5.

It should not surprise us to find

$$\sin\theta + \sin\phi \neq \sin(\theta + \phi),$$

because of our familiarity with the nonlinear graphs of sine functions. We find that there are situations where it is convenient to express $\sin(\theta \pm \phi)$ in terms of θ and ϕ separately. Figure 8-2a is used to develop such a relationship.

Let us place θ in standard position, and the initial side of ϕ on the terminal side of θ.

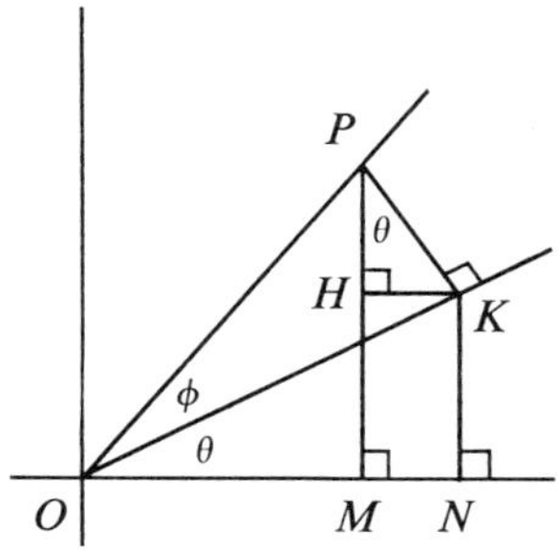

Fig. 8-2a

By definition,

$$\sin(\theta + \phi) = \frac{\overline{PM}}{\overline{OP}}.$$

Since $\overline{PM} = \overline{PH} + \overline{HM}$ and $\overline{HM} = \overline{KN}$, we may write

$$\sin(\theta + \phi) = \frac{\overline{KN} + \overline{PH}}{\overline{OP}},$$

or

$$\sin(\theta + \phi) = \frac{\overline{KN}}{\overline{OP}} + \frac{\overline{PH}}{\overline{OP}}.$$

We may multiply by 1 in the form of $\dfrac{\overline{OK}}{\overline{OK}}$ or $\dfrac{\overline{KP}}{\overline{KP}}$. Thus,

$$\sin(\theta + \phi) = \frac{\overline{KN}}{\overline{OP}} \cdot \frac{\overline{OK}}{\overline{OK}} + \frac{\overline{PH}}{\overline{OP}} \cdot \frac{\overline{KP}}{\overline{KP}},$$

which may be written

$$\sin(\theta + \phi) = \frac{\overline{KN}}{\overline{OK}} \cdot \frac{\overline{OK}}{\overline{OP}} + \frac{\overline{PH}}{\overline{KP}} \cdot \frac{\overline{KP}}{\overline{OP}}.$$

Since

$$\cos\theta = \frac{\overline{PH}}{\overline{KP}}, \qquad \sin\phi = \frac{\overline{KP}}{\overline{OP}},$$

and

$$\sin\theta = \frac{\overline{KN}}{\overline{OK}}, \qquad \cos\phi = \frac{\overline{OK}}{\overline{OP}},$$

we may write

$$\sin(\theta + \phi) = \sin\theta\cos\phi + \cos\theta\sin\phi. \qquad \text{(Eq. 8-2a)}$$

Similarly, we may derive the formula for expressing the cosine of the sum of two angles θ and ϕ in terms of functions of θ and ϕ. We use Fig. 8-2a and merely indicate the steps involved.

$$\cos(\theta + \phi) = \frac{\overline{OM}}{\overline{OP}} = \frac{\overline{ON} - \overline{MN}}{\overline{OP}},$$

$$\cos(\theta + \phi) = \frac{\overline{ON}}{\overline{OK}}\frac{\overline{OK}}{\overline{OP}} - \frac{\overline{HK}}{\overline{KP}}\frac{\overline{KP}}{\overline{OP}},$$

$$\cos(\theta + \phi) = \cos\theta\cos\phi - \sin\theta\sin\phi. \qquad \text{(Eq. 8-2b)}$$

Example 1

Find sin 75° from $\sin 75° = \sin(45° + 30°)$.

$$\begin{aligned}\sin 75° = \sin(45° + 30°) &= \sin 45° \cos 30° + \sin 30° \cos 45° \\ &= .707(.866) + .50(.707) \\ &= .612 + .353 \\ &= .965.\end{aligned}$$

Example 2

Verify that $\cos 180° = -1$ from the fact that $\cos 180° = \cos(120° + 60°)$.

$$\begin{aligned}\cos 180° = \cos(120° + 60°) &= \cos 120° \cos 60° - \sin 120° \sin 60° \\ &= -.50(.50) - .866(.866) \\ &= -.25 - .75 = -1.\end{aligned}$$

We can derive the formula for $\sin(\theta - \phi)$ by observing that

$$\sin(\theta - \phi) = \sin(\theta + (-\phi)) = \sin\theta \cos(-\phi) + \cos\theta \sin(-\phi).$$

Since $\cos(-\phi) = \cos\phi$ and $\sin(-\phi) = -\sin\phi$ (refer to Figure 6-3b), we have

$$\sin(\theta - \phi) = \sin\theta \cos\phi - \cos\theta \sin\phi. \qquad \text{(Eq. 8-2c)}$$

Show in a similar manner how the identity

$$\cos(\theta - \phi) = \cos\theta \cos\phi + \sin\theta \sin\phi \qquad \text{(Eq. 8-2d)}$$

can be derived from Eq. 8-2b.

■ $\cos(\theta - \phi) = \cos(\theta + (-\phi)) = \cos\theta\cos(-\phi) - \sin\theta\sin(-\phi)$.

Since $\cos(-\phi) = \cos\phi$ and $\sin(-\phi) = -\sin\phi$, we have

$\cos(\theta - \phi) = \cos\theta\cos\phi + \sin\theta\sin\phi$.

The fundamental identity $\tan\beta = \sin\beta/\cos\beta$ is used to arrive at the expression for $\tan(\theta \pm \phi)$ in terms of $\tan\theta$ and $\tan\phi$.

$$\tan(\theta \pm \phi) = \frac{\sin(\theta \pm \phi)}{\cos(\theta \pm \phi)} = \frac{\sin\theta\cos\phi \pm \cos\theta\sin\phi}{\cos\theta\cos\phi \mp \sin\theta\sin\phi}.$$

Dividing both numerator and denominator by $\cos\theta\cos\phi$ enables us to write the last expression in terms of tangent ratios.

$$\tan(\theta \pm \phi) = \frac{\dfrac{\sin\theta\cos\phi}{\cos\theta\cos\phi} \pm \dfrac{\cos\theta\sin\phi}{\cos\theta\cos\phi}}{\dfrac{\cos\theta\cos\phi}{\cos\theta\cos\phi} \mp \dfrac{\sin\theta\sin\phi}{\cos\theta\cos\phi}},$$

$$\tan(\theta \pm \phi) = \frac{\tan\theta \pm \tan\phi}{1 \mp \tan\theta\tan\phi}. \qquad \text{(Eq. 8-2e)}$$

Example 3

Find the $\tan(40° + 35°)$ using Eq. 8-2e.

$$\begin{aligned}\tan(40° + 35°) &= \frac{\tan 40° + \tan 35°}{1 - \tan 40° \tan 35°} \\ &= \frac{.8391 + .7002}{1 - (.8391)(.7002)} \\ &= \frac{1.5393}{.4125} = 3.7319.\end{aligned}$$

Before going on, review Example 1 of Section 7-5. In the following paragraph we pursue an algebraic analysis of a similar problem.

Consider the sum of two sinusoidal equations with the same period, $2\pi/b$,

$$g(t) = a_1 \sin(bt + c_1) \qquad \text{(Eq. 8-2f)}$$

and

$$h(t) = a_2 \sin(bt + c_2). \qquad \text{(Eq. 8-2g)}$$

Neither the amplitudes, $|a_1|$ and $|a_2|$, nor the initial phases, c_1 and c_2, are necessarily equal. By Eq. 8-2a,

$$g(t) = a_1(\sin bt \cos c_1 + \cos bt \sin c_1)$$

and

$$h(t) = a_2(\sin bt \cos c_2 + \cos bt \sin c_2).$$

Let $f(t)$ represent the sum of $g(t) + h(t)$ so that

$$\begin{aligned} f(t) &= g(t) + h(t) \\ &= a_1 \sin bt \cos c_1 + a_1 \cos bt \sin c_1 + a_2 \sin bt \cos c_2 + a_2 \cos bt \sin c_2 \\ &= (a_1 \sin c_1 + a_2 \sin c_2) \cos bt + (a_1 \cos c_1 + a_2 \cos c_2) \sin bt. \end{aligned}$$

Notice that the expressions in parentheses are constants. Figure 8-2b shows how they may be interpreted.

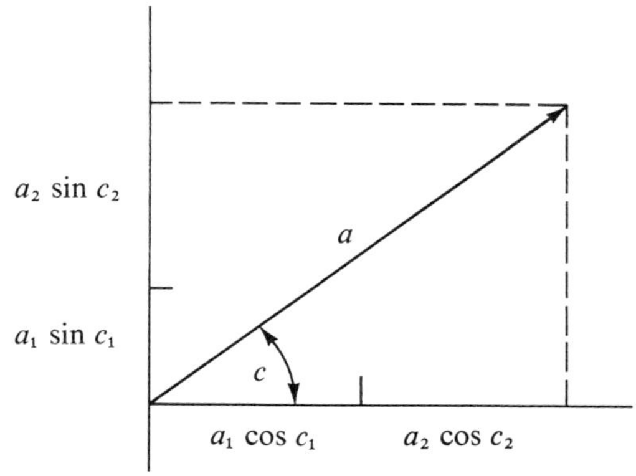

Fig. 8-2b

If we let

$$a \cos c = a_1 \cos c_1 + a_2 \cos c_2$$

and

$$a \sin c = a_1 \sin c_1 + a_2 \sin c_2,$$

the simplified results are

$$f(t) = a \sin c \cos bt + a \cos c \sin bt$$
$$= a \sin(bt + c). \qquad \text{(Eq. 8-2h)}$$

The result that appears in Eq. 8-2h tells us something about the sum of two sine waves with the same period and any relative phase shift. We may have suspected this when we graphed the result in Section 7-5. The equation for f is sinusoidal and has the same period $(2\pi/b)$ as the addends, $g(t)$ and $h(t)$.

From Fig. 8-2b we see that the constants of the sum are

$$a = \sqrt{(a_1 \cos c_1 + a_2 \cos c_2)^2 + (a_1 \sin c_1 + a_2 \sin c_2)^2} \qquad \text{(Eq. 8-2i)}$$

and

$$c = \arctan\left(\frac{a_1 \sin c_1 + a_2 \sin c_2}{a_1 \cos c_1 + a_2 \cos c_2}\right). \qquad \text{(Eq. 8-2j)}$$

Example 4

Find the sum of $g(t) = 2 \sin(\pi t + \pi/6)$ and $h(t) = 3 \sin(\pi t + \pi/4)$ in the form $f(t) = a \sin(\pi t + c)$.

i. $a\sqrt{(2 \cos \pi/6 + 3 \cos \pi/4)^2 + (2 \sin \pi/6 + 3 \sin \pi/4)^2}$
$= \sqrt{(2(.866) + 3(.707))^2 + (2(.500) + 3(.707))^2}$
$= \sqrt{14.84 + 9.74}$
$= 4.96;$

ii. $c = \arctan\left(\dfrac{2 \sin \pi/6 + 3 \sin \pi/4}{2 \cos \pi/6 + 3 \cos \pi/4}\right)$
$= \arctan \dfrac{3.12}{3.85} = \arctan .810$
$= .681 = .216\pi;$

iii. Therefore $f(t) = 4.96 \sin(\pi t + .216\pi)$.

8-2 Exercises

1. Find $\sin 105°$ by using $105° = 60° + 45°$.

---

2. Find $\cos 75°$ by using $75° = 30° + 45°$.

■ 1. $\sin(60° + 45°)$
$= \sin 60° \cos 45° + \cos 60° \sin 45°$
$= .866(.707) + .500(.707)$
$= .9658.$

---

3. Find $\sin 15°$ by using $15° = 45° - 30°$.

■ 2. $\cos(30° + 45°)$
$= \cos 30° \cos 45° - \sin 30° \sin 45°$
$= .866(.707) - .500(.707)$
$= .2588.$

---

4. Show that $\sin(270° - x) = -\cos x$.

■ 3. $\sin(45° - 30°)$
$= \sin 45° \cos 30° - \cos 45° \sin 30°$
$= .707(.866) - .707(.500)$
$= .2588.$

---

5. Show that $\sin(90° + x) = \cos x$.

■ 4. $\sin(270° - x)$
$= \sin 270° \cos x - \cos 270° \sin x$
$= -1 \cos x - (0) \sin x$
$= -1 \cos x.$

---

6. Given $\sin \alpha = 4/5$ (in the first quadrant) and $\cos \beta = -12/13$ (in the second quadrant), find $\sin(\alpha + \beta)$.

■ 5. $\sin(90° + x)$
$= \sin 90° \cos x + \cos 90° \sin x$
$= (1) \cos x + (0) \sin x$
$= \cos x.$

7. Find $\cos(\beta - \alpha)$ for the angles in Exercise 6.

■ 6. i.

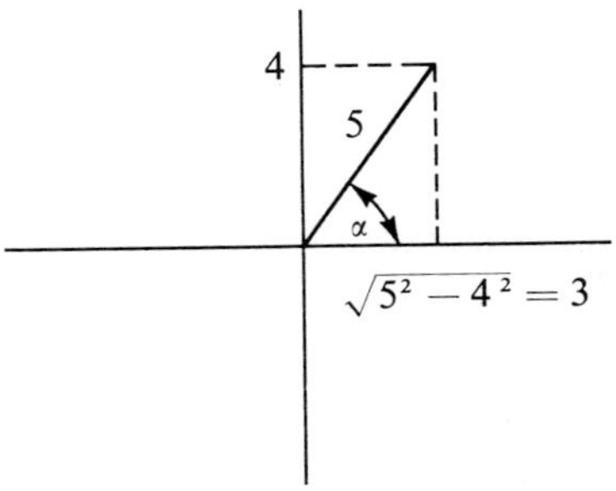

ii.

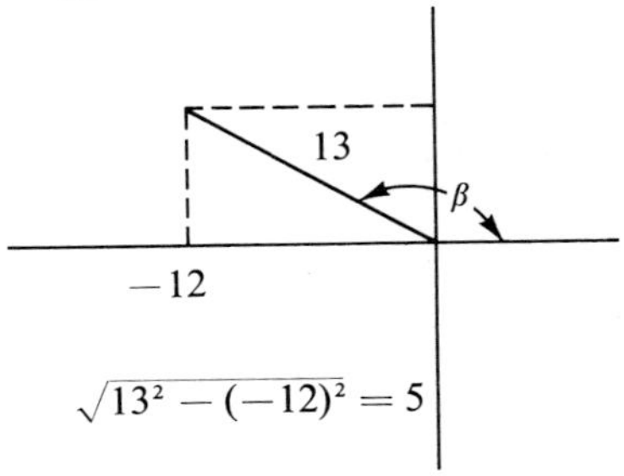

iii. $$\sin(\alpha+\beta) = \sin\alpha\cos\beta + \cos\alpha\sin\beta = \frac{4}{5}\left(-\frac{12}{13}\right) + \frac{3}{5}\left(\frac{5}{13}\right) = -.5077.$$

8. Express

$$\sin x \cos 2x + \sin 2x \cos x$$

as a single term.

■ 7. $$\cos(\beta - \alpha) = \cos\beta\cos\alpha + \sin\beta\sin\alpha = -\frac{12}{13}\left(\frac{3}{5}\right) + \frac{5}{13}\left(\frac{4}{5}\right) = -.2462.$$

9. Find the equation for the sum of $g(x) = \sin x$ and $h(x) = 2 \sin (x + 30°)$ in the form $f(x) = a \sin (x + c)$.

■ 8. $\sin x \cos 2x + \sin 2x \cos x$
$= \sin (x + 2x)$
$= \sin 3x.$

■ 9. i. $c = \arctan \left(\dfrac{\sin 0° + 2 \sin 30°}{\cos 0° + 2 \cos 30°}\right)$

$c = \arctan \left(\dfrac{0 + 2(.5)}{1 + 2(.866)}\right)$

$= \arctan \dfrac{1}{2.732}$

$= \arctan .366$

$= 20.1°;$

ii. $a = \sqrt{(\cos 0° + 2 \cos 30°)^2 + (\sin 0° + 2 \sin 30°)^2}$

$= \sqrt{(2.732)^2 + 1^2}$

$= 2.91;$

iii. Therefore, $f(x) = 2.91 \sin (x + 20.1°).$

8-2 Problems

1. The displacements, y_1 and y_2, of two waves traveling through the same medium are

$$f(t) = y_1 = A \sin 2\pi\left(\frac{t}{T} - \frac{a}{\lambda}\right)$$

and

$$g(t) = y_2 = A \sin 2\pi\left(\frac{t}{T} + \frac{a}{\lambda}\right).$$

Find an expression for the displacement, $y_1 + y_2$, of the combination of the waves.

2. Equation 8-2i is equivalent to the law of cosines. Show this by using the elementary identities of Section 8-1.

3.

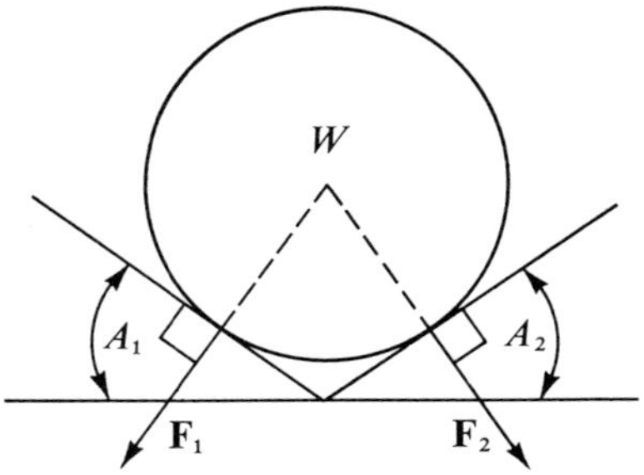

Vector analysis would show that the following conditions must exist for the sphere to be in equilibrium:

$$F_1 \cos A_1 + F_2 \cos A_2 = W$$

$$F_1 \sin A_1 - F_2 \sin A_2 = 0.$$

a. If $\angle A_1 = 45°$, $\angle A_2 = 25°$, and $W = 50$ pounds, find the magnitude of the forces $\mathbf{F}_1$ and $\mathbf{F}_2$.

b. A general solution for F_1 is

$$F_1 = \frac{W \sin A_2}{\sin(A_1 + A_2)}.$$

Find the general solution for F_2.

4.

Three single-phase alternators are in series. Find an expression for the sum of the voltages in the form

$$v_1 + v_2 + v_3 = V \sin(\omega t + \psi).$$

8-3 Double-Angle Identities

Useful trigonometric identities result from considering again Eqs. 8-2a through 8-2h when the two angles θ and ϕ are equal. If we let $\theta = \phi$ and substitute into

$$\sin(\theta + \phi) = \sin\theta \cos\phi + \cos\theta \sin\phi,$$

we get

$$\sin(\theta + \theta) = \sin\theta \cos\theta + \cos\theta \sin\theta,$$

or

$$\sin 2\theta = 2 \sin\theta \cos\theta. \qquad \text{(Eq. 8-3a)}$$

Similarly, the formula for the cosine of the sum of two angles may be modified to the form

$$\cos 2\theta = \cos^2\theta - \sin^2\theta. \qquad \text{(Eq. 8-3b)}$$

Alternate forms of Eq. 8-3b can be obtained by using the fundamental identity $\sin^2\theta + \cos^2\theta = 1$.

$$\cos 2\theta = \cos^2\theta - (1 - \cos^2\theta) = 2\cos^2\theta - 1, \qquad \text{(Eq. 8-3c)}$$

$$\cos 2\theta = (1 - \sin^2\theta) - \sin^2\theta = 1 - 2\sin^2\theta. \qquad \text{(Eq. 8-3d)}$$

The formula for the tangent of the sum of two angles may be modified to provide a double-angle formula;

$$\tan(\theta + \phi) = \frac{\tan\theta + \tan\phi}{1 - \tan\theta \tan\phi}$$

becomes

$$\tan 2\theta = \frac{2\tan\theta}{1 - \tan^2\theta}. \qquad \text{(Eq. 8-3e)}$$

Example 1

Simplify the expression $\cos^2 2x - \sin^2 2x$ using the double-angle formulas.

Using Eq. 8-3b and letting $2x = \theta$, we have

$$\begin{aligned}\cos^2 2x - \sin^2 2x &= \cos 2(2x)\\ &= \cos 4x.\end{aligned}$$

Example 2

Given that $\sin\theta = -4/5$, find $\cos 2\theta$.

$$\begin{aligned}\cos 2\theta &= 1 - 2\sin^2\theta \qquad \text{(Eq. 8-3d)}\\ &= 1 - 2\left(-\frac{4}{5}\right)^2\\ &= 1 - 2\left(\frac{16}{25}\right)\\ &= 1 - 1.28\\ &= -.28.\end{aligned}$$

8-3 Exercises

1. Verify each of the formulas of this section for $\theta = 45°$.

---

2. Simplify $\dfrac{\sin 2x}{1 + \cos 2x}$.

■ 1. Eq. 8-3a:

$$\begin{aligned}\sin 2(45°) &\stackrel{?}{=} 2\sin 45° \cos 45°\\ 1 &\stackrel{?}{=} 2(.707)(.707)\\ 1 &= 1,\end{aligned}$$

and so forth.
Notice that Eq. 8-3e, $\tan 2(45°) = \tan 90°$, is undefined.

---

3. Given $\sin\theta = 1/3$, find $\sin 2\theta$, $0° \leq \theta \leq 90°$.

■ 2. $\dfrac{\sin 2x}{1+\cos 2x}$

$$= \frac{2\sin x\cos x}{1+\cos^2 x-\sin^2 x}$$

$$= \frac{2\sin x\cos x}{2\cos^2 x}$$

$$= \frac{\sin x}{\cos x}$$

$$= \tan x.$$

---

4. Given $\cos\theta = 2/3$, find $\cos 2\theta$, $0° \leq \theta \leq 90°$.

■ 3. $\sin 2\theta = 2\sin\theta\cos\theta$

$$= 2\left(\frac{1}{3}\right)\left(\frac{2\sqrt{2}}{3}\right)$$

$$= \frac{4\sqrt{2}}{9}.$$

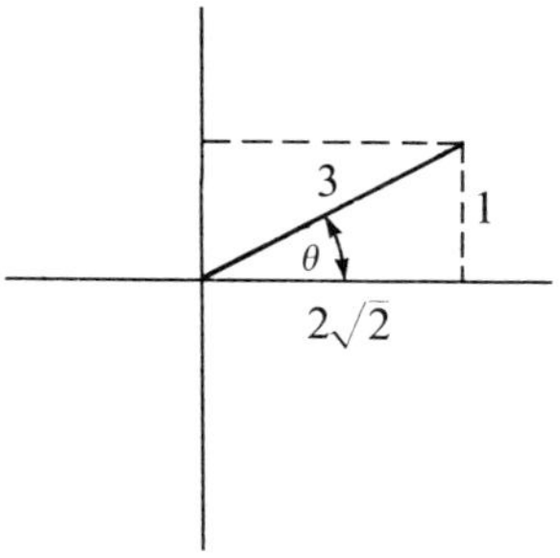

5. Simplify $\cos^4 x - \sin^4 x$.

■ 4. $\cos 2\theta = \left(\frac{2}{3}\right)^2 - \left(\frac{\sqrt{5}}{3}\right)^2$
$= \frac{4-5}{9} = -\frac{1}{9}$.

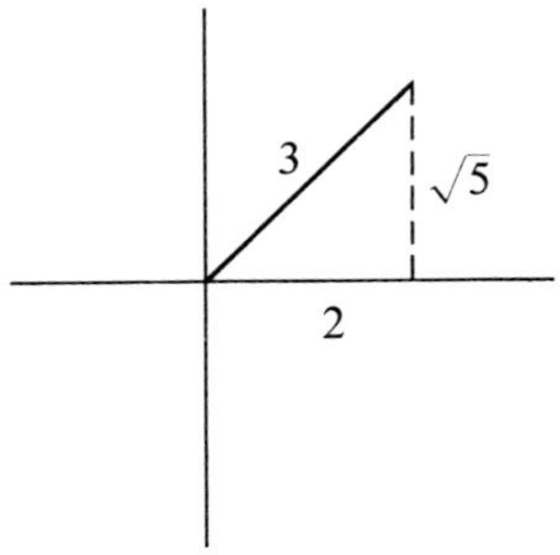

6. Simplify $\csc 2x + \cot 2x$.

■ 5. $\cos^4 x - \sin^4 x$
$= (\cos^2 x + \sin^2 x)$
$(\cos^2 x - \sin^2 x)$
$= 1(\cos 2x)$
$= \cos 2x$.

7. Simplify $\frac{1 - \tan^2 x}{1 + \tan^2 x}$.

■ 6. $\csc 2x + \cot 2x$

$$= \frac{1}{\sin 2x} + \frac{\cos 2x}{\sin 2x}$$

$$= \frac{1 + \cos 2x}{\sin 2x}$$

$$= \frac{1 + (2\cos^2 x - 1)}{2 \sin x \cos x}$$

$$= \frac{2\cos^2 x}{2 \sin x \cos x}$$

$$= \frac{\cos x}{\sin x} = \tan x.$$

---

8. Try to write an expression for sin $3x$ in terms of sin x. (*Hint*: $2x + x = 3x$).

■ 7. $\dfrac{1-\tan^2 x}{1+\tan^2 x}$

$$= \frac{1 - \dfrac{\sin^2 x}{\cos^2 x}}{\sec^2 x}$$

$$= \frac{1}{\sec^2 x} - \frac{\sin^2 x}{1}$$

$$= \cos^2 x - \sin^2 x$$

$$= \cos 2x.$$

---

9. Find the value of cos $3x$ in terms of cos x.

■ 8. $\sin 3x$

$$= \sin (2x + x)$$

$$= \sin 2x \cos x + \cos 2x \sin x$$

$$= 2 \sin x \cos^2 x + (1 - 2 \sin^2 x) \sin x$$

$$= 2 \sin x(1 - \sin^2 x) + \sin x - 2 \sin^3 x$$

$$= 2 \sin x - 2 \sin^3 x + \sin x - 2 \sin^3 x$$

$$= 3 \sin x - 4 \sin^3 x.$$

■ 9. $\cos 3x$

$$= \cos (2x + x)$$

$$= \cos 2x \cos x - \sin 2x \sin x$$

$$= (2 \cos^2 x - 1) \cos x - 2 \sin^2 x \cos x$$

$$= 2 \cos^3 x - \cos x - 2(1 - \cos^2 x) \cos x$$

$$= 4 \cos^3 x - 3 \cos x.$$

8-3 Problems

1. In a pure capacitive circuit, $v = V \sin \omega t$, $i = CV\omega \cos \omega t$, and $V = 10$. Plot i, v, and the power $= iv$ as functions of time.

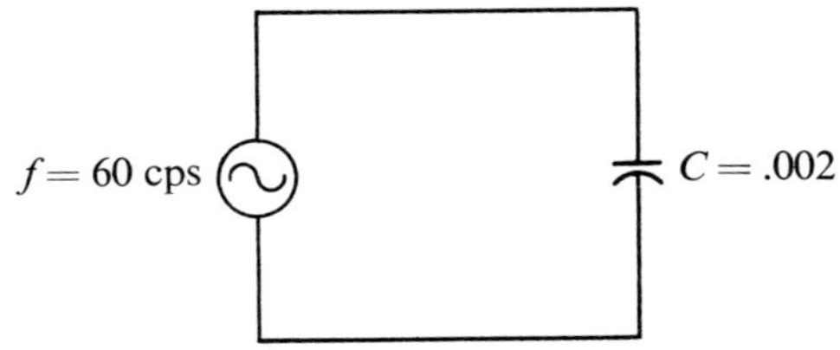

2. Standing waves are produced from incident and reflected waves. If the wave length is λ and the period is P, its displacement, y, at any position, x, at any time, t, is

$$y = A \sin 2\pi\left(\frac{x}{\lambda} - \frac{t}{P}\right) + A \sin 2\pi\left(\frac{x}{\lambda} + \frac{t}{P}\right).$$

This may be simplified to the form

$$y = Y \sin nx \cos mt.$$

Find the values of Y, n, and m in terms of the constants A, 2π, λ, and P.

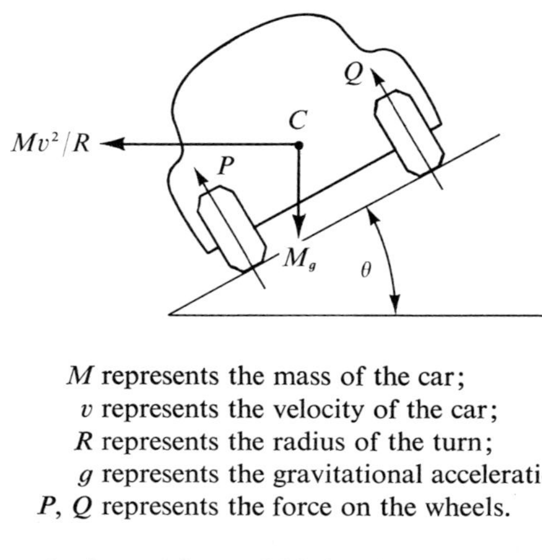

3. M represents the mass of the car;
v represents the velocity of the car;
R represents the radius of the turn;
g represents the gravitational acceleration;
P, Q represents the force on the wheels.

If there is to be *no* horizontal force of friction between the road and the wheels a curve must be banked at an angle θ, where

$$\frac{Mv^2}{R} = P \sin \theta + Q \sin \theta$$

and

$$Mg = P \cos \theta + Q \cos \theta.$$

a. Find the equation for θ by solving the above equations simultaneously.
b. What is the specific speed v (in fps) if $\theta = 15°$, $R = 2500$ feet, and $g = 32$ ft/sec^2? Convert your answer in fps to mph.

8-4 Conditional Trigonometric Equations

The emphasis thus far in Chapter 8 has been on trigonometric identities. In this section, we turn our attention to conditional equations; we shall see the principal reason for having given so much consideration to identities.

In the text, the very first subject matter concerning trigonometry deals with the solution of conditional equations. Sin $x = .3584$ is an example. The positive solutions to this equation less than or equal to 2π are [ ] and [ ].

■ $.117\pi$ (or 21°) and $.883\pi$ (or 159°)

This example is a **basic** trigonometric equation. It contains only one trigonometric ratio and the solution is ordinarily found with the direct aid of a table.

We now consider equations that have more than one trigonometric ratio, such as

$$2\sin^2 x - \cos^2 x + \frac{1}{4} = 0.$$

It is by using identities that we can transform such equations to equivalent basic equations.

There is no single method applicable to the solution of all trigonometric equations. The equations are in general of an algebraic form and the methods of algebra are used, but the unknowns are in terms of the sine, cosine, or other trigonometric ratios.

Example 1

Find the *principal* solutions (solutions in the interval zero to 2π) for $2\sin^2 x - \cos^2 x + 1/4 = 0$.

■ i. $2\sin^2 x - (1 - \sin^2 x) + \dfrac{1}{4} = 0$ (by Eq. 8-1k);

ii. $3\sin^2 x - \dfrac{3}{4} = 0;$

iii. $\sin^2 x = \dfrac{1}{4};$

iv. $\sin x = \pm\dfrac{1}{2}$

$$x = \arcsin \pm\frac{1}{2}$$

$$= \frac{\pi}{6}, \frac{5\pi}{6}, \frac{7\pi}{6}, \text{ and } \frac{11\pi}{6}, \text{ or in degrees}$$

$$x = 30^\circ, 150^\circ, 210^\circ, \text{ and } 330^\circ.$$

Example 2

For what principal angles are the sine and cosine equal?

i. $\sin x = \cos x;$

ii. $\dfrac{\sin x}{\cos x} = 1;$

iii. $\tan x = 1$ (by Eq. 6-2d);

iv. $x = \arctan 1$
$= 45^\circ, 225^\circ.$

Example 3

Find the principal roots of $2 \sin x + 5 = 8 \cos^2 x$.

i. $2\sin x + 5 = 8(1 - \sin^2 x)$ (by Eq. 8-1k);

ii. $8\sin^2 x + 2\sin x - 3 = 0;$
This is a quadratic equation in $\sin x$. Factoring (the quadratic formula could be used) gives

iii. $(2\sin x - 1)(4\sin x + 3) = 0;$

iv. $(2 \sin x - 1) = 0$

$$\sin x = \frac{1}{2}$$

$$x = \arcsin \frac{1}{2}$$

$$= 30°, 150°;$$

and

v. $(4 \sin x + 3) = 0$

$$\sin x = -\frac{3}{4}$$

$$x = \arcsin\left(-\frac{3}{4}\right)$$

$$\approx 228.6°, 311.4°.$$

Example 4

For what principal angles is the slope of the sine equal to the slope of the cosine?

i. $f(x) = \sin x$ and $g(x) = \cos x$;
ii. $f'(x) = \cos x$ and $g'(x) = -\sin x$;
iii. $\cos x = -\sin x$;

iv. $$1 = -\frac{\sin x}{\cos x}$$

$$1 = -\tan x$$

$$\tan x = -1$$

$$x = \arctan(-1) = 135°, 315°.$$

Many trigonometric equations are difficult to solve by analytical means. Finding approximate roots by graphing is often a useful alternative.

Example 5

Find graphically the principal roots of the equation $\sin x = x/2$.

i. Let $g(x) = \sin x$ and $h(x) = \dfrac{x}{2}$.

ii. Graph both functions on the same coordinate system.

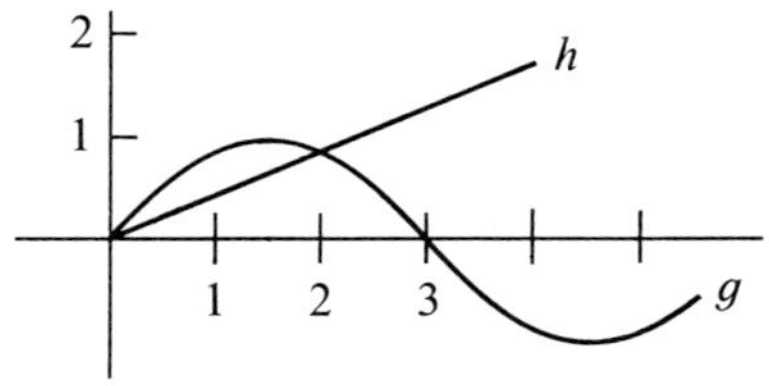

iii. The abscissas of the intersections are the roots, $x \approx 0$ and 1.9.

Example 6

Find graphically the principal roots to $\tan x - 1/x = 0$.

i. $\text{Tan}\ x - \dfrac{1}{x} = 0$ is equivalent to $\tan x = \dfrac{1}{x}$. Let $g(x) = \tan x$ and $h(x) = \dfrac{1}{x}$.

ii. Graph both functions on the same coordinate system.

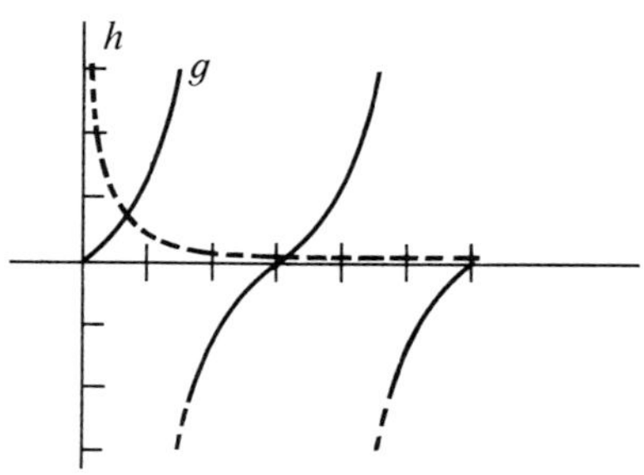

iii. $x \approx .86$ and 3.42.

8-4 Exercises

1. Solve for the principal roots of $\sec x = 2$.

---

2. Find the principal roots of $\csc x = \dfrac{2}{\sqrt{2}}$.

■ 1.
$$\sec x = 2$$
$$\frac{1}{\cos x} = 2$$
$$x = \arccos \frac{1}{2}$$
$$= 60°, 300°.$$

---

3. Find the principal roots of $\sec^2 x = \dfrac{1}{3}$.

■ 2.
$$\csc x = \frac{2}{\sqrt{2}}$$
$$\frac{1}{\sin x} = \frac{2}{\sqrt{2}}$$
$$x = \arcsin \frac{\sqrt{2}}{2}$$
$$= \arcsin .707$$
$$= 45°, 135°.$$

---

4. Find the principal roots of $2 \sin^2 x - \sin x - 1 = 0$.

■ 3.
$$\sec^2 x = \frac{1}{3}$$
$$\frac{1}{\cos^2 x} = \frac{1}{3}$$
$$\cos^2 x = 3$$
$$\cos x = \pm 1.7321$$
$$x = \arccos (\pm 1.7321).$$
(no solution)

---

5. Find the principal roots of $2\cos^2 x - \cos x = 0$.

■ 4. $2\sin^2 x - \sin x - 1 = 0$

$(2\sin x + 1)(\sin x - 1) = 0$

$2\sin x = -1$

$x = \arcsin\left(-\frac{1}{2}\right)$

$= 210°, 330°$

and

$\sin x = 1$

$x = \arcsin 1$

$= 90°.$

---

6. Find the principal roots of $\tan x - \cot x = 0$.

■ 5. $2\cos^2 x - \cos x = 0$

$\cos x(2\cos x - 1) = 0$

$\cos x = 0$

$x = \arccos 0$

$= 90°, 270°$

and

$\cos x = \frac{1}{2}$

$x = \arccos\frac{1}{2}$

$= 60°, 300°.$

---

7. Find the principal solutions to $5\sin x\cos x = -2\cos x$.

■ 6. $\tan x - \cot x = 0$

$\tan x - \frac{1}{\tan x} = 0$

$\tan^2 x - 1 = 0$

$\tan^2 x = 1$

$\tan x = \pm 1$

$x = 45°, 225°,$

$135°, 315°.$

---

8. Find the principal solutions for $\tan x = \cos x$.

■ 7. $5 \sin x \cos x + 2 \cos x = 0$

$\cos x(5 \sin x + 2) = 0$

$\cos x = 0$

$x = 90°, 270°$

and

$5 \sin x + 2 = 0$

$x = \arcsin\left(-\frac{2}{5}\right)$

$\approx 203.6°,$

$336.4°.$

---

9. Find the principal solutions for $\tan 3x = \cot 2x$.

■ 8. i. $\tan x = \cos x$

$\frac{\sin x}{\cos x} = \cos x$

$\sin x = \cos^2 x$

$\sin x = 1 - \sin^2 x$

$\sin^2 x + \sin x - 1 = 0;$

ii. $\sin x = \frac{-b \pm \sqrt{b - 4ac}}{2a}$

$= \frac{-1 + \sqrt{5}}{2}$

(sum only)

$= .618$

$x = \arcsin .618$

$= 38.2°, 141.8°.$

---

10. Find the simultaneous solution of

$w = 1 - \cos x$

$w = 1 + \sin x.$

■ 9. $\tan 3x = \cot 2x$

$\tan 3x = \tan (90° - 2x)$

$3x = 90° - 2x$

$5x = 90°$

$x = 18°.$

11. Find the principal solution for log tan $x = 1$.

■ 10. i. $w = 1 - \cos x$

$w = 1 + \sin x;$

ii. $\therefore 1 - \cos x = 1 + \sin x$

$$-\cos x = \sin x$$

$$-1 = \frac{\sin x}{\cos x}$$

$$\tan x = -1$$

$$x = \arctan(-1)$$

$$= 135°, 315°;$$

iii. $w = 1.707, .293.$

---

12. Where is the slope of

$$f(x) = \sin x$$

twice as great as it is at $x = 80°$?

■ 11. $\log \tan x = 1$

$$\tan x = 10$$

$$x = \arctan 10$$

$$= 84.3°, 264.3°.$$

■ 12. $f(x) = \sin x$

$$f'(x) = \cos x$$

$$f'(80°) = \cos 80° = .17365$$

$$f'(x) = \cos x = 2(.17365)$$

$$= .34730$$

$$x = \text{arc cos } .34730$$

$$= 69.67°.$$

8-4 Problems

1. When a light ray strikes the intersurface between two mediums, part of the ray reflects at the intersurface and part is refracted. The relationships that result are:

$$\frac{\sin \phi_1}{\sin \phi_2} = \frac{v_1}{v_2},$$

$$n_1 = \frac{c}{v_1},$$

$$n_2 = \frac{c}{v_2}.$$

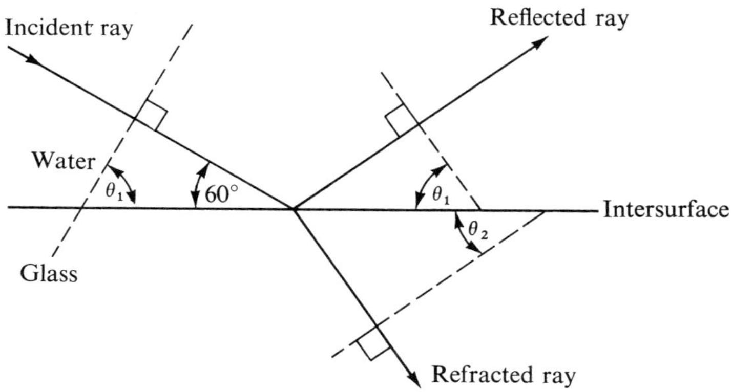

a. It is observed that the reflected ray and refracted ray form a 90° angle. Find the ratio v_1/v_2.

b. If $n_1 = 1.333$ (index of refraction for water), find n_2.

(*Note*: The index of refraction (n) for a given medium is constant and is equal to the velocity of the wave in a vacuum (c) divided by the velocity (v) of the wave in the given medium.)

2.

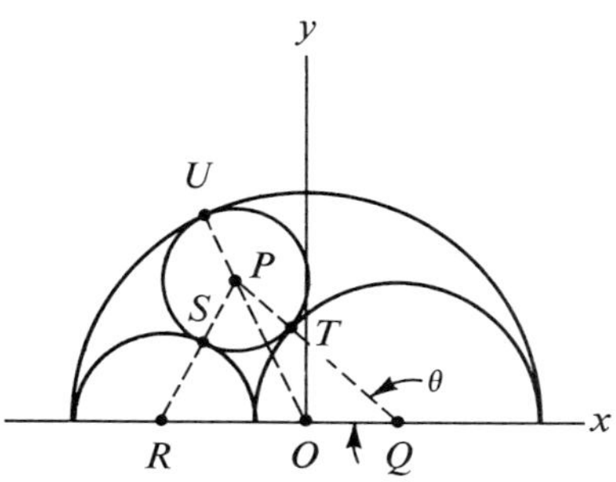

Radius $\overline{QT} = 3$

Radius $\overline{SR} = 2$

Radius $\overline{UO} = 5$

Find the radius of circle STU and the coordinates of its center. (*Hint*: $\angle\theta$ is common to triangles OQP and RQP.)

3. If $f(t) = \sin^2 t$, then the related area function is

$$F(t) = \frac{t_1}{2} - \frac{\cos t_1 \sin t_1}{2}$$

for $t = 0$ to $t = t_1$.

a. Draw the graph of f and calculate $F(2\pi)$.

b. What is the mean ordinate from $t = 0$ to $t = 2\pi$?

c. What is the square root of the mean ordinate found in (b)?

4. The area of the segment is one-fourth of the area of the entire circle. Find θ. (The solution to the equation which gives the answer may require graphical methods.)

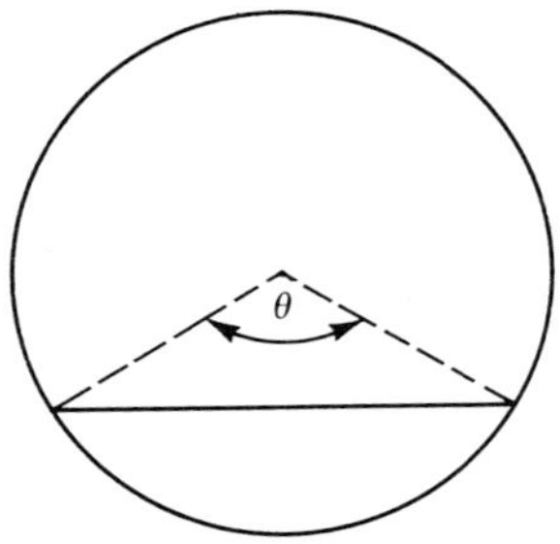

8 Supplementary Exercises and Problems

1. Find the value of the six trigonometric ratios for $\theta = 22°$. (8-1)
2. Find the value of the six trigonometric ratios for $\phi = 112°$. (8-1)
3. Find the secant, cosecant, and cotangent of .3141 radians. (8-1)
4. Find the secant, cosecant, and cotangent of 2.14 radians. (8-1)
5. Verify Equations 8-1g through 8-1n using $\theta = \pi/6$. (8-1)
6. Verify Equations 8-1g through 8-1n using $\theta = 2\pi/3$. (8-1)
7. Sketch the graph of $f(x) = \sec x$. (8-1)
8. Sketch the graph of $g(x) = \cot x$. (8-1)
9. From Equations 8-2c and 8-2d, develop the identity for $\cot(\theta \pm \phi)$ in terms of $\cot \theta$ and $\cot \phi$. (See the development of Equation 8-2e.) (8-2)
10. Handbooks of mathematics frequently list large numbers of trigonometric formulas and identities. Consult one or several handbooks to see how they are classified. (8-2)
11. If $h(x) = 2 \sin x$ and $g(x) = 3 \cos x$, find the values a and c when $h(x) + g(x) = a \sin(x + c)$. (8-2)
12. If $g(x) = 12 \sin(x + .20)$ and $h(x) = 5 \sin(x - .17)$, find the values a and c when $g(x) + h(x) = a \sin(x + c)$. (8-2)
13. Verify Equations 8-2a through 8-2e when $\theta = 57°$ and $\phi = 48°$. (8-2)
14. Interpolation of tangents for angles that approach 90° is very inaccurate. Use Equation 8-2e to calculate tan 89.35 with $\theta = 45.35°$ (to be interpolated) and $\phi = 44°$. (8-2)
15. Sin $x = .3000$. Find $\sin 2x$ using Equation 8-3a. (8-3)
16. Cos $2\theta = -.4000$. Find $\cos \theta$ using Equation 8-3c. (8-3)
17. Sin $x = .2000$. Find $\cos 2x$ using Equation 8-3d. (8-3)
18. Tan $\phi = 5.000$. Find $\tan 2\phi$ using Equation 8-3e. (8-3)
19. Tan $2x = .2500$. Find $\tan x$ using Equation 8-3e. (Quadratic in $\tan x$.) (8-3)
20. Sin $\theta = .4000$. Find $\cos 2\theta$ using Equation 8-3b. (8-3)

21. Simplify $\sin 2x - (\sin x + \cos x)^2$. (8-3)

22. Simplify: $\dfrac{\sin \phi}{1 - \cos \phi} + \dfrac{\sin \phi}{1 + \cos \phi}$. (8-3)

23. If $i = .04 \sin \omega t$ and $e_i = -12 \cos \omega t$, graph $p = ie_i$ when $\omega = 100\pi$. What is the period of p? (8-3)

24. Solve $2 \sin^2 x = \cos^2 x - 1/4$ for the principal roots. (8-4)

25. Solve for the principal roots of $2 \cos x + \sin^2 x = 1$. (8-4)

26. Find the principal roots of the equation

$$\tan x = \sqrt{2} \sin x. \quad (8\text{-}4)$$

27. Find the principal roots of

$$2 \sin^2 x + \cos^2 x = 4 \sin x - 2. \quad (8\text{-}4)$$

28. Find the principal roots of $\sin 2x = \tan x$. (8-4)

29. Find by graphical methods the principal roots of

$$x \cos 2x = 1. \quad (8\text{-}4)$$

30. Find the principal roots of $\sin x + \cos x = .5$. (8-4)

Chapter Nine Complex Numbers

9-1 Imaginary and Complex Numbers

When we were studying the solution of quadratic equations, we discovered that not all quadratic equations have real roots. In general, $ax^2 + bx + c = 0$ has real roots only if $b^2 - 4ac \geq 0$.

If we wish to include as roots the expressions that occur when $b^2 - 4ac < 0$, our concept of number has to be expanded.

Example 1

Solve $x^2 + 9 = 0$ for x.

$$x^2 + 9 = 0$$
$$x^2 = -9$$
$$x = \pm\sqrt{-9} = \pm\sqrt{9(-1)} = \pm 3\sqrt{(-1)}.$$

Example 2

Solve $x^2 + 4x + 5 = 0$ for x.

$$x^2 + 4x + 5 = 0$$
$$x = \frac{-4 \pm \sqrt{4^2 - 4(5)}}{2}$$
$$x = \frac{-4 \pm \sqrt{-4}}{2} = -2 \pm \sqrt{-1}.$$

Numbers in the form $b\sqrt{-1}$ are called **imaginary numbers**, and numbers in the form $a + b\sqrt{-1}$ are called **complex numbers**, where a and b are real numbers. The roots to the preceding examples illustrate both types of numbers.

For convenience and clarity the symbol j is introduced into the notation of imaginary and complex numbers. By definition,

$$j^2 = -1, \quad \text{or} \quad j = \sqrt{-1}. \qquad \text{(Def. 9-1a)}$$

So instead of writing $6 + \sqrt{-2}$, we write $6 + j\sqrt{2}$. Let us illustrate this change:

$$6 + \sqrt{-2} = 6 + \sqrt{(-1)(2)} = 6 + \sqrt{-1}\sqrt{2} = 6 + j\sqrt{2}.$$

The j symbol causes imaginary numbers to have the form jb and complex numbers to have the form $a + jb$.

The set of complex numbers is a large class of numbers to which the real numbers belong. Every real number can be written in complex form. For example, the number $2\sqrt{3}$ in complex form is $2\sqrt{3} + j0$.

To perform operations with imaginary and complex numbers we must be able to interpret integral powers of j. We can do this by using Def. 9-1a as follows:

$$j^1 = j$$

$$j^2 = -1$$

$$j^3 = j^2 j = -j$$

$$j^4 = j^2 j^2 = (-1)(-1) = 1$$

$$j^5 = j^4 j = j$$

$$j^6 = j^4 j^2 = j^2 = \square$$

$$j^7 = j^4 j^3 = j^3 = \square$$

$$j^8 = j^4 j^4 = \square.$$

■ -1; $-j$; 1.

If the list is extended in either direction, we discover that the only results are j, -1, $-j$, and 1. Every integral power of j is therefore represented by one of the following equations:

$$\begin{aligned} j^{4n} &= 1, \\ j^{4n+1} &= j, \\ j^{4n+2} &= -1, \\ j^{4n+3} &= -j, \quad n \text{ an integer.} \end{aligned} \qquad \text{(Eqs. 9-1a)}$$

Example 3

Simplify j^{10}.

$$j^{10} = j^{4(2)+2} = j^2 = -1.$$

Example 4

Simplify j^{-35}.

$$j^{-35} = j^{4(-9)+1} = j.$$

The inferences of mathematical definitions are not always clear until we encounter conflicts arising as a result of not using definitions correctly. Consider the following two demonstrations.

Demonstration 1

$$\begin{aligned} (-3\sqrt{-4})(-\sqrt{-9}) &= (-j6)(-j3) \\ &= j^2 18 \\ &= -18. \end{aligned}$$

Demonstration 2

$$\begin{aligned} (-3\sqrt{-4})(-\sqrt{-9}) &= 3\sqrt{-4}\sqrt{-9} \\ &= 3\sqrt{(-4)(-9)} \\ &= 3\sqrt{36} \\ &= 18. \end{aligned}$$

Only the first is correct! The second is incorrect, because the law of radicals, $\sqrt[n]{a}\,\sqrt[n]{b} = \sqrt[n]{ab}$, is not defined in terms of imaginary numbers. The mistake of performing an operation in the manner of the second demonstration can be avoided by introducing j early in computations where imaginary and complex numbers are permitted.

9-1 Exercises

In Exercises 1–8 express each number in terms of j.

1. $\sqrt{-36}$
2. $\sqrt{-32}$
3. $\sqrt{-48}$
4. $-\sqrt{-16}$
5. $\sqrt{-\dfrac{4}{9}}$
6. $\sqrt{-\dfrac{11}{4}}$
7. $\sqrt{-.01}$
8. $\sqrt{-\dfrac{2}{3}}$

■ 1. $\sqrt{-36} = \sqrt{(-1)(36)} = j6.$

■ 2. $\sqrt{-32} = \sqrt{(-1)(16)(2)} = j4\sqrt{2}.$

■ 3. $\sqrt{-48} = \sqrt{(-1)(16)(3)} = j4\sqrt{3}.$

■ 4. $-\sqrt{-16} = -\sqrt{(-1)(16)} = -j4.$

■ 5. $\sqrt{-\dfrac{4}{9}} = \sqrt{(-1)\dfrac{4}{9}} = j\dfrac{2}{3}.$

■ 6. $\sqrt{-\dfrac{11}{4}} = \sqrt{(-1)\dfrac{11}{4}} = j\dfrac{\sqrt{11}}{2}.$

■ 7. $\sqrt{-.01} = \sqrt{(-1)(.01)} = j(.1).$

In Exercises 9–14 express each number in terms of j, -1, $-j$, or 1.

9. j^5

■ 8. $\sqrt{-\frac{2}{3}} = \sqrt{(-1)\left(\frac{2}{3}\right)\left(\frac{3}{3}\right)}$
$= j\frac{\sqrt{6}}{3}.$

---

10. j^{-3}

■ 9. $j^5 = j^{4+1} = j.$

---

11. j^{12}

■ 10. $j^{-3} = j^{-4+1} = j.$

---

12. $-j^{14}$

■ 11. $j^{12} = j^{4(3)} = 1.$

---

13. j^{104}

■ 12. $-j^{14} = -j^{4(3)+2} = -j^2$
$= 1.$

---

14. j^{-9}

■ 13. $j^{104} = j^{4(26)} = 1.$

---

15. Simplify the imaginary expression $(-j3)(j0.5)^2$.

■ 14. $j^{-9} = j^{-4(3)+3} = j^3 = -j.$

---

16. Simplify the imaginary expression $(j5.2) + j(2.1)$.

■ 15. $(-j3)(j0.5)^2 = -j^3 0.75$
$= j0.75.$

---

17. $\dfrac{(j2.1)^2}{j}$

■ 16. $(j5.2) + (j2.1)$
$= j(5.2 + 2.1)$
$= j7.3.$

---

18. $\dfrac{1}{j}$

■ 17. $\dfrac{j^2 4.41}{j} = j4.41.$

■ 18. $\dfrac{1}{j} \cdot \dfrac{j}{j} = \dfrac{j}{-1} = -j.$

9-1 Problems

1. (Reactance) $= jX_L - jX_c$.
Find the reactance.

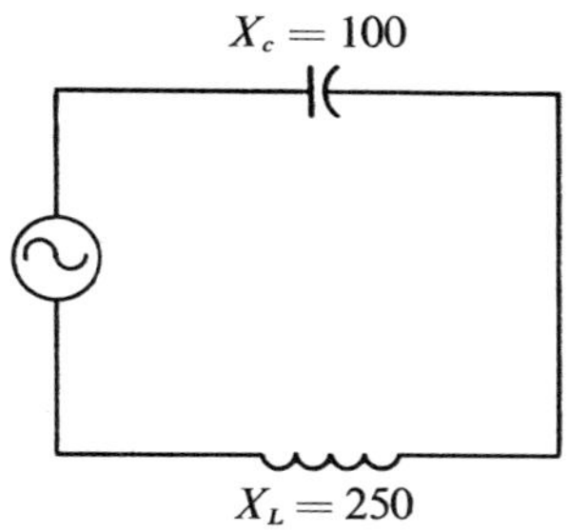

2. (Reactance) $= j(2\pi fL) - j\left(\dfrac{1}{2\pi fc}\right)$.
Find the reactance.

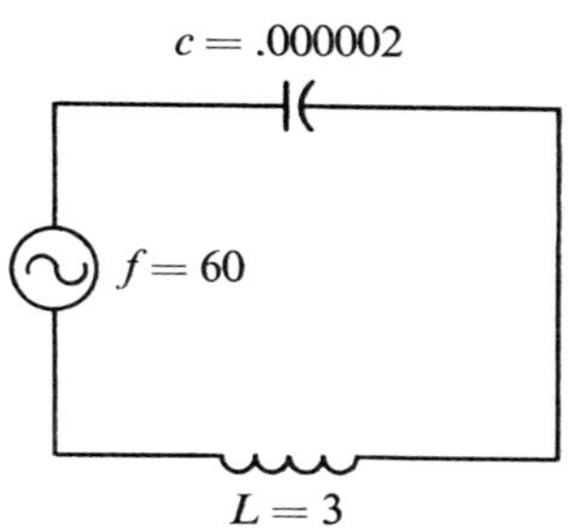

9-2 Fundamental Operations with Complex Numbers

Those who have studied elementary electricity know that the formula

$$R_t = R_1 + R_2 + \cdots \qquad \text{(Eq. 9-2a)}$$

is used to calculate series resistance and that the formula

$$\frac{1}{R_t} = \frac{1}{R_1} + \frac{1}{R_2} + \cdots \qquad \text{(Eq. 9-2b)}$$

is used to find parallel resistance.

If alternating-current circuits contain coils and capacitors in addition to resistors, the opposition to the flow of current is called impedance (Z). Formulas similar in format to Eqs. 9-2a and 9-2b are used to find impedances of series and parallel circuits.

$$Z_t = Z_1 + Z_2 + \cdots \qquad \text{(Series)} \qquad \text{(Eq. 9-2c)}$$

$$\frac{1}{Z_t} = \frac{1}{Z_1} + \frac{1}{Z_2} + \cdots \qquad \text{(Parallel)} \qquad \text{(Eq. 9-2d)}$$

The distinction between the formulas is not apparent until we discover that impedance is commonly represented by complex numbers, and algebraic operations with complex numbers are implied by the use of Z in place of R.

For convenience we will use Z to designate a complex number, in the same manner that x is often used to designate a real number. For example,

$$Z = a + jb$$

is a complex number which consists of the sum of the real number a and the imaginary number jb.

The operations with complex numbers need to be defined. Again, as in the case of vectors, the meanings of the words addition, subtraction, multiplication, and division are extended in terms of their application to complex numbers. The definitions are very much like algebraic manipulations of binomials with real coefficients.

Addition: If $Z_1 = a + jb$ and $Z_2 = c + jd$, then

$$Z_1 + Z_2 = (a + c) + j(b + d). \qquad \text{(Def. 9-2a)}$$

Example 1

If $Z_1 = 3 + j4$ and $Z_2 = 2 - j6$, then

$$\begin{aligned} Z_1 + Z_2 &= (3 + 2) + j(4 - 6) \\ &= 5 + j(-2) \\ &= 5 - j2. \end{aligned}$$

Subtraction: If $Z_1 = a + jb$ and $Z_2 = c + jd$, then

$$Z_1 - Z_2 = (a - c) + j(b - d). \qquad \text{(Def. 9-2b)}$$

Example 2

If $Z_1 = 8 + j4$ and $Z_2 = 5 - j6$, then

$$\begin{aligned} Z_1 - Z_2 &= (8 - 5) + j(4 - (-6)) \\ &= 3 + j10. \end{aligned}$$

Multiplication: If $Z_1 = a + jb$ and $Z_2 = c + jd$, then

$$Z_1Z_2 = (ac - bd) + j(ad + bc). \qquad \text{(Def. 9-2c)}$$

This result can be obtained if we simply multiply $(a + jb)$ and $(c + jd)$ as though they were binomials and replace j^2 by -1. Therefore, it is unnecessary to memorize this definition.

Example 3

$$\begin{aligned}(a + jb)(c + jd) &= ac + j^2bd + jad + jbc\\ &= (ac - bd) + j(ad + bc).\end{aligned}$$

Example 4

If $Z_1 = 3 + j5$ and $Z_2 = 4 - j2$, then

$$\begin{aligned}Z_1Z_2 &= (3 + j5)(4 - j2)\\ &= 12 - j^210 - j6 + j20\\ &= \boxed{}\,.\end{aligned}$$

■ $22 + j14$

Example 5

If $Z_1 = 1.2 + j1.3$ and $Z_2 = 1.2 - j1.3$, then

$$\begin{aligned}Z_1Z_2 &= (1.2 + j1.3)(1.2 - j1.3)\\ &= 1.44 - j^21.69 + j1.56 - j1.56\\ &= 1.44 + 1.69\\ &= 3.13.\end{aligned}$$

Example 5 is particularly important. Notice that the product is a real number. In general, for every nonzero complex number Z_1 there is another unique complex number Z_2 such that Z_1Z_2 is a real number. Each is called the **complex conjugate** of the other.

For example, Z_1 and Z_2 of the preceding example are conjugates. If we can identify the particular connection between conjugates in the example, then we should be able to specify the conjugate of $2 + j3$. It is [ ].

■ $2 - j3$

Finding the conjugate of a complex number is easy, since for any complex number $a + jb$ the conjugate is simply $a - jb$. This fact is important to our consideration of the division operation.

Division: To divide two complex numbers, we consider them in fractional form and multiply both numerator and denominator of the fraction by the conjugate of the denominator. If $Z_1 = a + jb$ and $Z_2 = c + jd$, then

$$Z_1 \div Z_2 = \frac{Z_1}{Z_2} = \frac{a + jb}{c + jd}$$

$$= \frac{(a + jb)(c - jd)}{(c + jd)(c - jd)}$$

$$= \frac{(ac + bd) + j(bc - ad)}{c^2 + d^2}$$

$$= \frac{(ac + bd)}{(c^2 + d^2)} + j\left(\frac{bc - ad}{c^2 + d^2}\right). \qquad \text{(Def. 9-2d)}$$

Definition 9-2d may look complicated, but in specific computations, division is really not difficult.

Example 6

If $Z_1 = 3 + j2$ and $Z_2 = 4 - j5$, then

$$\frac{Z_1}{Z_2} = \frac{(3 + j2)(4 + j5)}{(4 - j5)(4 + j5)} = \frac{(12 - 10) + j(8 + 15)}{16 + 25}$$

$$= \frac{2}{41} + j\frac{23}{41}.$$

9-2 Exercises

1. $(2 + j4) + (5 + j3)$
$= \boxed{}$

2. $(7 - j2) + (3 - j2)$
$= \boxed{}$

■ 1. $(2 + 5) + j(4 + 3)$
$= 7 + j7.$

3. $(8 - j5) - (7 + j6)$
$= \boxed{}$

■ 2. $(7 + 3) + j(-2 - 2)$
$= 10 - j4.$

4. $(-6 + j7) - (8 + j4)$
$= \boxed{}$

■ 3. $(8 - 7) + j(-5 - 6)$
$= 1 - j11.$

5. $(3 + j5)(2 - j7)$
$= \boxed{}$

■ 4. $(-6 - 8) + j(7 - 4)$
$= -14 + j3.$

6. $(2 + j3)(j6) = \boxed{}$

■ 5. $6 - j^2 35 + j10 - j21$
$= (6 + 35) + j(10 - 21)$
$= 41 - j11.$

7. $(0.3 + j0.4)^2 = \boxed{}$

■ 6. $j^2 18 + j12 = -18 + j12.$

8. $(2 - j3)^3 = \boxed{}$

■ 7. $(0.3)^2 + (j0.4)^2 + 2j(0.3)(0.4)$
$= 0.09 - 0.16 + j0.24$
$= -0.07 + j0.24.$

9. What is the conjugate of

a. $2 - j3$?

b. $-\frac{7}{10} + j\frac{4}{3}$?

c. $1.24 + j\pi$?

■ 8. $(2 - j3)^2(2 - j3)$
$= (4 + j^2 9 - j6 - j6)(2 - j3)$
$= (-5 - j12)(2 - j3)$
$= -10 + j^2 36 - j24 + j15$
$= -46 - j9.$

---

10. $\frac{2 - j3}{4 + j2} =$ ☐

■ 9. a. $2 + j3$;
b. $-\frac{7}{10} - j\frac{4}{3}$;
c. $1.24 - j\pi$.

---

11. $\frac{1 - j}{1 + j} =$ ☐

■ 10. $\frac{(2 - j3)(4 - j2)}{(4 + j2)(4 - j2)} = \frac{2 - j16}{16 + 4}$
$= \frac{1}{10} - j\frac{4}{5}.$

---

12. $\frac{1}{-3 + j7} =$ ☐

■ 11. $\frac{(1 - j)(1 - j)}{(1 + j)(1 - j)}$
$= \frac{1 + j^2 - j1 - j1}{1 + 1}$
$= \frac{-j2}{2} = -j.$

---

13. Solve for x and show the check of both roots:

$$x^2 + 2x + 4 = 0.$$

■ 12. $\frac{(-3 - j7)}{(-3 + j7)(-3 - j7)}$
$= \frac{-3 - j7}{9 + 49}$
$= \frac{-3}{58} - j\frac{7}{58}.$

■ 13. i. $x^2 + 2x + 4 = 0,$
$x = \frac{-2 \pm \sqrt{4 - 16}}{2}$
$= -1 \pm j\sqrt{3};$

ii. Check $-1+j\sqrt{3}$

$(-1+j\sqrt{3})^2$
$+2(-1+j\sqrt{3})$
$+4$
$=-2-j2\sqrt{3}-2$
$+j2\sqrt{3}+4$
$=0;$

iii. Check $-1-j\sqrt{3}$

$(-1-j\sqrt{3})^2$
$+2(-1-j\sqrt{3})$
$+4$
$=-2+j2\sqrt{3}-2$
$-j2\sqrt{3}+4$
$=0.$

9-2 Problems

1. Find series impedance when $Z_1 = 100 - j190$ and $Z_2 = 75 + j40$. (See Eq. 9-2c.)
2. Solve $\frac{1}{Z_t} = \frac{1}{Z_1} + \frac{1}{Z_2}$ for Z_t.
3. If $Z_1 = 25 + j35$ and $Z_2 = 30 - j65$, find the parallel impedance. Use the results of the preceding problem.
4. $Z_1 = 2Z_2 - Z_3$. Find Z_2 if $Z_1 = 480 - j500$ and $Z_3 = 200 + j350$.
5. If $f(Z) = 4Z^2 + Z$,
   a. find $f(2 + j3)$,
   b. find $f(-1 - j)$.

9-3 Complex Numbers and Vectors

The manner in which complex numbers are commonly graphed is suggestive of the manner in which ordered pairs of real numbers are graphed. We graph $Z = a + jb$ on a coordinate plane by graphing a along the horizontal axis and b along the vertical axis. Every complex number may in this way be represented geometrically by a point.

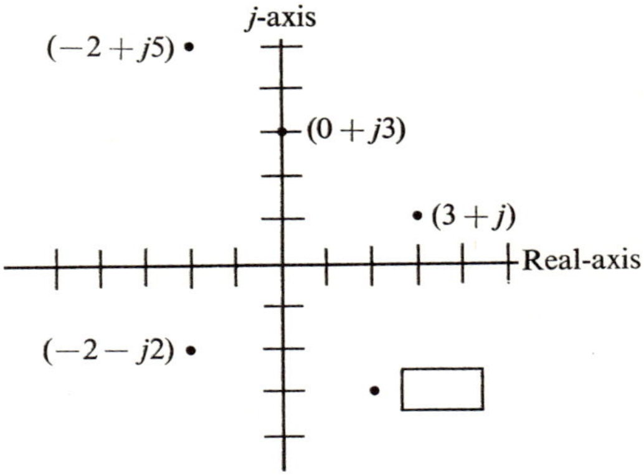

■ $2 - j3$

When the coordinate system is used in this way, it is called the **complex plane**; the horizontal axis is called the **real-axis** and the vertical axis, the **imaginary** or ***j*-axis**.

A comparison of Sections 6-7 and 9-2 shows that the fundamental operations defined in terms of two-dimensional vectors (addition, subtraction, and scalar multiplication) are analogous to similar operations with complex numbers.

Table 9-3a

Two-dimensional vectors	Complex numbers
$\mathbf{r}_1 = (x_1, y_1)$	$Z_1 = a_1 + jb_1$
$\mathbf{r}_2 = (x_2, y_2)$	$Z_2 = a_2 + jb_2$
$\mathbf{r}_1 \pm \mathbf{r}_2 = (x_1 \pm x_2, y_1 \pm y_2)$	$Z_1 \pm Z_2 = (a_1 \pm a_2) + j(b_1 \pm b_2)$
$k\mathbf{r}_1 = (kx_1, ky_1)$	$kZ_1 = ka_1 + jkb_1$

Because of the approach used in graphing complex numbers and the definition given for operations with complex numbers, two-dimensional vector problems are frequently done in terms of complex numbers. Graphically, the initial end of the vector is at the origin of the complex plane, and the terminal end is at the point designated by the complex number.

Example 1

What complex number Z is equivalent to the vector $\mathbf{r} = (-3.0, 4.0)$?

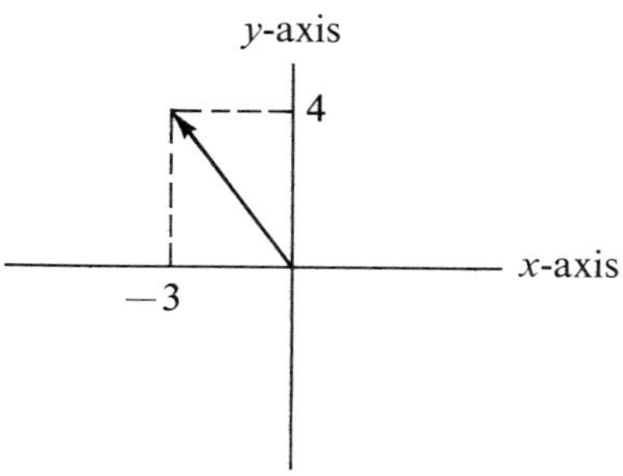

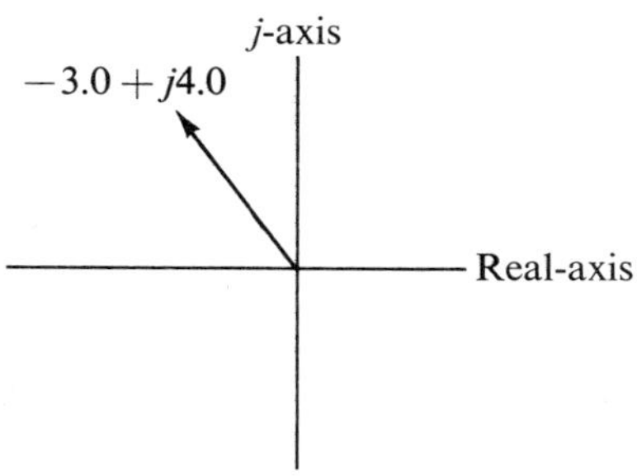

$$\mathbf{r} = (-3.0, 4.0) \longrightarrow Z = -3.0 + j4.0.$$

Example 2

Find the sum (resultant) of two vector forces, $Z_1 = 5.3 + j2.1$ and $Z_2 = 1.8 + j6.2$.

$$\begin{aligned} Z_1 + Z_2 &= (5.3 + 1.8) + j(2.1 + 6.2) \\ &= 7.1 + j8.3. \end{aligned}$$

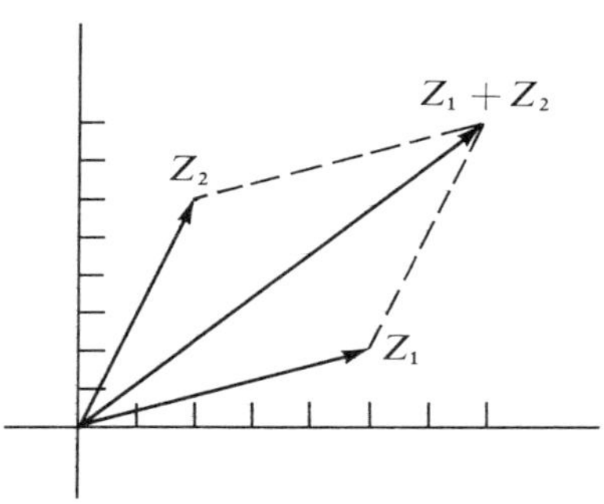

It should be clear that the rectangular form of the vector $\mathbf{r} = (x, y)$ can be written in the equivalent complex form $Z = a + jb$, where $x = a$ and $y = b$.

To give us the proper perspective, we must emphasize that the analogy of complex numbers to vectors applies only to two-dimensional vectors and that the operations defined for complex numbers do not necessarily have a counterpart in vector operations.

A considerable number of advantages, developed in the next sections, result from introducing a second form for complex numbers. Examine Fig. 9-3a. If we write a and b in terms of r and θ, then $a = r\cos\theta$, $b = r\sin\theta$, and

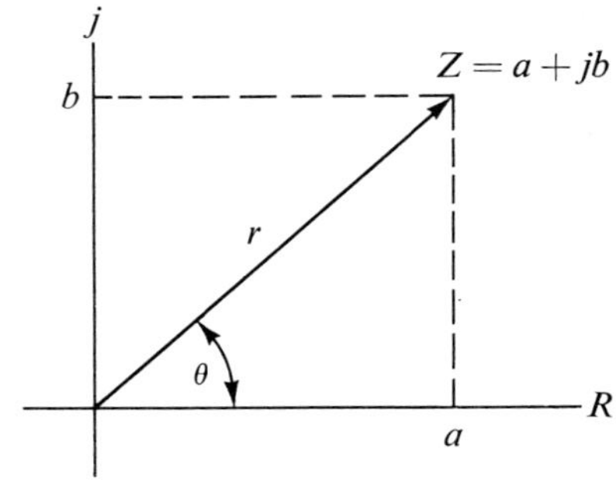

Fig. 9-3a

$$Z = r\cos\theta + jr\sin\theta$$
$$= r(\cos\theta + j\sin\theta). \qquad \text{(Eq. 9-3a)}$$

As in similar circumstances,

$$r = \sqrt{a^2 + b^2} \quad \text{and} \quad \theta = \arctan\frac{b}{a}.$$

Example 3

Express $Z = -1 - j\sqrt{3}$ in the equivalent form $Z = r(\cos\theta + j\sin\theta)$.

i. $\theta = \arctan\dfrac{-\sqrt{3}}{-1}$

$= \arctan 1.7321 = 240°$;

ii. $r = \sqrt{(-1)^2 + (-\sqrt{3})^2}$

$= \sqrt{4} = 2$;

iii. $Z = 2(\cos 240° + j\sin 240°)$.

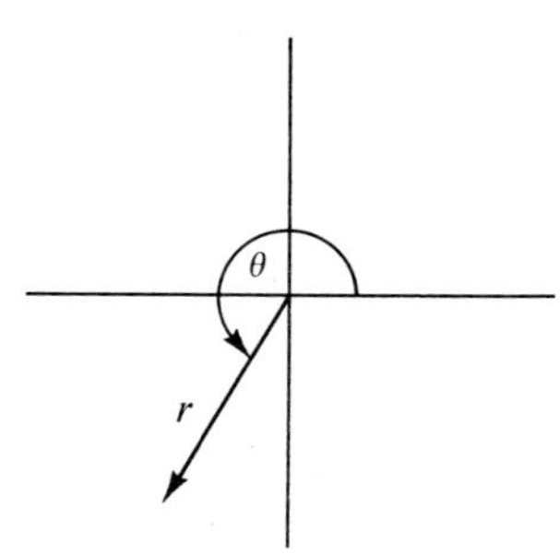

$Z = r(\cos\theta + j\sin\theta)$ is called the **polar form** of the complex number $Z = a + jb$. This name was also used in Chapter 6. To avoid confusion, let us agree to modify the name by calling $Z = r(\cos\theta + j\sin\theta)$ the **complex polar form** of the complex number Z.

9-3 Exercises

1. Graph the complex numbers

$$Z_1 = 3 + j2,$$
$$Z_2 = -5 + j3,$$
$$Z_3 = -4 - j2.$$

---

2. Graph the complex vectors

$$Z_1 = 5 + j2,$$
$$Z_2 = -7 - j8,$$
$$Z_3 = -2 + j6.$$

■ 1.

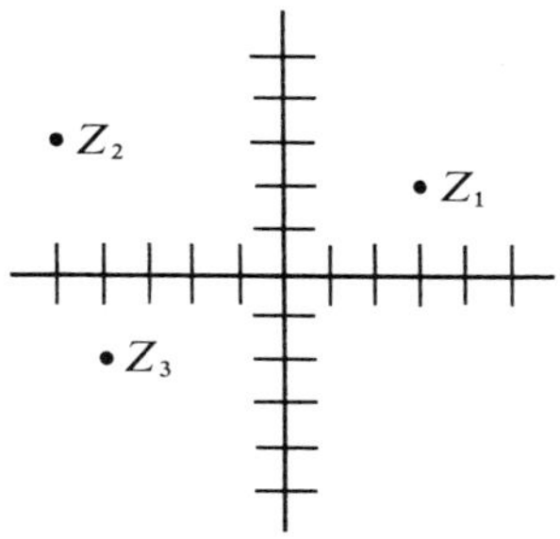

---

3. Illustrate graphically the sum of the complex vectors $Z_1 = 2 - j2$ and $Z_2 = 3 + j4$ and then check algebraically.

■ 2.

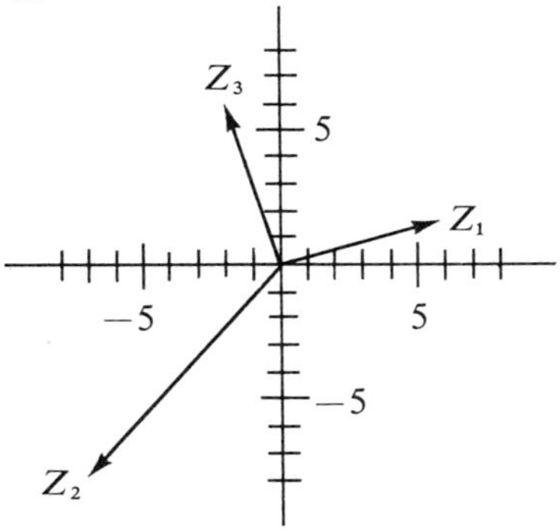

4. Illustrate graphically the difference $Z_1 - Z_2$ when $Z_1 = 3 + j5$ and $Z_2 = -3 + j2$. Check the results algebraically.

■ 3. i.

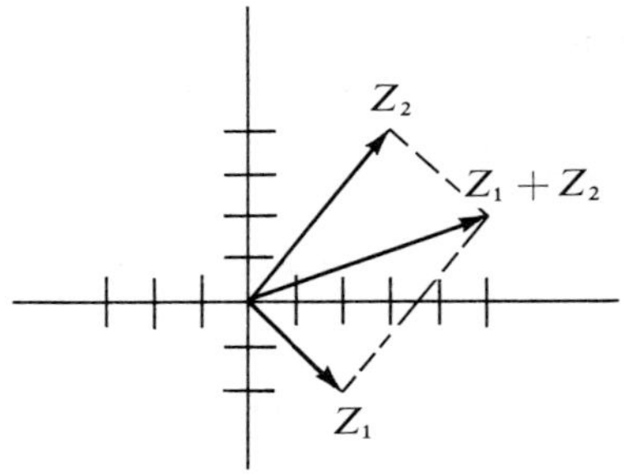

ii. $(2 - j2) + (3 + j4)$
$= 5 + j2.$

---

5. What is the magnitude and direction of the vector represented by $3 - j2$?

■ 4. i.

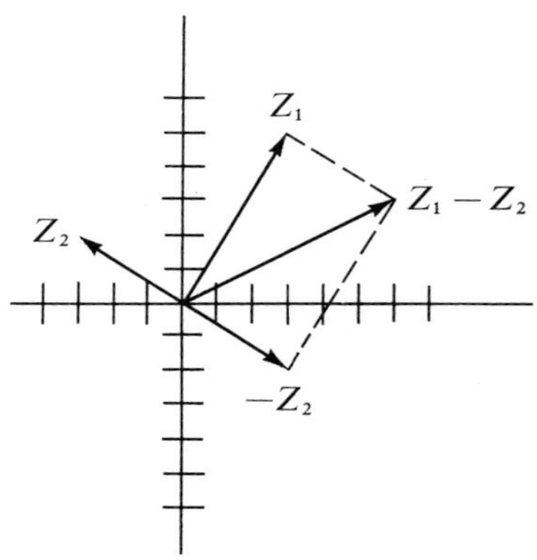

ii. $Z_1 - Z_2$
$= (3 + j5)$
$\quad - (-3 + j2)$
$= 6 + j3.$

---

6. Express $Z = 3 + j5$ in complex polar form.

■ 5. $r = \sqrt{a^2 + b^2}$
$= \sqrt{3^2 + (-2)^2}$
$= \sqrt{13} \approx 3.6;$
$\theta = \arctan b/a$
$= \arctan -2/3$
$= 360° - 33.7° = 326.3°.$

7. Express $Z = -2 + j3$ in complex polar form.

■ 6. $|Z| = r = \sqrt{a^2 + b^2}$
$= \sqrt{9 + 25} \approx 5.83;$
$\theta = \arctan 5/3$
$= 59.0°;$
$Z = 5.83(\cos 59°$
$+ j \sin 59°).$

---

8. Express $Z = 0 + j6$ in complex polar form.

■ 7. $|Z| = \sqrt{4 + 9} \approx 3.61;$
$\theta = \arctan 3/-2$
$= 180° - 56.3°$
$= 123.7°;$
$Z = 3.61(\cos 123.7°$
$+ j \sin 123.7°).$

---

9. What complex number Z represents the vector with magnitude 12 and direction 40°?

■ 8. $|Z| = 6;$
$\theta = 90°;$
$Z = 6(\cos 90° + j \sin 90°).$

---

10. Express $Z = 4(\cos 30° + j \sin 30°)$ in the complex (rectangular) form $a + jb$.

■ 9. $a = 12 \cos 40° = 9.19;$
$b = 12 \sin 40° = 7.71;$
$Z = a + jb$
$= 9.19 + j7.71.$

---

11. Express
$$Z = 7(\cos 316° + j \sin 316°)$$
in complex rectangular form.

■ 10. $a = 4 \cos 30° = 3.46;$
$b = 4 \sin 30° = 2.00;$
$Z = 3.46 + j2.00.$

---

12. Express
$$Z = \sqrt{3}(\cos 121° + j \sin 121°)$$
in complex rectangular form.

■ 11. $a = 7 \cos 316°$
$= 7 \cos 44° = 5.04;$
$b = 7 \sin 316°$
$= -7 \sin 44° = -4.87;$
$Z = 5.04 - j4.87.$

■ 12. $a = \sqrt{3} \cos 121°$
$= -1.73 \cos 59°$
$= -.892;$
$b = \sqrt{3} \sin 121°$
$= 1.73 \sin 59° = 1.48;$
$Z = -.892 + j1.48.$

9-3 Problems

1. Two vector forces, Z_1 and Z_2, are held in equilibrium by a third force Z_3. Find Z_3 if $Z_1 = 1000 + j2000$ and $Z_2 = -800 + j2200$. (*Hint*: $Z_1 + Z_2 + Z_3 = 0$.)
2. If $3.4Z_1 = 5.3Z_2$ and $Z_2 = 0 - j42$, find Z_1.

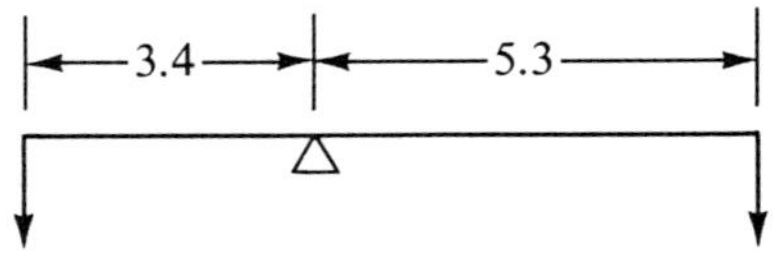

3. If $Z = X_L + X_c + R$, $X_L = 0 + j200$, $X_c = 0 - j300$ and $R = 50 + j0$, find Z.

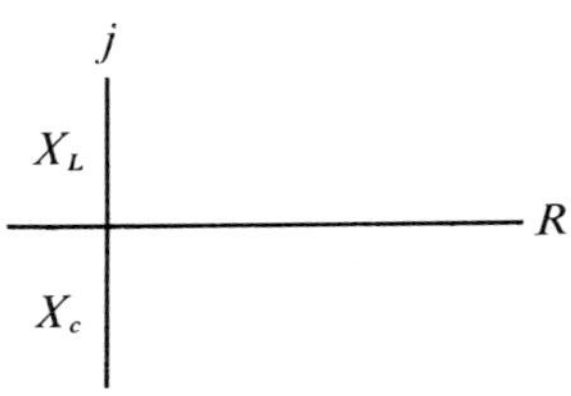

4. $aZ_1 + bZ_2 = 0$; $a = 5.0$; $b = 1.5$; $Z_1 = 100(\cos 30° + j \sin 30°)$. Find Z_2.
5. $Z_1 = 1.6(\cos \pi/3 + j \sin \pi/3)$; $Z_2 = 12(\cos \pi/4 + j \sin \pi/4)$; find $Z = Z_1Z_2$.

9-4 Multiplication and Division in Complex Polar Form

There is an advantage in considering the operations of multiplication and division of complex numbers in complex polar form. The rather involved work that follows leads to results that will simplify all subsequent calculations of products and quotients with complex numbers.

First, let us consider the product of Z_1Z_2 when

$$Z_1 = r_1(\cos \theta_1 + j \sin \theta_1)$$

and

$$Z_2 = r_2(\cos \theta_2 + j \sin \theta_2).$$

$$\begin{aligned}
Z_1Z_2 &= r_1(\cos\theta_1 + j\sin\theta_1)r_2(\cos\theta_2 + j\sin\theta_2)\\
&= r_1r_2(\cos\theta_1\cos\theta_2 + j\cos\theta_1\sin\theta_2 + j\sin\theta_1\cos\theta_2\\
&\quad + j^2\sin\theta_1\sin\theta_2)\\
&= r_1r_2(\cos\theta_1\cos\theta_2 - \sin\theta_1\sin\theta_2 + j(\sin\theta_1\cos\theta_2\\
&\quad + \cos\theta_1\sin\theta_2))\\
&= r_1r_2(\cos(\theta_1+\theta_2) + j\sin(\theta_1+\theta_2)). \qquad \text{(Eq. 9-4a)}
\end{aligned}$$

The perfectly general results appearing in Eq. 9-4a tell us the product of two complex numbers in polar form is obtained by simply *multiplying* r_1 and r_2 and *adding* angles θ_1 and θ_2.

Example 1

If $Z_1 = 3(\cos 70° + j\sin 70°)$ and $Z_2 = 4(\cos 80° + j\sin 80°)$, find $Z_1 \cdot Z_2$.

$$\begin{aligned}
Z_1 \cdot Z_2 &= 3\cdot 4(\cos(70° + 80°) + j\sin(70° + 80°))\\
&= 12(\cos 150° + j\sin 150°).
\end{aligned}$$

Next let us look at the division process. Again

$$Z_1 = r_1(\cos\theta_1 + j\sin\theta_1)$$

and

$$Z_2 = r_2(\cos\theta_2 + j\sin\theta_2).$$

$$\begin{aligned}
\frac{Z_1}{Z_2} &= \frac{r_1(\cos\theta_1 + j\sin\theta_1)(\cos\theta_2 - j\sin\theta_2)}{r_2(\cos\theta_2 + j\sin\theta_2)(\cos\theta_2 - j\sin\theta_2)}\\
&= \frac{r_1}{r_2}\,\frac{(\cos\theta_1\cos\theta_2 + \sin\theta_1\sin\theta_2) + j(\sin\theta_1\cos\theta_2 - \cos\theta_1\sin\theta_2)}{\cos^2\theta_2 + \sin^2\theta_2}\\
&= \frac{r_1}{r_2}(\cos(\theta_1 - \theta_2) + j\sin(\theta_1 - \theta_2)). \qquad \text{(Eq. 9-4b)}
\end{aligned}$$

The general results stated in Eq. 9-4b can be used to shorten division of complex numbers by simply dividing magnitudes and subtracting angles, being careful to observe the order of operation.

Example 2

If $Z_1 = \sqrt{8}(\cos 40° + j \sin 40°)$ and $Z_2 = \sqrt{2}(\cos 10° + j \sin 10°)$, find $Z_1 \div Z_2$.

$$Z_1 \div Z_2 = \frac{\sqrt{8}}{\sqrt{2}} (\cos(40° - 10°) + j \sin(40° - 10°))$$
$$= 2(\cos 30° + j \sin 30°).$$

Example 3

If $Z_1 = 10^6(\cos 50° + j \sin 50°)$ and $Z_2 = 10^4(\cos 120° + j \sin 120°)$, find $Z_1 \div Z_2$.

$$Z_1 \div Z_2 = \frac{10^6}{10^4} (\cos(50° - 120°) + j \sin(50° - 120°))$$
$$= 10^2(\cos 290° + j \sin 290°).$$

9-4 Exercises

1. Find Z_1Z_2 and Z_1/Z_2 when

$$Z_1 = 3(\cos 40° + j \sin 40°)$$

and

$$Z_2 = 4(\cos 20° + j \sin 20°).$$

---

2. Find Z_1Z_2 and Z_1/Z_2 when

$$Z_1 = 2(\cos \pi + j \sin \pi)$$

and

$$Z_2 = 7\left(\cos \frac{\pi}{2} + j \sin \frac{\pi}{2}\right).$$

■ 1. i. Z_1Z_2
$= 3 \cdot 4(\cos(40° + 20°) + j \sin(40° + 20°))$
$= 12(\cos 60° + j \sin 60°)$;

ii. $\frac{Z_1}{Z_2} = \frac{3}{4} (\cos(40° - 20°) + j \sin(40° - 20°))$
$= .75(\cos 20° + j \sin 20°)$.

3. Find Z_1Z_2 and Z_1/Z_2 when
$Z_1 = 0.1(\cos 300° + j \sin 300°)$
and
$Z_2 = 0.01(\cos 400° + j \sin 400°)$.

■ 2. i. $Z_1Z_2 = 14\left(\cos \frac{3\pi}{2} + j \sin \frac{3\pi}{2}\right)$;

ii. $\frac{Z_1}{Z_2} = \frac{2}{7}\left(\cos \frac{\pi}{2} + j \sin \frac{\pi}{2}\right)$.

---

4. Find Z^2 when
$Z = 2.1(\cos 3.14 + j \sin 3.14)$.

■ 3. i. $Z_1Z_2 = 0.001(\cos 700° + j \sin 700°)$;

ii. $\frac{Z_1}{Z_2} = 10(\cos(-100°) + j \sin(-100°)) = 10(\cos 260° + j \sin 260°)$.

---

5. If $Z_2 = 1/Z_1$ and
$Z_1 = \sqrt{3}\left(\cos \frac{3\pi}{5} + j \sin \frac{3\pi}{5}\right)$,
find Z_2.

■ 4. $Z^2 = ZZ$
$= (2.1)(2.1)(\cos(3.14 + 3.14) + j \sin(3.14 + 3.14))$
$= 4.41(\cos 6.28 + j \sin 6.28)$.

■ 5. $Z_2 = \frac{1(\cos 0 + j \sin 0)}{\sqrt{3}(\cos 3\pi/5 + j \sin 3\pi/5)}$

$= \frac{\sqrt{3}}{3}\left(\cos\left(\frac{-3\pi}{5}\right) + j \sin\left(\frac{-3\pi}{5}\right)\right)$

$= \frac{\sqrt{3}}{3}\left(\cos \frac{7\pi}{5} + j \sin \frac{7\pi}{5}\right)$.

9-4 Problems

1. If $Z^2 = .81(\cos 72° + j \sin 72°)$, find Z.
2. If $I = E/Z$, $E = 12(\cos 0° + j \sin 0°)$, and $Z = 1000(\cos 20° + j \sin 20°)$, find I.

3. If $Z_1 = 100(\cos 35° + j \sin 35°)$ and $Z_2 = 80(\cos(-15°) + j \sin(-15°))$, find

$$Z_t = \frac{Z_1 Z_2}{Z_1 + Z_2}.$$

4. If $P = IE$, $I = 2.5(\cos \pi/5 + j \sin \pi/5)$, and $E = 100(\cos \pi/8 + j \sin \pi/8)$, find P.
5. $I = 200(\cos 2\pi ft + j \sin 2\pi ft)$ and $E = 10(\cos(2\pi ft + \pi/12) + j \sin(2\pi ft + \pi/12))$. Find $P = IE$.

9-5 Powers and Roots in Complex Polar Form

The procedure used for the multiplication of three or more complex factors is a very simple extension of the work done in Section 9-4.

$$\begin{aligned} &Z_1 Z_2 \ldots Z_n \\ &\quad = r_1 r_2 \ldots r_n(\cos(\theta_1 + \theta_2 + \cdots + \theta_n) + j \sin(\theta_1 + \theta_2 + \cdots + \theta_n)). \end{aligned}$$

(Eq. 9-5a)

Example 1

Find the product of $Z_1 = 2(\cos 15° + j \sin 15°)$,

$Z_2 = 3(\cos 21° + j \sin 21°)$, and $Z_3 = 7(\cos 50° + j \sin 50°)$.

$$\begin{aligned} Z_1 Z_2 Z_3 &= 2 \cdot 3 \cdot 7(\cos(15° + 21° + 50°) + j \sin(15° + 21° + 50°)) \\ &= 42(\cos 86° + j \sin 86°). \end{aligned}$$

If all the complex numbers in Eq. 9-5a are equal, then for integral values of n,

$$\begin{aligned} Z^n &= (r(\cos \theta + j \sin \theta))^n \\ &= r^n(\cos n\theta + j \sin n\theta). \end{aligned}$$

(Eq. 9-5b)

Example 2

Find the fourth power of $Z = 2(\cos 21° + j \sin 21°)$.

$$\begin{aligned} Z^4 &= 2^4(\cos 4(21°) + j \sin 4(21°)) \\ &= 16(\cos 84° + j \sin 84°). \end{aligned}$$

If $Z = r^{1/2}(\cos \theta/2 + j \sin \theta/2)$, then by using Eq. 9-5b,

$$\begin{aligned} Z^2 &= (r^{1/2})^2 \left(\cos 2\left(\frac{\theta}{2}\right) + j \sin 2\left(\frac{\theta}{2}\right)\right) \\ &= r(\cos \theta + j \sin \theta). \end{aligned}$$

This should clearly illustrate that the square root of a complex number is found by taking the square root of the magnitude r and halving the angle θ.

Example 3

Find $\sqrt{Z}$ when $Z = 32(\cos 36° + j \sin 36°)$.

$$\begin{aligned} \sqrt{Z} = Z^{1/2} &= (32(\cos 36° + j \sin 36°))^{1/2} \\ &= 32^{1/2}\left(\cos \frac{36°}{2} + j \sin \frac{36°}{2}\right) \\ &= 4\sqrt{2}(\cos 18° + j \sin 18°). \end{aligned}$$

The approach used to justify the procedure for taking the square root can be extended to calculating cube roots, fourth roots, and so on.

Example 4

Find a fifth root of $Z = 32(\cos 36° + j \sin 36°)$.

$$\begin{aligned} \sqrt[5]{Z} = Z^{1/5} &= (32(\cos 36° + j \sin 36°))^{1/5} \\ &= 32^{1/5}\left(\cos \frac{36°}{5} + j \sin \frac{36°}{5}\right) \\ &= 2(\cos 7.2° + j \sin 7.2°). \end{aligned}$$

In the two preceding examples, we purposely overlooked the possibility of the existence of more than one root for a complex number. In general, there are *n* distinct *n*th roots for every complex number.

$$
\begin{aligned}
\text{Consider again } Z &= 32(\cos 36^\circ + j\sin 36^\circ), \quad k = 0 \\
&= 32(\cos(36^\circ + 360^\circ) + j\sin(36^\circ + 360^\circ)), \quad k = 1 \\
&= 32(\cos(36^\circ + 2(360^\circ)) + j\sin(36^\circ + 2(360^\circ))), \quad k = 2 \\
&= 32(\cos(36^\circ + 3(360^\circ)) + j\sin(36^\circ + 3(360^\circ))), \quad k = 3 \\
&\;\vdots \\
&= 32(\cos(36^\circ + k(360^\circ)) + j\sin(36^\circ + k(360^\circ))).
\end{aligned}
$$

Now if we reconsider the square root of Z, we obtain two distinct roots, one for $k = 0$ and the other for $k = 1$. The results for greater values of k are *not* unique.

The five distinct fifth roots of Z result from k having the values 0, 1, 2, 3, and 4. No other unique roots result from greater values of k.

Example 5

Calculate the three roots of Z when $Z = 32(\cos 36^\circ + j\sin 36^\circ)$.

i. $k = 0;\quad \sqrt[3]{Z} = 32^{1/3}\left(\cos\dfrac{36^\circ}{3} + j\sin\dfrac{36^\circ}{3}\right)$

$\approx 3.18(\cos 12^\circ + j\sin 12^\circ);$

ii. $k = 1;\quad \sqrt[3]{Z} = 32^{1/3}\left(\cos\dfrac{36^\circ + 360^\circ}{3} + j\sin\dfrac{36^\circ + 360^\circ}{3}\right)$

$\approx 3.18(\cos 132^\circ + j\sin 132^\circ);$

iii. $k = 2;\quad \sqrt[3]{Z} = 32^{1/3}\left(\cos\dfrac{36^\circ + 2(360^\circ)}{3} + j\sin\dfrac{36^\circ + 2(360^\circ)}{3}\right)$

$\approx 3.18(\cos 252^\circ + j\sin 252^\circ).$

Since we have considered both integer powers and roots of complex polar numbers, we can now say that Eq. 9-5b is valid for all *rational values* of n. Equation 9-5b is known as Demoivre's Theorem.

Example 6

If $Z = 4.2(\cos 72° + j \sin 72°)$, find $Z^{3/2}$.

i. $k = 0$; $Z^{3/2} = (4.2(\cos 72° + j \sin 72°))^{3/2}$

$$= 4.2^{3/2}\left(\cos \frac{3}{2}(72°) + j \sin \frac{3}{2}(72°)\right)$$

$$\approx 8.61(\cos 108° + j \sin 108°);$$

ii. $k = 1$; $Z^{3/2} = (4.2(\cos(72° + 360°) + j \sin(72° + 360°)))^{3/2}$

$$= 4.2^{3/2}\left(\cos \frac{3}{2}(432°) + j \sin \frac{3}{2}(432°)\right)$$

$$\approx 8.61(\cos 648° + j \sin 648°)$$

$$\approx 8.61(\cos 288° + j \sin 288°).$$

9-5 Exercises

1. Find the 5th power of $Z = 1.2(\cos 10° + j \sin 10°)$.

2. Find the 12th power of $Z = 1 + j$.

■ 1. $Z^5 = 1.2^5(\cos 5(10°) + j \sin 5(10°)) \approx 2.49(\cos 50° + j \sin 50°)$.

3. Find the three cube roots of $Z = 27(\cos 135° + j \sin 135°)$.

■ 2. $Z = 1 + j$

$$= \sqrt{2}\left(\cos \frac{\pi}{4} + j \sin \frac{\pi}{4}\right);$$

$$Z^{12} = \left(\sqrt{2}\left(\cos \frac{\pi}{4} + j \sin \frac{\pi}{4}\right)\right)^{12}$$

$$= (2^{1/2})^{12}\left(\cos 12\left(\frac{\pi}{4}\right) + j \sin 12\left(\frac{\pi}{4}\right)\right)$$

$$= 2^6(\cos 3\pi + j \sin 3\pi)$$

$$= 64(\cos 3\pi + j \sin 3\pi).$$

---

4. Find the four fourth roots of $Z = 81(\cos 240° + j \sin 240°)$.

■ 3. $Z^{1/3}$

$= 27^{1/3}\left(\cos \dfrac{135° + k360°}{3} + j \sin \dfrac{135° + k360°}{3}\right).$

i. $k = 0;$
$Z^{1/3} = 3(\cos 45° + j \sin 45°).$

ii. $k = 1;$
$Z^{1/3} = 3(\cos 165° + j \sin 165°).$

iii. $k = 2;$
$Z^{1/3} = 3(\cos 285° + j \sin 285°).$

---

5. Find the three complex cube roots of (-1).

■ 4. $Z^{1/4}$

$= 81^{1/4}\left(\cos \dfrac{240° + k360°}{4} + j \sin \dfrac{240° + k360°}{4}\right).$

i. $k = 0;$
$Z^{1/4} = 3(\cos 60° + j \sin 60°).$

ii. $k = 1;$
$Z^{1/4} = 3(\cos 150° + j \sin 150°).$

iii. $k = 2;$
$Z^{1/4} = 3(\cos 240° + j \sin 240°).$

iv. $k = 3;$
$Z^{1/4} = 3(\cos 330° + j \sin 330°).$

---

6. If $Z = 4.3 + j2.8$, find only one value for $Z^{0.3}$.

■ 5. All real numbers belong to the set of complex numbers. Therefore, let

$Z = (-1) = -1 + j0$
$= 1(\cos 180° + j \sin 180°)$

and

$Z^{1/3}$

$$= \left(\cos \frac{180° + k360°}{3} + j \sin \frac{180° + k360°}{3}\right).$$

i. $k = 0$;
$Z^{1/3} = \cos 60° + j \sin 60°.$

ii. $k = 1$;
$Z^{1/3} = \cos 180° + j \sin 180°.$

iii. $k = 2$;
$Z^{1/3} = \cos 300° + j \sin 300°.$

■ 6. i. $Z = 4.3 + j2.8$
$r = \sqrt{(4.3)^2 + (2.8)^2} \approx 5.13;$
$\theta = \arctan 2.8/4.3 \approx \arctan .6512 \approx 33.1°.$

ii. $Z^{0.3} = (5.13(\cos 33.1° + j \sin 33.1°))^{0.3} = 1.63(\cos 9.93° + j \sin 9.93°).$

9-5 Problems

1. Calculate the three complex cube roots of 64.
2. Calculate $P_{app.} = I^2R$, when $I = .10(\cos 30° + j \sin 30°)$ and $R = 1000$.
3. If $E^2 = 15(\cos 53° + j \sin 53°)$, find E.
4. $E = \dfrac{I}{Z}$.

   $I = .12(\cos 15° + j \sin 15°)$.
   Find E.

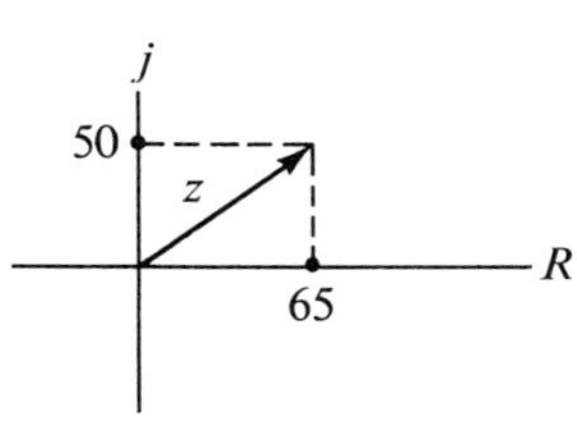

5. Find the complex zeros of the function $f(Z) = Z^2 + 4$.

9-6 Solution of AC Circuit Problems by Using Complex Numbers

Complex numbers are a very convenient and important tool for the solution of sinusoidal alternating-current circuit problems. Current, voltage, and impedance are not vector quantities in the same sense that force, velocity, and acceleration are, because they do not have a physical direction. However, because of certain phase relationships, angles as well as magnitudes are assigned to the quantities of current, voltage, and impedance; and operations with complex numbers are used.

Consider an AC circuit having a coil (L), capacitor (C), and resistor (R) in series. If the current and voltage about each component of this circuit are

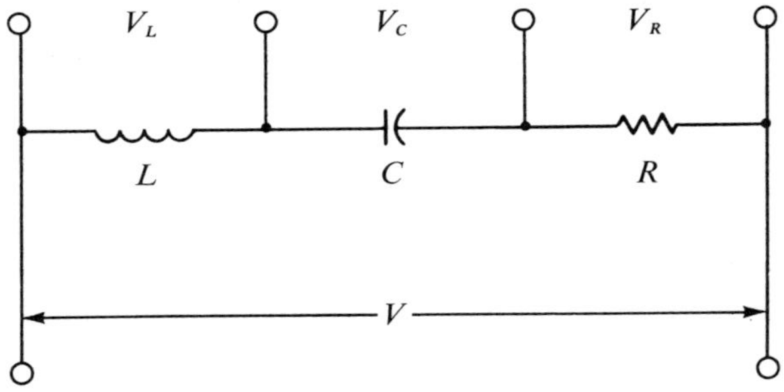

examined separately, the current and voltage V_R are found to be in phase. The voltage V_C lags the current by 1/4 cycle and the voltage V_L leads the current by 1/4 cycle. The net effect of these separate phase shifts is determined by constructing a complex vector diagram, where R is plotted in phase with current on the positive real axis, X_L (the coil's reactance) is plotted on the positive j-axis, and

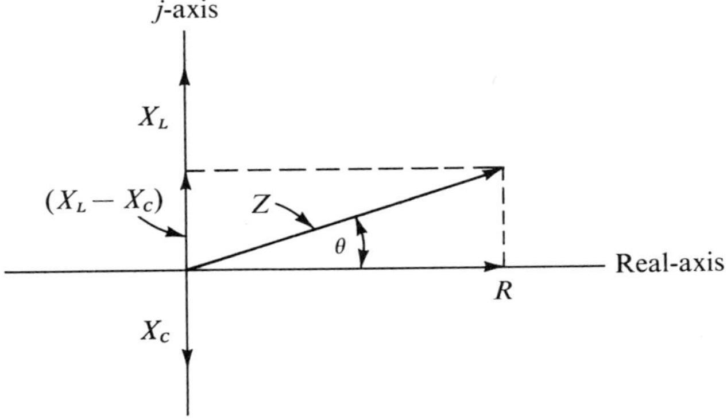

X_C (the capacitor's reactance) is plotted on the negative j-axis. The voltages across the separate components are

$$V_R = IR, \qquad V_L = IjX_L, \qquad V_C = I(-j)X_C. \qquad \text{(Eqs. 9-6a)}$$

By Kirchhoff's second law, the total voltage drop is

$$\begin{aligned} V &= IR + IjX_L - IjX_C \\ &= I(R + j(X_L - X_C)). \end{aligned} \qquad \text{(Eq. 9-6b)}$$

The expression, $R + j(X_L - X_C)$, is the complex expression for impedance Z. Its magnitude is

$$|Z| = \sqrt{R^2 + (X_L - X_C)^2},$$

and the angle

$$\theta = \arctan\left(\frac{X_L - X_C}{R}\right)$$

is the phase constant between the current I and voltage V.

Example 1

Find the impedance (Z) and phase constant between the voltage (V) and current (I).

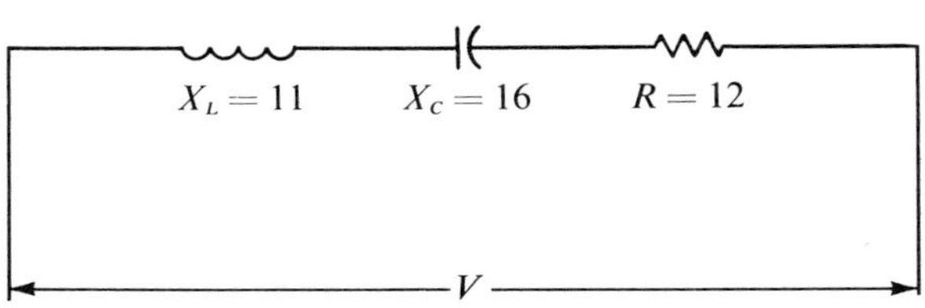

$$\begin{aligned} Z &= R + j(X_L - X_C) \\ &= 12 + j(11 - 16) \\ &= 12 - j5; \\ |Z| &= \sqrt{12^2 + 5^2} = 13; \\ \theta &= \arctan \frac{X_L - X_C}{R} \\ &= \arctan -\frac{5}{12} \\ &= \arctan(-.4167) = -22.62°. \end{aligned}$$

Suppose the complex numbers representing current and voltage are represented on the complex plane. The current vector could be placed at any position. Once it is drawn, the position of the voltage vector is determined by the value $\theta = \arctan(X_L - X_C)/R$. For our illustration, assume X_C is greater than X_L.

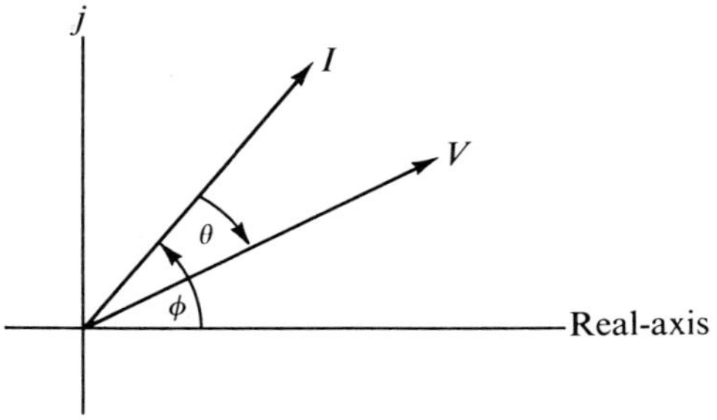

$$\begin{aligned} I &= I_1 + jI_2 \\ &= |I|(\cos\phi + j\sin\phi); \\ V &= V_1 + jV_2 \\ &= |V|(\cos(\phi - \theta) + j\sin(\phi - \theta)). \end{aligned}$$

If both I and V are known, Z is calculated by Ohm's law as

$$Z = \frac{V}{I} \qquad \text{(Eq. 9-6c)}$$

$$= \frac{V_1 + jV_2}{I_1 + jI_2}$$

$$= \frac{|V|\ (\cos(\phi - \theta) + j\sin(\phi - \theta))}{|I|(\cos\phi + j\sin\phi)}$$

$$= \left|\frac{V}{I}\right| (\cos(-\theta) + j\sin(-\theta)).$$

Example 2

Let $I = 3 + j4$ and $V = 12 + j2$. Calculate $|I|$, $|V|$, Z, and $|Z|$.

$$|I| = \sqrt{3^2 + 4^2} = 5.00;$$

$$|V| = \sqrt{12^2 + 2^2} \approx 12.17;$$

$$Z = \frac{12 + j2}{3 + j4} = \frac{12.17(\cos 9.5^\circ + j \sin 9.5^\circ)}{5(\cos 53.1^\circ + j \sin 53.1^\circ)}$$

$$= 2.43\ (\cos(9.5^\circ - 53.1^\circ) + j \sin(9.5^\circ - 53.1^\circ))$$

$$= 2.43\ (\cos(-43.6^\circ) + j \sin(-43.6^\circ))$$

$$= 1.76 - j1.67;$$

$$|Z| = \sqrt{(1.76)^2 + (1.67)^2}$$

$$= 2.43.$$

9-6 Problems

1. Given the values $R = 10\Omega$, $X_C = 8\Omega$, and $X_L = 15\Omega$, find Z and the phase constant between the current and voltage. Which leads?
2. Suppose $I = 9 + j4$ and $V = 18 + j15$. Find Z and the phase constant between the current and voltage.
3. Both X_L and X_C are variables in an AC circuit where the frequency is variable, but R does not change. What is the phase relationship between I and V in a series circuit when $X_L = X_C$?
4. Consider a network of parallel branches of resistance, inductive reactance, and capacitive reactance. Calculate Z_1 and Z_2 in complex form.

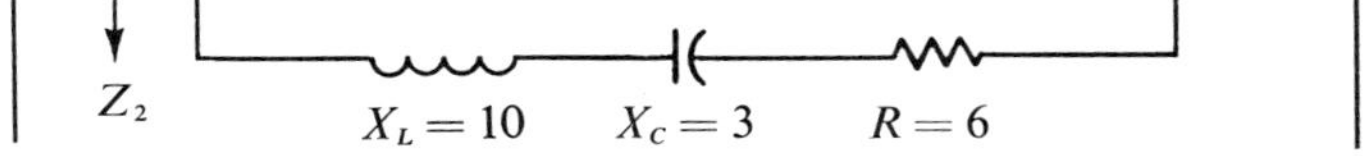

5. The network impedance (Z) of Problem 4 is computed by the formula

$$\frac{1}{Z} = \frac{1}{Z_1} + \frac{1}{Z_2}.$$

   a. Rearrange the formula for Z.
   b. Compute Z in complex form.

9 Supplementary Problems and Exercises

1. Write in imaginary form: (9-1)

   a. $\sqrt{-12}$,
   b. $\sqrt{-0.16}$,
   c. $\sqrt{(1.8)^2 - 4(1)(2.1)}$,
   d. $\sqrt{-8a^2}$.

2. Express in terms of j, -1, $-j$ or 1: (9-1)

   a. j^3,
   b. j^{24},
   c. j^{-1},
   d. j^{-6}.

3. Determine which are real numbers and which are imaginary numbers: (9-1)

   a. $(j4)^2$,
   b. $(-j)^3$,
   c. $\left(\frac{1}{j}\right)^{-3}$,
   d. $\left(\frac{-1}{j^5}\right)^2$.

Simplify the expressions in Problems 4–18.

4. $(6 - j2) + (3 + j4)$ (9-2)
5. $(12 + j5) + (-2 + j6)$ (9-2)
6. $4(3 + j5)$ (9-2)
7. $\frac{2}{3}\left(-12 + j\frac{3}{4}\right)$ (9-2)
8. $(1 - j)(-1 + j)$ (9-2)
9. $(3.8 + j4.1)(2 - j5)$ (9-2)
10. $(16 + j47) - (31 - j41)$ (9-2)
11. $(2.3 - j6) + (0 + j7)$ (9-2)
12. $\left(4 + j\frac{2}{3}\right)^2$ (9-2)
13. $(-200 + j120)^2$ (9-2)
14. $\frac{5 + j11}{2 + j2}$ (9-2)
15. $\frac{1}{400 + j80}$ (9-2)
16. $(2 + j3)(-4 + j2) + (2 + j3)(4 - j2)$ (9-2)
17. $\frac{15 - j20}{3 - j4}$ (9-2)

18. $\dfrac{1000 - j500}{-j200}$ (9-2)

19. Graph the complex numbers (9-3)

$$Z_1 = 5 + j6,$$
$$Z_2 = -3 - j4,$$
$$Z_3 = 1 - j6,$$
$$Z_4 = -2 + j5.$$

20. Graph the complex vectors (9-3)

$$Z_1 = 3 + j4,$$
$$Z_2 = -5 + j12,$$
$$Z_3 = 4 - j7.$$

21. What is the magnitude r and angle θ of each Z in Problem 20? (9-3)
22. Write the complex polar form of each Z given in Problem 20. (9-3)
23. Graph the complex polar numbers (9-3)

$$Z_1 = 3(\cos 40^\circ + j \sin 40^\circ),$$
$$Z_2 = 5(\cos 210^\circ + j \sin 210^\circ).$$

24. Change $120 + j75$ to complex polar form. (9-3)
25. Change $27(\cos 38^\circ + j \sin 38^\circ)$ to complex rectangular form. (9-3)
26. Change $42(\cos 5\pi/3 + j \sin 5\pi/3)$ to complex rectangular form. (9-3)
27. Write $-23 + j15$ in complex polar form. (9-3)
28. $(12 + j3) + Z = 0$. Find Z. (9-3)
29. $(6.5 - j4.1) + Z = 0$. Find Z. (9-3)
30. Write the number one in complex polar form. (9-3)
31. Find Z_1Z_2 and Z_1/Z_2 when $Z_1 = 4.5(\cos 108^\circ + j \sin 108^\circ)$ and $Z_2 = 2.5(\cos(-40^\circ) + j \sin(-40^\circ))$. (9-4)
32. Convert $Z_1 = 12 + j8$ and $Z_2 = -4 + j6$ to complex polar form and then find Z_1Z_2. (9-4)
33. $1/Z = 0.04(\cos 0.2\pi + j \sin 0.2\pi)$. Find Z. (9-4)
34. $Z = 10(\cos 0.34 + j \sin 0.34)$. Find $1/Z$. (9-4)
35. $Z_1 = 2500(\cos 3\pi/8 + j \sin 3\pi/8)$ and $Z_2 = 3500(\cos \pi/4 + j \sin \pi/4)$. Calculate $Z = Z_1Z_2/(Z_1+Z_2)$. (9-4)
36. $Z = 1.2(\cos 18^\circ + j \sin 18^\circ)$. Calculate Z^3. (9-5)
37. $Z = 2.4(\cos \pi/12 + j \sin \pi/12)$. Calculate Z^2. (9-5)
38. $Z = 1.2 + j1.6$. Calculate Z^3. (9-5)
39. $Z = 3 - j2$. Calculate Z^4. (9-5)
40. $Z = 5(\cos 60^\circ + j \sin 60^\circ)$. Calculate $Z^{1/2}$. (9-5)
41. Find the three cube roots of $15(\cos 45^\circ + j \sin 45^\circ)$. (9-5)
42. Graph the results of Problem 41, assuming the complex numbers represent vectors. (9-5)
43. Calculate the three complex cube roots of 10. (9-5)
44. $Z = 2.0 + j3.6$. Calculate $Z^{1/3}$. (9-5)
45. Find the complex zeros of the equation $f(Z) = 2Z^2 + 5$. (9-5)

Chapter Ten Plane Analytic Geometry

10-1 Distance Formula and Midpoints

Plane analytic geometry is the application of algebraic methods to the study of plane geometry. The fundamental elements of algebra and the fundamental elements of geometry are linked through the use of the coordinate system.

We have already made use of this link to bring more understanding to the work with linear and quadratic functions in Chapters 1, 2, and 3. For example, in plane geometry a *line* is an undefined entity expressing the notion of an unturning path extending in opposite directions indefinitely. In analytic geometry a line is designated by the graph of the solution set of an equation in the general form $f(x) = mx + b$, where m is the constant slope and b the y-intercept.

In this chapter we shall extend the study of analytic geometry to a group of plane curves called *conic sections* and to other ideas that are fundamental to many engineering and technical applications.

First we must consider the matter of *distance* on the coordinate plane. **Distance** is defined to be the shortest absolute length separating two geometric configurations.

The formula for calculating the distance (d) between two points on a coordinate system is

$$d = \sqrt{(x_2 - x_1)^2 + (y_2 - y_1)^2}, \qquad \text{(Eq. 10-1a)}$$

where (x_1, y_1) and (x_2, y_2) are the coordinates of the two points. An examination of Fig. 10-1a will show that Eq. 10-1a is simply an application of the Pythagorean theorem.

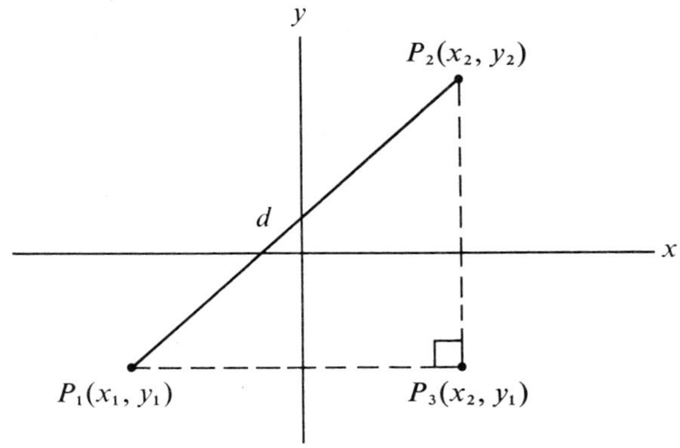

Fig. 10-1a

Point P_3 is given the coordinates (x_2, y_1) so that $P_1P_2P_3$ is a right triangle. The legs of the triangle are $|\overline{P_1P_3}| = |x_2 - x_1|$ and $|\overline{P_2P_3}| = |y_2 - y_1|$.

Example 1

Calculate the distance between points $P_1(-5, -3)$ and $P_2(7, 12)$.

$$\begin{aligned} |\overline{P_1P_2}| &= \sqrt{(-5-7)^2 + (-3-12)^2} \\ &= \sqrt{144 + 225} \\ &\approx 19.2. \end{aligned}$$

The distance from a point to a line is the length of the perpendicular line segment from the point to the line.

The technique of using a compass and straightedge to locate the line segment representing the distance from a point P to a line L is illustrated in Fig. 10-1b.

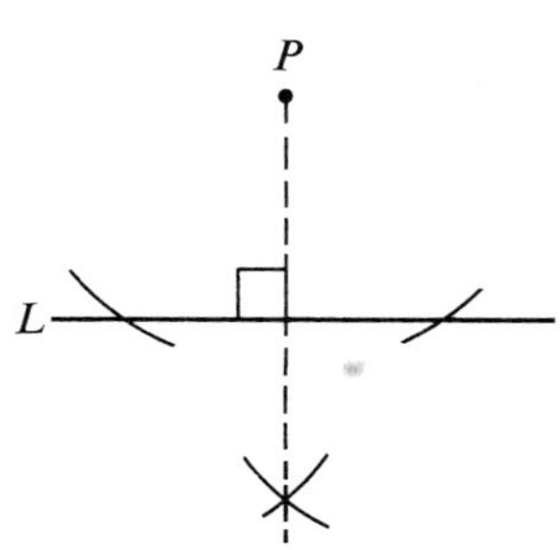

Fig. 10-1b

In analytic geometry the distance from a point P to a line L can be calculated if the coordinates of P and the equation of L are known. The procedure, briefly outlined, is: (1) find the equation of the line through P and perpendicular to L using Eqs. 1-3b and 1-4a, (2) find the simultaneous solution to the equations of Step 1 and the equation of L, and (3) use the results of Step 2 and point P to find the desired distance.

If the equation of a line L has the specific form $Ax + By + C = 0$ and point P has the coordinates (x_0, y_0), then using the steps outlined in the preceding paragraph, the distance d from the point to the line may be written in the very concise form

$$d = \left| \frac{Ax_0 + By_0 + C}{\sqrt{A^2 + B^2}} \right|. \qquad \text{(Eq. 10-1b)}$$

Example 2

How far is $P_1(1, 3)$ from $2x + 6y + 4 = 0$?

$$d = \left| \frac{2 \cdot 1 + 6 \cdot 3 + 4}{\sqrt{2^2 + 6^2}} \right| = \frac{24}{\sqrt{40}} = \frac{24}{2\sqrt{10}} = \frac{12}{\sqrt{10}}$$

$$= \frac{6\sqrt{10}}{5}$$

$$\approx 3.79.$$

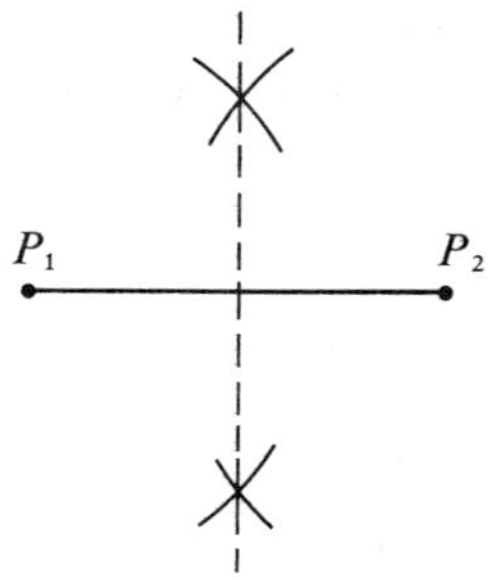

Fig. 10-1c

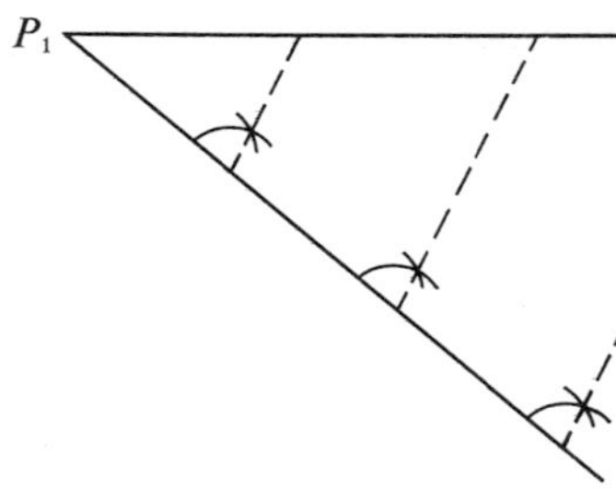

Fig. 10-1d

Figures 10-1c and 10-1d illustrate the compass and straightedge geometric techniques of bisecting and trisecting a line segment. From an analytic point of view, not only can such points as the midpoint be located on a line segment in a coordinate system, but the precise coordinates of the points can be calculated as well.

Example 3

Find the coordinates of the midpoint of $\overline{P_1P_2}$ in Fig. 10-1e. So that the

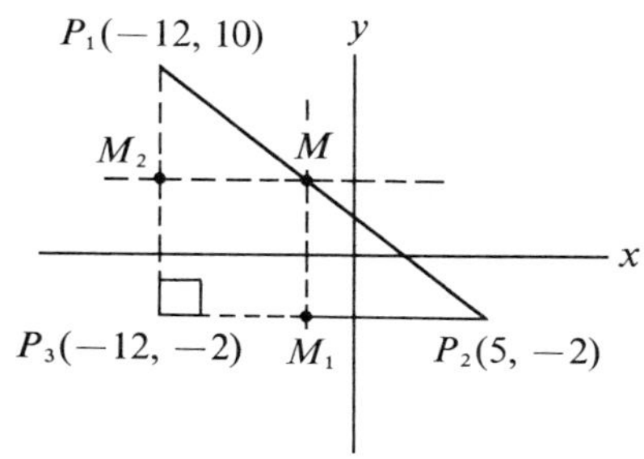

Fig. 10-1e

procedure for determining the coordinates of the midpoint will be understood, we choose P_3 with coordinates $(-12, -2)$ to form the right triangle $P_1P_2P_3$. If M_1 is the midpoint of $\overline{P_3P_2}$ and M_2 is the midpoint of $\overline{P_1P_3}$, then the abscissa of point M_1 and the ordinate of M_2 designate the coordinates of the midpoint M of $\overline{P_1P_2}$.

$$M_1 = \left(\frac{-12+5}{2}, -2\right)$$

$$= \left(-\frac{7}{2}, -2\right);$$

$$M_2 = \left(-12, \frac{10+(-2)}{2}\right)$$

$$= (-12, 4).$$

Therefore the midpoint of $\overline{P_1P_2}$ is $(-7/2, 4)$.

In general, the midpoint of $P_1(x_1, y_1)$ and $P_2(x_2, y_2)$ is

$$M(\overline{P_1P_2}) = \left(\frac{x_1+x_2}{2}, \frac{y_1+y_2}{2}\right). \qquad \text{(Eq. 10-1c)}$$

10-1 Exercises

1. Determine the lengths of the four line segments illustrated.

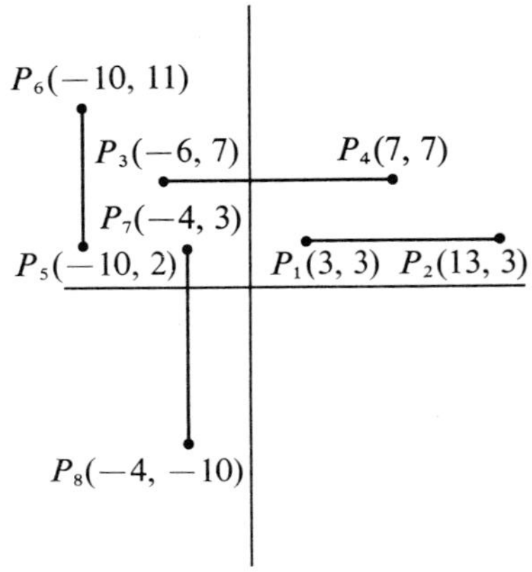

2. Find the distance between $P_1(-5, 7)$ and $P_2(8, 2)$.

■ 1. $|\overline{P_1P_2}| = |3 - 13| = 10;$
$|\overline{P_3P_4}| = |-6 - 7| = 13;$
$|\overline{P_5P_6}| = |2 - 11| = 9;$
$|\overline{P_7P_8}| = |-10 - 3| = 13.$

3. Find the distance between $P_1(1.23, 3.47)$ and $P_2(-3.16, 1.76)$.

■ 2. $\overline{P_1P_2}$
$= \sqrt{(-5 - 8)^2 + (7 - 2)^2}$
$= \sqrt{169 + 25}$
$\approx 13.9.$

4. Represent the distance between $P_1(x, y)$ and $P_2(h, k)$.

■ 3. $\overline{P_1P_2}$
$= \sqrt{(1.23 + 3.16)^2}$
$+ \sqrt{(3.47 - 1.76)^2}$
$\approx \sqrt{19.27 + 2.92}$
$\approx 4.71.$

5. Find the distance from $P_1(3, 5)$ to the line L whose equation is $3x - 2y + 3 = 0$.

■ 4. $|\overline{P_1P_2}|$
$= \sqrt{(x - h)^2 + (y - k)^2}.$

---

6. Find the distance between $P_1(-4, -2)$ and line L whose equation is $x = y$.

■ 5. $d = \left|\dfrac{3(3) - 2(5) + 3}{\sqrt{3^2 + (-2)^2}}\right|$

$\approx \dfrac{2}{3.606} \approx .554.$

---

7. Calculate the distance between $P_1(0.1, 0.01)$ and the graph of the equation $0.2x + 0.3y + 0.02 = 0$.

■ 6. $L: x = y \leftrightarrow x - y + 0 = 0;$

$d = \left|\dfrac{(-4) + 2}{\sqrt{1^2 + (-1)^2}}\right|$

$= \dfrac{2}{\sqrt{2}} \approx 1.414.$

---

8. How far is it from the graph of $3x + 2y + 6 = 0$ to the point of intersection of

$$2x - 5y + 7 = 0$$

and

$$3x + 5y - 4 = 0?$$

■ 7. $d =$

$\left|\dfrac{0.2(0.1) + 0.3(0.01) + 0.02}{\sqrt{(0.2)^2 + (0.3)^2}}\right|$

$\approx \dfrac{.0430}{.3606} \approx .119.$

---

9. Determine the midpoints

$M(\overline{P_1P_2})$, $M(\overline{P_2P_3})$, and $M(\overline{P_1P_3})$.

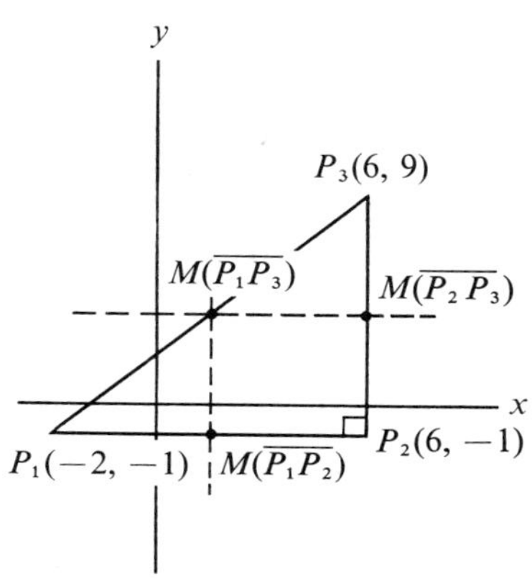

■ 8. i. $\left.\begin{aligned} 2x - 5y &= -7 \\ 3x + 5y &= 4 \end{aligned}\right]$ System of equations

$\therefore x = -3/5;$

$y = 29/25.$

ii. $P(-3/5, 29/25);$

$L: 3x + 2y + 6 = 0;$

$d = \left|\dfrac{3(-3/5) + 2(29/25) + 6}{\sqrt{3^2 + 2^2}}\right|$

$\approx \dfrac{6.52}{3.61} \approx 1.81.$

10. Find the coordinates of the midpoint of the line segment whose endpoints are $(-3, -2)$ and $(5, 3)$.

■ 9. i. $M(\overline{P_1P_2})$

$$= \left(\frac{-2+6}{2}, \frac{-1-1}{2}\right)$$

$= (2, -1)$;

ii. $M(P_2P_3)$

$$= \left(\frac{6+6}{2}, \frac{9-1}{2}\right)$$

$= (6, 4)$;

iii. $M(P_1P_3)$

$$= \left(\frac{-2+6}{2}, \frac{9-1}{2}\right)$$

$= (2, 4)$.

■ 10. $M\left(\frac{-3+5}{2}, \frac{-2+3}{2}\right)$, or $M(1, 1/2)$.

10-1 Problems

1. Town A is 3 miles east and 4 miles north of town B, and town C is 5 miles west and 6 miles south of town B. What is the straight-line distance from town A to town C?
2. The center of a rectangle whose sides are of lengths $2a$ and $2b$ is at the origin of the coordinate system. If the sides are parallel to the coordinate axes, what are the coordinates of the vertices of the rectangle?
3. The center of a square whose side has a length of $2a$ is at the origin of the coordinate system. Determine the coordinates of the vertices if they all lie on the coordinate axes.
4. Find the area of the circle whose center is $(2, 1)$ if the circle passes through the point $(5, -2)$.
5. Find the coordinates of the point one-third of the distance from $(2, 1)$ to $(6, 12)$.
6. Find the area of the triangle whose vertices are $(-4, -2)$, $(3, 5)$, and $(3, -2)$.

7. An aerial topographic survey map superimposed on a grid is illustrated in Fig. 10-1f. What is the distance from the radio tower to the transmission line?

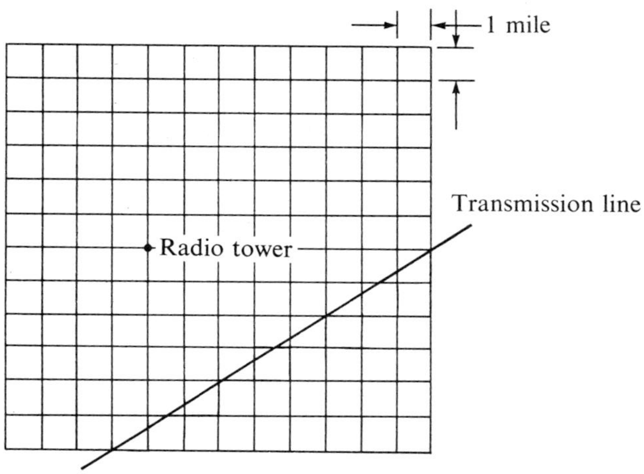

Fig. 10-1f

8. Find the center of the circle that circumscribes the triangle whose vertices are $(-3, 5)$, $(-3, 7)$, and $(4, 2)$.

10-2 Areas of Polygons on a Coordinate System

The area of a triangle placed on a coordinate system can be calculated by knowing the coordinates of the vertices. The orientation of the triangle has a great deal to do with how complicated the computation will be.

If at least one side of a triangle is parallel to either the x-axis or y-axis of the coordinate system, calculating the area is very easy.

Example 1

Find the area of triangle $P_1P_2P_3$.

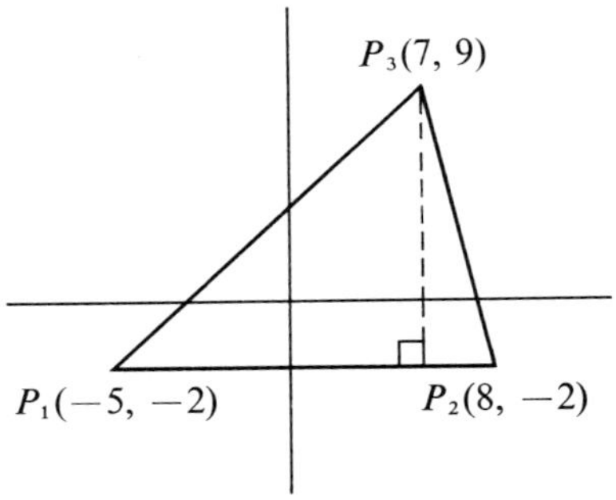

Fig. 10-2a

Let $|\overline{P_1P_2}| = |-5-8| = 13$ be the base. Point A has the coordinates $(x, -2)$. Therefore the altitude is $|9-(-2)| = 11$, and the area of $P_1P_2P_3 = (1/2)(13)(11) = 71.5$.

Consider the problem of calculating the area of a triangle plotted on a coordinate system where no sides of the triangle are parallel to either axis. Equation 10-1a can be used to determine the length of a base and Eq. 10-1b can be used to calculate the length of the corresponding altitude. This procedure works in all cases, but more efficient techniques exist.

From Fig. 10-2b we shall proceed to derive a general equation that can be used not only for all triangles, but for all polygons. The triangular area $P_1P_2P_3$

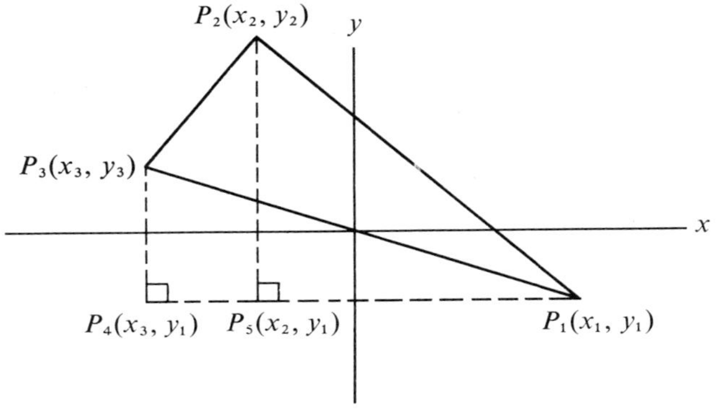

Fig. 10-2b

is equal to the sum of the triangular area $P_1P_2P_5$ plus the trapezoidal area $P_2P_3P_4P_5$ minus the triangular area of $P_1P_3P_4$. In terms of the coordinates of the vertices,

$$\text{(area) } P_1P_2P_5 = \frac{1}{2}(x_1 - x_2)(y_2 - y_1)$$

$$= \frac{1}{2}(x_1y_2 - x_1y_1 - x_2y_2 + x_2y_1);$$

$$\text{(area) } P_2P_3P_4P_5 = \frac{1}{2}(x_2 - x_3)((y_3 - y_1) + (y_2 - y_1))$$

$$= \frac{1}{2}(x_2y_3 - x_2y_1 + x_2y_2 - x_2y_1 - x_3y_3 + x_3y_1 - x_3y_2 + x_3y_1);$$

and

$$\text{(area) } P_1P_3P_4 = \frac{1}{2}(x_1 - x_3)(y_3 - y_1)$$

$$= \frac{1}{2}(x_1y_3 - x_1y_1 - x_3y_3 + x_3y_1).$$

Therefore,

$$\text{(area) } P_1P_2P_3 = P_1P_2P_5 + P_2P_3P_4P_5 - P_1P_3P_4$$

$$= \frac{1}{2}(x_1y_2 - x_1y_3 + x_2y_3 - x_2y_1 + x_3y_1 - x_3y_2),$$

or finally,

$$\text{(area) } P_1P_2P_3 = \frac{1}{2}(x_1(y_2 - y_3) + x_2(y_3 - y_1) + x_3(y_1 - y_2)). \qquad \text{(Eq. 10-2a)}$$

In order for Eq. 10-2a to yield correct results, the triangle to which it is applied must have vertices numbered in a *counterclockwise* order!

As a means of checking our general results, use Eq. 10-2a to calculate the area of the first example of this section (Fig. 10-2a).

$$\blacksquare\ (\text{Area})P_1P_2P_3 = \frac{1}{2}(-5(-2-9) + 8(9-(-2)) + 7(-2-(-2)))$$

$$= \frac{1}{2}(55 + 88) = 71.5.$$

Example 2

Use Eq. 10-2a to calculate the area of the triangle whose vertices are $P_1(-8, -3)$, $P_2(6, -5)$, and $P_3(4, 7)$.

$$(\text{Area})P_1P_2P_3 = \frac{1}{2}((-8)(-5-7) + 6(7-(-3)) + 4(-3-(-5)))$$

$$= \frac{1}{2}(-8(-12) + 6(10) + 4(2))$$

$$= \frac{1}{2}(164) = 82.$$

Figure 10-2c illustrates the manner in which any polygon may be subdivided into triangular areas by drawing from only one vertex all possible diagonals. If the vertices of the polygon are numbered in counterclockwise order, then each triangle has numbered vertices that occur in counterclockwise order, even though not consecutively numbered. To each triangle we may apply

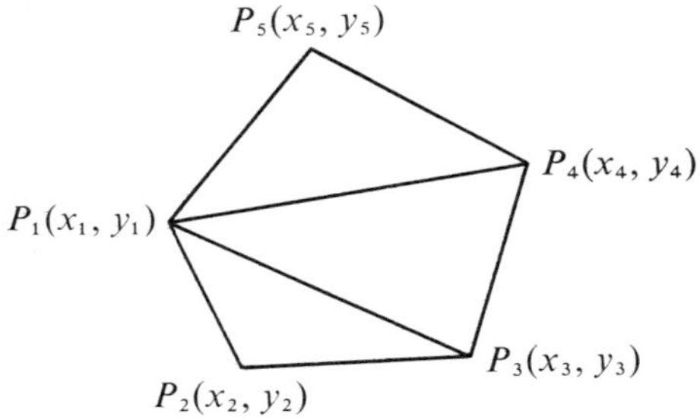

Fig. 10-2c

Eq. 10-2a and ultimately find the total area. We would discover, however, that many of the terms in the computation are opposites and that deleting these terms results in the equation

$$(\text{Area})\ P_1P_2P_3P_4P_5 = \frac{1}{2}(x_1(y_2 - y_5) + x_2(y_3 - y_1) + x_3(y_4 - y_2)$$

$$+ x_4(y_5 - y_3) + x_5(y_1 - y_4)). \quad \text{(Eq. 10-2b)}$$

This equation is extremely difficult to remember unless we observe the pattern which reoccurs in each term. Each abscissa x_i is in numerical order and multiplies the difference of the ordinates of the vertices immediately ahead and behind y_i.

Example 3

Find the area of Fig. 10-2d.

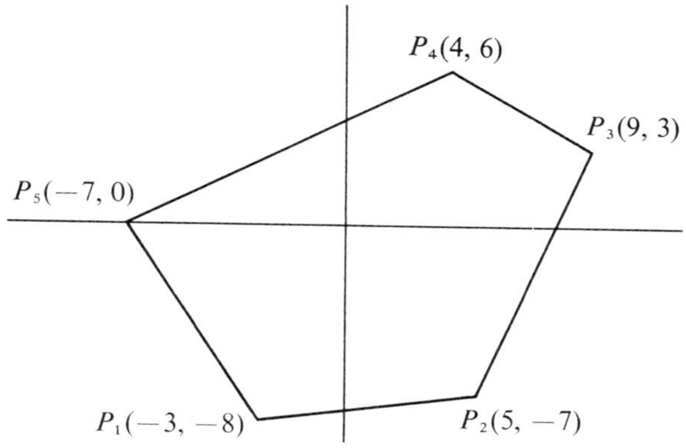

Fig. 10-2d

$$(\text{Area})P_1P_2 \ldots P_5 = \frac{1}{2}(-3(-7-0) + 5(3-(-8)) + 9(6-(-7))$$

$$+ 4(0-3) - 7(-8-6))$$

$$= \frac{1}{2}(21 + 55 + 117 - 12 + 98)$$

$$= 139.5.$$

For a polygon of n sides the area formula can be generalized to

$$\text{(area) } P_1P_2 \ldots P_n = \frac{1}{2}(x_1(y_2 - y_n) + x_2(y_3 - y_1) + \cdots + x_i(y_{i+1} - y_{i-1}) + \cdots + x_{n-1}(y_n - y_{n-2}) + x_n(y_1 - y_{n-1})). \quad \text{(Eq. 10-2c)}$$

10-2 Exercises

1. Calculate the area:

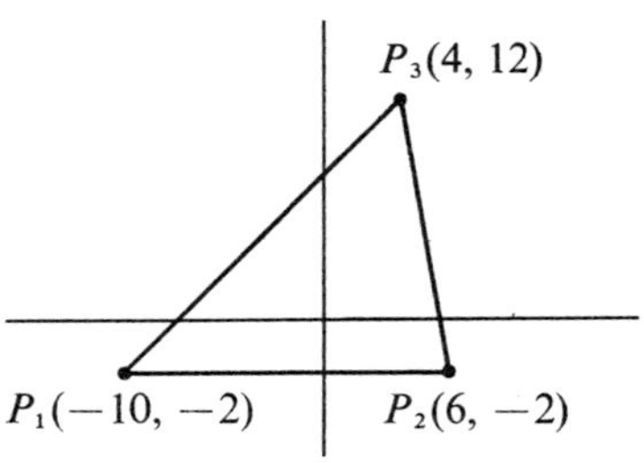

2. Calculate the area:

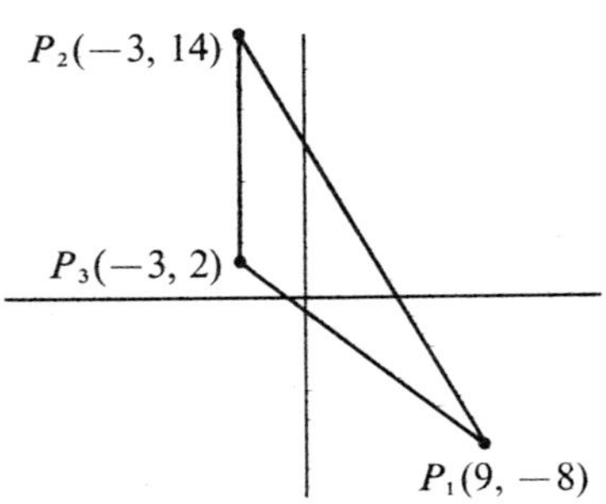

■ 1. i. $|\overline{P_1P_2}| = |-10 - 6| = 16;$

ii. Altitude from P_3: $|12 - (-2)| = 14;$

iii. (Area) $= (1/2)(16)(14) = 112.$

3. Calculate the area:

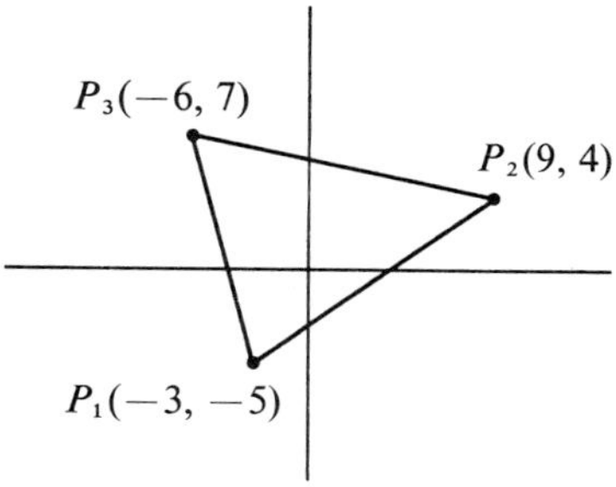

■ 2. i. $|\overline{P_2P_3}| = |14 - 2|$
$= 12;$

ii. Altitude from P_1:
$|9 - (-3)| = 12;$

iii. (Area) $= (1/2)(12)(12)$
$= 72.$

4. Calculate the area:

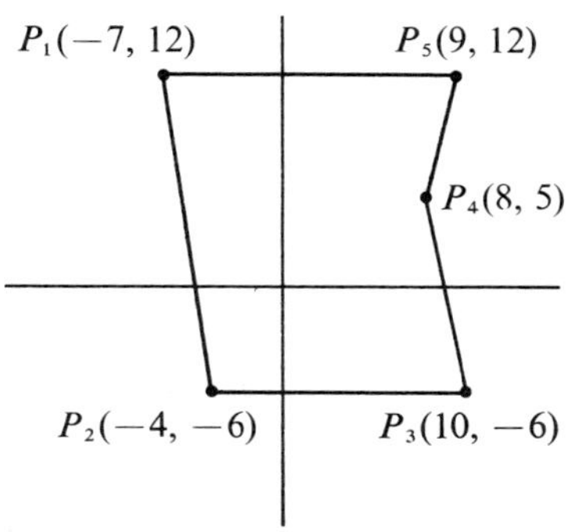

■ 3. Area $P_1P_2P_3$

$$= \frac{1}{2}(-3(4 - 7) + 9(7 - (-5)) - 6(-5 - 4))$$
$$= \frac{1}{2}(9 + 108 + 54)$$
$$= 85.5.$$

5. Calculate the area of ABC when $r_1 = 100$, $r_2 = 85$, $\theta = 62°$, and $\phi = 28°$.

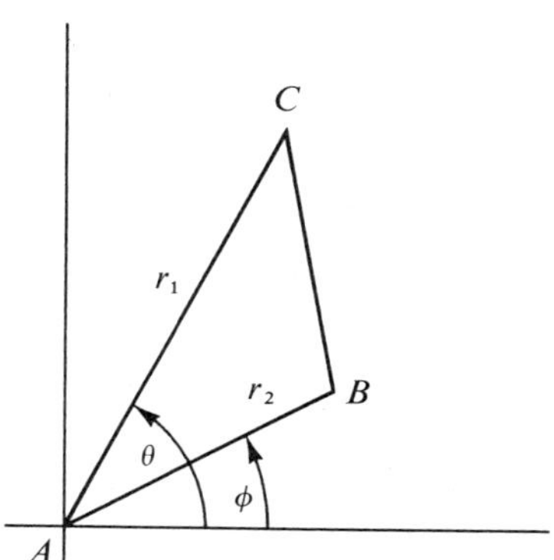

■ 4. (Area)$P_1 \ldots P_5$

$$= \frac{1}{2}(-7(-6 - 12) - 4(-6 - 12) + 10(5 - (-6)) + 8(12 - (-6)) + 9(12 - 5))$$
$$= \frac{1}{2}(126 + 72 + 110 + 144 + 63)$$
$$= 257.5.$$

■ 5. i. $A(0, 0)$;

ii. $B(85 \cos 28°, 85 \sin 28°)$, or $B(75.05, 39.91)$;

iii. $C(100 \cos 62°, 100 \sin 62°)$, or $C(46.95, 88.29)$;

iv. (Area)ABC

$$= \frac{1}{2}(0(39.91 - 88.29) + 75.05(88.29 - 0) + 46.95(0 - 39.91))$$

$$= \frac{1}{2}(6625.9 - 1873.7)$$

$$= \frac{1}{2}(4752.2) = 2376.1.$$

10-2 Problems

1. If each of the grids represents one square mile, calculate the area of $ABCD$ in Fig. 10-2e.

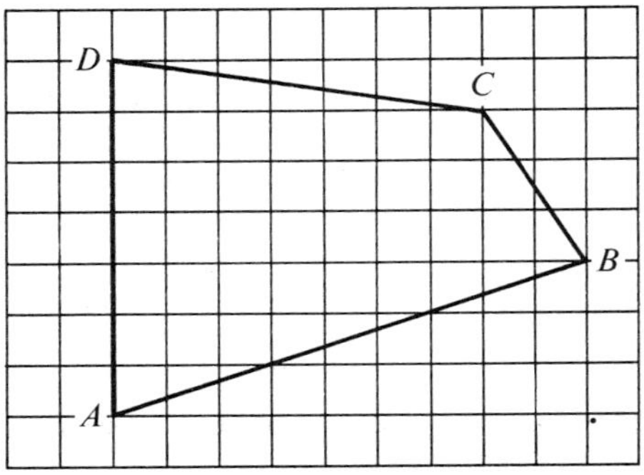

Fig. 10-2e

2. Find the area of $ABCD$ in Fig. 10-2f when $\overline{AB} = 72$ feet, $\overline{AC} = 110$ feet, and $\overline{AD} = 60$ feet.

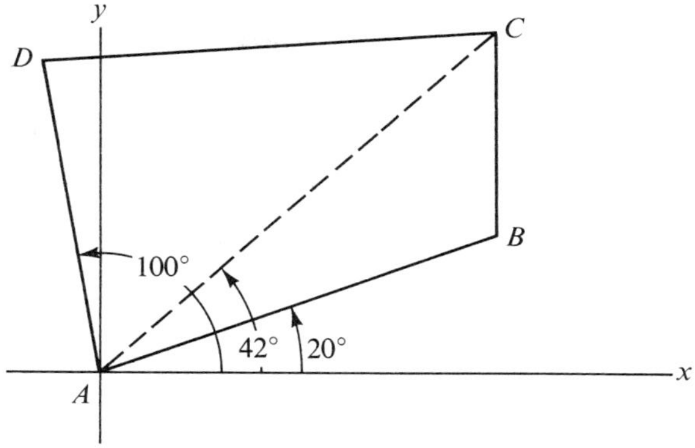

Fig. 10-2f

3. Find the area of the traverse $ABCD$ in Fig. 10-2g if $\overline{AB} = 180$ feet, $\overline{BC} = 210$ feet, and $\overline{CD} = 360$ feet.

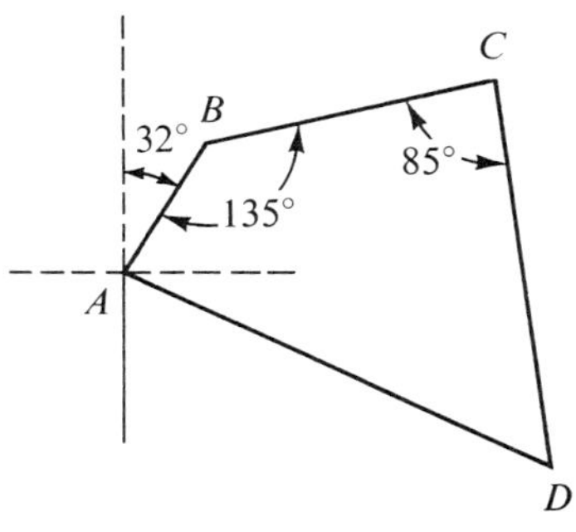

Fig. 10-2g

10-3 Circles on the Coordinate Plane

If a point $P(x, y)$ moves in such a manner that it is always the same distance from a fixed point, then the point P moves in a closed curve called a circle. Let r be the distance from P to the fixed point, and let the coordinates of the fixed point be (h, k). With these constants, we can write an equation that produces the related pairs of numbers (x, y) that are the points on the circle.

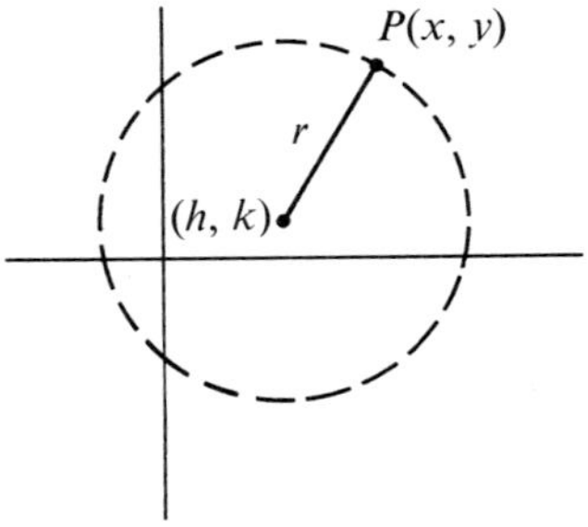

Fig. 10-3a

From the distance formula, we can write the equation

$$r^2 = (x - h)^2 + (y - k)^2. \qquad \text{(Eq. 10-3a)}$$

The fixed point (h, k) is, of course, the center of the circle. Should the coordinates (h, k) be the origin of the coordinate system, Eq. 10-3a simply becomes

$$r^2 = x^2 + y^2. \qquad \text{(Eq. 10-3b)}$$

Example 1

Find the equation of a circle whose center is at (3, 4) and whose radius is 5.

$$r^2 = (x - h)^2 + (y - k)^2,$$

$$5^2 = (x - 3)^2 + (y - 4)^2,$$

$$25 = x^2 - 6x + 9 + y^2 - 8y + 16$$

$$x^2 + y^2 - 6x - 8y = 0.$$

Example 2

If the equation of a circle is $x^2 + y^2 + 6x - 10y + 12 = 0$, where is its center, and what is its radius?

We must write the equation in the form

$$(x-h)^2+(y-k)^2=r^2,$$

or

$$x^2-2xh+h^2+y^2-2yk+k^2=r^2.$$

$$(x^2+6x+\square)+(y^2-10y+\square)=-12+\square.$$

In order to make x^2+6x a perfect square, we must add 9, and to make y^2-10y a perfect square, we must add 25. We now have

$$x^2+6x+9+y^2-10y+25=-12+9+25$$

$$(x+3)^2+(y-5)^2=22,$$

so $h=-3$, $k=5$, and $r=\sqrt{22}$.

The equation of a circle has in general the form

$$Ax^2+By^2+Cxy+Dx+Ey+F=0, \qquad \text{(Eq. 10-3c)}$$

where $A=B=1$ and $C=0$; in other words,

$$x^2+y^2+Dx+Ey+F=0. \qquad \text{(Eq. 10-3d)}$$

Although every circle has an equation of this *general form*, it is incorrect to assume every equation in this form is a circle.

Example 3

Investigate the equation $x^2+y^2-4x+6y+15=0$.

Since the coefficients of x^2 and y^2 are equal, we may suppose that this is a circle.

$$x^2-4x+4+y^2+6y+9=-15+13$$

$$(x-2)^2+(y+3)^2=-2.$$

Since the left side of the equation is positive for every (x, y) and the right side is negative, we see that the equation has no *real* solutions and the equation has no graph on the real coordinate plane.

A general solution to Eq. 10-3d will provide us with several useful equations that will make determining the center and radius of a circle quite simple.

i. $x^2 + y^2 + Dx + Ey + F = 0;$

ii. $x^2 + Dx + \boxed{} + y^2 + Ey + \boxed{} = \boxed{} + \boxed{} - F;$

iii. $\left(x + \frac{D}{2}\right)^2 + \left(y + \frac{E}{2}\right)^2 = \frac{D^2}{4} + \frac{E^2}{4} - F = \frac{D^2 + E^2 - 4F}{4}.$

■ $(D/2)^2$; $(E/2)^2$; $(D/2)^2$; $(E/2)^2$.

A comparison of the results of step (iii) and Eq. 10-3a shows that

$$h = -\frac{D}{2},$$

$$k = -\frac{E}{2},$$

and

$$r = \frac{\sqrt{D^2 + E^2 - 4F}}{2}. \qquad \text{(Eq. 10-3e)}$$

The equation for r clearly reveals the limitations on the values of constants D, E, and F in order for the solution set of Eq. 10-3d to be a circle on the coordinate system: $D^2 + E^2 \geq 4F$.

Thus, we have two methods for finding the coordinates of the center and the radius when given a circle in general form. We may proceed to complete the square as in Example 2, or we may use Eqs. 10-3e.

An equation such as $x^2 + y^2 = 9$ gives a relationship between x and y, but it is not a *functional relationship*. If we solve for y in terms of x, we get $y = \pm\sqrt{9 - x^2}$. This gives us two values of y for each value of x, where $-3 < x < 3$. We may, if we wish, split this relationship into two equations that *are* each functional:

$$y = f(x) = \sqrt{9 - x^2}$$

$$y = g(x) = -\sqrt{9 - x^2}.$$

The graph of f is the top half of the circle, and the graph of g is the bottom half of the same circle.

10-3 Exercises

In Exercises 1–3, write the equation of the circle in general form.

1. Center (0, 0), radius 5.

2. Center (2, −4), radius 3.

3. Center (−7, 3), radius 10.

4. Write in general form the equation of the circle whose center is on the graph of $y = 2x$, whose radius is 5 if the circle is tangent to the x-axis.

5. Write in general form the equation of the circle tangent to both axes, whose radius is 5 and which is in the fourth quadrant.

■ 1. $x^2 + y^2 - 25 = 0.$

■ 2. $(x - 2)^2 + (y + 4)^2 - 9 = 0;$
$x^2 + y^2 - 4x + 8y + 11 = 0.$

■ 3. $(x + 7)^2 + (y - 3)^2 - 100 = 0;$
$x^2 + y^2 + 14x - 6y - 42 = 0.$

■ 4.

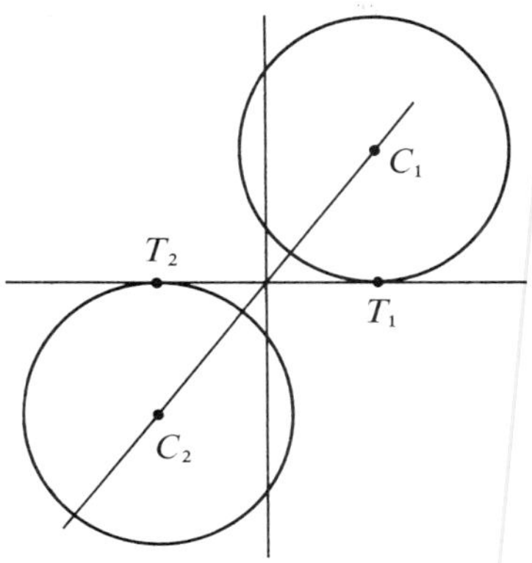

There are two possible solutions.

Part 1

i. $y = 2x$
$5 = 2x;\ x = 5/2$
$C_1(5/2, 5)$.

ii. $\left(x - \frac{5}{2}\right)^2 + (y - 5)^2 - 25 = 0$
$x^2 + y^2 - 5x - 10y + \frac{25}{4} = 0.$

Part 2

i. $y = 2x$
$-5 = 2x;\ x = -5/2$
$C_2(-5/2, -5)$.

ii. $\left(x + \frac{5}{2}\right)^2 + (y + 5)^2 - 25 = 0$
$x^2 + y^2 + 5x + 10y + \frac{25}{4} = 0.$

---

6. Find the center and radius of the graph of

$$x^2 + y^2 - 2x + 4y - 2 = 0.$$

■ 5.

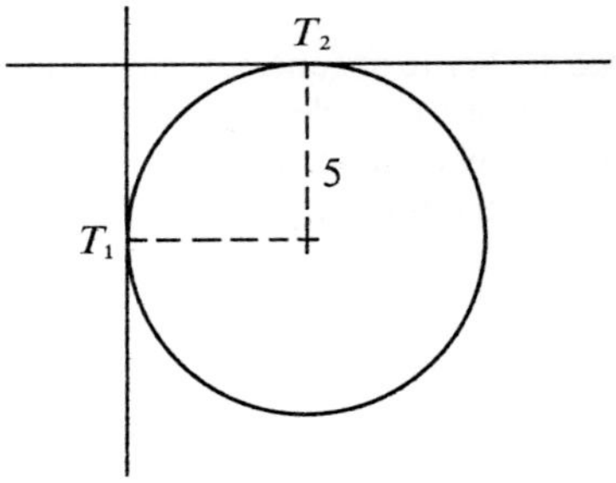

$(x - 5)^2 + (y + 5)^2 - 25 = 0;$

$x^2 + y^2 - 10x + 10y + 25 = 0.$

---

7. Find the center and radius of the graph of

$$x^2 + y^2 + 8x - 10y - 8 = 0.$$

■ 6. $x^2 + y^2 - 2x + 4y - 2 = 0;$
$x^2 - 2x + 1 + y^2 + 4y + 4$
$= 2 + 1 + 4$
$(x-1)^2 + (y+2)^2 = 7.$
Center: $(1, -2)$
Radius: $\sqrt{7}$

---

8. Find the center and radius of the graph of

$$x^2 + y^2 - 2x - 8 = 0.$$

■ 7. $x^2 + y^2 + 8x - 10y - 8 = 0;$
$x^2 + 8x + 16 + y^2 - 10y$
$+ 25 = 8 + 16 + 25$
$(x+4)^2 + (y-5)^2 = 49.$
Center: $(-4, 5)$
Radius: 7

---

9. Where does $y = x - 1$ intersect

$$x^2 + y^2 - x - 3y = 0?$$

■ 8. $r = \dfrac{\sqrt{D^2 + E^2 - 4F}}{2}$

$= \dfrac{\sqrt{(-2)^2 + (0)^2 - 4(-8)}}{2}$

$= 3;$

$h = -\dfrac{D}{2} = -\dfrac{-2}{2} = 1;$

$k = -\dfrac{E}{2} = -\dfrac{0}{2} = 0.$

---

10. Find the equation of the circle passing through $(-2, -2)$, $(10, -8)$, and $(7, 1)$.

■ 9. i. $x^2 + y^2 - x - 3y = 0$
$y = x - 1;$

ii. $x^2 + (x-1)^2 - x$
$- 3(x-1) = 0$

$x^2 - 3x + 2 = 0$
$(x-2)(x-1) = 0;$

iii. $x = 2$
$y = 2 - 1 = 1;$ $(2, 1)$

iv. $x = 1$
$y = 1 - 1 = 0;$ $(1, 0)$

■ 10. i. The general form of the equation is $x^2 + y^2 + Dx + Ey + F = 0$. With the three known values (x, y), we write a system of equations in terms of D, E, and F.

ii. $$\begin{aligned} -2D - 2E + F &= -8 \\ 10D - 8E + F &= -164 \\ 7D + E + F &= -50; \end{aligned}$$

iii. $$\left(\begin{array}{ccc|c} -2 & -2 & 1 & -8 \\ 10 & -8 & 1 & -164 \\ 7 & 1 & 1 & -50 \end{array}\right),$$

$$\left(\begin{array}{ccc|c} -2 & -2 & 1 & -8 \\ 0 & -6 & 2 & -68 \\ 0 & 0 & 5 & -20 \end{array}\right);$$

iv. $F = -4$
$E = 10$
$D = -8$;

v. $x^2 + y^2 - 8x + 10y - 4 = 0$.

10-3 Problems

1. The equation of the circle is $x^2 + y^2 - 24x + 8y + 111 = 0$. Find the coordinates of the vertices of the inscribed square such that the diagonals are parallel to the axes.

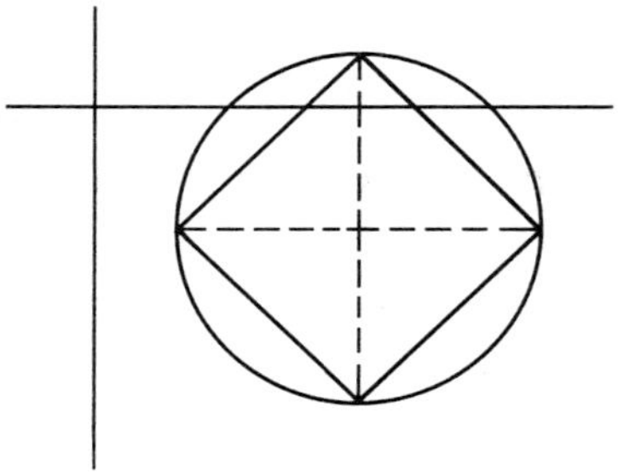

2. The equation of the circle is $x^2 + y^2 = 25$. The chord $\overline{P_1P_2}$ lies on the line whose equation is $y = (2/3)x - 4$. Find the midpoint of the chord.

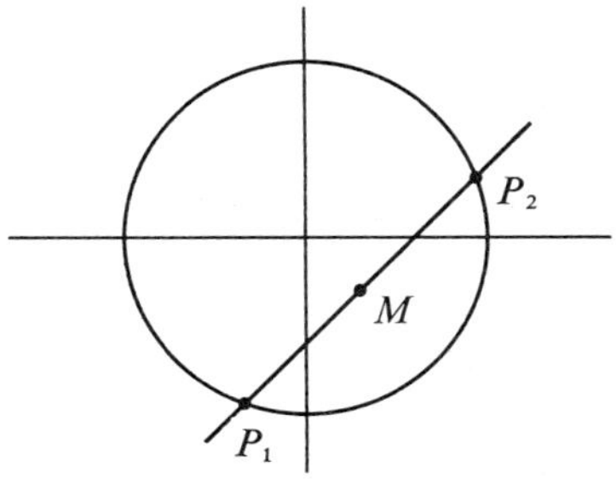

3. Given two equations of the form $x^2 + y^2 + Dx + Ey + F = 0$, how many times would their graphs intersect? Investigate the possibilities, and illustrate each with sketches of circles.
4. Do the two circles

$$x^2 + y^2 - 2x - 4y - 4 = 0$$

and

$$x^2 + y^2 - 6x - 8y + 9 = 0$$

intersect? If so, where?
5. Since longitudinal and transversal seismic waves travel at different rates, it is possible to determine the distance that a disturbance occurs from a particular station. Two stations are separated by 300 miles. Station A reports a disturbance 280 miles from its center, and station B reports the same disturbance 120 miles from its center. Where is the center of disturbance with relation to the two stations?
6. Three municipalities are located at points (12, 9), (−15, 0) and (−12, −20). What is the minimum range of a transmitter which is ideally located so that all three cities can receive transmissions of equal intensity?

10-4 Analytic Properties of Parabolas

In Chapter 3 we studied quadratic functions in the form $y = f(x) = ax^2 + bx + c$. Rewriting the coefficients as

$$-Ey = Ax^2 + Dx + F,$$

or

$$Ax^2 + Dx + Ey + F = 0, \qquad \text{(Eq. 10-4a)}$$

we see that the functions whose graphs we called *parabolas* in Chapter 3 belong to the same general class of equations introduced in the preceding section (Eq. 10-3c).

It is our purpose in this section to rediscover parabolas from another point of view.

If a point on a coordinate system moves so that it is equidistant from a fixed point, called a **focus**, and a fixed line, called a **directrix**, then the path of the moving point is a **parabola**. The methods of analytic geometry enable us to determine the equation of the locus of points.

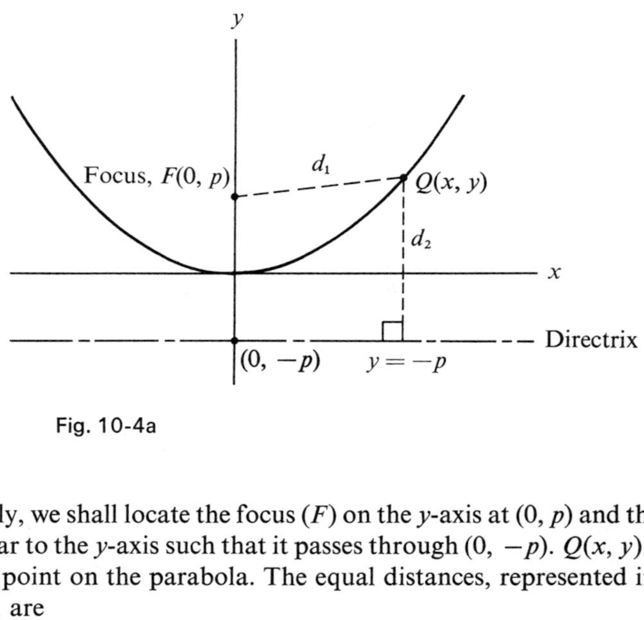

Fig. 10-4a

Initially, we shall locate the focus (F) on the y-axis at $(0, p)$ and the directrix perpendicular to the y-axis such that it passes through $(0, -p)$. $Q(x, y)$ represents any general point on the parabola. The equal distances, represented in terms of coordinates, are

$$d_1 = \sqrt{(x-0)^2 + (y-p)^2}$$

and

$$d_2 = |y + p|.$$

Since $d_1 = d_2$,

$$|y + p| = \sqrt{(x - 0)^2 + (y - p)^2}.$$

Squaring both sides and carrying out algebraic simplifications gives

$$y^2 + 2py + p^2 = x^2 + y^2 - 2py + p^2$$

$$x^2 - 4py = 0. \qquad \text{(Eq. 10-4b)}$$

Equation 10-4b is the most elementary form for an equation of a parabola because of the specific position chosen for the focus and directrix. Notice that the y-axis is the axis of symmetry and that the point midway between the focus and directrix, called the **vertex**, corresponds to the origin of the graph.

Example 1

Find the focus, vertex, and equation of the directrix of the parabola $2x^2 - 9y = 0$.

i. $2x^2 - 9y = 0$,

$$x^2 - \frac{9}{2}y = 0;$$

ii. $\therefore \quad -4p = -9/2$

$$p = 9/8;$$

iii. Vertex: $(0, 0)$,
Focus: $(0, 9/8)$,
Directrix: $y = -9/8$.

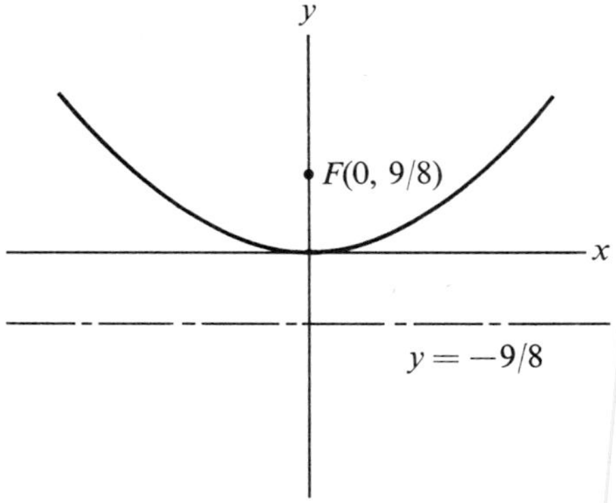

Example 2

Find the vertex, focus, and equation of the directrix of the parabola $y = -2/3x^2$.

i. $y = -\frac{2}{3}x^2$

$\frac{2}{3}x^2 + y = 0$

$x^2 + \frac{3}{2}y = 0;$

ii. $-4p = 3/2$

$p = -3/8;$

iii. Vertex: (0, 0),
Focus: (0, −3/8),
Directrix: $y = 3/8$.

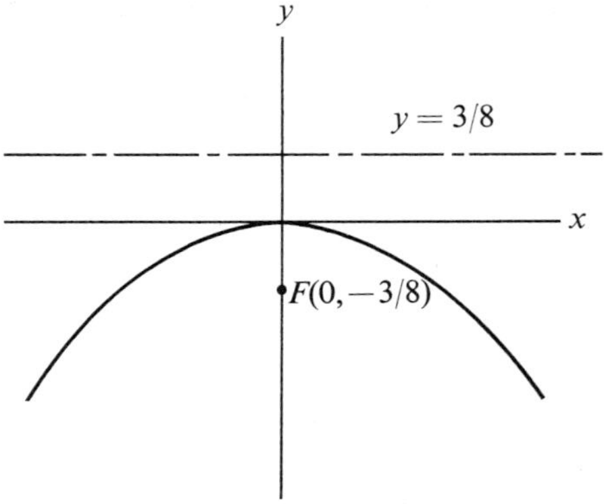

Notice that this parabola "opens downward."

Example 3

Write the equation of the parabola with focus (0, 6) and directrix $y = -6$.

Because of the positions of the focus and directrix, the equation is of the form given by Eq. 10-4b.

i. $p = 6;$

ii. therefore, $x^2 - 4(6)y = 0$ or $x^2 - 24y = 0$.

If the x and y of the equation of a parabola are interchanged, Eq. 10-4b becomes

$$y^2 - 4px = 0. \qquad \text{(Eq. 10-4c)}$$

The focus is on the x-axis at $(p, 0)$ and the equation of the directrix is $x = -p$. The parabola opens to the right if p is greater than zero, and to the left if p is less than zero.

Example 4

Find the focus, vertex, and equation of the directrix of the parabola $y^2 = 20x$.

i. $y^2 - 20x = 0$;

ii. $-4p = -20$ and $p = 5$;

iii. Vertex: $(0, 0)$,

Focus: $(5, 0)$,

Directrix: $x = -5$.

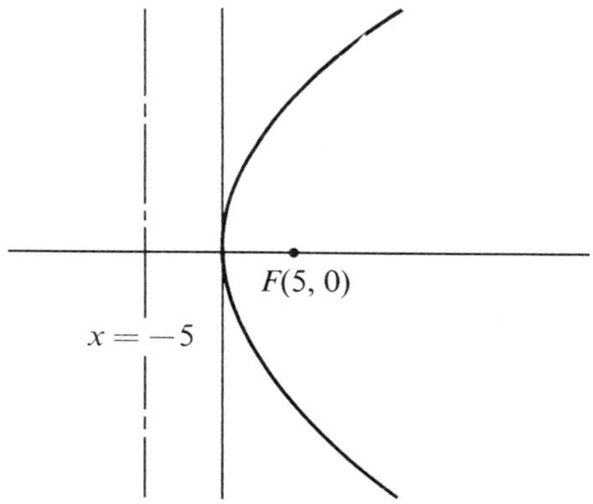

If, in Eq. 10-4a, D and F are unequal to zero, then the parabola has some ordinate other than $x = 0$ as an axis of symmetry and a vertex at some point other than $(0, 0)$. You should recall from Section 3-4 the methods for finding the axis of symmetry and vertex of a parabola.

Example 5

Find the vertex and axis of symmetry for the graph of $x^2 - 2x - y + 3 = 0$.

i. $x^2 - 2x - y + 3 = 0$ is equivalent to $y = x^2 - 2x + 3$.

ii. By Eq. 3-4b, the axis of symmetry is

$$\bar{x} = \frac{-b}{2a} = -\frac{(-2)}{2(1)} = 1.$$

iii. One is the abscissa of the vertex and

$$y = (1)^2 - 2(1) + 3 = 2$$

is the ordinate.

The problem of locating the focus and directrix for the equation given in Example 5 is a bit more complicated but may be done by **translating** the coordinate axis.

Figure 10-4b represents the results of Example 5. If x^* and y^* are considered the coordinate axes for the parabola, then $y = x^2 - 2x + 3$ in the (x, y) system will have the form $(x^*)^2 - 4py^* = 0$ in the (x^*, y^*) system, and the constant p will tell us where the focus and directrix are in the (x^*, y^*) system.

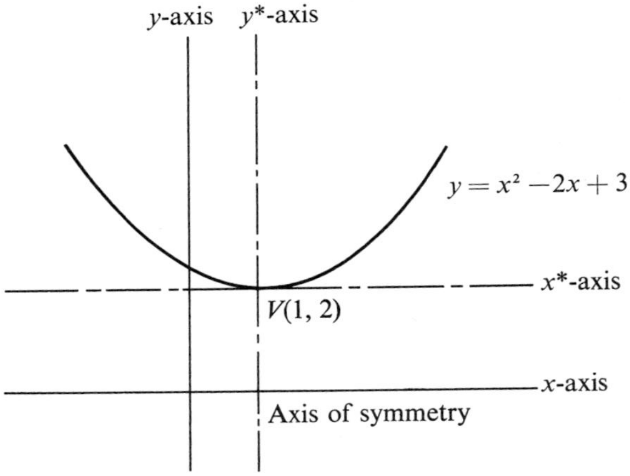

Fig. 10-4b

Figure 10-4c shows us how the points in both systems are generally related.

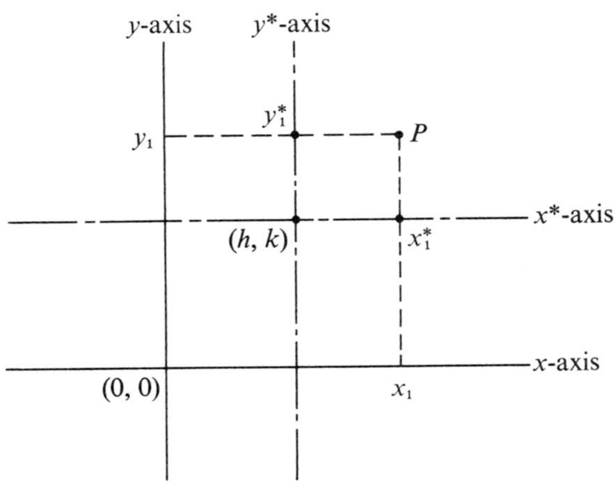

Fig. 10-4c

For every general point P, we have

$$x_1 = h + x_1^*,$$
$$y_1 = k + y_1^*. \qquad \text{(Eqs. 10-4d)}$$

Returning to Fig. 10-4b, the origin (h, k) required for our *translated* coordinate system is $(1, 2)$; in other words, $h = 1$ and $k = 2$.

Example 6

Find the focus and equation of the directrix for Example 5. (Refer to Fig. 10-4b.)

i. $y = x^2 - 2x + 3$ in the (x, y) system.

ii. Substituting $1 + x^*$ for x and $2 + y^*$ for y,

$$2 + y^* = (1 + x^*)^2 - 2(1 + x^*) + 3 \text{ or}$$
$$y^* = \boxed{}.$$

iii. $y^* = (x^*)^2$ is equivalent to $(x^*)^2 - y^* = 0$.
Compared with Eq. 10-4b, this means that $-4p = -1$ or $p = 1/4$. So in the (x^*, y^*) system the focus is $(0, 1/4)$ and the directrix has the equation $y^* = -1/4$.

iv. In the (x, y) system the focus is $(1 + 0, 2 + 1/4)$ or $(1, 9/4)$ and the equation of the directrix is $y = 2 - 1/4$ or $y = 7/4$.

■ $y^* = (x^*)^2 + 2x^* + 1 - 2 - 2x^* + 3 - 2 = (x^*)^2$.

10-4 Exercises

1. Graph $x^2 - 8y = 0$ after finding the focus, vertex, and equation of the directrix.

---

2. Graph $y^2 = -x$ after finding the focus, vertex, and directrix.

■ 1. $-4p = -8$ or $p = 2$;

Focus: (0, 2),

Vertex: (0, 0),

Directrix: $y = -2$.

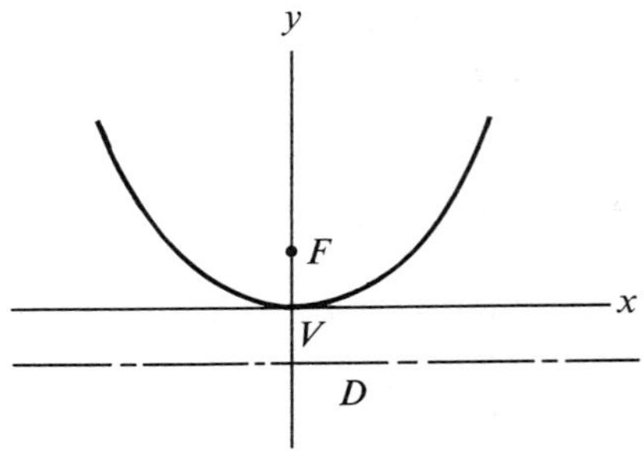

---

3. Graph $y^2 = 8x$ after finding the focus, vertex, and directrix.

■ 2. $y^2 + x = 0$

$\therefore\ -4p = 1$ or

$p = -1/4$;

Focus: $(-1/4, 0)$,

Vertex: (0, 0),

Directrix: $x = 1/4$.

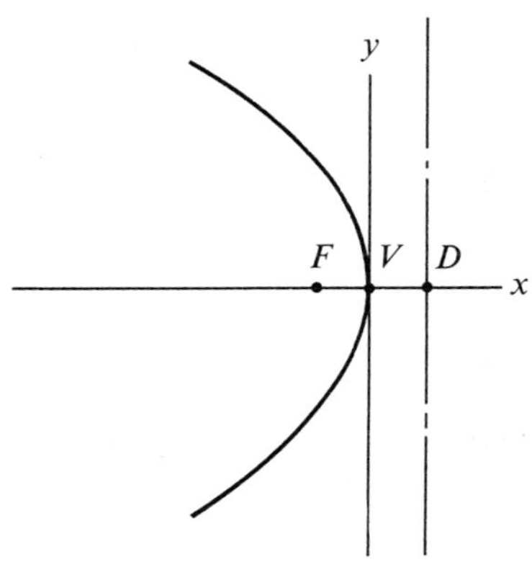

4. Graph $-y = 4x^2$ after finding the vertex, focus, and directrix.

■ 3. $y^2 - 8x = 0$

$-4p = -8$ or $p = 2$;

Focus: (2, 0),
Vertex: (0, 0),
Directrix: $x = -2$.

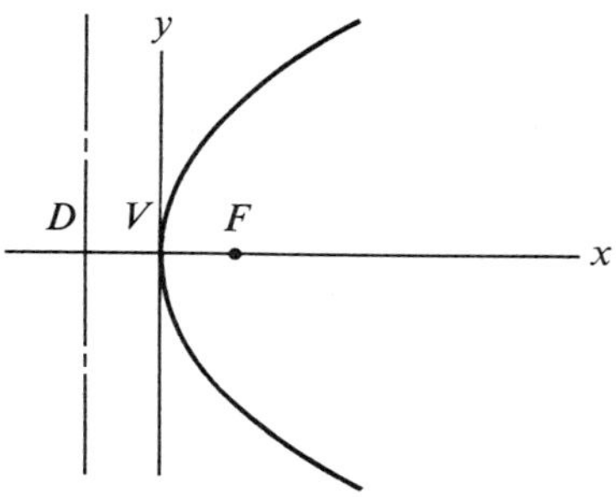

5. Write the equation for the parabola when the focus is $(-3, 0)$ and the directrix is $x = 3$.

■ 4. $4x^2 + y = 0$

$x^2 + \frac{1}{4} y = 0$

$-4p = \frac{1}{4}$ or $p = -\frac{1}{16}$;

Focus: (0, −1/16),
Vertex: (0, 0),
Directrix: $y = 1/16$.

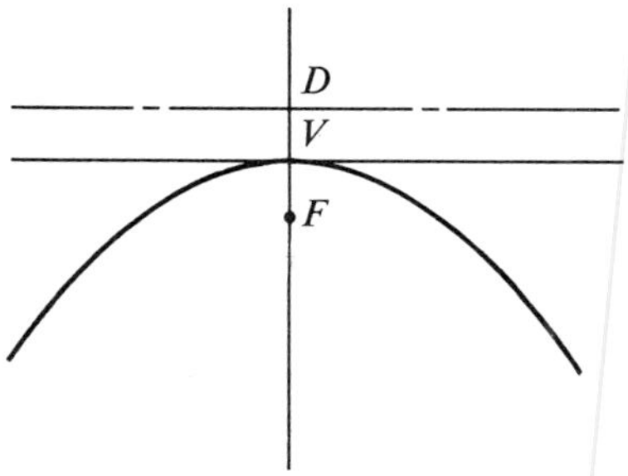

---

6. Write the equation for the parabola when the focus is (0, 2) and the directrix is $y = -2$.

■ 5. $y^2 - 4px = 0$ and $p = -3$.
Therefore
$y^2 - 4(-3)x = 0$ or $y^2 + 12x = 0$.

---

7. Write the equation for the parabola when the focus is (0, 3) and the vertex (0, 0).

■ 6. $x^2 - 4py = 0$ and $p = 2$.
Therefore $x^2 - 4(2)y = 0$ or $x^2 - 8y = 0$.

---

8. Find the equation of the parabola with vertex (0, 0) that passes through point (12, 15).

■ 7. $x^2 - 4py = 0$ and $p = 3$.
Therefore $x^2 - 4(3)y = 0$ or $x^2 - 12y = 0$.

■ 8. a. $x^2 - 4py = 0$ and $(12)^2 - 4p(15) = 0$.
Therefore $p = 12/5$ and $5x^2 - 48y = 0$ is one equation for the given parabola.

b. $y^2 - 4px = 0$ and $(15)^2 - 4p(12) = 0$.
Therefore $p = 75/16$ and $4y^2 - 75x = 0$ is another equation for the given parabola.

10-4 Problems

1. A cable hanging between two points of equal elevation approximates the shape of a parabola. If the points are 40 feet apart and the middle of the cable is one foot lower than the end points, find the equation of the cable's parabolic shape.
2. Making use of the equation in Problem 1, where would the cable be two feet above the lowest point?
3. If a cylindrical vessel is partly filled with a liquid and whirled about the axis of the cylinder, every plane through the axis cuts the surface of the liquid in a parabola. If the liquid is 8 inches up the side of a cylinder 12 inches in diameter, and the vertex is one inch from the bottom, how far is the vertex from the focus?

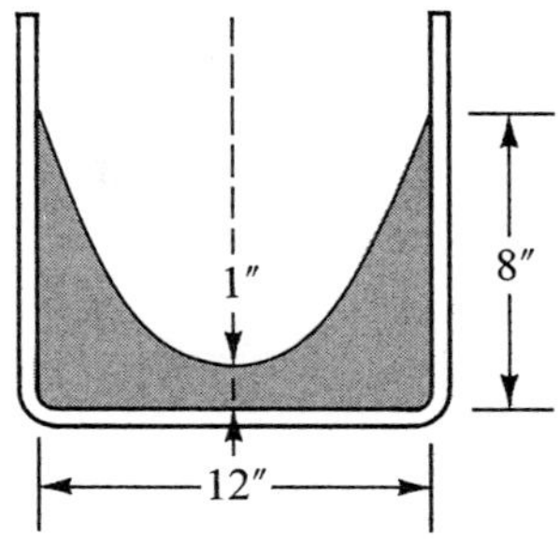

4. In reference to Problem 3, the faster the cylinder is whirled, the higher the liquid goes up the sides of the vessel and the nearer the vertex goes to the bottom. What, therefore, is the effect of speed on the distance between the focus and vertex?

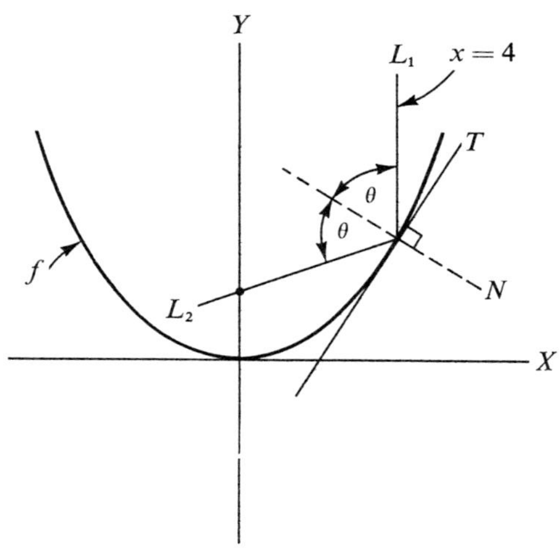

5. $$f(x) = \frac{1}{4}x^2, \qquad f'(x) = \frac{1}{2}x.$$

T is perpendicular to N and tangent to f.

a. Find the slope of L_2.
b. Find the y-intercept of the equation of L_2.
c. Find the coordinates of the focus of the graph of the function f.

Note: The results of Problem 5 illustrate a very useful property of parabolic curves. If L_1 and L_2 are incident and reflected rays of wave energy, then waves of energy leaving the focus of the parabola all reflect parallel to the axis of symmetry.

6. Find the focus and equation of the directrix for the parabola $x^2 - 5x - y = -2$.

10-5 Ellipses

An **ellipse** is a plane curve that results from the movement of a point in such a way that the *sum* of its *distances* from *two fixed points is constant*.

Determining the equation of an ellipse is simplified if we elect to place the two fixed points, called **foci**, on one axis of the coordinate system so that the origin (0, 0) is midway between them. If the foci are located at $(c, 0)$ and $(-c, 0)$ and the constant sum of the two distances, d_1 and d_2, is $2a$, then $d_1 + d_2 = 2a$,

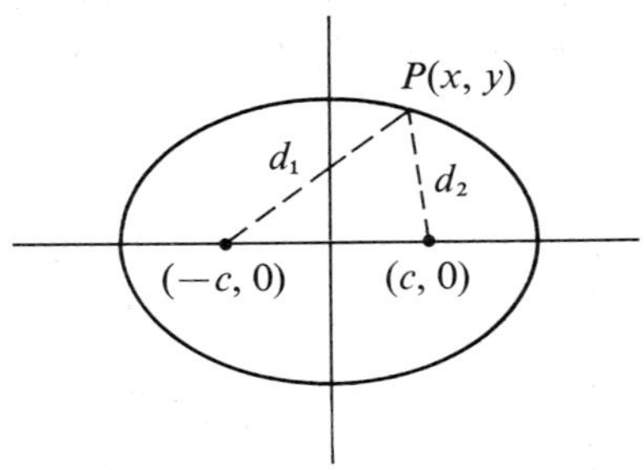

Fig. 10-5a

$a > 0$. The distances in terms of coordinates are

$$d_1 = \sqrt{(x + c)^2 + y^2}$$

and

$$d_2 = \sqrt{(x - c)^2 + y^2}.$$

Therefore,

$$\sqrt{(x + c)^2 + y^2} + \sqrt{(x - c)^2 + y^2} = 2a.$$

To simplify the preceding equation, we proceed in the manner developed in Section 4-4.

$$\sqrt{(x - c)^2 + y^2} = 2a - \sqrt{(x + c)^2 + y^2}$$

$$(x - c)^2 + y^2 = 4a^2 - 4a\sqrt{(x + c)^2 + y^2} + (x + c)^2 + y^2$$

$$x^2 - 2cx + c^2 + y^2 = 4a^2 - 4a\sqrt{(x + c)^2 + y^2} + x^2 + 2cx + c^2 + y^2$$

$$a^2 + cx = a\sqrt{(x+c)^2 + y^2}$$

$$a^4 + 2a^2cx + c^2x^2 = a^2(x^2 + 2cx + c^2 + y^2)$$

$$a^4 + 2a^2cx + c^2x^2 = a^2x^2 + 2a^2cx + a^2c^2 + a^2y^2$$

$$a^4 - a^2c^2 = a^2x^2 - c^2x^2 + a^2y^2$$

$$(a^2 - c^2)a^2 = (a^2 - c^2)x^2 + a^2y^2. \qquad \text{(Eq. 10-5a)}$$

The final steps in the simplification require a little more than simple algebraic manipulation. We would like to substitute b^2 for the constant difference $a^2 - c^2$. This is permissible since $d_1 + d_2 = 2a$ must be greater than the distance between foci. In other words,

$$2a > |c - (-c)| = 2|c|,$$

or

$$a^2 > c^2,$$

which is equivalent to saying

$$a^2 - c^2 > 0.$$

Therefore, the positive constant b^2 may be substituted for $a^2 - c^2$. Finally,

$$b^2a^2 = b^2x^2 + a^2y^2$$

or

$$1 = \frac{x^2}{a^2} + \frac{y^2}{b^2}. \qquad \text{(Eq. 10-5b)}$$

By examining both Fig. 10-5a and Eq. 10-5b, we can make the following observations:

1. The x and y axes are both lines of symmetry.
2. When $y = 0$, $x = \pm a$. The points $V_1(-a, 0)$ and $V_2(a, 0)$ are called **vertices**. The line segment $\overline{V_1V_2}$ is called the **major axis** and has the length $|a - (-a)| = 2a$.

3. When $x = 0$, $y = \pm b$. $P_1(0, b)$ and $P_2(0, -b)$ are the end points of the **minor axis.** The length of the minor axis is $|b - (-b)| = 2|b|$.
4. Since, from previous discussion, $a^2 - c^2 = b^2$, we can conclude that $a^2 > b^2$. This tells us that the major axis is always greater in length that the minor axis. Notice that the major axis contains the focus points.
5. In finding solutions to Eq. 10-5b, real values of y correspond only to x's in the interval $-a \leq x \leq a$.

Figure 10-5b graphically summarizes some of the preceding observations.

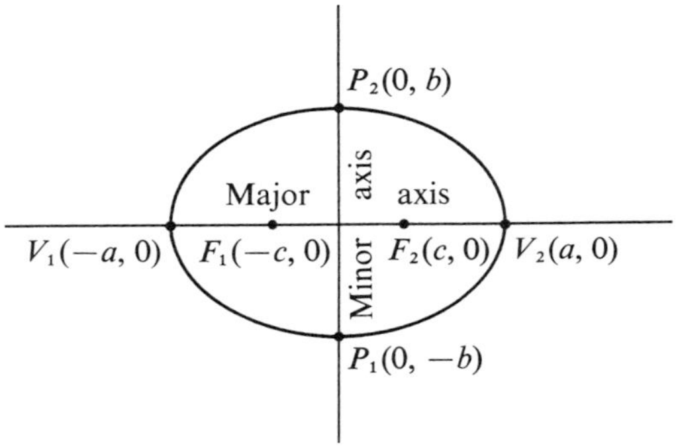

Fig. 10-5b

Example 1

Find the equation of the ellipse that has a major axis of length 10 and foci at $F_1(-3, 0)$ and $F_2(3, 0)$.

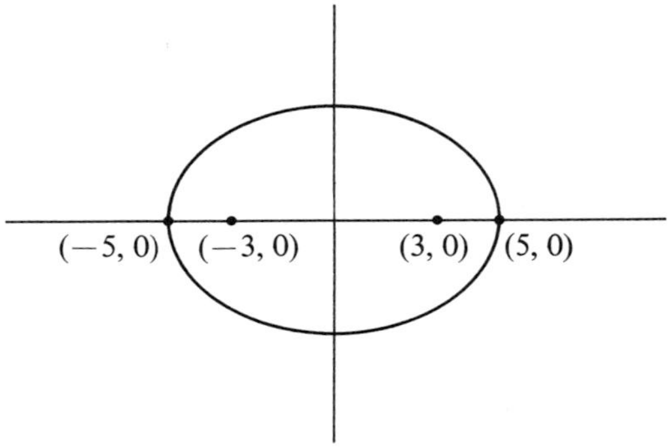

i. $2a = 10$
$a^2 = 25.$

ii. $c = 3$
$c^2 = 9.$

iii. $b^2 = a^2 - c^2$
$= 25 - 9 = 16.$

iv. $\frac{x^2}{a^2} + \frac{y^2}{b^2} = 1$

$\frac{x^2}{25} + \frac{y^2}{16} = 1,$ or $16x^2 + 25y^2 - 400 = 0.$

If x and y are interchanged in Eq. 10-5b, so that

$$\frac{y^2}{a^2} + \frac{x^2}{b^2} = 1, \qquad \text{(Eq. 10-5c)}$$

then the major axis with the foci will be positioned on the y-axis.

Example 2

If the foci of an ellipse are at (0, 2) and (0, −2) and the ellipse passes through (3, 0), what is its equation?

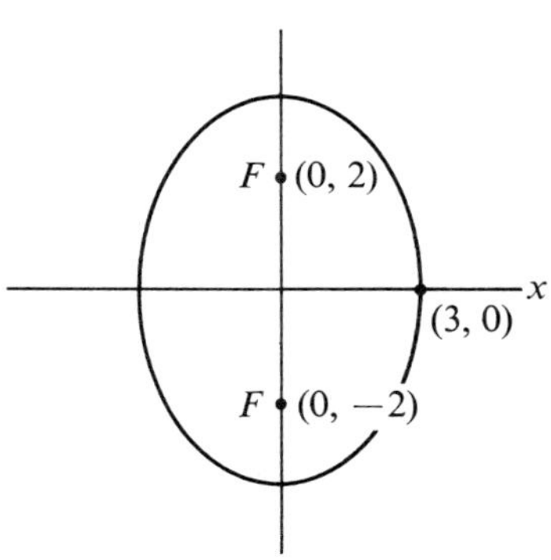

i. $c = 2, \quad c^2 = 4;$

ii. $b = 3, \quad b^2 = 9;$

iii. $a^2 - c^2 = b^2$
$a^2 = b^2 + c^2 = 9 + 4 = 13;$

iv. $\dfrac{y^2}{a^2} + \dfrac{x^2}{b^2} = 1$

$\dfrac{y^2}{13} + \dfrac{x^2}{9} = 1, \quad \text{or} \quad 13x^2 + 9y^2 - 117 = 0.$

10-5 Exercises

1. Graph

$$\frac{x^2}{10} + \frac{y^2}{5} = 1$$

after first finding the foci, vertices, and end points of the minor axis.

2. Graph $\dfrac{x^2}{9} + \dfrac{y^2}{16} = 1$ after first finding the foci, vertices, and end points of the minor axis.

■ 1. i. $a^2 = 10, \quad b^2 = 5,$
$a^2 - c^2 = b^2,$
$c^2 = a^2 - b^2 = 5.$

ii. $F_1(-\sqrt{5}, 0),$
$F_2(\sqrt{5}, 0);$
$V_1(-\sqrt{10}, 0),$
$V_2(\sqrt{10}, 0);$
$P_1(0, -\sqrt{5}),$
$P_2(0, \sqrt{5}).$

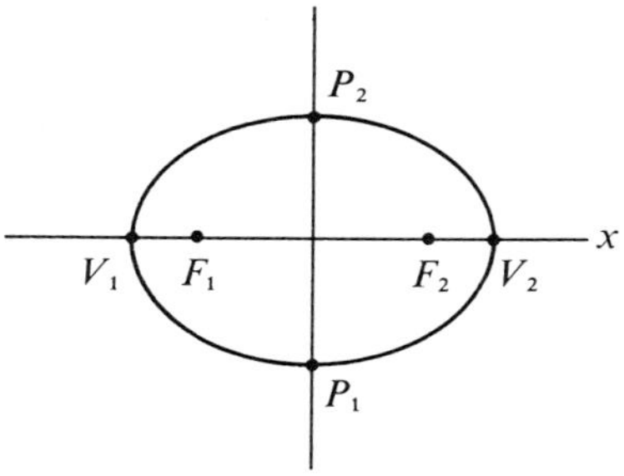

---

3. Graph
$$8x^2 + 2y^2 - 32 = 0$$
after determining the lengths of the major and minor axes and the position of the foci.

■ 2. i. $a^2 = 16$,
$b^2 = 9$,
$c^2 = a^2 - b^2 = 7$.

ii. $F_1(0, \sqrt{7})$, $F_2(0, -\sqrt{7})$;
$V_1(0, 4)$, $V_2(0, -4)$;
$P_1(-3, 0)$, $P_2(3\ 0)$.

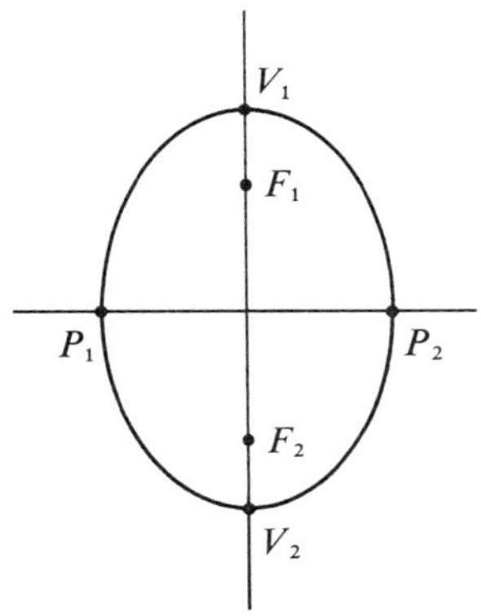

---

4. Graph
$$2x^2 + 5y^2 = 10$$
after determining the lengths of the axes and the position of the foci.

■ 3. i. $8x^2 + 2y^2 = 32$,
$$\frac{x^2}{4} + \frac{y^2}{16} = 1.$$

ii. $a^2 = 16$, major axis: 8;
$b^2 = 4$, minor axis: 4;
$c^2 = a^2 - b^2 = 12$.

iii. $F_1(0, 2\sqrt{3})$,
$F_2(0, -2\sqrt{3})$;
$V_1(0, 4)$, $V_2(0, -4)$;
$P_1(-2, 0)$, $P_2(2, 0)$.

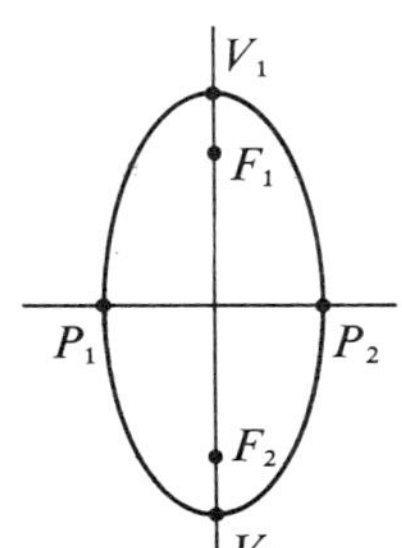

5. Find the equation of the ellipse that has symmetry about the coordinate axes, one focus at (3, 0) and a major axis of length 8.

■ 4. i. $2x^2 + 5y^2 = 10$,

$$\frac{x^2}{5} + \frac{y^2}{2} = 1.$$

ii. $a^2 = 5$,

major axis: $2\sqrt{5}$;

$b^2 = 2$,

minor axis: $2\sqrt{2}$;

$c^2 = a^2 - b^2 = 3$.

iii. $F_1(-\sqrt{3}, 0)$, $F_2(\sqrt{3}, 0)$;

$V_1(-\sqrt{5}, 0)$, $V_2(\sqrt{5}, 0)$;

$P_1(0, -\sqrt{2})$, $P_2(0, \sqrt{2})$.

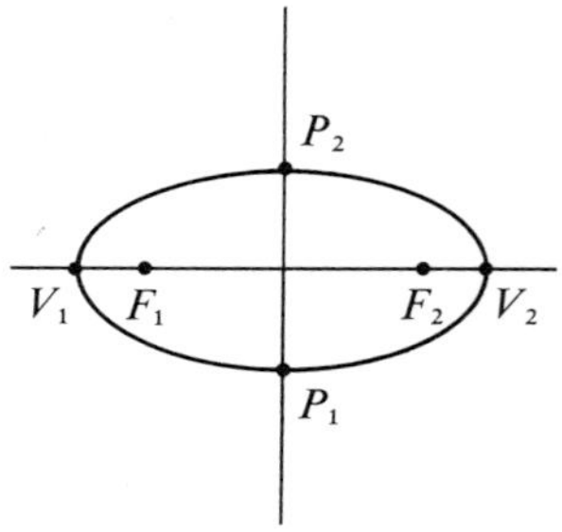

6. If a vertex is at the point (10, 0) and a focus is at the point (6, 0) of an ellipse, what is the equation of the ellipse when it is symmetrical about the coordinate axes?

■ 5.

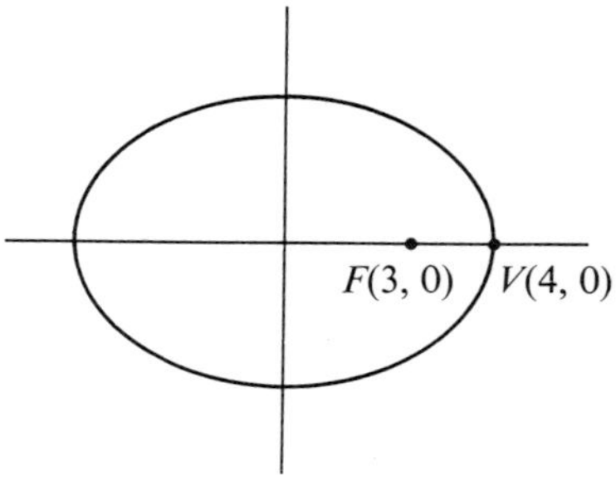

i. $a = 4$, $a^2 = 16$;

$c = 3$, $c^2 = 9$;

$b^2 = a^2 - c^2 = 7$.

ii. $\dfrac{x^2}{a^2} + \dfrac{y^2}{b^2} = 1,$

$\dfrac{x^2}{16} + \dfrac{y^2}{7} = 1.$

---

7. If an ellipse has symmetry about the coordinate system and passes through (4, 1) and (−2, 2), what is its equation?

■ 6.

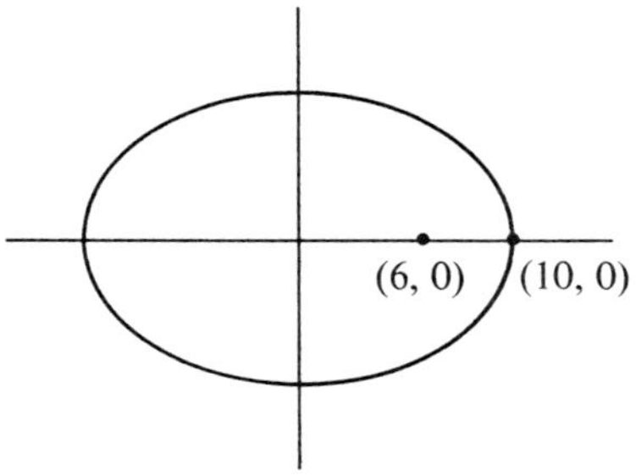

i. $a = 10, \quad a^2 = 100;$

$c = 6, \quad c^2 = 36;$

$b^2 = a^2 - c^2 = 64.$

ii. $\dfrac{x^2}{a^2} + \dfrac{y^2}{b^2} = 1,$

$\dfrac{x^2}{100} + \dfrac{y^2}{64} = 1.$

■ 7.

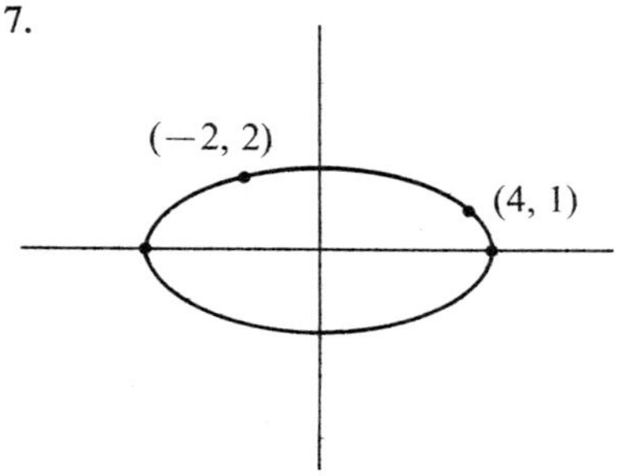

i. $\dfrac{x^2}{a^2}+\dfrac{y^2}{b^2}=1;$

ii. $\left.\begin{aligned}\dfrac{4}{a^2}+\dfrac{4}{b^2}&=1\\ \dfrac{16}{a^2}+\dfrac{1}{b^2}&=1\end{aligned}\right]$ System of equations;

iii. $\left.\begin{aligned}\dfrac{16}{a^2}+\dfrac{16}{b^2}&=4\\ \dfrac{16}{a^2}+\dfrac{1}{b^2}&=1\end{aligned}\right];$

iv. $\therefore \dfrac{15}{b^2}=3, \quad b^2=5;$

and $\dfrac{4}{a^2}+\dfrac{4}{5}=1, \quad a^2=20;$

v. $\dfrac{x^2}{20}+\dfrac{y^2}{5}=1.$

10-5 Problems

1. An elliptical table top is to be manufactured so that the major axis is 48 inches and the minor axis is 36 inches. The pattern is to be drawn by a method that requires knowing the position of the foci and using a taut string in the manner illustrated.

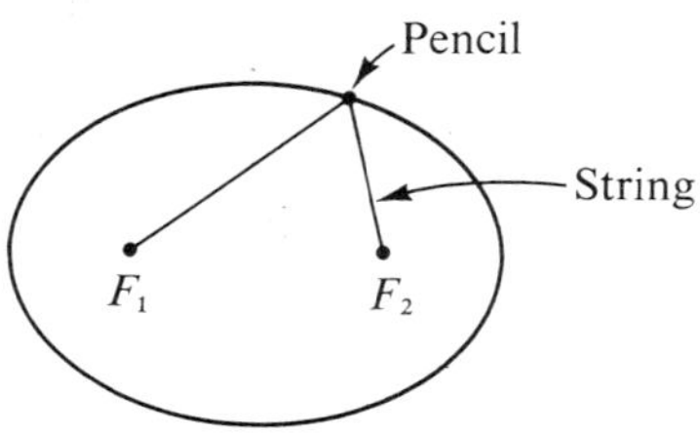

Determine the positions of the foci and the proper length of string.

2. Table tops of a shape *similar* to that in Problem 1 are to be manufactured in slightly different sizes. Calculate the positions of the foci and the string length for a table that has only 3/4 the area of the table in Problem 1. What are the principal dimensions?
3. The top view of the retaining side of a dam is in the shape of half an ellipse.

What is the equation of the ellipse? (Fig. 10-5c)

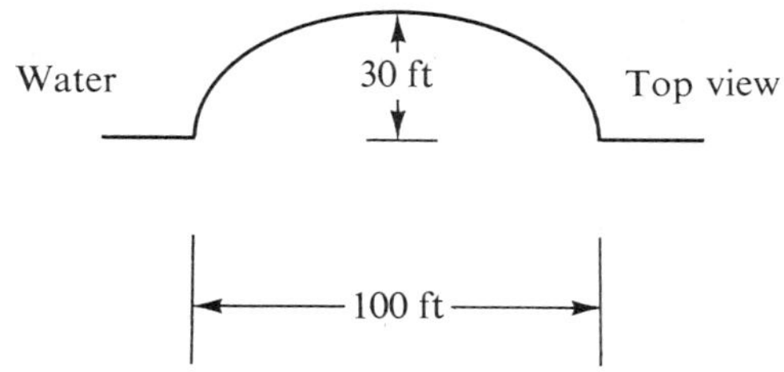

Fig. 10-5c

4. Earth satellites assume elliptical orbits with the earth's center at one focus. A Soviet communications satellite, Molniya 1D, when launched in 1966, had a perigee of 300 miles above the earth's surface and an apogee of 24,670 miles above the earth's surface. Prepare a scale drawing of the earth and one of the satellite's orbits. What are the dimensions of its major and minor axes?

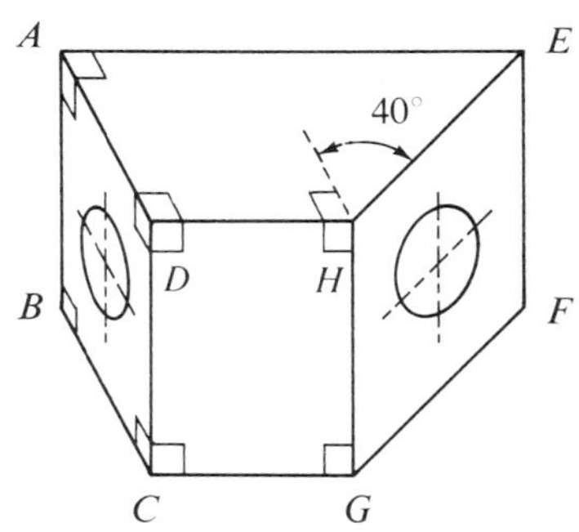

5. A circular hole 0.75 inches in diameter is drilled through the block perpendicular to plane *ABCD*. The plane *EFGH* intersects the hole in the form of an ellipse. Calculate all the principal measurements of the ellipse and determine its equation assuming the intersection of the major and minor axes is the origin.

10-6 Hyperbolas

Like the ellipse, the curve of a hyperbola begins with two fixed points called foci. The hyperbola is the locus of points in a plane whose distances from the foci have a *constant difference*.

Let us consider briefly the concept of constant difference. Consider the variable distances $|d_1|$ and $|d_2|$ that have a constant difference of $2a > 0$. Four of an unlimited number of possibilities are illustrated in Fig. 10-6a. As we view Fig. 10-6a, it should be apparent that at least one of the distances, $|d_1|$ or $|d_2|$, must be greater than $2a$.

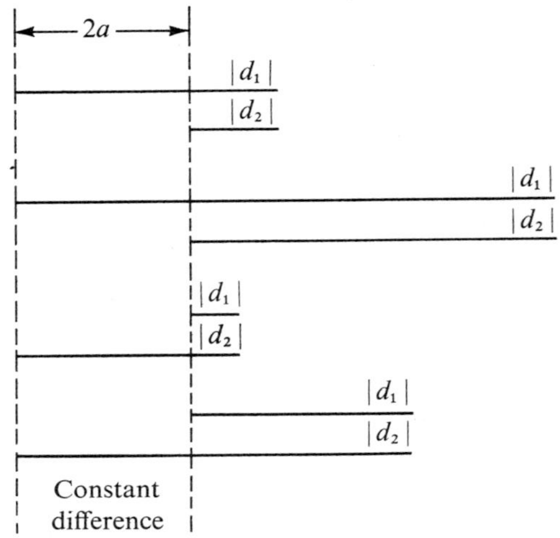

Fig. 10-6a

Figure 10-6b shows the general appearance of a hyperbola when the two

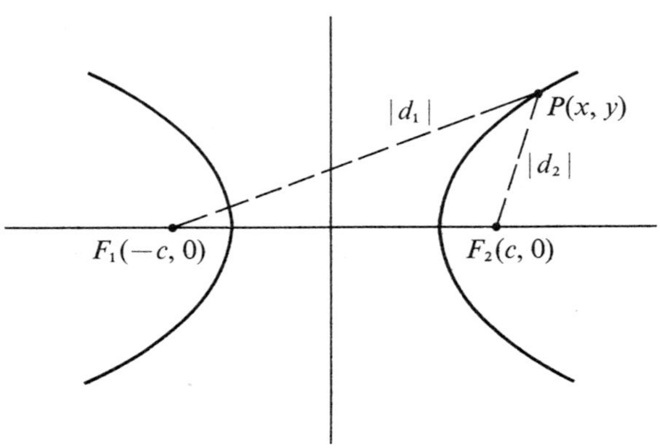

Fig. 10-6b

fixed points, F_1 and F_2, called foci, are placed on the x-axis at $(-c, 0)$ and $(c, 0)$, and $\|d_1| - |d_2\| = 2a$. Unlike the other plane curves of this chapter, the hyperbola has two unattached parts.

In order to determine the equation of the hyperbola, we must state in terms of coordinates the condition that

$$||d_1| - |d_2|| = 2a.$$

Since subtraction is an ordered operation (not commutative), we need to use care in dealing with absolute differences. The outer absolute value sign can be eliminated by assuming $|d_1| > |d_2|$, which appears to be true only for the part of the graph on the right side of the y-axis. It happens that if we assumed $|d_2| > |d_1|$, the final results are unaltered.

The two distances written in terms of their coordinates are (referring to Fig. 10-6b)

$$|d_1| = \sqrt{(x + c)^2 + y^2}$$

and

$$|d_2| = \sqrt{(x - c)^2 + y^2}.$$

Therefore

$$\begin{aligned}
\sqrt{(x + c)^2 + y^2} - \sqrt{(x - c)^2 + y^2} &= 2a \\
\sqrt{(x + c)^2 + y^2} &= 2a + \sqrt{(x - c)^2 + y^2} \\
x^2 + 2cx + c^2 + y^2 &= 4a^2 + 4a\sqrt{(x - c)^2 + y^2} \\
&\quad + x^2 - 2cx + c^2 + y^2 \\
4cx - 4a^2 &= 4a\sqrt{(x - c)^2 + y^2} \\
cx - a^2 &= a\sqrt{(x - c)^2 + y^2} \\
c^2x^2 - 2a^2cx + a^4 &= a^2(x^2 - 2cx + c^2 + y^2) \\
c^2x^2 - 2a^2cx + a^4 &= a^2x^2 - 2a^2cx + a^2c^2 + a^2y^2 \\
c^2x^2 - a^2x^2 - a^2y^2 &= a^2c^2 - a^4 \\
(c^2 - a^2)x^2 - a^2y^2 &= a^2(c^2 - a^2). \qquad \text{(Eq. 10-6a)}
\end{aligned}$$

The final steps in the simplification are made by letting $b^2 = c^2 - a^2$. This is clearly permissible when we observe that by letting $y = 0$ in Eq. 10-6a, $x^2 = a^2$ or $x = \pm a$. These x-intercepts are called the **vertices** of the hyperbola. Since $c > a$ or $c^2 > a^2$, which is equivalent to saying $c^2 - a^2 > 0$, it is permissible to represent $c^2 - a^2$ by the positive constant b^2. Making this substitution, we have

$$b^2x^2 - a^2y^2 = a^2b^2,$$

or

$$\frac{x^2}{a^2} - \frac{y^2}{b^2} = 1. \qquad \text{(Eq. 10-6b)}$$

Example 1

Write the equation of the hyperbola that passes through $(-1, 0)$ and $(1, 0)$ and has foci located at $(-3, 0)$ and $(3, 0)$.

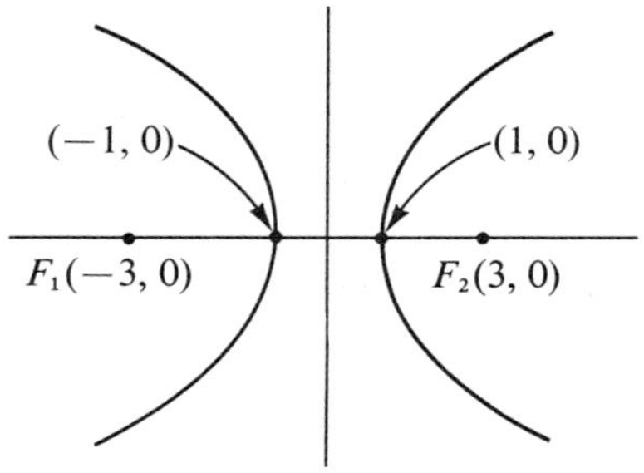

Fig. 10-6c

i. $c = 3, \quad c^2 = 9;$
$a = 1, \quad a^2 = 1;$
$c^2 - a^2 = b^2 = 8.$

ii. $\dfrac{x^2}{a^2} - \dfrac{y^2}{b^2} = 1$

$\dfrac{x^2}{1} - \dfrac{y^2}{8} = 1, \quad \text{or} \quad 8x^2 - y^2 - 8 = 0.$

In the process of graphing hyperbolas, it is helpful to know the positions of the vertices and focus points. Greater assistance comes from knowing the location of two straight line graphs called **asymptotes** that are uniquely determined for each hyperbola. For *large* values of $|x|$ and correspondingly large values of $|y|$, the difference of one between x^2/a^2 and y^2/b^2 in Eq. 10-6b is very insignificant. If we replace the difference by zero, then

$$\frac{x^2}{a^2} - \frac{y^2}{b^2} = 0,$$

or

$$\left(\frac{x}{a}+\frac{y}{b}\right)\cdot\left(\frac{x}{a}-\frac{y}{b}\right)=0. \qquad \text{(Eq. 10-6c)}$$

The graph of the solution set for Eq. 10-6c is equivalent to the graphs of the system

$$\frac{x}{a}+\frac{y}{b}=0,$$

$$\frac{x}{a}-\frac{y}{b}=0,$$

or

$$y=-\frac{b}{a}x, \qquad \text{(Eq. 10-6d)}$$

$$y=\frac{b}{a}x. \qquad \text{(Eq. 10-6e)}$$

Furthermore, the graphs of Eqs. 10-6d and 10-6e approximate the graph of the hyperbola from which they are derived.

Example 2

Graph the hyperbola $12x^2-3y^2-4=0$ using, as aids, the locations of the vertices and asymptotes.

i. $12x^2-3y^2-4=0$

$$\frac{12x^2}{4}-\frac{3y^2}{4}=1$$

$$\frac{x^2}{1/3}-\frac{y^2}{4/3}=1;$$

ii. $a^2=1/3$

$a=\pm\sqrt{3}/3\approx\pm.577$

Vertices: $(\pm.577, 0)$;

iii. Asymptotes:

$$y=\frac{b}{a}x=\frac{\sqrt{4/3}}{\sqrt{1/3}}x=2x.$$

$$y=-\frac{b}{a}x=\boxed{}.$$

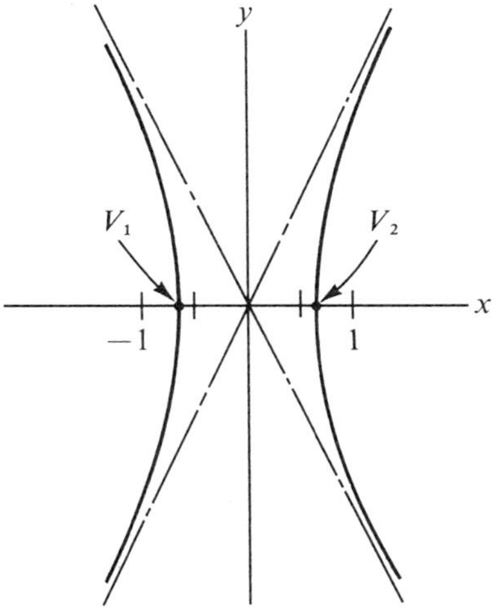

Fig. 10-6d

■ $-2x$

Interchanging the x and y in Eq. 10-6b gives the equation

$$\frac{y^2}{a^2} - \frac{x^2}{b^2} = 1. \qquad \text{(Eq. 10-6f)}$$

The effect of making this change is a ninety-degree rotation of Fig. 10-6b about the origin of the coordinate system.

Example 3

Sketch the graph of

$$\frac{y^2}{4} - \frac{x^2}{9} = 1.$$

i. The vertices are (0, 2) and (0, −2).

ii. The asymptotes result from letting $\frac{y^2}{4} - \frac{x^2}{9} = 0$,

so that $\frac{y}{2} + \frac{x}{3} = 0$, or $y = -\frac{2}{3}x$

and $\frac{y}{2} - \frac{x}{3} = 0$, or $y = \frac{2}{3}x$.

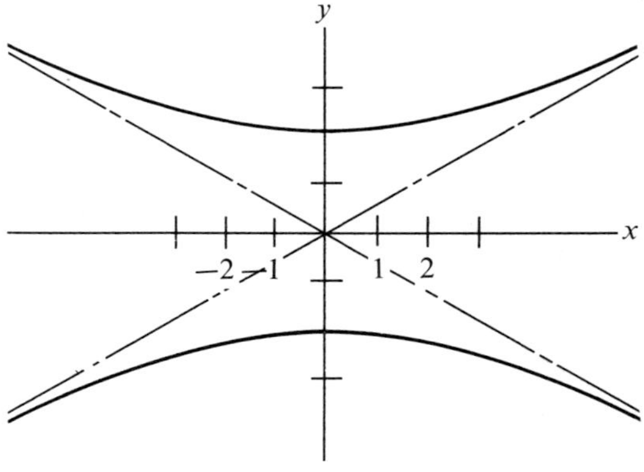

Fig. 10-6e

Notice that for Eq. 10-6f,

$$y = \frac{a}{b}x \quad \text{and} \quad y = -\frac{a}{b}x$$

are the asymptotes.

10-6 Exercises

1. Graph

$$\frac{x^2}{16} - \frac{y^2}{25} = 1$$

by finding the vertices and plotting the asymptotes.

---

2. Graph the hyperbola $4x^2 - y^2 = 40$ by finding the vertices and asymptotes.

■ 1. $a^2 = 16, \quad a = 4;$
$b^2 = 25, \quad b = 5;$
Vertices: $(-4, 0), (4, 0);$
Asymptotes:

$$y = \frac{b}{a}x = \frac{5}{4}x,$$

$$y = -\frac{b}{a}x = -\frac{5}{4}x.$$

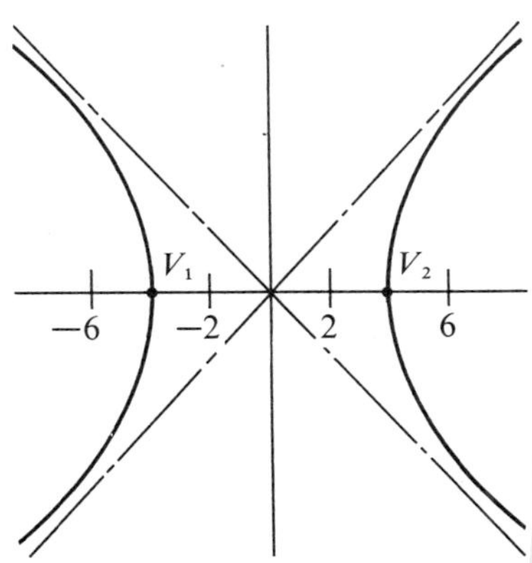

3. Sketch the graph of

$$\frac{y^2}{16} - \frac{x^2}{32} = 1;$$

illustrate the positions of the vertices, foci, and asymptotes.

■ 2. $4x^2 - y^2 = 40,$

$$\frac{x^2}{10} - \frac{y^2}{40} = 1;$$

$a^2 = 10, \quad a = \sqrt{10};$
$b^2 = 40, \quad b = 2\sqrt{10};$
Vertices: $(-\sqrt{10}, 0),$
$(\sqrt{10}, 0);$

Asymptotes:

$$y = \frac{b}{a}x = \frac{2\sqrt{10}}{\sqrt{10}}x = 2x,$$

$$y = -\frac{b}{a}x = -2x.$$

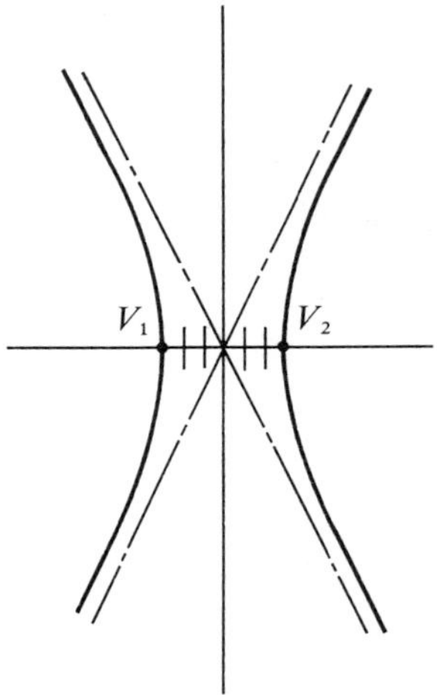

4. Write the equation of the hyperbola, given the following:

Foci: $(-\sqrt{6}, 0), (\sqrt{6}, 0);$
Vertices: $(-\sqrt{2}, 0), (\sqrt{2}, 0).$

■ 3. $\frac{y^2}{16} - \frac{x^2}{32} = 1$

$a^2 = 16, \quad a = 4;$

$b^2 = 32, \quad b = 4\sqrt{2} \approx 5.66;$

$c^2 = a^2 + b^2$

$= 48, \; c = 4\sqrt{3} \approx 6.93;$

$$y = \frac{a}{b}x = \frac{4}{4\sqrt{2}}\,x \approx .71x;$$

$$y = -\frac{a}{b}\,x \approx -.71x.$$

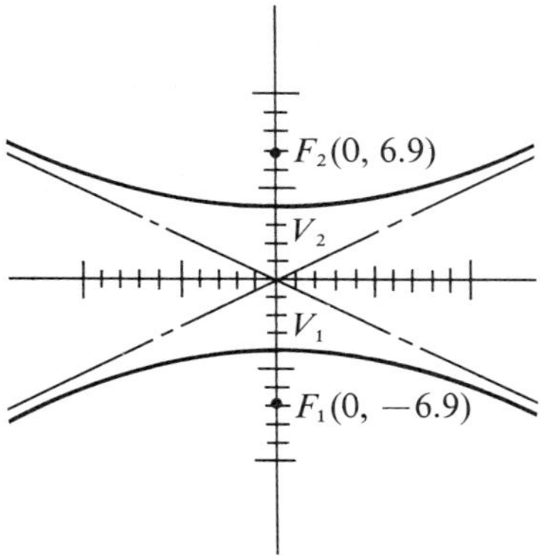

---

5. Write the equation of the following hyperbola:

Vertices:

$(0, -\sqrt{5}),\quad (0, \sqrt{5});$

Asymptotes:

$$y = \frac{\sqrt{2}}{2}\,x,$$

$$y = -\frac{\sqrt{2}}{2}\,x.$$

■ 4. $a = \sqrt{2},\quad a^2 = 2;$
$c = \sqrt{6},\quad c^2 = 6;$
$b^2 = c^2 - a^2 = 4;$

$$\therefore \frac{x^2}{2} - \frac{y^2}{4} = 1.$$

---

6. Find the equation of the hyperbola which passes through points $P_1(3\sqrt{2}, 2)$ and $P_2(2\sqrt{3}, 2\sqrt{3}/3)$ and has symmetry about the coordinate system.

■ 5. i. $a = \sqrt{5},\quad a^2 = 5;$

ii. $\dfrac{a}{b} = \dfrac{\sqrt{2}}{2}$

$$b = \frac{2a}{\sqrt{2}}$$

$$b^2 = \frac{4a^2}{2} = 2a^2 = 10;$$

iii. $\dfrac{y^2}{5} - \dfrac{x^2}{10} = 1.$

7. The graph of the equation $xy=9$ is a hyperbola whose axes of symmetry have the equations $y=x$ and $y=-x$.

   a. Graph $xy=9$.

   b. Find the vertices.

   c. What do you think are the equations of the asymptotes?

■ 6. i.

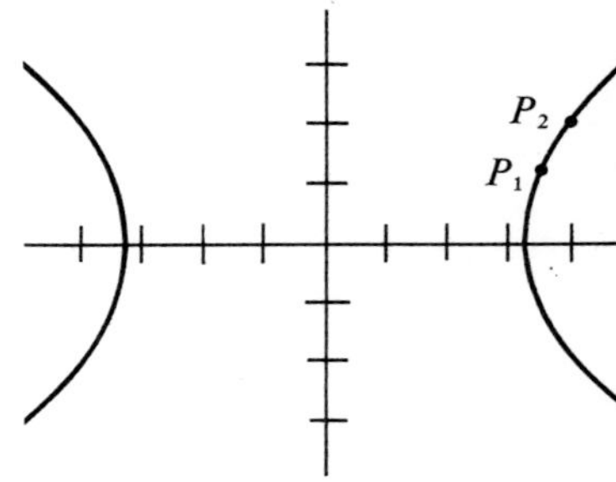

$$\frac{x^2}{a^2}-\frac{y^2}{b^2}=1;$$

ii. $$\left.\begin{array}{l}\dfrac{(3\sqrt{2})^2}{a^2}-\dfrac{2^2}{b^2}=1\\[2ex]\dfrac{(2\sqrt{3})^2}{a^2}-\dfrac{(2\sqrt{3}/3)^2}{b^2}=1\end{array}\right]$$

System of equations,

$$\left.\begin{array}{l}\dfrac{18}{a^2}-\dfrac{4}{b^2}=1\\[2ex]\dfrac{12}{a^2}-\dfrac{4}{3b^2}=1\end{array}\right],$$

$$\left.\begin{array}{l}\dfrac{6}{a^2}-\dfrac{4}{3b^2}=\dfrac{1}{3}\\[2ex]\dfrac{12}{a^2}-\dfrac{4}{3b^2}=1\end{array}\right],$$

$$\therefore \frac{12}{a^2}-\frac{6}{a^2}=1-\frac{1}{3}$$

$$a^2=9,$$

and

$$\frac{18}{9}-\frac{4}{b^2}=1$$

$$b^2=4;$$

iii. $$\frac{x^2}{9}-\frac{y^2}{4}=1.$$

■ 7. a.

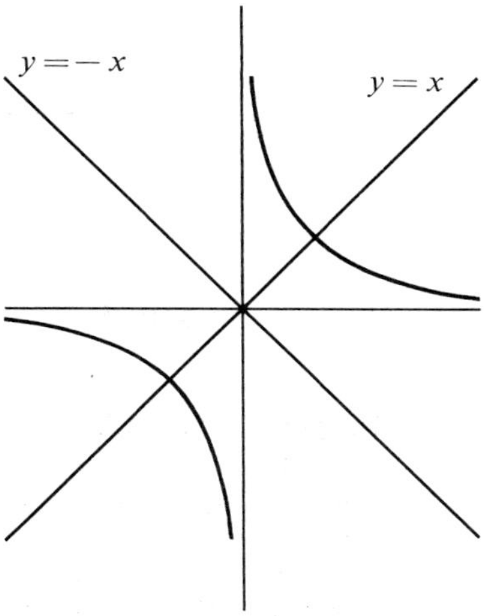

b. $\left.\begin{array}{r} y=x \\ xy=9 \end{array}\right]$ System.

Therefore $x^2=9$ or $x=\pm 3$ and $y=\pm 3$. Consequently the vertices are $(-3, -3)$ and $(3, 3)$.

c. The asymptotes are the axes of the coordinate system.

10-6 Problems

1. The equation $\frac{x^2}{16}-\frac{y^2}{16}=1$ in the (X, Y) coordinate system has the form $x^*y^*=k$ in the (X^*, Y^*) system (Fig. 10-6f) when the angle of rotation of the (X^*, Y^*) system is 45°. The equations relating the coordinates of a point in both systems are

$$x=x^* \cos\theta - y^* \sin\theta$$

and

$$y=x^* \sin\theta + y^* \cos\theta.$$

Find the value k.

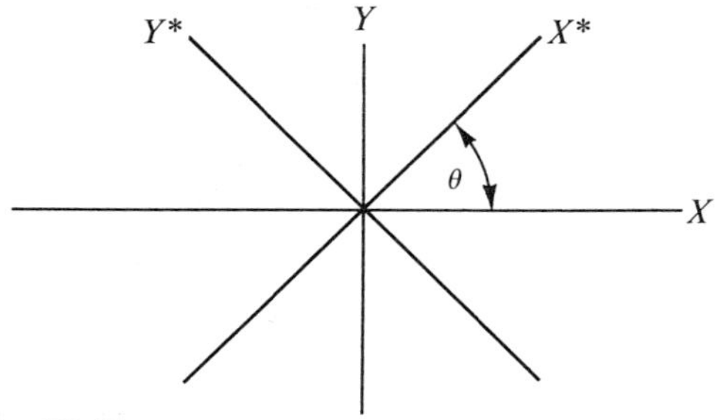

Fig. 10-6f

2. The graphs of equations in the form $xy = k$ are hyperbolas. The simple relationship between the variables is called **inverse variation**. Try to find physical applications of this type of relationship.
3. Graph $xy = -5$.
4. LORAN is a LOng Range Aid to Navigation system that is used by ships at sea. A radio signal is transmitted from a stationary master transmitter and is received by ships at sea and is also received by a second stationary slave transmitter that sends out a signal in response to the master signal. A ship at sea measures only the *difference* in time between the master and the slave signal. Assume that the master and slave transmitting stations are on the same latitude and separated by 2000 miles.

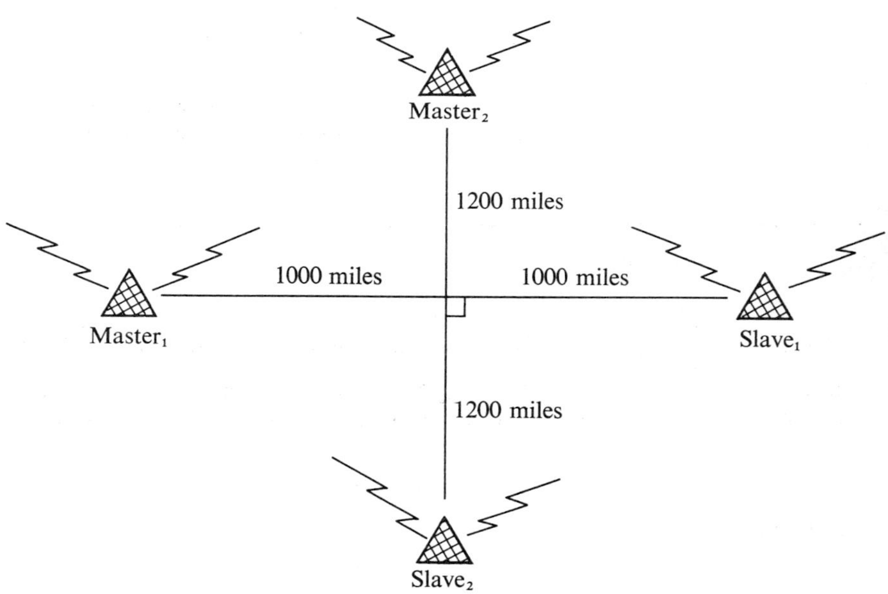

Fig. 10-6g

a. What is known about the position of a ship when the two signals occur at a .0043 second interval? (Velocity of radio waves is 186,000 miles per second.)

b. Sufficient information is not given in part (a) to isolate the location of the ship. In actuality a second master and slave system provides the means for narrowing down the possibilities. Assume that the two systems are located in the manner illustrated and that the interval of time between the signals of the second system is .0032 seconds. Where is the ship relative to the stations?

10 Supplementary Exercises and Problems

1. Find the distance from $P_1(-6, 7)$ to $P_2(-6, -8)$. (10-1)
2. Find the distance from P_1 $(2.4, -16.1)$ to $P_2(0, 7.4)$. (10-1)
3. Find the perimeter of the triangle whose vertices are $(-4, 3)$, $(5, 3)$, and $(11, -5)$. (10-1)
4. Find the length of each diagonal of the quadrilateral whose vertices are $(6, 2)$, $(0, 7)$, $(-5, 0)$, and $(2, -4)$. (10-1)
5. The center of a circle has coordinates $(5, 5)$. If the circle passes through $P_1(6, -7)$, does it also pass through $P_2(-3, -4)$? (10-1)
6. Find a point on the x-axis equidistant from $(2, 9)$ and $(10, 3)$. (10-1)
7. Find the midpoint of the line segment which joins $P_1(-6, 3)$ and $P_2(12, 14)$. (10-1)
8. Find the point equidistant from $(-1, 9)$, $(5, 3)$, and $(3, -3)$. (10-1)
9. Find three points which divide into four equal parts the line segment with endpoints $P_1(-6, -3)$ and $P_2(10, 4)$. (10-1)
10. Find the distance from the point $(12, 15)$ to the straight line passing through $P_1(2, 5)$ and $P_2(6, -1)$. (10-1)
11. The line segment drawn from $P_1(-5, 1)$ to $P_2(1, 3)$ is extended 10 units beyond P_2. Find the coordinates of the end point. (10-1)
12. Find the altitude a for the triangle ABC. (10-1)

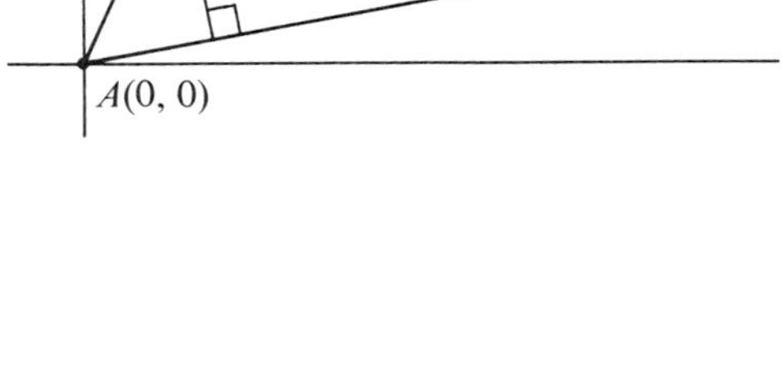

13. Graph and calculate the area of the triangle whose vertices are (−5, 2), (37, 16), and (−5, 94). (10-2)
14. Graph and calculate the area of the triangle whose vertices are (60, −9), (60, 30) and (−50, −20). (10-2)
15. Calculate the area.

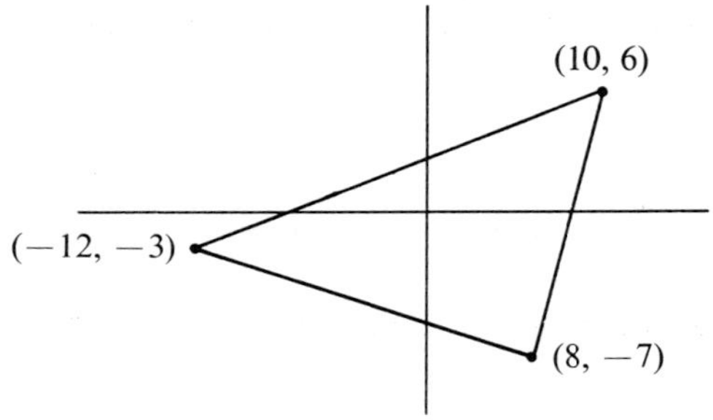

16. Calculate the area. (10-2)

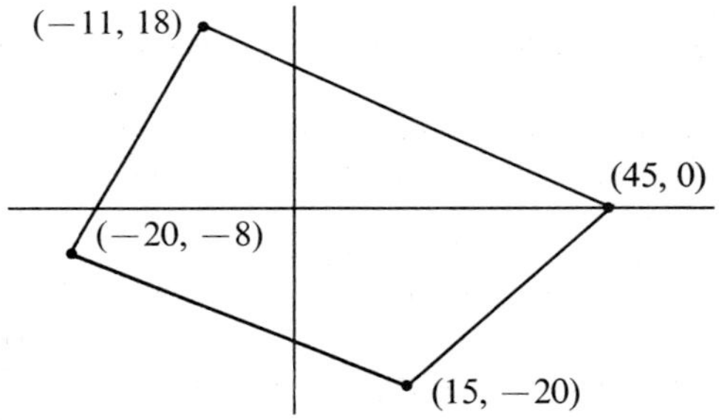

(10-2)

17. Calculate the area. (10-2)

Point	Standard angle	Distance to point from origin
A	45°	108 feet
B	110°	93 feet
C	202°	105 feet
D	255°	48 feet
E	331°	90 feet

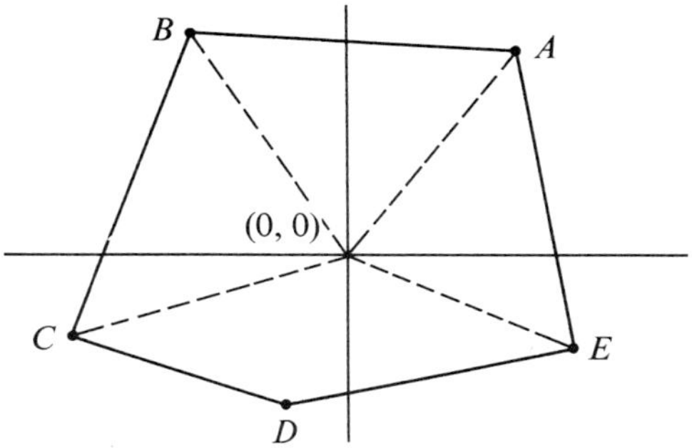

18. Calculate the area enclosed by the graphs of $f(x) = 3.0x + 27$, $g(x) = \frac{8x}{10} - 1$, and $h(x) = -\frac{x}{4} + 27$. (10-2)
19. Find the equation of the circle whose center is at the origin and whose radius is 12. (10-3)
20. Find the equation of the circle with the center at the origin and which passes through the point (12, 5). (10-3)
21. If the ends of the diameter of a circle are the points $P_1(4, -2)$ and $P_2(-8, 10)$, find the equation of the circle. (10-3)
22. Find the equation of the circle with the center (9, 15) and tangent to the x-axis. (10-3)
23. Prepare a graph of the equation $x^2 + y^2 = 5$. (10-3)
24. Find the coordinates of the center of the circle $x^2 + y^2 - 4x - 8y + 4 = 0$. Calculate the radius and draw the graph. (10-3)
25. Find the radius and coordinates of the center of the graph of $x^2 + y^2 - 8x + 7 = 0$. (10-3)
26. Find the radius and coordinates of the center of the graph of $y^2 - 6y = 6x - x^2$. (10-3)
27. Find the equation of the circle with center (1, 2) and tangent to $y = \frac{3}{4}x + 5$. (10-3)
28. Find the point of tangency in Problem 27. (10-3)
29. Find the equation of the circle passing through the point (4, 2) and concentric with the circle $x^2 + y^2 + 5x - 6y - 6 = 0$. (10-3)
30. Graph on the same coordinate system $x^2 - 4py = 0$ for $p = 1/2, 1, 2$. What effect does changing p have on the shape of the parabola? (10-4)
31. Find the focus and directrix of the parabola $y^2 = 4x$. (10-4)
32. Find the focus and directrix of the parabola $x^2 + 25y = 0$. (10-4)
33. Find the focus and directrix of the parabola $5.6y - 1.8x^2 = 0$. (10-4)
34. Write the equation of the parabola with vertex (0, 0) and focus (0, 16). (10-4)

35. A parabolic mirror is 20 centimeters in diameter. The distance c is 1 centimeter. How far is the focus from the vertex? (10-5)

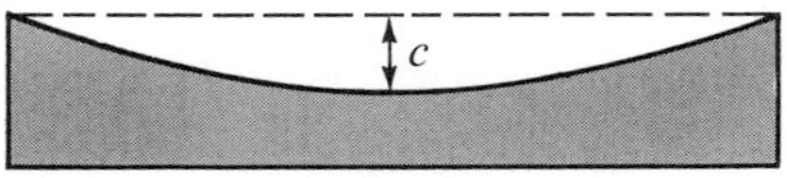

36. Determine the vertex, focus, and equation of the directrix of the parabola $2x^2 - 10x + 5y = 0$. (10-5)

37. Find the y-intercepts, x-intercepts, and foci of the ellipse $\frac{x^2}{36} + \frac{y^2}{16} = 1$. (10-6)

38. Find the major axis, minor axis, and foci of the ellipse $\frac{x^2}{15} + \frac{y^2}{25} = 1$. (10-6)

39. Find the foci, major axis, and minor axis of the ellipse $2x^2 + y^2 = 100$. (10-6)

40. Find the foci, major axis, and minor axis of the ellipse $x^2 + 4y^2 = 1$. (10-6)

41. If the major axis is 8 and the foci are at (2, 0) and $(-2, 0)$, find the equation of the ellipse. (10-6)

42. If the minor axis is 12 and the foci are at (0, 3) and $(0, -3)$, find the equation of the ellipse. (10-6)

43. For every point P on the ellipse, the tangent line $\overline{LM}$ through P is perpendicular to $\overline{PQ}$ and angle F_1PQ equals angle QPF_2. If the ellipse has the equation $\frac{x^2}{16} + \frac{y^2}{4} = 1$, what is the slope of $\overline{LM}$ for $P(2, y)$? (10-6)

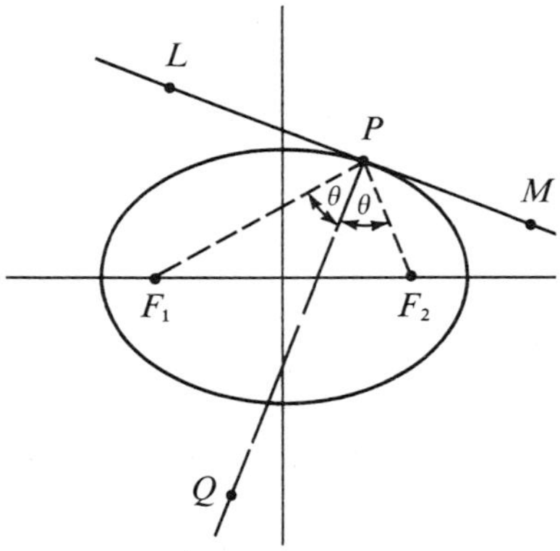

44. Graph the hyperbola $9x^2 - 4y^2 = 36$ and find its foci, vertices, and asymptotes. (10-6)

45. Graph the hyperbola $45y^2 - 4x^2 - 180 = 0$ and find its foci, vertices, and asymptotes. (10-6)

46. Find the equation of the hyperbola that has foci (6, 0) and (−6, 0) and has a distance of 8 between vertices. (10-6)
47. Find the equation of the hyperbola which has vertices on the x-axis equidistant from the origin and passes through (3, 1) and (9, 5). (10-6)
48. Find the equation of the locus of a point which moves in a plane so that the difference of its distances from (0, 60) and (0, −60) is 40. (10-6)
49. Find the equation of the hyperbola passing through (4, 6) with asymptotes $y = \sqrt{3}x$ and $y = -\sqrt{3}x$. (10-6)
50. Discover a method for drawing hyperbolas with string, pencil, and tacks at the foci. (10-6)

Chapter Eleven Differential Calculus

11-1 Limits

The invention of calculus is one of the great achievements in the history of mathematics. Calculus is one of the principal mathematical tools that aid scientists and engineers in solving the complex technical problems of our day. Technicians who want an appreciation and understanding of the ways in which scientists and engineers communicate should consider the very elementary introduction to calculus given in this book essential to their occupational preparation.

Early in the course of this text some of the fundamental notions of calculus were introduced. In particular, we gave attention to the concepts of slope functions (f') and area functions (F), which are at the heart of calculus. We cannot make further progress in terms of achieving greater mathematical understanding of these related functions until we learn something of the fundamental concept of **limit**. The following examples use familiar situations to help us toward an understanding of the mathematical concept of limit.

Radioactive elements have a characteristic known as *half-life*. Half-life is the constant period of time, peculiar to each radioactive substance, required for the rate of activity to "decay" to half the initial rate. For example, the half-life of carbon eleven is 20.5 minutes. After each 20.5 minutes, the level of activity is halved.

What is the theoretical life of the radioactivity? What is the ultimate level (limit) of activity?

■ The life of the activity is *infinite*. The ultimate level or *limit* of the activity is *zero*.

We have often been faced with the problem of writing 1/3 as a decimal fraction. Statements such as $1/3 = .33$ or $1/3 = .333$ are not entirely true. We know that there is less error in the second statement than in the first. By placing more digits of three after the decimal point, we can get closer decimal approximations of 1/3, but the expression will never be exact as long as the number of digits of three is finite. We can say that

$$\frac{3}{10} + \frac{3}{10^2} + \frac{3}{10^3} + \cdots + \frac{3}{10^n}$$

approaches 1/3 as a limit as n assumes larger and larger integral values. This may be written

$$\lim_{n \to \infty} \left(\frac{3}{10} + \frac{3}{10^2} + \frac{3}{10^3} + \cdots + \frac{3}{10^n} \right) = \frac{1}{3}.$$

This is an exact statement because the *limit* is exactly 1/3.

Consider the regular polygon of n sides inscribed in a circle. The perimeter (P) of the polygon approaches the circumference (C) of the circle when the number (n) of sides becomes larger and larger. But for any finite n, no matter how

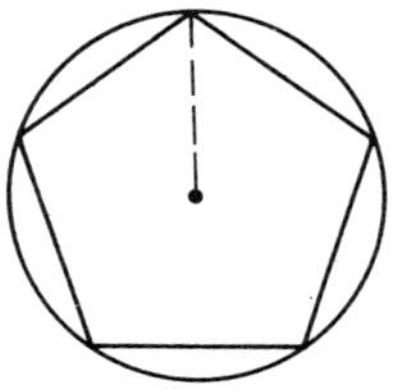

Fig. 11-1a

large, the perimeter of the inscribed polygon is less than the circumference of the circle. The *limit* of P as n approaches infinity is equal to C. This may be written

$$\lim_{n \to \infty} P_n = C.$$

The general problems that we must now look at involve finding the **limits of functions.**

The particular functions we may consider are numerous in type, and the particular limits we seek may vary.

Example 1

If $f(x) = 2x + 3$, find $\lim_{x\to 3} f(x)$.

If we select values of x close to 3, we gain some insight into the desired limit.

$$\begin{aligned} f(2.9) &= 2(2.9) + 3 = 8.8 \\ f(2.99) &= 2(2.99) + 3 = 8.98 \\ f(2.999) &= 2(2.999) + 3 = 8.998 \end{aligned}$$

We may surmise that the desired limit is 9. In this particular exercise, the $\lim_{x\to 3} f(x) = f(3)$. However, it is frequently impossible to find a limit in this way.

Example 2

If $f(x) = \dfrac{(x^2 - 9)}{x - 3}$, find $\lim\limits_{x\to 3} f(x)$.

If we try to evaluate $f(3)$, we meet with no success. The fraction 0/0 is said to be *indeterminant*, and so we must use other devices. Now if $x \neq 3$,

$$\frac{x^2 - 9}{x - 3} = x + 3;$$

so

$$\lim_{x\to 3} \frac{x^2 - 9}{x - 3} = \lim_{x\to 3} (x + 3).$$

We are interested in values of x close to, but not equal to, 3. Therefore, the desired limit is 6.

Example 3

If $f(x) = \dfrac{3x^2}{x^2 + x}$, find $\lim\limits_{x\to 0} f(x)$.

Again $f(0)$ results in the indeterminant form 0/0. If $x \neq 0$, then

$$\frac{3x^2}{x^2+x} = \frac{3}{1+(1/x)}.$$

The limit of

$$\frac{3}{1+(1/x)}$$

is dependent upon the limit of $(1/x)$ as x approaches zero. $\text{Lim}_{x\to 0}\,(1/x)$ does not exist; in other words, the nearer x approaches zero, the larger $(1/x)$ becomes. Therefore

$$\lim_{x\to 0} \frac{3}{1+(1/x)} = 0.$$

In Examples 2 and 3, the functions are not defined at the limit points. The next example shows that it is possible to have a function that exists at the limit point, but yet has no real limit.

Example 4

If $f(x) = 2x\sqrt{x-2}$, find $\lim_{x\to 0} f(x)$.

Now $f(0) = 0$, but since $f(x)$ is not defined for values near zero, the function does not have a limit as $x \to 0$. Thus we must be careful not to claim to find a limit simply because the function can be evaluated at the limit point.

We may discover that finding the limits of some functions is beyond our level of mathematical ability. In these cases we may develop some insight by performing calculations near the limit.

Example 5

If $f(x) = (\sin x)/x$, x in radians, then find $\lim_{x\to 0} f(x)$.

Direct substitution is inconsequential. Calculations near zero provide the following results:

x	$\sin x$	$\dfrac{\sin x}{x}$
.500000	.479425	.958851
.100000	.099833	.998334
.050000	.049979	.999958
.010000	.009999	.999983
.001000	.000999	.999999

We may guess that $\lim_{x\to 0}(\sin x)/x = 1$, which, in fact, is the correct result.

11-1 Exercises

1. Find the following limits, if they exist:

   a. $\lim_{x\to 0} \dfrac{x}{k}$,

   b. $\lim_{x\to 0} \dfrac{k}{x}$,

   c. $\lim_{x\to 0} \dfrac{kx}{x}$.

2. If $f(x) = \dfrac{x^2 + x - 2}{x - 1}$, use a calculator to find $f(.9)$, $f(.99)$, $f(.999)$. Is $f(1)$ defined? What is $\lim_{x\to 1} f(x)$?

■ 1. a. $\lim_{x\to 0} \dfrac{x}{k} = 0$;

b. $\lim_{x\to 0} \dfrac{k}{x}$ is undefined;

c. $\lim_{x\to 0} \dfrac{kx}{x} = \lim_{x\to 0} k, \quad x \neq 0$

$= k.$

3. If $f(x) = x^2 + 1$, find $\lim_{x\to 2} f(x)$.

■ 2. i. $f(.9) = 2.900$,

$f(.99) = 2.990$,

$f(.999) = 2.999$;

ii. $f(1)$ is undefined or indeterminate;

iii. $\lim_{x\to 1} f(x) = 3$.

---

4. If $f(x) = \dfrac{2}{x^2 + 3}$, find $\lim_{x \to 2} f(x)$.

■ 3. $\lim_{x \to 2}(x^2 + 1) = 5.$

---

5. If $f(x) = \dfrac{x^2 - 9}{x - 3}$, find $\lim_{x \to 3} f(x)$.

■ 4. $\lim_{x \to 2} \dfrac{2}{x^2 + 3} = \dfrac{2}{7}.$

---

6. If $f(h) = \dfrac{(2 + h)^2 - 4}{h}$, find $\lim_{h \to 0} f(h)$.

■ 5.
$$\begin{aligned} f(x) &= \frac{x^2 - 9}{x - 3} \\ &= \frac{(x + 3)(x - 3)}{(x - 3)} \\ &= x + 3, \quad x \neq 3; \\ \therefore \lim_{x \to 3} \frac{x^2 - 9}{x - 3} &= \lim_{x \to 3} (x + 3) \\ &= 6. \end{aligned}$$

---

7. If $f(x) = (x + 2)\sqrt{x^2 - 5}$, find $\lim_{x \to -2} f(x)$.

■ 6.
$$\begin{aligned} f(h) &= \frac{(2 + h)^2 - 4}{h} \\ &= \frac{4 + 4h + h^2 - 4}{h} \\ &= 4 + h, \quad h \neq 0; \\ \therefore \lim_{h \to 0} f(h) &= 4. \end{aligned}$$

---

8. If $f(x) = \dfrac{1 - (1 + x)^3}{x}$, find $\lim_{x \to 0} f(x)$.

■ 7. The function $f(x)$ has no real limit near -2.

■ 8.
$$\begin{aligned} f(x) &= \frac{1 - (1 + x)^3}{x} \\ &= \frac{1 - (1 + 3x + 3x^2 + x^3)}{x} \\ &= -3 - 3x - x^2, \\ &\quad x \neq 0; \\ \therefore \lim_{x \to 0} f(x) &= -3. \end{aligned}$$

11-1 Problems

1. A casting is placed in a very large cooling reservoir whose temperature remains almost constant at 60° F. The temperature of the casting falls at a constant rate of 20 per cent per hour of the temperature difference. If the casting is originally 1200° F, what is its temperature after 1 hour? 2 hours? 3 hours? How long will it continue to cool? What is its ultimate temperature approaching?
2. When a constant DC voltage is applied to a series RL circuit, the current approaches the value $i = E/R$ as determined by Ohm's law, but theoretically never reaches it. Instead, for each small unit of time, the current increases by a constant fractional part of the remaining difference separating it from the limit E/R. Suppose that for each .01 second, the constant fractional part is .9 for a certain combination of resistance (R) and inductance (L). Show in a table how the current is increasing.
3. Use a calculator to find the approximate limit of $(1 + 1/m)^m$ as m approaches some large integral value.
4. Approximate the value of $\overline{PM}/\overset{\frown}{NM}$ as θ approaches zero.

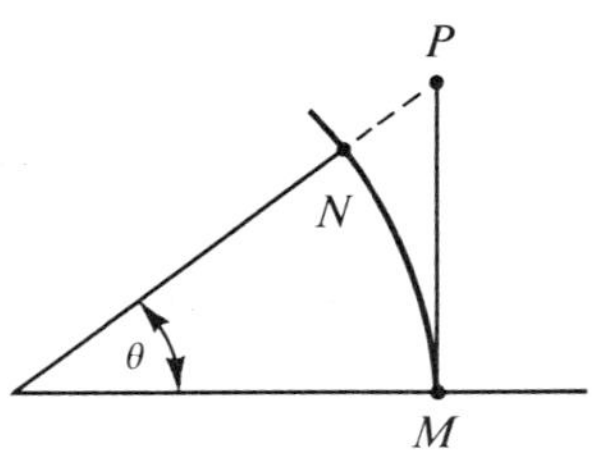

11-2 Definition of Derivative

The straight line L which intersects the curve C at points P and Q is called a *secant line* (Fig. 11-2a). Suppose that we permit point Q to move along C

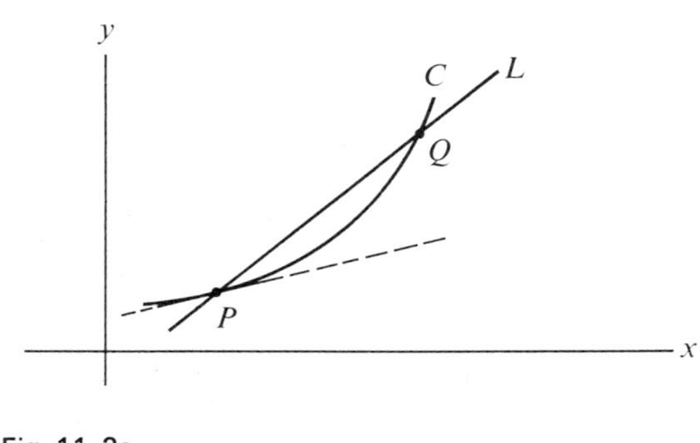

Fig. 11-2a

toward point P. The *limit* as Q approaches P, but does not become P, is the line *tangent* to C at P. If C is the graph of some function f such that $f(x) = y$, then we should recognize, from earlier work, that the tangent at P is equivalent to the *slope* and also the *rate of change* of y with respect to x at P. We can write an algebraic expression that states precisely these ideas by using the methods of analytic geometry and the notion of limit.

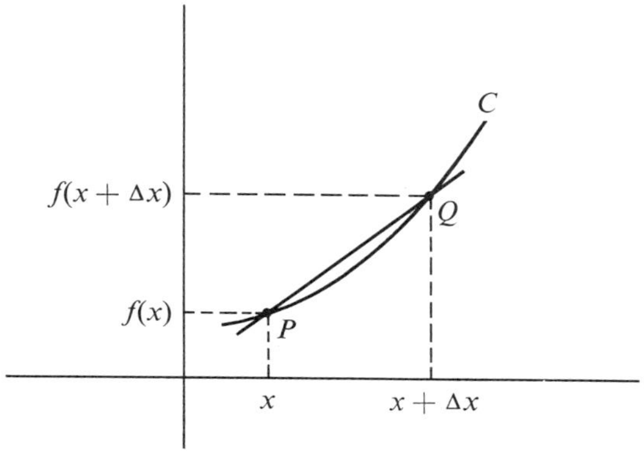

Fig. 11-2b

Let P have the coordinates $(x, f(x))$ and Q the coordinates $(x + \Delta x, f(x + \Delta x))$, where Δx is a small positive value (Fig. 11-2b). Then

$$f'(x) = \lim_{\Delta x \to 0} \frac{f(x + \Delta x) - f(x)}{(x + \Delta x) - x}. \qquad \text{(Eq. 11-2a)}$$

Until now, all the related functions f and f' that we have studied were discovered either by means of plane geometry or graphical approximations. For example, in Section 1-2 we used a geometric approach to show that if $f(x) = 2x$, then $f'(x) = 2$. In Section 3-5 we used approximate graphic methods to discover that if

$$f(x) = \frac{x^2}{3} + \frac{x}{2} - 2,$$

then

$$f'(x) = \frac{2}{3}x + \frac{1}{2}.$$

Equation 11-2a can be used to reaffirm both of these results.

Example 1

Use Eq. 11-2a to show that if $f(x) = 2x$, then $f'(x) = 2$.

$$f'(x) = \lim_{\Delta x \to 0} \frac{f(x + \Delta x) - f(x)}{(x + \Delta x) - x}$$

$$= \lim_{\Delta x \to 0} \frac{2(x + \Delta x) - 2x}{\Delta x}$$

$$= \lim_{\Delta x \to 0} \frac{2\Delta x}{\Delta x}$$

$$= \lim_{\Delta x \to 0} 2 = 2.$$

Example 2

Use Eq. 11-2a to show that if

$$f(x) = \frac{x^2}{3} + \frac{x}{2} - 2,$$

then

$$f'(x) = \frac{2}{3}x + \frac{1}{2}.$$

$$f'(x) = \lim_{\Delta x \to 0} \frac{f(x + \Delta x) - f(x)}{(x + \Delta x) - x}$$

$$= \lim_{\Delta x \to 0} \frac{\left(\frac{(x + \Delta x)^2}{3} + \frac{(x + \Delta x)}{2} - 2\right) - \left(\frac{x^2}{3} + \frac{x}{2} - 2\right)}{\Delta x}$$

$$= \lim_{\Delta x \to 0} \frac{\frac{x^2}{3} + \frac{2x(\Delta x)}{3} + \frac{(\Delta x)^2}{3} + \frac{x}{2} + \frac{\Delta x}{2} - 2 - \frac{x^2}{3} - \frac{x}{2} + 2}{\Delta x}$$

$$= \lim_{\Delta x \to 0} \frac{\frac{2x(\Delta x)}{3} + \frac{(\Delta x)^2}{3} + \frac{\Delta x}{2}}{\Delta x}$$

$$= \lim_{\Delta x \to 0} \left(\frac{2x}{3} + \frac{\Delta x}{3} + \frac{1}{2}\right) = \frac{2x}{1} + \frac{1}{2}.$$

Equation 11-2a is a precise definition of what is called the **derivative of a function**. The process of finding the derivative is called **differentiation**. Differentiating a function f gives the related function f' that describes the slope or rate of change of f.

Several symbols are commonly used to indicate derivatives:

Function	Derivative
$y = f(x)$	$y' = f'(x)$
$y = f(x)$	$\frac{dy}{dx} = \frac{df(x)}{dx}$
$y =$ (expression)	$\dot{y} =$ (derivative of expression)

We shall use the notation f and f' for a function and its derivative.

Example 3

If $f(x) = 3x^2 - 5x$, find the equation for f' and the slope of f at $x = 1/2$.

i. $$f'(x) = \lim_{\Delta x \to 0} \frac{f(x+\Delta x) - f(x)}{(x+\Delta x) - x}$$

$$= \lim_{\Delta x \to 0} \frac{3(x+\Delta x)^2 - 5(x+\Delta x) - 3x^2 + 5x}{\Delta x}$$

$$= \lim_{\Delta x \to 0} \frac{3x^2 + 6x(\Delta x) + 3(\Delta x)^2 - 5x - 5\Delta x - 3x^2 + 5x}{\Delta x}$$

$$= \lim_{\Delta x \to 0} \frac{6x(\Delta x) + 3(\Delta x)^2 - 5\Delta x}{\Delta x}$$

$$= \lim_{\Delta x \to 0} (6x + 3\Delta x - 5)$$

$$= 6x - 5;$$

ii. $f'(1/2) = 6(1/2) - 5 = -2.$

Example 4

If the position of a freely falling object is determined by the equation $f(t) = 16t^2$, find the equation for its speed as a function of time.

$$\text{(Speed)} = f'(t) = \lim_{\Delta t \to 0} \frac{f(t+\Delta t) - f(t)}{(t+\Delta t) - t}$$

$$= \lim_{\Delta t \to 0} \frac{16(t+\Delta t)^2 - 16t^2}{(t+\Delta t) - t}$$

$$= \lim_{\Delta t \to 0} \frac{16t^2 + 32t(\Delta t) + 16(\Delta t)^2 - 16t^2}{\Delta t}$$

$$= \lim_{\Delta t \to 0} \frac{32t(\Delta t) + 16(\Delta t)^2}{\Delta t}$$

$$= \lim_{\Delta t \to 0} (32t + 16\Delta t) = 32t.$$

11-2 Exercises

Find the derivative of each of the following functions by using Eq. 11-2a.

1. $f(x) = x.$

2. $f(x) = x^2.$

■ 1. $$f'(x) = \lim_{\Delta x \to 0} \frac{f(x+\Delta x) - f(x)}{(x+\Delta x) - x}$$

$$= \lim_{\Delta x \to 0} \frac{x + \Delta x - x}{\Delta x}$$

$$= \lim_{\Delta x \to 0} \frac{\Delta x}{\Delta x}$$

$$= \lim_{\Delta x \to 0} 1 = 1.$$

3. $f(x) = x^3.$

■ 2. $$f'(x) = \lim_{\Delta x \to 0} \frac{f(x+\Delta x) - f(x)}{(x+\Delta x) - x}$$

$$= \lim_{\Delta x \to 0} \frac{(x+\Delta x)^2 - x^2}{\Delta x}$$

$$= \lim_{\Delta x \to 0} \frac{x^2 + 2x(\Delta x) + \Delta x^2 - x^2}{\Delta x}$$

$$= \lim_{\Delta x \to 0} (2x + \Delta x) = 2x.$$

---

4. $f(x) = x^4.$

■ 3. $$f'(x) = \lim_{\Delta x \to 0} \frac{f(x + \Delta x) - f(x)}{(x + \Delta x) - x}$$
$$= \lim_{\Delta x \to 0} \frac{(x + \Delta x)^3 - x^3}{\Delta x}$$
$$= \lim_{\Delta x \to 0} \frac{x^3 + 3x^2\Delta x + 3x(\Delta x)^2 + (\Delta x)^3 - x^3}{\Delta x}$$
$$= \lim_{\Delta x \to 0} (3x^2 + 3x(\Delta x) + (\Delta x)^2)$$
$$= 3x^2.$$

---

5. $f(x) = a.$

■ 4. $$f'(x) = \lim_{\Delta x \to 0} \frac{f(x + \Delta x) - f(x)}{(x + \Delta x) - x}$$
$$= \lim_{\Delta x \to 0} \frac{(x + \Delta x)^4 - x^4}{\Delta x}$$
$$= \lim_{\Delta x \to 0} \left(\frac{x^4 + 4x^3(\Delta x) + 6x^2(\Delta x)^2}{\Delta x} + \frac{4x(\Delta x)^3 + (\Delta x)^4 - x^4}{\Delta x}\right)$$
$$= \lim_{\Delta x \to 0} (4x^3 + 6x^2(\Delta x) + 4x(\Delta x)^2 + (\Delta x)^3)$$
$$= 4x^3.$$

---

6. $f(x) = 7x.$

■ 5. $$f'(x) = \lim_{\Delta x \to 0} \frac{f(x + \Delta x) - f(x)}{(x + \Delta x) - x}$$
$$= \lim_{\Delta x \to 0} \frac{a - a}{\Delta x}$$
$$= 0.$$

---

7. $g(x) = 3x^2.$

■ 6. $$f'(x) = \lim_{\Delta x \to 0} \frac{7(x + \Delta x) - 7x}{\Delta x}$$
$$= \lim_{\Delta x \to 0} 7 = 7.$$

8. $f(x) = ax^2$.

■ 7. $g'(x) = \lim_{\Delta x \to 0} \frac{g(x + \Delta x) - g(x)}{(x + \Delta x) - x}$

$= \lim_{\Delta x \to 0} \frac{3(x + \Delta x)^2 - 3x^2}{\Delta x}$

$= \lim_{\Delta x \to 0} \frac{3x^2 + 6x(\Delta x) + 3(\Delta x)^2 - 3x^2}{\Delta x}$

$= \lim_{\Delta x \to 0} (6x + 3\Delta x) = 6x.$

---

9. $h(x) = x^2 + 4$.

■ 8. $f'(x) = \lim_{\Delta x \to 0} \frac{a(x + \Delta x)^2 - ax^2}{\Delta x}$

$= \lim_{\Delta x \to 0} \frac{a(x^2 + 2x(\Delta x) + (\Delta x)^2 - x^2)}{\Delta x}$

$= a(2x).$

---

10. $f(x) = 3x^2 + 7x$.

■ 9. $h'(x) = \lim_{\Delta x \to 0} \frac{h(x + \Delta x) - h(x)}{(x + \Delta x) - x}$

$= \lim_{\Delta x \to 0} \frac{(x + \Delta x)^2 + 4 - (x^2 + 4)}{\Delta x}$

$= \lim_{\Delta x \to 0} \frac{x^2 + 2x(\Delta x) + (\Delta x)^2 + 4 - x^2 - 4}{\Delta x}$

$= \lim_{\Delta x \to 0} (2x + \Delta x) = 2x.$

---

11. $f(x) = x^3 + 2x$.

■ 10. $f'(x) =$

$\lim_{\Delta x \to 0} \frac{3(x + \Delta x)^2 + 7(x + \Delta x) - (3x^2 + 7x)}{\Delta x}$

$= \lim_{\Delta x \to 0} \frac{3x^2 + 6x(\Delta x) + 3(\Delta x)^2 + 7x}{\Delta x}$

$+ \frac{7\Delta x - 3x^2 - 7x}{\Delta x}$

$= \lim_{\Delta x \to 0} (6x + 3(\Delta x) + 7)$

$= 6x + 7.$

12. $f(x) = a(x^2 + 2x)$.

■ 11. $f'(x) =$

$$\lim_{\Delta x \to 0} \frac{(x + \Delta x)^3 + 2(x + \Delta x) - (x^3 + 2x)}{\Delta x}$$

$$= \lim_{\Delta x \to 0} \left(\frac{x^3 + 3x^2(\Delta x) + 3x(\Delta x)^2 + (\Delta x)^3}{\Delta x} + \frac{2x + 2\Delta x - x^3 - 2x}{\Delta x} \right)$$

$$= \lim_{\Delta x \to 0} (3x^2 + 3x(\Delta x) + (\Delta x)^2 + 2)$$

$$= 3x^2 + 2.$$

13. What is the slope of g at $x = 3/4$ in Exercise 7?

■ 12. $$f'(x) = \lim_{\Delta x \to 0} \left(\frac{a((x + \Delta x)^2 + 2(x + \Delta x))}{\Delta x} - \frac{a(x^2 + 2x)}{\Delta x} \right)$$

$$= \lim_{\Delta x \to 0} \left(\frac{ax^2 + 2ax(\Delta x) + a(\Delta x)^2}{\Delta x} + \frac{2ax + 2a\,\Delta x}{\Delta x} - \frac{ax^2 + 2ax}{\Delta x} \right)$$

$$= \lim_{\Delta x \to 0} (2ax + a(\Delta x) + 2a)$$

$$= 2ax + 2a = a(2x + 2).$$

14. What is the slope of f at $x = (-1/3)$ in Exercise 11?

■ 13. $g'(x) = 6x$

$g'(3/4) = 6(3/4) = 9/2.$

15. What is the rate of change of $h(x)$ with respect to x at $x = -2.5$ in Exercise 9?

■ 14. $f'(x) = 3x^2 + 2$

$f'(-1/3) = 3(-1/3)^2 + 2$

$= (1/3) + 2 = 7/3.$

■ 15. $h'(x) = 2x$

$h'(-2.5) = 2(-2.5)$

$= -5.$

11-2 Problems

1. A cable is hanging with the approximate curve of a parabola whose equation is $f(x) = (2.2 \times 10^{-3})x^2$, where the lowest point is midway between the poles. What is the angle at which the line meets the poles if they are separated by 200 feet?
2. Represent graphically $f(x) = x^2/3 - 5$ and the derivative f' on the same coordinate system.
3. In electronics, simple RC circuits electronically transform input signals so that the output represents the rate of change (derivative) of the input. If the graph of the input looks like Fig. 11-2c, *sketch* the differentiated output.

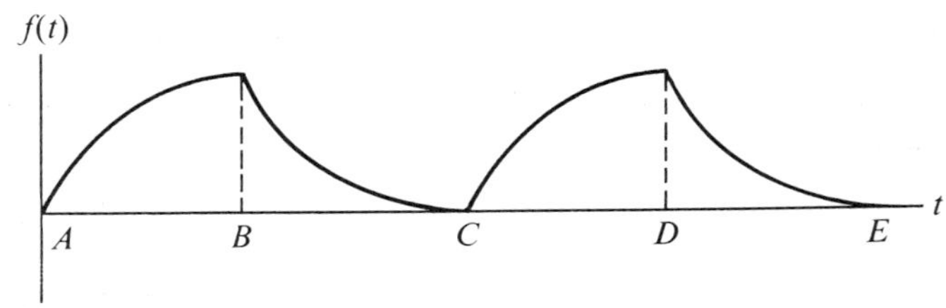

Fig. 11-2c

4. Prepare two graphs that illustrate the distinction between

$$f'(x) = \lim_{\Delta x \to 0} \frac{f(x + \Delta x) - f(x)}{x + \Delta x - x}$$

and

$$\bar{y} = \frac{f(x + \Delta x) - f(x)}{x + \Delta x - x}$$

when $f(x) = x^2/2 - 2x$, $x = 3$, and $\Delta x = 2$.

11-3 Derivative Formulas

The brief experiences we have had using Eq. 11-2a to calculate the derivatives of polynomial functions are probably sufficient to show us that the process can be abbreviated.

We shall use the examples and exercises of Section 11-2 to suggest the abbreviated methods.

In Exercise 11-2 we made the following calculations which are listed here for our review.

Function	*Derivative*
$f(x) = x$	$f'(x) = 1, (x^0)$
$f(x) = x^2$	$f'(x) = 2x$
$f(x) = x^3$	$f'(x) = 3x^2$
$f(x) = x^4$	$f'(x) = 4x^3$

The pattern appearing in the preceding list suggests that if

$$f(x) = x^n,$$

then

$$f'(x) = nx^{n-1}. \qquad \text{(Eq. 11-3a)}$$

The applications of Eq. 11-3a are not limited to positive-integer values of n, as the development of the equation may imply. The exponent n may be any rational number.

Example 1

If $f(x) = x^{-.4}$, find the derivative f'.

$$f(x) = x^{-.4}$$

$$f'(x) = -.4x^{-.4-1} = -.4x^{-1.4}.$$

Exercise 5 of Section 11-2 was actually a perfectly general approach to finding the derivative of a constant function. Our experience and intuition alone should tell us that if

$$f(x) = k,$$

then

$$f'(x) = 0. \qquad \text{(Eq. 11-3b)}$$

Here are the results of Exercises 6, 7, 8, and 12 of Section 11-2.

Function	*Derivative*
$f(x) = 7x$	$f'(x) = 7$
$f(x) = 3x^2$	$f'(x) = 6x = 3(2x)$
$f(x) = ax^2$	$f'(x) = 2ax = a(2x)$
$f(x) = a(x^2 + 2x)$	$f'(x) = a(2x + 2)$

These examples illustrate how the derivative of a constant times a function of x is equal to the constant times the derivative of the function of x. That is, if

$$\boldsymbol{f(x) = kg(x),}$$

then

$$\boldsymbol{f'(x) = kg'(x).} \qquad \text{(Eq. 11-3c)}$$

The function $f(x) = 3x^2 + 7x$ can be thought of as the sum of the two functions $g(x) = 3x^2$ and $h(x) = 7x$. The derivative of f was calculated by means of Eq. 11-2a and found to be $f'(x) = 6x + 7$. We should notice that $6x = g'(x)$ and $7 = h'(x)$. Exercises 9, 11, and 12 all reinforce the conclusion that if

$$\boldsymbol{f(x) = g(x) \pm h(x),}$$

then

$$\boldsymbol{f'(x) = g'(x) \pm h'(x).} \qquad \text{(Eq. 11-3d)}$$

Example 2

Find the derivative of

$$f(x) = \frac{1}{4x^2} + \frac{1}{3x}$$

using the formulas of this section.

$$f(x) = \frac{1}{4x^2} + \frac{1}{3x} = \frac{1}{4}x^{-2} + \frac{1}{3}x^{-1}$$

$$f'(x) = \frac{1}{4}(-2)x^{-2-1} + \frac{1}{3}(-1)x^{-1-1}$$

$$= -\frac{1}{2}x^{-3} - \frac{1}{3}x^{-2} = -\left(\frac{1}{2x^3} + \frac{1}{3x^2}\right).$$

The equations of this section may be written in equivalent forms using other notation:

a. $\dfrac{d(x^n)}{dx} = nx^{n-1}.$ (Eq. 11-3a)

b. $\dfrac{d(k)}{dx} = 0.$ (Eq. 11-3b)

c. $\dfrac{d(ku)}{dx} = k\dfrac{d(u)}{dx},\ u = g(x).$ (Eq. 11-3c)

d. $\dfrac{d(u \pm v)}{dx} = \dfrac{d(u)}{dx} \pm \dfrac{d(v)}{dx},$

$u = g(x),\ v = h(x).$ (Eq. 11-3d)

Differentiation of polynomial functions using formulas is really quite easy. At the same time, it is easy to lose sight of the meaning and purpose of differentiating a function. We can avert this possibility by always keeping in our minds the definition of derivatives stated concisely in Eq. 11-2a.

11-3 Exercises

1. If $f(x) = 12x^2$, find $f'(1/3)$.

---

2. If $g(x) = -0.6(10^{-3})$, find g'.

■ 1. $f(x) = 12x^2$,
$f'(x) = 12(2x) = 24x$,
$f'(1/3) = 24(1/3) = 8.$

3. If $f(r) = (6.2)(10^3)\pi r^2$, find $f'(10^{-3})$.

■ 2. $g(x) = -0.6(10^{-3})$,
$g'(x) = 0$.

---

4. Find the slope of
$$f(x) = 0.2x^4 - 0.3x^2$$
at $x = 1$.

■ 3. $f'(r) = (6.2)(10^3)\pi(2r)$,
$f'(10^{-3}) = (6.2 \times 10^3)\pi$
$(2)(10^{-3})$
$= 12.4\pi$.

---

5. If $h(x) = 0.7x^{1.3}$, find $h'(1)$.

■ 4. $f'(x) = 0.2(4)x^3 - 0.3(2)x$,
$f'(1) = .8(1)^3 - 0.6(1)$
$= 0.8 - 0.6 = 0.2$.

---

6. If $g(t) = \frac{1}{5.3t^2}$, find the slope of g at $t = 2$.

■ 5. $h'(x) = 0.7(1.3)x^{1.3-1}$
$= 0.91x^{.3}$,
$h'(1) = 0.91$.

---

7. If $f(p) = \frac{(p-4)^2}{3}$, find f'.

■ 6. $g(t) = \frac{1}{5.3t^2}$
$= \frac{1}{5.3} t^{-2}$,
$g'(t) = -\frac{2}{5.3} t^{-3}$
$= -\frac{2}{5.3t^3}$,
$g'(2) = -\frac{2}{5.3(2)^3}$
$= -\frac{1}{5.3(2)^2}$
$\approx .047$.

---

8. Find h' when $h(t) = \left(\log \frac{E}{R}\right)t$.

■ 7. $f(p) = \frac{p^2 - 8p + 16}{3}$,
$f'(p) = \frac{2p - 8}{3}$.

9. Find f' when $f(T) = K\left(\frac{T}{10}\right)^4$.

■ 8. $h'(t) = \left(\log \frac{E}{R}\right)$.

10. Where is the slope of $f(x) = 3x^2 - 2x + 2$ equal to one?

■ 9. $f'(T) = \left(\frac{4K}{10^4}\right)T^3$.

■ 10. i. $f'(x) = 6x - 2$;
ii. $6x - 2 = 1$; $x = 3/6 = 1/2$;
iii. $\therefore f'(1/2) = 1$.

11-3 Problems

1. The equation $f(t) = \alpha t^2 + \omega_0 t + \theta$ gives the angular distance a wheel rotates in t seconds. The derivative of f is an equation of the angular velocity of the wheel. Find the equation for angular velocity.
2. If $T = \frac{p^{0.94}}{330.3}$ and T is a function of p, find T'.
3. If $B = \frac{-\mu_0 I}{2\pi r}$ and $B = f(r)$, find $f'(r)$.
4. If $R = \left(\frac{R_1}{T_1}\right)T^{1.2}$ and $R = f(T)$, find $f'(T)$.
5. Curve $GBDF$ has the equation $f(x) = (1/2)x^2 - 12$. $\overline{CE}$ is tangent to the curve at $D(4, f(4))$. Find the length of $\overline{CD}$.

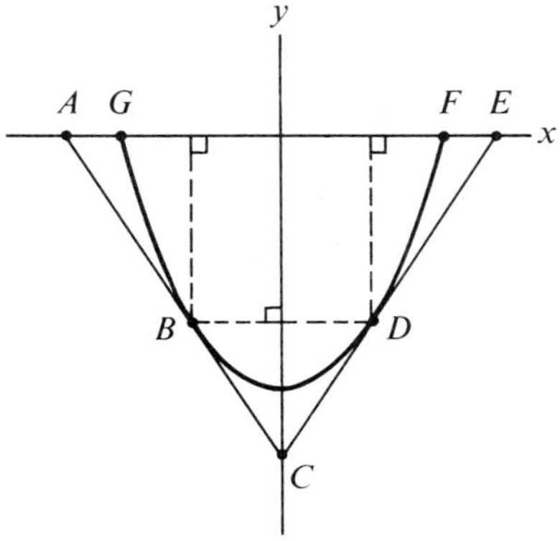

6. The volume of a spherical segment is

$$V = \pi h^2\left(r - \frac{h}{3}\right).$$

Determine the formula for the rate of change of V as h changes.

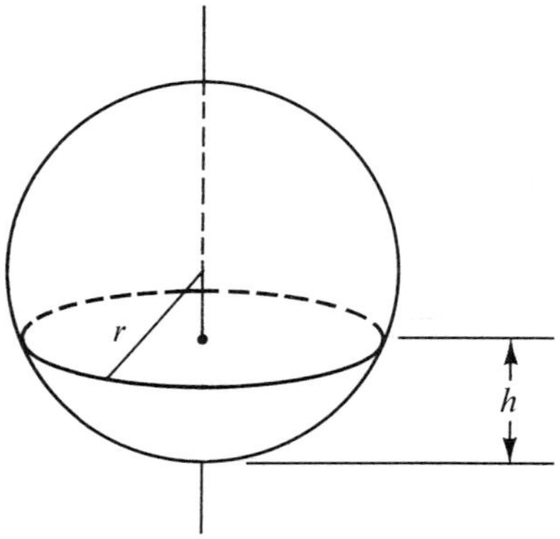

7. A circle is growing in such a way that the radius increases by 3/2 inches per second. What is the rate at which the area of the circle is changing?

11-4 Derivatives of Composite Functions

Many algebraic functions may be thought of as the composition of two or more other functions. For example, if $f(x) = 3x^2$ and $g(x) = 5x - 1$, then $f(g(x)) = 3(5x - 1)^2$.

Many of the formulas of science are composite functions. In electrical circuits, power is a function of current, and current may be a function of voltage. Volume may be a function of the radius, and radius, in turn, a function of temperature. Position may be a function of speed, and speed a function of time.

Taking the derivative of a composite function sometimes requires a special formula. If f is a composite function such that

$$f(x) = g(h(x)),$$

then

$$f'(x) = g'(h(x)) \cdot h'(x). \qquad \text{(Eq. 11-4a)}$$

Example 1

If $f(x) = (3x - 1)^2$, find $f'(x)$.

i. Let us think of the function f as the composite of $h(x) = 3x - 1$ and $g(h(x)) = (3x - 1)^2$.

ii. Then $h'(x) = 3$ and $g'(h(x)) = 2(3x - 1)$.

iii. Therefore
$$\begin{aligned} f'(x) &= g'(h(x))h'(x) \\ &= 2(3x - 1)(3) \\ &= 18x - 6. \end{aligned}$$

The function f in Example 1 can be differentiated in another manner using only the formulas of Section 11-3.

$$\begin{aligned} f(x) &= (3x - 1)^2 \\ &= 9x^2 - 6x + 1, \\ f'(x) &= \boxed{} \end{aligned}$$

■ $18x - 6$

The results are, of course, the same, but the second method of calculation is so much less complicated than the first that there appears to be little reason for giving further consideration to Eq. 11-4a. There are, however, composite functions that can be differentiated only by using Eq. 11-4a.

Example 2

If $f(x) = (x^2 - 1)^{1/2}$, find $f'(x)$.

The binomial $(x^2 - 1)$ has no perfect square root, so we must proceed using Eq. 11-4a.

i. Let $h(x) = x^2 - 1$ and $g(h(x)) = (x^2 - 1)^{1/2}$.

ii. Then $h'(x) = 2x$ and $g'(h(x)) = \frac{1}{2}(x^2 - 1)^{-1/2}$.

iii. Therefore $f'(x) = g'(h(x))h'(x)$

$$= \frac{1}{2}(x^2 - 1)^{-1/2}(2x)$$

$$= \frac{x}{\sqrt{x^2 - 1}}.$$

Equation 11-4a frequently appears in another form. If $y = f(u)$ and $u = h(x)$ so that $y = f(h(x))$, then

$$\frac{dy}{dx} = \frac{dy}{du} \cdot \frac{du}{dx}, \qquad \text{(Eq. 11-4b)}$$

where dy/dx is equivalent to $f'(x)$, dy/du is equivalent to $g'(h(x))$, and du/dx is equivalent to $h'(x)$. The name **chain rule** is given to the formula for differentiating composite functions because u is the "link" between y and x.

Example 3

If $y = u^2 - u$ and $u = 4x + 1$, find dy/dx.

i. $\dfrac{dy}{du} = 2u - 1;$

ii. $\dfrac{du}{dx} = 4;$

iii. $\dfrac{dy}{dx} = \dfrac{dy}{du} \cdot \dfrac{du}{dx}$

$$= (2u - 1)(4)$$

$$= 8u - 4$$

$$= 8(4x + 1) - 4 = 32x + 4.$$

The preceding example was chosen so that the results could be checked by using only the formula of Section 11-3.

$$\begin{aligned} y = u^2 - u &= (4x+1)^2 - (4x+1) \\ &= 16x^2 + 8x + 1 - 4x - 1 \\ &= 16x^2 + 4x \\ y' &= \boxed{} \end{aligned}$$

■ $32x + 4$

11-4 Exercises

1. Use Eq. 11-4a to differentiate $f(x) = (2x+1)^2$ and then check the results by using only the formulas of Section 11-3.

2. Differentiate $f(x) = (x^2 - x)^4$.

■ 1. i. Let $h(x) = 2x + 1$ and $g(h(x)) = (2x+1)^2$. Then $h'(x) = 2$ and $g'(h(x)) = 2(2x+1)$. Therefore,

$$\begin{aligned} f'(x) &= g'(h(x))h'(x) \\ &= 2(2x+1)(2) \\ &= 8x + 4. \end{aligned}$$

ii.

$$\begin{aligned} f(x) &= (2x+1)^2 \\ &= 4x^2 + 4x + 1 \\ f'(x) &= 8x + 4. \end{aligned}$$

3. What is the slope of $f(x) = \sqrt{1 + x^2}$ at $x = 2$?

■ 2. i. Let $h(x) = x^2 - x$, and $g(h(x)) = (x^2 - x)^4$.

ii. Then $h'(x) = 2x - 1$ and $g'(h(x)) = 4(x^2 - x)^3$.

iii. $f'(x) = g'(h(x))h'(x)$
$= 4(x^2 - x)^3(2x - 1)$.

4. If $v = f(t) = \sqrt{t/m}$, find $f'(t)$.

■ 3. i. $h(x) = 1 + x^2$

$g(h(x)) = (1 + x^2)^{1/2}$;

ii. $h'(x) = 2x$

$g'(h(x)) = \frac{1}{2}(1 + x^2)^{-1/2}$;

iii. $f'(x) = \frac{1}{2\sqrt{1 + x^2}} \cdot \frac{2x}{1}$

$= \frac{x}{\sqrt{1 + x^2}}$;

iv. $f'(2) = \frac{2}{\sqrt{1 + 4}}$

$= \frac{2\sqrt{5}}{5}$

$\approx .894$.

5. If $D = f(x)$ and

$$f(x) = \left(\frac{x^4}{16} - 2x + 5\right)^{1/2},$$

where is the slope of f equal to zero?

■ 4. i. $h(t) = t/m$

$g(h(t)) = (t/m)^{1/2}$;

ii. $h'(t) = 1/m$

$g'(h(t)) = (1/2)(t/m)^{-1/2}$

$= (1/2)(m/t)^{1/2}$;

iii. $f'(t) = (1/2)(m/t)^{1/2}(1/m)$

$= \frac{1}{2\sqrt{tm}}$.

6. If $Q = f(H)$ and $f(H) = CL(H + h)^{1.5}$, find $f'(H)$.

■ 5. i. $h(x) = \frac{x^4}{16} - 2x + 5$

$g(h(x)) = \left(\frac{x^4}{16} - 2x + 5\right)^{1/2}$;

ii. $h'(x) = \frac{x^3}{4} - 2$

$g'(h(x)) = \frac{1}{2}\left(\frac{x^4}{16} - 2x + 5\right)^{-1/2}$;

$$\text{iii.}\quad f'(x) = \frac{\left(\frac{x^3}{4} - 2\right) \cdot \frac{1}{2}}{\sqrt{\frac{x^4}{16} - 2x + 5}}$$

$$= 0$$

$$\frac{x^3}{4} - 2 = 0$$

$$x^3 = 8$$

$$x = 2.$$

■ 6. i. $k(H) = H + h$
$g(k(H)) = (H + h)^{1.5}$;
ii. $k'(H) = 1$
$g'(k(H)) = 1.5(H + h)^{.5}$;
iii. $f'(H) = CLg'(k(H))\,k'(H)$
$= CL(1.5)(H + h)^{.5}$.

11-4 Problems

1. $\widehat{AB}$ is one-quarter of a circle with a radius of 4 inches. $\overline{CE}$ is tangent to $\widehat{AB}$ at D.

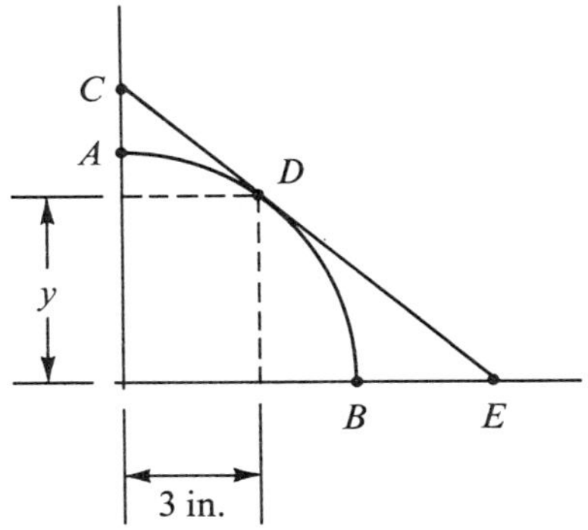

a. Find the angle that the tangent makes with $\overline{BE}$.
b. Find the length of line segment $\overline{CE}$.

2. If the curve AC is hyperbolic and has the equation

$$\frac{y^2}{4} - \frac{x^2}{9} = 1$$

and if line segments $\overline{AB}$ and $\overline{BC}$ are tangent at A and C respectively, find $\angle ABC$.

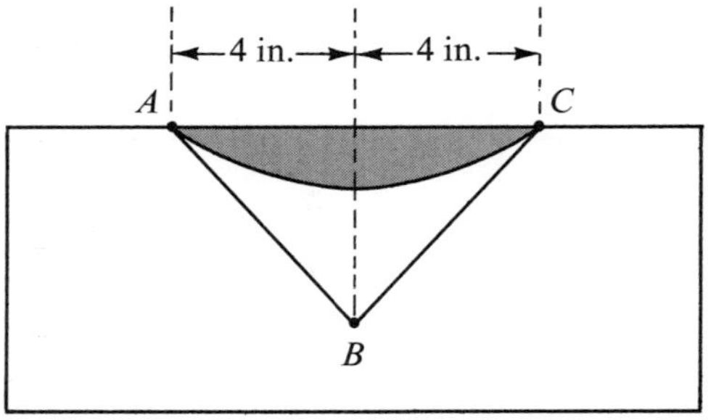

3. Thermisters are semi-conductors that have the interesting characteristic of decreasing in electrical resistance as the temperature increases. They are frequently used for temperature measurement because the current (i) flowing through a thermister is a function of resistance (r), which in turn is a function of temperature (t). If

$$i = \frac{5.5}{r} \quad \text{and} \quad r = \frac{6 \times 10^3}{t},$$

find di/dt.

4. A uniformly loaded beam held rigid at one end is deflected y inches at a distance x inches from the rigid end according to the formula

$$y = f(x) = \frac{1}{EI}\left(-\frac{W}{24}(L-x)^4 - \frac{WL^4}{6}x + \frac{WL^4}{24}\right).$$

If $I = 72$, $E = 30(10^6)$, $W = 100$, and $L = 48$, compute the slope of the beam at $x = 24$ inches.

5. A point is moving along the positive x-axis so that $x = 2t$. Consequently the distance $r = \sqrt{(2t)^2 + d^2}$. Find the expression that gives the rate at which the distance r is increasing.

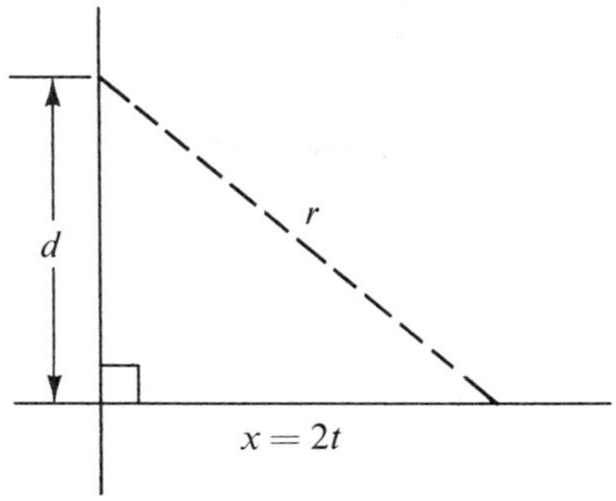

11-5 Derivatives of Products and Quotients

The purpose of this section is to derive and use the formulas for differentiating products and quotients.

We found in Section 11-3 that the derivative of a sum (or difference) of two functions is simply the sum (or difference) of the derivatives of the two functions (Eq. 11-3d). The derivative of a product or quotient is not quite so simple.

If

$$\boldsymbol{f(x) = g(x) \cdot h(x),}$$

then

$$\boldsymbol{f'(x) = g(x) \cdot h'(x) + h(x) \cdot g'(x).} \qquad \text{(Eq. 11-5a)}$$

Our mathematical background is adequate for deriving Eq. 11-5a. In doing so we will review again the fundamental concept of differentiation and also capture a little of the spirit of mathematical analysis.

In general terms, the function we are differentiating is $f(x) = g(x)h(x)$.

$$f'(x) = \lim_{\Delta x \to 0} \frac{f(x + \Delta x) - f(x)}{x + \Delta x - x}$$

$$= \lim_{\Delta x \to 0} \frac{g(x + \Delta x)h(x + \Delta x) - g(x)h(x)}{\Delta x}.$$

To the numerator of the last expression we add zero in the form

$$g(x + \Delta x)h(x) - g(x + \Delta x)h(x).$$

$$f'(x) = \lim_{\Delta x \to 0} \frac{g(x+\Delta x)h(x+\Delta x) - g(x+\Delta x)h(x) + g(x+\Delta x)h(x) - g(x)h(x)}{\Delta x}$$

$$= \lim_{\Delta x \to 0} \left(\frac{g(x+\Delta x)(h(x+\Delta x) - h(x))}{\Delta x} + \frac{h(x)(g(x+\Delta x) - g(x))}{\Delta x} \right).$$

Observe that

$$\lim_{\Delta x \to 0} \frac{h(x+\Delta x) - h(x)}{\Delta x} = h'(x), \quad \lim_{\Delta x \to 0} \frac{g(x+\Delta x) - g(x)}{\Delta x} = g'(x),$$

and $\lim_{\Delta x \to 0} g(x + \Delta x) = g(x)$. Therefore $f'(x) = g(x)h'(x) + h(x)g'(x)$.

Example 1

If $f(x) = (4x + 2)(3 - 2x)$, find f'.

i. Let $f(x) = g(x)h(x)$, where $g(x) = 4x + 2$ and $h(x) = 3 - 2x$.

ii. $h'(x) =$ ☐ and $g'(x) =$ ☐.

iii. $f'(x) = g(x)h'(x) + h(x)g'(x)$
$= (4x + 2)(-2) + (3 - 2x)(4)$
$= -8x - 4 + 12 - 8x$
$= 8 - 16x.$

■ -2; 4.

Example 1 was chosen to illustrate the use of Eq. 11-5a because it can be differentiated in another manner requiring no new techniques.

$$\begin{aligned} f(x) &= (4x + 2)(3 - 2x) \\ &= 12x - 4x - 8x^2 + 6 \\ &= 8x - 8x^2 + 6, \\ f'(x) &= 8 - 16x. \end{aligned}$$

In many problems we do not have the choice of using alternate methods.

Example 2

If $f(t) = 2t^2(3t - 5)^{1.3}$, find f'.

i. Let $g(t) = 2t^2$, so that $g'(t) = 4t$.

ii. Let $h(t) = (3t - 5)^{1.3}$, so that $h'(t) = 1.3(3t - 5)^{0.3}(3)$ or $3.9(3t - 5)^{0.3}$.

iii. $$\begin{aligned} f'(t) &= g(t)h'(t) + h(t)g'(t) \\ &= 2t^2(3.9)(3t - 5)^{0.3} + (3t - 5)^{1.3}(4t) \\ &= 7.8t^2(3t - 5)^{0.3} + 4t(3t - 5)^{1.3}. \end{aligned}$$

We do not need to return to the definition of a derivative to derive a formula for differentiating a quotient. Since

$$f(x) = \frac{g(x)}{h(x)} = g(x)(h(x))^{-1},$$

we may begin with Eq. 11-5a. Eq. 11-5a requires that we differentiate both factors of the function f. The derivative of $g(x)$ is $g'(x)$. The derivative of $(h(x))^{-1}$ is (using the chain rule) $(-1)(h(x))^{-2}h'(x)$. Substituting these values into the appropriate places in Eq. 11-5a, we have

$$\begin{aligned} f'(x) &= g(x)(-1)(h(x))^{-2}h'(x) + (h(x))^{-1}g'(x) \\ &= \frac{-g(x)h'(x)}{(h(x))^2} + \frac{g'(x)}{h(x)} \\ &= \frac{-g(x)h'(x) + h(x)g'(x)}{(h(x))^2}, \end{aligned}$$

or finally,

$$f'(x) = \frac{h(x)g'(x) - g(x)h'(x)}{(h(x))^2}. \qquad \text{(Eq. 11-5b)}$$

Example 3

$$\text{If } f(x) = \frac{3x+1}{2x^2-5}, \quad \text{find } f'.$$

i. Let $g(x) = 3x + 1$ so that $g'(x) = 3$.

ii. Let $h(x) = 2x^2 - 5$ so that $h'(x) = 4x$.

iii. Therefore
$$\begin{aligned} f'(x) &= \frac{h(x)g'(x) - g(x)h'(x)}{(h(x))^2} \\ &= \frac{(2x^2-5)3 - (3x+1)4x}{(2x^2-5)^2} \\ &= \frac{6x^2 - 15 - 12x^2 - 4x}{(2x^2-5)^2} \\ &= \frac{-6x^2 - 4x - 15}{(2x^2-5)^2}. \end{aligned}$$

11-5 Exercises

1. Find the derivative of

$$f(x) = 4x(3x^2 - 2),$$

a. using Eq. 11-5a,
b. using only the methods of Section 11-3.

---

2. Find the derivative of

$$f(x) = \frac{9x^2 + 4x}{3x}$$

a. using Eq. 11-5b,
b. using only the methods of Section 11-3.

■ 1. a. By Eq. 11-5a

i. $g(x) = 4x$, $g'(x) = 4$;

ii. $h(x) = (3x^2 - 2)$, $h'(x) = 6x$;

iii.
$$\begin{aligned} f'(x) &= g(x)h'(x) + h(x)g'(x) \\ &= 4x(6x) + (3x^2-2)4 \\ &= 24x^2 + 12x^2 - 8 \\ &= 36x^2 - 8. \end{aligned}$$

b. By Section 11-3

i. $f(x) = 4x(3x^2 - 2)$
$= 12x^3 - 8x;$

ii. $f'(x) = 36x^2 - 8.$

---

3. If $f(x) = (3x - 1)(x^2 - 1)^{3/2}$, find f'.

■ 2. a. By Eq. 11-5b

i. $g(x) = 9x^2 + 4x,$
$g'(x) = 18x + 4;$

ii. $h(x) = 3x, \quad h'(x) = 3;$

iii. $f'(x)$

$$= \frac{h(x)g'(x) - g(x)h'(x)}{(h(x))^2}$$

$$= \frac{3x(18x + 4) - (9x^2 + 4x)3}{(3x)^2}$$

$$= \frac{54x^2 + 12x - 27x^2 - 12x}{9x^2}$$

$$= \frac{27x^2}{9x^2} = 3.$$

b. By Section 11-3

i. $f(x) = \dfrac{9x^2 + 4x}{3x}$

$= 3x + \dfrac{4}{3};$

ii. $f'(x) = 3.$

---

4. Find the derivative of

$$f(x) = (x^2 + 3)\left(\frac{1}{x}\right)$$

a. using Eq. 11-5a,
b. using Eq. 11-5b.

■ 3. i. $g(x) = (3x - 1),$
$g'(x) = 3;$

ii. $h(x) = (x^2 - 1)^{3/2},$

$h'(x) = \dfrac{3}{2}(x^2 - 1)^{1/2}(2x)$

$= 3x(x^2 - 1)^{1/2};$

iii. $f'(x) = g(x)h'(x)$
$+ h(x)g'(x)$
$= (3x - 1)(3x)$
$(x^2 - 1)^{1/2}$
$+ (x^2 - 1)^{3/2}(3)$
$= (9x^2 - 3x)$
$(x^2 - 1)^{1/2}$
$+ 3(x^2 - 1)^{3/2}.$

5. If $f(x) = \dfrac{1}{x+1}$, find f'.

■ 4. a. By Eq. 11-5a

i. $f(x) = (x^2+3)(x)^{-1}$;

ii. $g(x) = x^2+3$,
$g'(x) = 2x$;

iii. $h(x) = x^{-1}$,
$h'(x) = -x^{-2}$;

iv. $f'(x) = g(x)h'(x) + h(x)g'(x)$

$$= (x^2+3)(-x^{-2}) + (x^{-1})(2x)$$

$$= 2 - \frac{x^2+3}{x^2}$$

$$= 2 - \frac{x^2}{x^2} - \frac{3}{x^2}$$

$$= 1 - \frac{3}{x^2}.$$

b. By Eq. 11-5b

i. $f(x) = \dfrac{x^2+3}{x}$;

ii. $g(x) = x^2+3$,
$g'(x) = 2x$;

iii. $h(x) = x$, $h'(x) = 1$;

iv. $f'(x)$

$$= \frac{h(x)g'(x) - g(x)h'(x)}{(h(x))^2}$$

$$= \frac{x(2x) - (x^2+3)}{x^2}$$

$$= 1 - \frac{3}{x^2}.$$

■ 5. i. $f(x) = (x+1)^{-1}$;

ii. $f'(x) = (-1)(x+1)^{-2}(1)$

$$= -\frac{1}{(x+1)^2}.$$

11-5 Problems

1. The variable r is a function of t in the equation

$$r = \frac{t^2 - 1}{t^3 + t}.$$

Find the equation for the rate of change of r with respect to t.

2. If $v = \sqrt{2GMh/(Rh + R^2)}$ and $v = f(h)$, find f'.
3. If $P = 2\pi\sqrt{(k^2 + L^2)/Lg}$ and $P = f(L)$, find f'.
4. A point is moving along the path of the parabola $y = f(t) = 2t^2 + 5$. If the velocity in the x direction is $v_x = 7.5/t$, find the velocity of the point at $t = 2.5$. (*Hint*: The v_x component is known, and the y component, v_y, is found by differentiating $f(t)$.)

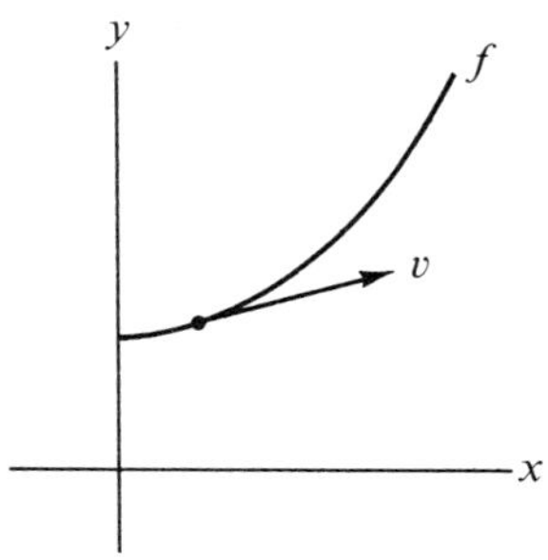

5. An object in a force field is in a position determined by the equations

$$x = \frac{3}{\sqrt{t^2 + 1}} \quad \text{and} \quad y = \frac{3t}{\sqrt{t^2 + 1}}$$

at any time t. What are the magnitude and direction of the velocity of the object at $t = 1$? (*Hint*: The derivatives of x and y, both with respect to t, give the x and y components of the velocity.)

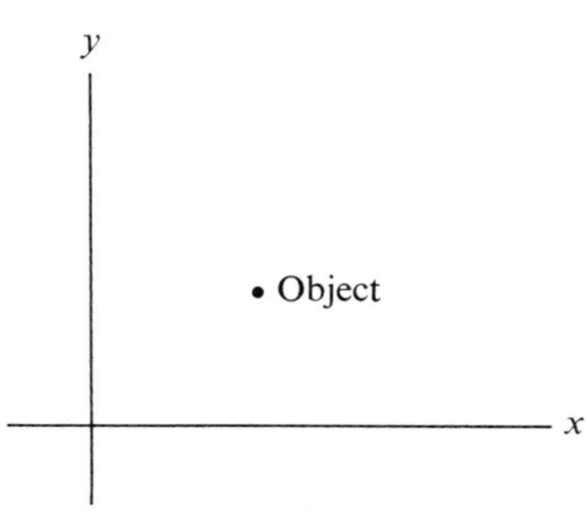

6. Suppose f is a function of x and $f(a) < 0$ and $f(b) > 0$. Then for some x_0 in the interval from a to b, $f(x_0) = 0$. Let x_1 be a *first* approximation of the root x_0.

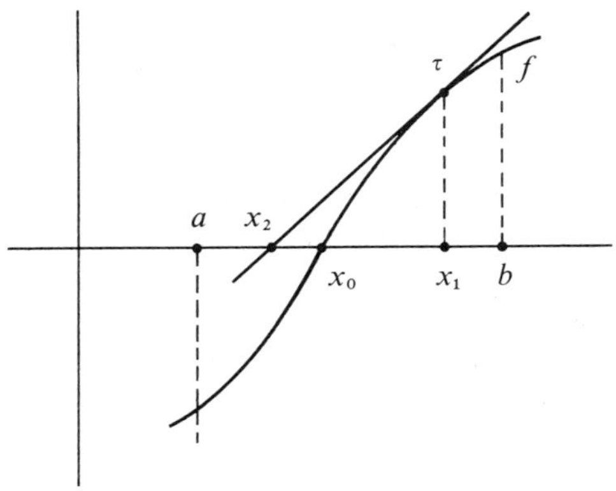

Under suitable conditions the line drawn tangent to the graph of f at x_1 intersects the x-axis at a point x_2 which is closer to x_0 than x_1. From the illustration we can see that

$$f'(x_1) = \frac{f(x_1)}{x_1 - x_2}.$$

Solving for x_2, we get

$$x_2 = x_1 - \frac{f(x_1)}{f'(x_1)},$$

which is known as **Newton's formula.** Repeated use of Newton's formula, where the value of x_2 becomes the new x_1, results in an approximation of x_0 to any degree of accuracy desired.

Find a root for $f(x) = x^3 - 2x - 5$ in the interval between 2 and 3 by Newton's method.

11-6 Maxima and Minima

One very important use of differentiation is in finding the maximum or minimum of a function. The task of an engineer is often to design the strongest component with the least amount of material. The plant superintendent wishes to maximize production and minimize costs. The merchant wishes to maximize profits and minimize overhead. If the variables in these types of problems can be expressed as functions, then differential calculus is a very useful tool for finding the values of the variables that correspond to maximum or minimum conditions.

In Section 3-4 we discovered techniques for finding the minimum or maximum point of a quadratic function. The method depended upon the symmetry of the graph of the quadratic. Symmetry is not a necessary characteristic of a function when derivatives are used to find maximum or minimum points. The derivative of a function f is the function f' relating the slope of f to the domain of f. The zeros of f', which are equivalent to the points where the slope is zero, are ordinarily relative maxima or minima of the function f.

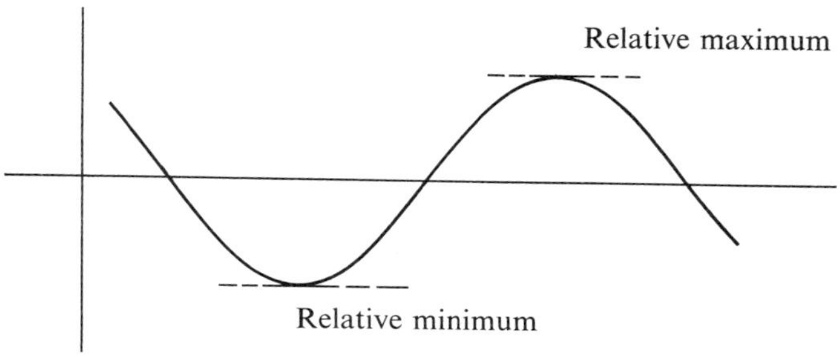

Fig. 11-6a

Example 1

Find the relative maximum or minimum points of the function $f(x) = x^3 - 6x^2 - 15x$.

i. $f'(x) = 3x^2 - 12x - 15;$

ii. $3x^2 - 12x - 15 = 0$

$(3x + 3)(x - 5) = 0$

$x = -1 \text{ and } 5;$

iii. $f(-1) = (-1)^3 - 6(-1)^2 - 15(-1)$

$= 8 \qquad (-1, 8)$

$f(5) = 5^3 - 6(5)^2 - 15(5)$

$= -100 \qquad (5, -100).$

iv. The graph (Fig. 11-6b) illustrates the positions of $(-1, 8)$ and $(5, -100)$ relative to the other points of f.

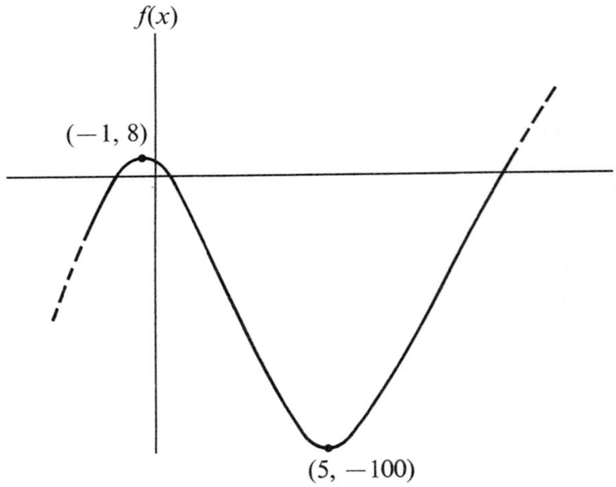

Fig. 11-6b

Example 2

A uniform bar weighing 2 pounds per foot is to be used to lift a 100-pound load, when the load is .5 feet from the fulcrum. What length of bar will require the minimum force, and what is the force?

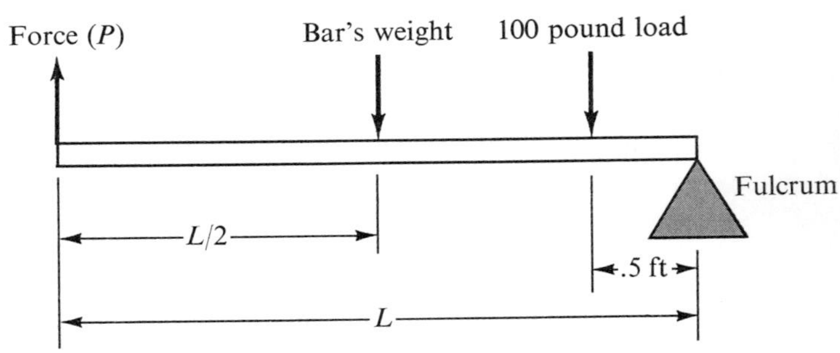

Fig. 11-6c

Let the required force be P and the length of the bar be L. As L is increased, the force required to lift the 100 pounds decreases, but additional force is required to lift the bar. If a function which expresses P in terms of L can be found, then the minimum of this function represents the desired solution. Equating the counterclockwise moments with the clockwise moment, we have

$$PL = 100(.5) + 2L\left(\frac{L}{2}\right)$$

and

$$P = \frac{50}{L} + L.$$

Recognizing that P is a function of L, we may write

$$P = f(L) = 50L^{-1} + L.$$

Next, we differentiate f and find the zeros of f'.

$$f'(L) = -50L^{-2} + 1$$

$$\frac{-50}{L^2} + 1 = 0$$

$$-50 + L^2 = 0$$

$$L^2 = 50$$

$$L = \pm 7.07 \text{ feet.}$$

The negative value has no significance in this real problem, so $L = 7.07$ feet.

$$P = f(7.07) = \frac{50}{7.07} + 7.07$$

$$= 14.1 \text{ pounds.}$$

All the relative maxima and minima for any given function are called **critical points**.

One problem always arises after having located the critical points. Are the points maxima or minima? Perhaps the simplest technique for making this known is checking the ordinate of a point *near* each critical point. Its value will reveal the nature of the critical point.

11-6 Exercises

Find all the relative maxima or minima for each of the exercises.

1. $f(x) = 8x^2 - 4x + \frac{1}{3}$.

---

2. $f(x) = 5 - 2x - 3x^2 - x^3.$

■ 1. i. $f'(x) = 16x - 4;$

ii. $16x - 4 = 0$

$x = 4/16 = 1/4;$

iii. $f(1/4) = 8(1/4)^2 - 4(1/4) + (1/3) = (1/2) - 1 + (1/3) = -1/6;$

iv. $(1/4, -1/6)$ is a minimum point.

---

3. $f(x) = -x + 2\sqrt{x^2 + 30}.$

■ 2. i. $f'(x) = -2 - 6x - 3x^2;$

ii. $-3x^2 - 6x - 2 = 0$

$3x^2 + 6x + 2 = 0$

$$x = \frac{-6 \pm \sqrt{6^2 - 4(3)2}}{6}$$

$= -.423$ and $-1.58;$

iii. $f(-.423) = 5.38$

$(-.423, 5.38)$, maximum point;

iv. $f(-1.58) = 4.62$

$(-1.58, 4.62)$, minimum point.

---

4. $f(t) = 4t^2 + \dfrac{128}{t}.$

■ 3. i. $f(x) = -x + 2(x^2 + 30)^{1/2},$

$f'(x) = -1 + \frac{2}{2}(x^2 + 30)^{-1/2}(2x)$

$$= \frac{2x}{\sqrt{x^2 + 30}} - 1;$$

ii. $$\frac{2x}{\sqrt{x^2 + 30}} - 1 = 0$$

$2x = \sqrt{x^2 + 30}$

$4x^2 = x^2 + 30$

$3x^2 = 30$

$x = \pm\sqrt{10} \approx \pm 3.16;$

iii. $f(3.16) = 15.8$
$(3.16, 15.8)$,
maximum point;
$f(-3.16) = 9.49$
$(-3.16, 9.49)$,
minimum point.

---

5. Find two positive numbers that have a product of 75 and a sum that is minimum.

■ 4. i. $f(t) = 4t^2 + 128t^{-1}$

$$f'(t) = 8t - 128t^{-2}$$
$$= 8t - \frac{128}{t^2};$$

ii. $8t - \dfrac{128}{t^2} = 0$

$$8t^3 = 128$$
$$t^3 = 16$$
$$t = 2\sqrt[3]{2} \approx 2.52;$$

iii. $f(2.52) = 76.2$
$(2. 52, 76.2)$,
minimum point.

■ 5. i. $\dfrac{75}{x} \cdot x = 75;$

ii. $(\text{sum}) = f(x) = \dfrac{75}{x} + x;$

iii. $f'(x) = -\dfrac{75}{x^2} + 1$

$$\frac{-75}{x^2} + 1 = 0$$
$$x^2 = 75$$
$$x = 5\sqrt{3};$$

iv. The two numbers are equal to $5\sqrt{3}$.

11-6 Problems

1. Design an open top box with the largest possible volume from a piece of sheet metal 12 inches square by cutting out squares from each corner and turning up the sides.

2. What relationship exists between the diameter and height of the cylindrical container with the greatest volume and least surface area?
3. What are the coordinates of point Q when it is nearest point $P(1, 2)$ as Q moves along the graph of $f(x) = x^2/4$?

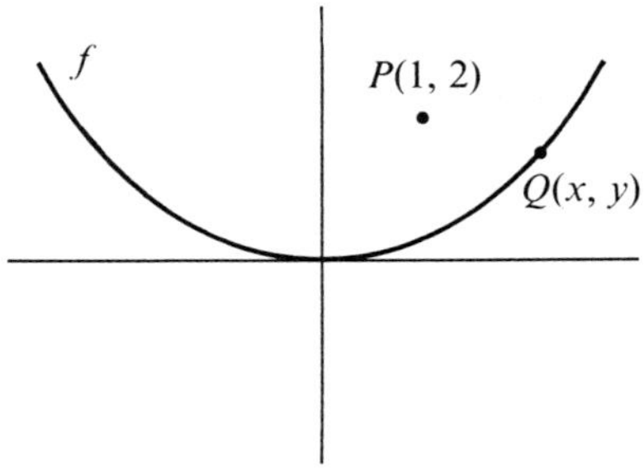

4. Where does the graph of $xy = 4$ in the first quadrant come closest to the origin?
5. Three mining sites are to be joined by a "Y" network of roads. How far from A must the junction be located in order to minimize the road construction?

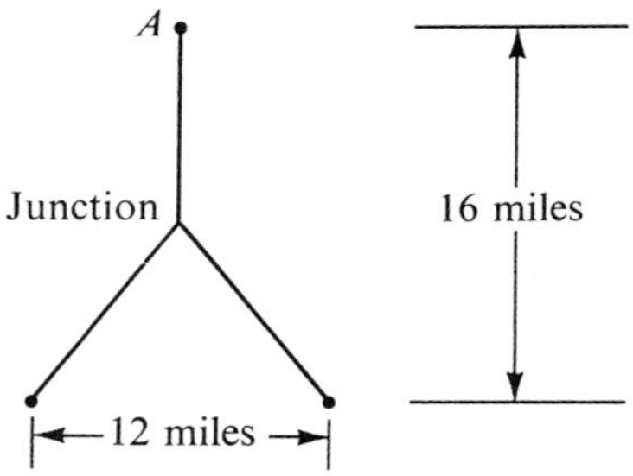

6. A cone can be shaped from sheet metal by cutting out a sector of a circle and bringing together the edges where the sector was removed. Find the angle of the sector that produces the cone with maximum volume.

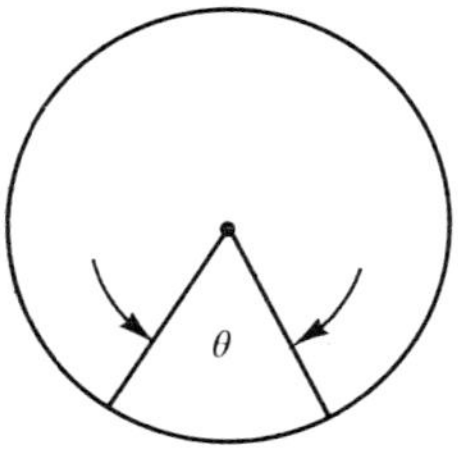

7. The strength of a rectangular beam is proportional to the product of its width and the square of its depth. What are the relative dimensions of the strongest rectangular beam that can be sawn from a circular log?

8. Flight A and B both pass over point P. Flight A is 500 miles west of P and is flying east at 300 mph. Flight B is 250 miles south of P and is flying north at 220 mph. What is the minimum passing distance?
9. Calculate R for maximum power (P) at the speaker. Use the equations

$$IR + 8I = 6 \quad \text{and} \quad P = I^2 R.$$

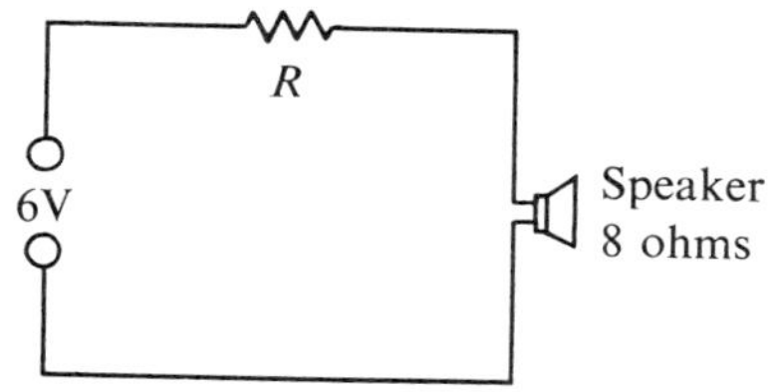

10. Overproduction raises storage cost and underproduction increases the number of production setups. If the total costs (c) of a year's operation are represented by the mathematical model

$$c = \frac{bn}{x} + pn + \frac{ix}{2},$$

where

$n = 12{,}000$ (number of items sold per year),
$b = \$3{,}000$ (cost per setup),
$i = \$4$ (cost per storage per item per year),
$p =$ (cost per item),

and

$x =$ (number of items to be manufactured per setup),

find x for minimum costs.

11-7 Derivatives of Trigonometric Functions

In order to consider an analytical approach to the derivative of sine and cosine functions, we must give attention to the development of two trigonometric identities.

Recall from Section 8-2 that

$$\sin(u + v) = \sin u \cos v + \cos u \sin v$$

and

$$\sin(u - v) = \sin u \cos v - \cos u \sin v.$$

Subtracting the respective sides of these two identities gives

$$\sin(u + v) - \sin(u - v) = \boxed{}. \quad \text{(Eq. 11-7a)}$$

■ $2 \cos u \sin v$

If we let

$$u + v = a$$

and

$$u - v = b, \quad \text{(Eqs. 11-7b)}$$

then the simultaneous solution for u and v is

$$u = \frac{a + b}{2}$$

and

$$v = \frac{a - b}{2}. \quad \text{(Eqs. 11-7c)}$$

Substituting Eqs. 11-7b and 11-7c into Eq. 11-7a gives

$$\boxed{\sin a - \sin b = 2 \cos\left(\frac{a + b}{2}\right) \sin\left(\frac{a - b}{2}\right).} \quad \text{(Eq. 11-7d)}$$

An analogous sequence of steps starting with Eqs. 8-2b and 8-2d results in the identity

$$\boxed{\cos b - \cos a = 2 \sin\left(\frac{a + b}{2}\right) \sin\left(\frac{a - b}{2}\right).} \quad \text{(Eq. 11-7e)}$$

We may now consider the derivative of $f(x) = \sin x$;

$$f'(x) = \lim_{\Delta x \to 0} \frac{f(x+\Delta x) - f(x)}{(x+\Delta x) - x}$$

$$= \lim_{\Delta x \to 0} \frac{\sin(x+\Delta x) - \sin x}{\Delta x}.$$

Now we use Eq. 11-7d. Let

$$a = x + \Delta x,$$

$$b = x,$$

$$\frac{a+b}{2} = \frac{2x+\Delta x}{2},$$

and

$$\frac{a-b}{2} = \frac{\Delta x}{2}.$$

Then

$$f'(x) = \lim_{\Delta x \to 0} \frac{2\cos\left(\frac{2x+\Delta x}{2}\right)\sin\left(\frac{\Delta x}{2}\right)}{\Delta x}$$

$$= \lim_{\Delta x \to 0} \cos\left(x + \frac{\Delta x}{2}\right) \frac{\sin \frac{\Delta x}{2}}{\frac{\Delta x}{2}}.$$

From Example 5, Section 11-1, we know that

$$\lim_{\Delta x \to 0} \frac{\sin \frac{\Delta x}{2}}{\frac{\Delta x}{2}} = 1$$

and

$$\lim_{\Delta x \to 0} \cos\left(x + \frac{\Delta x}{2}\right) = \cos x.$$

Therefore, if

$$f(x) = \sin x,$$

then

$$f'(x) = \cos x. \qquad \text{(Eqs. 11-7f)}$$

An analogous sequence of steps shows that if

$$f(x) = \cos x,$$

then

$$f'(x) = -\sin x. \qquad \text{(Eqs. 11-7g)}$$

When a trigonometric relation is not simply a function of x, we must use the chain rule, Eq. 11-4a.

Example 1

If $f(x) = \sin 2\pi x$, find the equation for f'.

i. Let $f(x) = g(h(x))$, where $g(h(x)) = \sin h(x)$ and $h(x) = 2\pi x$.

ii. $g'(h(x)) = \cos(h(x)) = \cos 2\pi x$
$h'(x) = 2\pi$.

iii. $f'(x) = h'(x)g'(h(x))$
$= 2\pi \cos 2\pi x$.

Example 2

If $f(t) = E/R \sin(120\pi t + \pi/5)$, find f'.

i. Let $f(t) = g(h(t))$ where $h(t) = 120\pi t + \pi/5$ and $g(h(t)) = E/R \sin(h(t))$.

ii. $g'(h(t)) = E/R \cos(h(t)) = E/R \cos(120\pi t + \pi/5)$
$h'(t) = 120\pi$.

iii. $f'(t) = E/R\,(120\pi) \cos(120\pi t + \pi/5)$.

11-7 Exercises

Additional trigonometric derivatives are developed in the exercises.

1. Find the derivative of

$$f(x) = 12 \sin x.$$

---

2. Find the derivative of

$$g(x) = \frac{1}{2} \cos x.$$

■ 1. $f'(x) = 12 \cos x.$

---

3. Use the chain rule (Eq. 11-4a) to differentiate

$$f(x) = \sin^2 x.$$

■ 2. $g'(x) = -\frac{1}{2} \sin x.$

---

4. Use the chain rule to differentiate

$$f(t) = 2 \sin \omega t.$$

■ 3. i. Let $f(x) = g(h(x))$, where $g(h(x)) = (h(x))^2$ and $h(x) = \sin x$.

ii. $g'(h(x)) = 2(h(x)) = 2 \sin x,$
$h'(x) = \cos x,$

iii. $f'(x) = 2 \sin x \cos x = \sin 2x.$

---

5. Use the chain rule to differentiate

$$f(x) = (\sin x)^{-1}.$$

■ 4. i. Let $f(t) = g(h(t))$, where $h(t) = \omega t$ and $g(h(t)) = 2 \sin (h(t))$.

ii. $h'(t) = \omega,$
$g'(h(t)) = 2 \cos(h(t)) = 2 \cos \omega t.$

iii. $f'(t) = 2\omega \cos \omega t.$

---

6. Differentiate $f(x) = \sec x$. (*Hint*: $\sec x = (\cos x)^{-1}$.)

■ 5. i. Let $f(x) = g(h(x))$, where $h(x) = \sin x$ and $g(h(x)) = (h(x))^{-1}$.

ii. $h'(x) = \cos x$, $g'(h(x)) = -(h(x))^{-2}$
$$= -\frac{1}{\sin^2 x}.$$

iii. $f'(x) = -\frac{1}{\sin^2 x} \cos x$
$$= -\cot x \csc x.$$
(*Note:* $f(x) = (\sin x)^{-1} = \csc x$.)

---

7. Use the product rule (Eq. 11-5a) to differentiate
$$f(x) = \sin x \cos x.$$

■ 6. i. $f(x) = \sec x = (\cos x)^{-1}$.

ii. Let $f(x) = g(h(x))$, where $h(x) = \cos x$ and $g(h(x)) = (h(x))^{-1}$.

iii. $h'(x) = -\sin x$, $g'(h(x)) = -(h(x))^{-2}$
$$= -\frac{1}{\cos^2 x}.$$

iv. $f'(x) = \frac{1}{\cos^2 x} \sin x$
$$= \tan x \sec x.$$

---

8. Use the quotient rule (Eq. 11-5b) to differentiate
$$f(t) = \frac{2 \cos t}{3t}.$$

■ 7. i. $f(x) = g(x)h(x)$, where $g(x) = \sin x$ and $h(x) = \cos x$.

ii. $g'(x) = \cos x$, $h'(x) = -\sin x$.

iii. $f'(x) = -\sin^2 x + \cos^2 x = 1 - 2 \sin^2 x$.

---

9. Use the quotient rule to differentiate
$$f(x) = \frac{\sin x}{\cos x}.$$

■ 8. i. $f(t) = \frac{2 \cos t}{3t} = \frac{g(t)}{h(t)}$.

ii. $g'(t) = -2 \sin t$, $h'(t) = 3$.

iii. $f'(t)$

$$= \frac{-3t(2 \sin t) - 3(2 \cos t)}{9t^2}$$

$$= \frac{-2t \sin t - 2 \cos t}{3t^2}.$$

---

10. Differentiate $f(x) = \cot x$.

$\left(Hint: \cot x = \dfrac{\cos x}{\sin x}.\right)$

■ 9. i. $f(x) = \dfrac{\sin x}{\cos x} = \dfrac{g(x)}{h(x)}$.

ii. $g'(x) = \cos x$,

$h'(x) = -\sin x$.

iii. $f'(x) = \dfrac{\cos^2 x + \sin^2 x}{\cos^2 x}$

$= \dfrac{1}{\cos^2 x} = \sec^2 x$.

$\left(Note: \dfrac{\sin x}{\cos x} = \tan x.\right)$

---

11. If $f(t) = Y \sin (2\pi ft + \Psi)$, find the equation for f'.

■ 10. i. $f(x) = \dfrac{\cos x}{\sin x} = \dfrac{g(x)}{h(x)}$.

ii. $g'(x) = -\sin x$,
$h'(x) = \cos x$.

iii. $f'(x)$

$$= \frac{-\sin^2 x - \cos^2 x}{\sin^2 x}$$

$= -\csc^2 x$.

■ 11. i. $f(t) = g(h(t))$, where $h(t) = (2\pi ft + \Psi)$ and $g(h(t)) = Y \sin (h(t))$.

ii. $h'(t) = 2\pi f$,

$g'(h(t)) = Y \cos (h(t))$

$= Y \cos (2\pi ft + \Psi)$.

iii. $f'(t) = 2\pi f Y \cos (2\pi ft + \Psi)$.

11-7 Problems

1. Summarize the results of the examples and exercises of this section by completing the following chart.

Function (f)	Derivative (f')
$f(x) = \sin\ x$	
$f(x) = \cos x$	
$f(x) = \tan x$	
$f(x) = \csc x$	
$f(x) = \sec x$	
$f(x) = \cot x$	

2. What is the value of x when the slope of $g(x) = \sin x$ is greatest? What is the greatest slope?
3. What is the value of t when the rate of change of $f(t) = \sin \omega t$ is greatest? What is the greatest slope?
4. At what $t > 0$ does the first maximum of $f(t) = 12 \sin (4\pi t)$ occur?
5. Find the critical points in the first period of $f(t) = 16 \cos (50\pi t + \pi/12)$.
6. The periodic displacement of a point is given by the equation $f(t) = 6.5 \cos 20\pi t$. The equation for f' represents the velocity of the point. What is the velocity at $t = .02$?
7. What is the slope of $f(t) = \tan t$ at $t = \pi/6$? $t = \pi/3$?
8. If

$$i = 0.2 \sin 120\ t$$

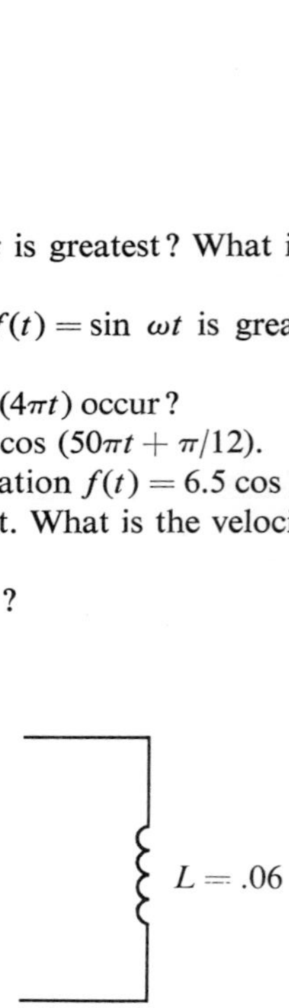

and

$$e = L \frac{di}{dt},$$

find e at $t = .01$.

$$\left(Note:\ i = f(t), \frac{di}{dt} = f'(t).\right)$$

9. Find the derivative of $f(t) = r^2 - (s \sin t)^2$.
10. The position of a piston operating off a crank shaft is

$$f(t) = s \sin \omega t + \sqrt{r^2 - s^2 \cos^2 \omega t} - \sqrt{r^2 - s^2}.$$

Find the equation for f', which is equivalent to the velocity of the piston.

11 Supplementary Exercises and Problems

1. If $f(x) = x$, find $\lim_{x \to 0} f(x)$. (11-1)
2. If $f(x) = x^2 + 2x + 3$, find $\lim_{x \to 0} f(x)$. (11-1)
3. If $f(x) = (x^2 + 2x)/x$, find $\lim_{x \to 0} f(x)$. (11-1)
4. If $f(x) = (x + 5x)/x$, find $\lim_{x \to 0} f(x)$. (11-1)
5. What do we call the limit of $\overline{AB}$ as d approaches zero? (11-1)

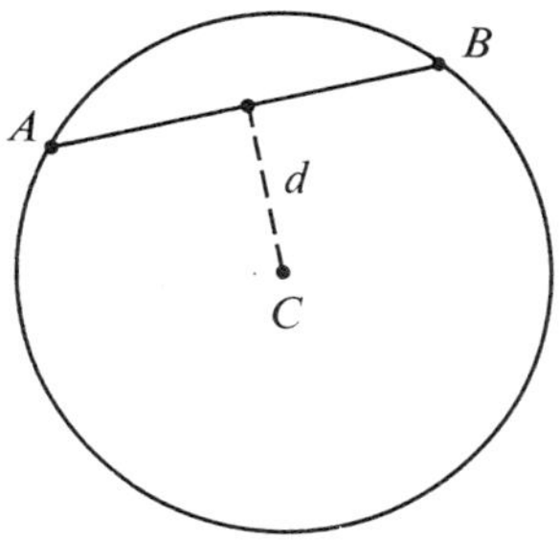

6. F_1 and F_2 are foci of an ellipse L. What is the limit of L as $\overline{F_1F_2}$ approaches zero? (11-1)

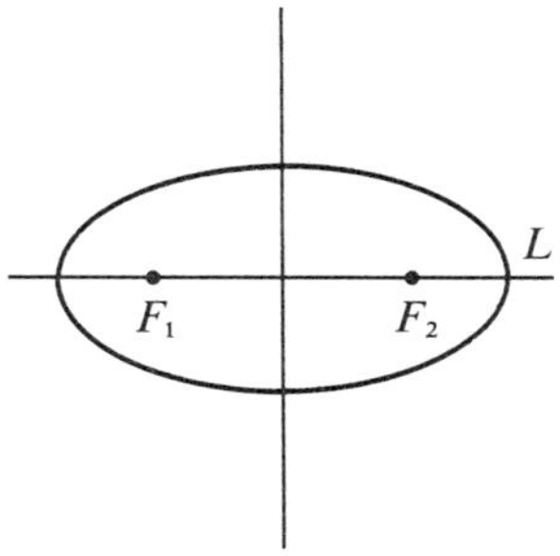

7. L is a parabola with focus F, vertex V, and directrix D. What is the limit of L as F approaches D? (11-1)

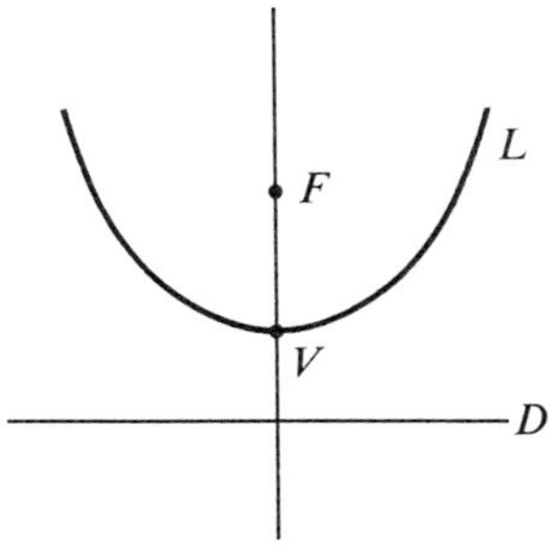

8. If $f(x) = (x + 5)/x$, find $\lim_{x \to 0} f(x)$. (11-1)
9. If $f(x) = x^2/(x + 4)$, find $\lim_{x \to 0} f(x)$. (11-1)
10. If $f(x) = \sin x$, find $\lim_{x \to 0} f(x)$. (11-1)
11. If $f(x) = k - x$, find $\lim_{x \to k} f(x)$. (11-1)
12. If $f(x) = 2x^2$, find $\lim_{x \to a}(f(x) - f(a))$. (11-1)
13. Use Eq. 11-2a to find f' when $f(x) = 5x - 2$. Evaluate $f'(3)$ and $f'(10)$. (11-2)
14. Calculate the derivative of $f(t) = (t^2/2) + 3t$. (11-2)
15. Using the results of Problem 14, for what values of t is the slope of f equal to -1 and 1? (11-2)
16. Angular distance is represented by the equation $f(t) = 20\pi t + \pi$. The derivative $f'(t)$ represents angular velocity. Find f'. (11-2)
17. Use Eq. 11-2a to find f' when $f(x) = 1/(x + 1)$. For what x is $f'(x)$ undefined? (11-2)
18. Using the abbreviated methods of Section 11-3, find $f'(.05)$ when $v = f(t) = 45t - 30$. (11-3)
19. Find $g'(1.2)$ when $g(t) = 4(t^2 - 3t)$. (11-3)
20. Find $h'(0)$ when $h(t) = (\pi t^2 - t)/4$. (11-3)
21. Find $f'(1/8)$ when $f(s) = 16s^{3/4} - 6.5$. (11-3)
22. Over what interval of x's is $-5 \leq f'(x) \leq -1$ when $f(x) = 1/x$? (11-3)
23. Over what interval of x's is $|f'(x)| \leq 1$ when $f(x) = -3.2x^2 - 1.2x + 4$? (11-3)
24. Find i at $t = .001$ using the formula $i = C(dv/dt)$, when $C = .002$, $v = f(t) = 140t + 20$, and $dv/dt = f'(t)$. (11-3)
25. Sketch the derivative of the function illustrated. (11-3)

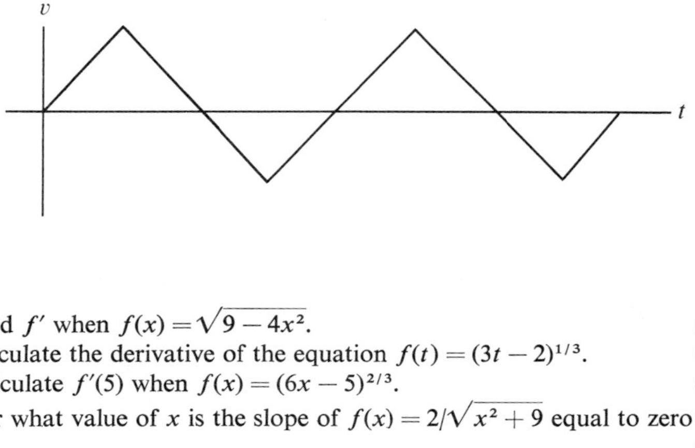

26. Find f' when $f(x) = \sqrt{9 - 4x^2}$. (11-4)
27. Calculate the derivative of the equation $f(t) = (3t - 2)^{1/3}$. (11-4)
28. Calculate $f'(5)$ when $f(x) = (6x - 5)^{2/3}$. (11-4)
29. For what value of x is the slope of $f(x) = 2/\sqrt{x^2 + 9}$ equal to zero? (11-4)
30. Find f' when $f(C) = 1/2\pi\sqrt{LC}$. (11-4)
31. Find dV/dt when $V = (4/3)\pi r^3$ and $r = 10t$. (11-4)
32. If $i_p = k(v_g + v_p/c)^{3/2}$, find the mutual conductance

$$g_m = \frac{di_p}{dv_g}. \qquad (11\text{-}4)$$

33. Find the derivative of $f(x) = (x^2 + 3)(x^2 + 7)$. (11-5)
34. Find the derivative of $g(t) = (t + 3)/(t - 3)$. (11-5)
35. Differentiate $h(v) = 1/v^{1.5}$. (11-5)
36. Differentiate $f(x) = \sqrt{x^2 - 5/x^2}$. (11-5)
37. Differentiate $g(w) = 2w/\sqrt{w^2 + 9}$. (11-5)
38. Find the critical point(s) of $f(x) = -4x^2 + 5x + 2$. (11-6)
39. Find the critical point(s) of $g(x) = x^3 - 4x^2 + 2x$. (11-6)
40. What is the maximum volume of a container that can be constructed from a sheet of cardboard 24 inches by 15 inches by cutting out corners and folding up the sides in the manner illustrated? (11-6)

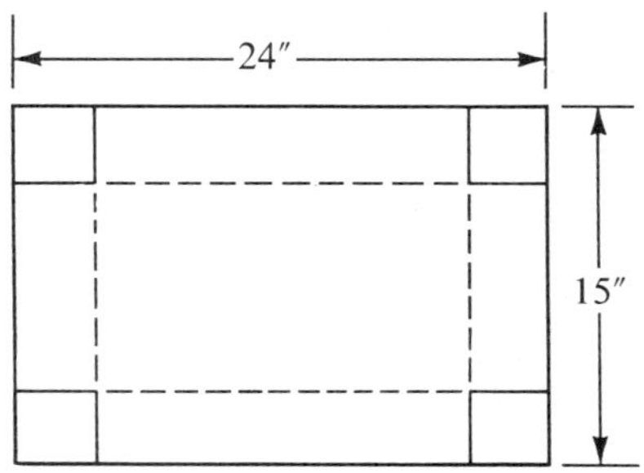

41. The operating cost of a certain machine is proportional to the cube of the cutting speed. The cost is \$12 per hour when the cutting speed is 2000 inches per hour plus a fixed charge of \$5 per hour. Determine the most economical speed for an operation that requires 40,000 machining inches.
42. A manufacturer has to produce 50,000 parts a year and wishes to select the most economical lot size to be manufactured during equally spaced intervals of time over the year. The annual cost of storage varies directly as the lot size and is \$1000 when the entire annual requirement is manufactured in one lot. The cost of setting up and dismantling for each run is \$40. What is the most economical lot size? (11-6)
43. Find the derivative of $f(x) = \cos(x/2)$. (11-7)
44. Differentiate $f(x) = \sin(6 - x)$. (11-7)
45. Differentiate $g(x) = \cos^2 x$. (11-7)
46. Differentiate $g(x) = \sin^2 3x$. (11-7)
47. Find the derivative of $f(t) = 110 \sin 2\pi ft$. (11-7)
48. Find the derivative of $h(t) = \sin 10t \cos 10t$. (11-7)
49. Differentiate $g(t) = t^2 \sin 2t$. (11-7)
50. Find a minimum and maximum point for the equation $f(x) = \sin x \cos x$. (11-7)

Chapter Twelve Integral Calculus

12-1 Indefinite Integral—Antiderivative

The *derivative* of an equation expressing the relationship between time and distance results in an equation that gives the functional relationship between time and velocity. Assume that we know an equation for velocity in terms of time and desire the related equation of distance in terms of time.

The problem involves first knowing whether or not the processes of finding derivatives, studied in Chapter 11, can be reversed. For example, if the equation for velocity is $f(t) = 32t$, can we find the equation for distance as a function of time? In other words, can we find a function whose derivative is $f(t) = 32t$?

By experimentation, we discover that the equation $F(t) = 16t^2$ meets the requirements because $F'(t) = \boxed{}$.

■ $32t$

The act of finding a function whose derivative is known is called finding the **antiderivative**.

Example 1

Find the antiderivative of $f(x) = 3$.
Our experience with differentiating functions should suggest that $F(x) = 3x$ is the antiderivative of $f(x) = 3$ because $F'(x) = \boxed{}$.

■ 3

But further thought will remind us that $F(x) = 3x + c$ also has the derivative $F'(x) = 3$, for any constant c.

Generally we may say that $F(x) + c$ is equal to the antiderivative of $f(x)$ whenever the derivative of $F(x) + c$ is equal to $f(x)$. The symbol that is equivalent to $F(x) + c$ in terms of $f(x)$ is called the **indefinite integral** of $f(x)$ and is written

$$\int f(x)\,dx = F(x) + c. \qquad \text{(Eq. 12-1a)}$$

Example 2

If $f(x) = 4x + 3$, find the indefinite integral $\int f(x)\,dx$.

$$\begin{aligned}\int f(x)\,dx &= \int (4x + 3)\,dx \\ &= 2x^2 + 3x + c.\end{aligned}$$

An explanation for why the symbols $\int$ and dx appear in the notation is delayed until Section 12-4.

The value of the constant c is indefinite unless we know at least one solution to the indefinite integral.

Example 3

The pressure on a long rectangular observation window on a reservoir is a function of height and is given by the formula

$$f(h) = 624 - 62.4h.$$

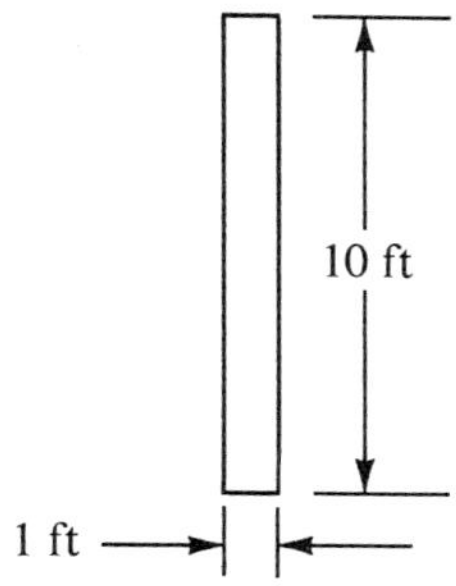

The total force on the window is the integral of f. If the force is zero when $h = 0$, find the equation for total force.

i. (Force) $= F(h) = \int f(h)\, dh$

$$= \int (624 - 62.4h)\, dh$$
$$= 624h - 31.2h^2 + c.$$

ii. Since $F(0) = 0$, $c = 0$ and the formula for total force becomes

$$F(h) = 624h - 31.2h^2.$$

12-1 Exercises

Use an experimental approach to find the indefinite integrals of the following functions. Check the results by differentiation.

1. $f(x) = 1$.

2. $f(x) = 5x$.

■ 1. $\int f(x)\, dx = \int (1)\, dx = x + c.$

3. $f(x) = x/4$.

■ 2. $\int f(x)\, dx = \int (5x)\, dx = \frac{5}{2} x^2 + c.$

4. $f(x) = 6x + 5$.

■ 3. $\int f(x)\, dx = \int (x/4)\, dx = \frac{x^2}{8} + c.$

5. $f(w) = 12w$.

■ 4. $\int f(x)\, dx = \int (6x + 5)\, dx = 3x^2 + 5x + c.$

---

6. $f(t) = t^2.$

■ 5. $\int f(w)\,dw = \int (12w)\,dw$
$= 6w^2 + c.$

(Notice that the indefinite integral notation is altered to accommodate the variable w.)

---

7. $f(x) = 3x^2 + x.$

■ 6. $\int f(t)\,dt = \int (t^2)\,dt$
$= \frac{t^3}{3} + c.$

---

8. $f(x) = 4x^3 + x^2 - 3x.$

■ 7. $\int f(x)\,dx = \int (3x^2 + x)\,dx$
$= x^3 + \frac{x^2}{2} + c.$

---

9. $f(x) = \frac{1}{x^2}.$

■ 8. $\int f(x)\,dx$
$= \int (4x^3 + x^2 - 3x)\,dx$
$= x^4 + \frac{x^3}{3} - \frac{3x^2}{2} + c.$

---

10. $f(x) = \sqrt{x}.$

■ 9. $\int f(x)\,dx = \int \left(\frac{1}{x^2}\right) dx$
$= \int x^{-2}\,dx$
$= -x^{-1} + c$
$= -\frac{1}{x} + c.$

■ 10. $\int f(x)\,dx = \int (\sqrt{x})\,dx$
$= \int (x^{1/2})\,dx$
$= \frac{2x^{3/2}}{3} + c.$

12-1 Problems

1. The velocity of an experimental vehicle is $f(t) = 20t$ (feet per second). What is the related equation for distance with respect to time? If at $t = 0$ the distance is zero, how far does the vehicle travel in 10 seconds?
2. The equation that relates time (t) and electrical charge (q) is the integral of the equation relating current (i) and time. If $i = f(t) = t^{-2}$, find the equation for q as a function of time.
3. *Marginal cost* (M) is the rate of change of the cost of producing an item. Total cost (C) is the integral of M. When $M = .00012x^2 + .12x + 10$, where x is the number of items, find the equation for total cost. If the initial costs are \$20,000 before the first item is manufactured, what is the total cost after producing 1000 items?
4. When force (F) is a function of displacement (s), then the integral of force is an expression for energy (E) as a function of displacement. When $F = -ks$, find the equation for E, given $E = 0$, when $s = 0$.

12-2 Integration Formulas

For every derivative formula given in Section 11-3 there is a corresponding integration formula. Our purpose in this section is to introduce some formulas of integration that will say specifically what we must do to integrate elementary polynomial functions. Although Eq. 12-1a clearly specifies the relationship that exists between $f(x)$ and the indefinite integral of $f(x)$, it does not specify in what manner the indefinite integral is found.

We may be sure that $F(x) + c = \int f(x)\,dx$ is the indefinite integral of $f(x)$ if the derivative of $F(x) + c$ is equal to $f(x)$. It is from this basis that we proceed to develop some of the simpler integration formulas.

The integration formula corresponding to Eq. 11-3a is

$$\int x^n\,dx = \frac{x^{n+1}}{n+1} + c, \quad n \neq -1. \qquad \text{(Eq. 12-2a)}$$

This can be confirmed by differentiating

$$F(x) = \frac{x^{n+1}}{n+1}. \quad F'(x) = \boxed{}.$$

■ $(n+1)\dfrac{x^{(n+1)-1}}{n+1} = x^n.$

Example 1

Integrate $f(x) = x^{1/2}$ using Eq. 12-2a.

$$\begin{aligned}\int f(x)\,dx &= \int x^{1/2}\,dx \\ &= \frac{x^{1/2+1}}{(1/2)+1} + c \\ &= \frac{2x^{3/2}}{3} + c.\end{aligned}$$

We can see that the integral for $f(x) = x^{-1}$ using Eq. 12-2a is meaningless. In more advanced work the integral of $f(x) = x^{-1}$ is shown to be $\ln x + c$. Equation 12-2a does, however, apply to all other rational powers of x.

Let $n = 0$, so that $f(x) = x^0$. Then $\int x^0\,dx =$ [ ].

■ $\dfrac{x^{0+1}}{0+1} + c = x + c.$

These results are ordinarily summarized in the formula

$$\int dx = x + c, \qquad \text{(Eq. 12-2b)}$$

recognizing, of course, that $x^0 = 1$ for all x.

The equation

$$\int af(x)\,dx = a\int f(x)\,dx \qquad \text{(Eq. 12-2c)}$$

corresponds to Eq. 11-3c.

Example 2

Find the indefinite integral of the equation $f(x) = 5x^2$.

$$\int (5x^2)\,dx = 5\int x^2\,dx \qquad \text{(By Eq. 12-2c)}$$

$$= 5\left(\frac{x^3}{3}\right) + c. \qquad \text{(By Eq. 12-2a)}$$

Since the antiderivative of $f(x) = g(x) \pm h(x)$ is $F(x) = G(x) \pm H(x)$, it follows that

$$\int (g(x) \pm h(x))\,dx = \int g(x)\,dx \pm \int h(x)\,dx. \qquad \text{(Eq. 12-2d)}$$

Example 3

Find the indefinite integral of $f(x) = 4x^2 - x + 2$.

$$\int (4x^2 - x + 2)\,dx = \int (4x^2)\,dx + \int (-x)\,dx + \int (2)\,dx \qquad \text{(By Eq. 12-2d)}$$

$$= 4\int x^2\,dx - \int x\,dx + 2\int dx \qquad \text{(By Eq. 12-2c)}$$

$$= \frac{4x^3}{3} - \frac{x^2}{2} + 2x + c. \qquad \text{(By Eq. 12-2a)}$$

We now have the tools for integrating polynomial functions. The problems of understanding the mechanics for integrating more complicated algebraic functions and other logarithmic, exponential, and trigonometric functions require special considerations beyond the scope of this text. It is, however, possible for us to make use of *tables of integration formulas* and integrate functions not considered in this chapter. The important fact, in the case of using tables, is that we realize what it is that integration accomplishes. It is to this point that we address ourselves in the remaining sections.

12-2 Exercises

1. Find the indefinite integral of

$$f(x) = 12x^2.$$

---

2. Find the indefinite integral of

$$g(t) = t^{1/2} - 1.$$

■ 1. $$\int f(x)\,dx = \int 12x^2\,dx = 12 \int x^2\,dx = 12\frac{x^3}{3} + c = 4x^3 + c.$$

---

3. Find the indefinite integral of

$$f(w) = \frac{w^2 - w}{2}.$$

■ 2. $$\int g(t)\,dt = \int (t^{1/2} - 1)\,dt = \int t^{1/2}\,dt - \int dt = \frac{2t^{3/2}}{3} - t + c.$$

---

4. Find the indefinite integral of

$$f(t) = \frac{6.2(10^4)}{t^2}.$$

■ 3. $$\int f(w)\,dw = \int \frac{w^2 - w}{2}\,dw = \frac{1}{2}\left(\int w^2\,dw - \int w\,dw\right) = \frac{1}{2}\left(\frac{w^3}{3} - \frac{w^2}{2}\right) + c = \frac{w^3}{6} - \frac{w^2}{4} + c.$$

---

5. Find the indefinite integral of

$$f(x) = \frac{2\pi}{x}.$$

■ 4. $$\int f(t)\,dt = \int \frac{6.2(10^4)}{t^2}\,dt = 6.2(10^4) \int t^{-2}\,dt = 6.2(10^4)(-t^{-1}) + c = \frac{-6.2(10^4)}{t} + c.$$

6. Given $v = \frac{1}{C}\int i\,dt$,

$i = f(t) = .03t^{0.6}$,

$C = .0006$,

$v = 0$ when $t = 0$.

Find v at $t = .2$.

■ 5. $\int f(x)\,dx = \int (2\pi x^{-1})\,dx$
$= 2\pi \int x^{-1}\,dx$
$= 2\pi \ln x + c.$

We remember that Eq. 12-2a holds for all $n \neq -1$. If you remembered from the reading that $\int x^{-1}\,dx = \ln x + c$, that is very good. Tables of indefinite integrals give the same information.

---

7. Integrate $\int (x-2)^2\,dx$.

■ 6. i. $v = \frac{1}{C}\int i\,dt$

$$= \frac{1}{.0006}\int (.03t^{0.6})\,dt$$

$$= 50\left(\frac{t^{1.6}}{1.6}\right) + c$$

$$= 31.25t^{1.6} + c.$$

ii. Since $v = 0$ when $t = 0$, c must equal zero.

iii. $v = 31.25(.2)^{1.6}$
$\approx 31.25(.07614)$
$\approx 2.379.$

---

8. Starting with $f(x) = c_0$, perform the integrating process twice.

■ 7. $\int (x-2)^2\,dx$

$$= \int (x^2 - 4x + 4)\,dx$$

$$= \frac{x^3}{3} - 2x^2 + 4x + c.$$

---

9. The equation of the slope of a function F is $f(x) = 5x + 4$. Find the equation of F if $F(2) = 15$.

■ 8. i. $\int c_0\,dx = c_0 x + c_1.$

ii. $\int (c_0 x + c_1)\,dx$

$$= \frac{c_0}{2}x^2 + c_1 x + c_2.$$

■ 9. i. $F(x) = \int f(x)\,dx$

$$= \int (5x + 4)\,dx$$

$$= \frac{5x^2}{2} + 4x + c.$$

ii. $\frac{5(2)^2}{2} + 4(2) + c = 15$

$$10 + 8 + c = 15$$

$$c = -3.$$

iii. $F(x) = \frac{5x^2}{2} + 4x - 3.$

12-2 Problems

1. Given: $v = \int a\,dt$,
   $a = f(t) = 9.8$ (meters/sec^2),
   $v = 2.0$ (meters/sec) when $t = 0.0$ (seconds).
   Find v, when $t = 10$ (seconds).
2. The equation of the slope of a curve is $f'(x) = 1/2x - 3$. Find the equation of the curve if it passes through the point (2, 4).
3. An airplane has an acceleration of 12 feet per second per second and is airborne at a velocity of 150 feet per second. Find the length of runway used in take-off. (*Hint*: Velocity is the integral of acceleration in terms of time, and distance is the integral of velocity in terms of time.)
4. Graphical approximations of integrals are an important part of some technical work. Sketch the integral of the function f (Fig. 12-2a), knowing that it passes through the point (0, 2).

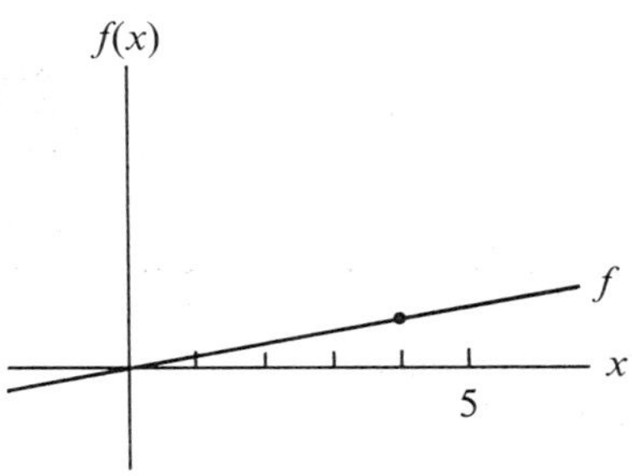

Fig. 12-2a

5. If the function f in Problem 4 is linear, determine its equation and the solution to $\int f(x)\,dx$.

6. Draw the integral of the periodic voltage illustrated in Fig. 12-2b.

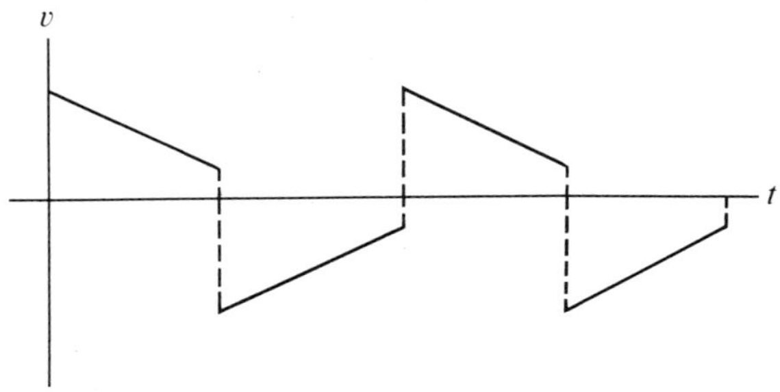

Fig. 12-2b

7. In beam design, Fig. 12-2c is called a *shear diagram*. The integral of the shear diagram is the *moment diagram*. Sketch the moment diagram given that it begins at the origin of the graph.

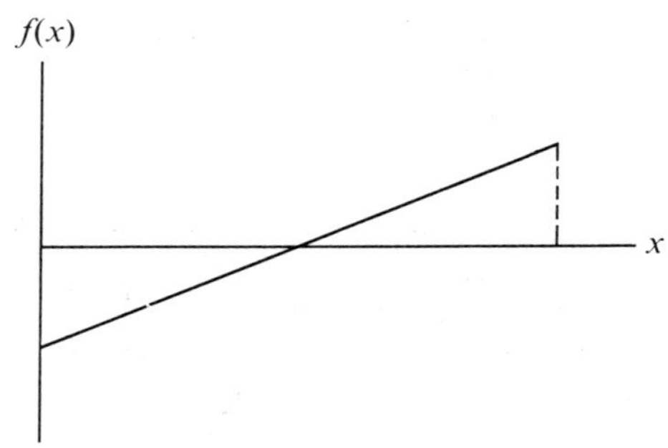

Fig. 12-2c

8. If the shear diagram in Fig. 12-2c has an equation in the general form $f(x) = wx - k$, what is the general form of the equation for the moment diagram?

12-3 Area under a Curve—Definite Integral

The discussion of area enclosed by the graph of a function, the x-axis, $x = a$, and $x = b$ appeared early in the text. We should know that the area bounded in this manner is a function of x (the independent variable), and may

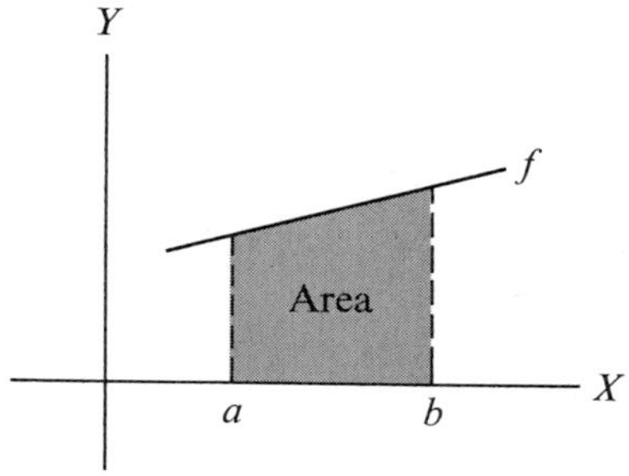

Fig. 12-3a

have physical significance if the function to which it is related has physical meaning. A review of Sections 1-7 and 3-6 will help us recall some of the physical relationships that exist between a function f and its related area function F.

f function	Related F function
(time, acceleration)	(time, velocity)
(time, velocity)	(time, distance)
(displacement, force)	(displacement, work)
(time, current)	(time, charge)
(length, weight)	(length, bending moment)
(time, current)	(time, voltage)

If f is a linear function, the related area function F may be determined by means of formulas for the areas of triangles and trapezoids.

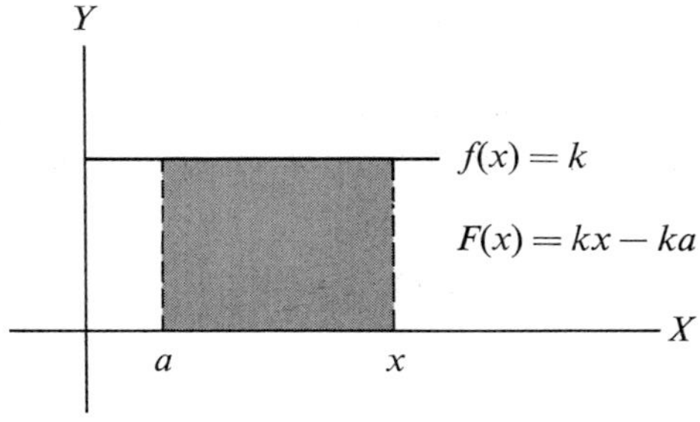

Fig. 12-3b

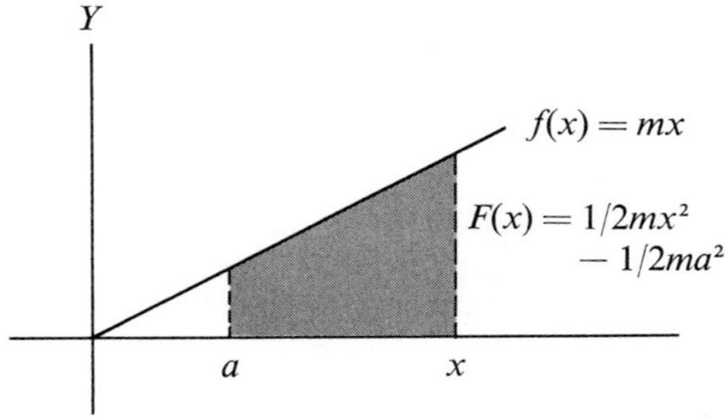

Fig. 12-3c

We have delayed, until now, discussing the precise methods for determining area functions when the graphs of the related functions are nonlinear.

Let $A(x)$ be the area under the graph of f between the ordinates at a and x as illustrated in Fig. 12-3d. Consider the derivative of $A(x)$ in terms of Eq. 11-2a;

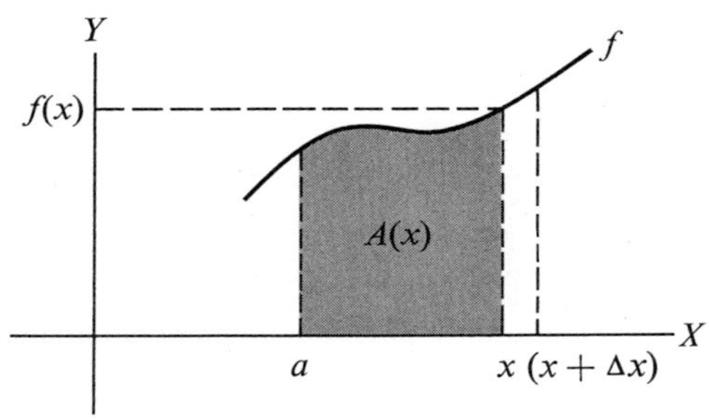

Fig. 12-3d

$$A'(x) = \lim_{\Delta x \to 0} \frac{A(x + \Delta x) - A(x)}{(x + \Delta x) - x}.$$

The difference $A(x + \Delta x) - A(x)$ is represented in Fig. 12-3d as the area between the ordinates x and $x + \Delta x$ and under the graph of f. For very small values of Δx,

$$A(x + \Delta x) - A(x) \approx f(x)\, \Delta x,$$

whereas

$$\lim_{\Delta x \to 0} \frac{A(x + \Delta x) - A(x)}{\Delta x} = \lim_{\Delta x \to 0} \frac{f(x)\, \Delta x}{\Delta x}.$$

Therefore

$$A'(x) = \lim_{\Delta x \to 0} \frac{f(x)\,\Delta x}{\Delta x} = f(x).$$

Looking back at Figs. 12-3b and 12-3c, we see that this is precisely the relationship that exists between $F(x)$ and $f(x)$, namely, $F'(x) = f(x)$.

These results introduce us to the very important relationship that exists between indefinite integrals and areas under a graph:

$$A(x) = F(x) + c = \int f(x)\,dx. \qquad \text{(Eq. 12-3a)}$$

The area $A(x)$ and $F(x)$ differ only by a constant c.

Since areas are obviously discrete values, we should be able to determine specifically the constant c. Using Eq. 12-3a and referring to Fig. 12-3d, we have

$$A(x) = F(x) + c.$$

When $x = a$, then

$$A(a) = F(a) + c = 0,$$

or

$$c = -F(a).$$

If we specify the right-hand abscissa as b, we may finally write

$$\int_a^b f(x)\,dx = F(b) - F(a) \qquad \text{(Eq. 12-3b)}$$

to designate the area illustrated in Fig. 12-3e. Eq. 12-3b is known as the *fundamental theorem of calculus* and defines the **definite integral** of $f(x)$ where $F(a)$ and $F(b)$ are the *antiderivatives* of $f(x)$ evaluated at the *lower* and *upper limits* a and b.

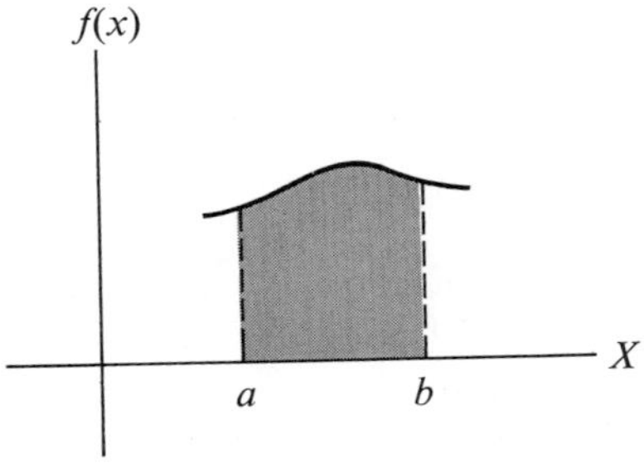

Fig. 12-3e

Example 1

If $a = 1$, $b = 3$, and $f(x) = (1/3)x^2 + 2$, find the area illustrated in Fig. 12-3f.

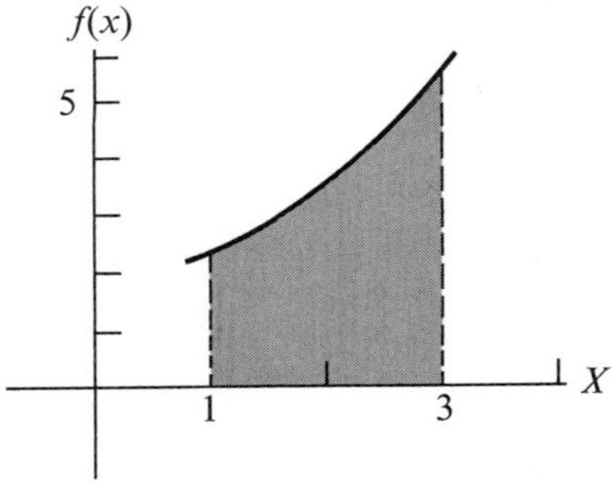

Fig. 12-3f

i. $\int_1^3 \left(\frac{1}{3}x^2 + 2\right)dx = F(3) - F(1);$

ii. $F(x) = \frac{x^3}{9} + 2x$

$$F(3) = \frac{(3)^3}{9} + 2(3) = 9$$

$$F(1) = \frac{(1)^3}{9} + 2(1) = \frac{19}{9};$$

iii. $F(3) - F(1) = 9 - \frac{19}{9} = \frac{62}{9}.$

Example 2

The force (P) on a spring is a function of displacement (x). If $P = 6x$, find the work (W) required to displace the spring from $x = 2$ to $x = 3$.

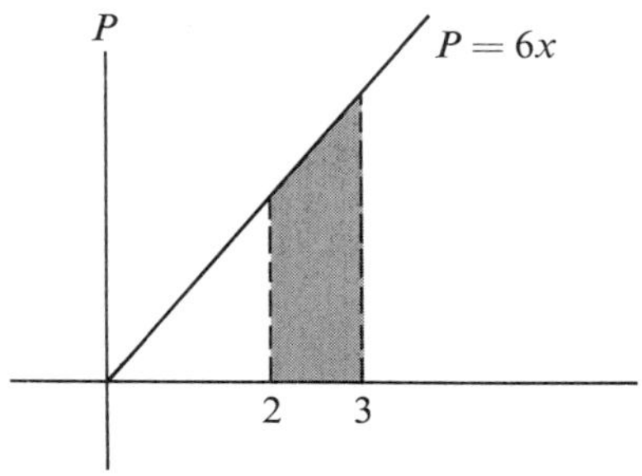

Fig. 12-3g

The work required is equivalent to the area designated in Fig. 12-3g. Therefore

$$W = \int_2^3 6x\,dx$$
$$= 3x^2]_2^3$$
$$= 3(3)^2 - 3(2)^2 = 27 - 12 = 15.$$

Notice the form $3x^2]_2^3$ that is used for evaluating the definite integral. It permits us to write the integral of $6x$ before substituting the limits 3 and 2. Without this form, we are compelled to use steps like those appearing in Example 1.

12-3 Exercises

In Exercises 1–7, evaluate the definite integrals and sketch the areas they represent.

1. $\int_1^2 3x\,dx = \boxed{}$

2. $\int_0^2 x^3\,dx =$ ☐

■ 1.

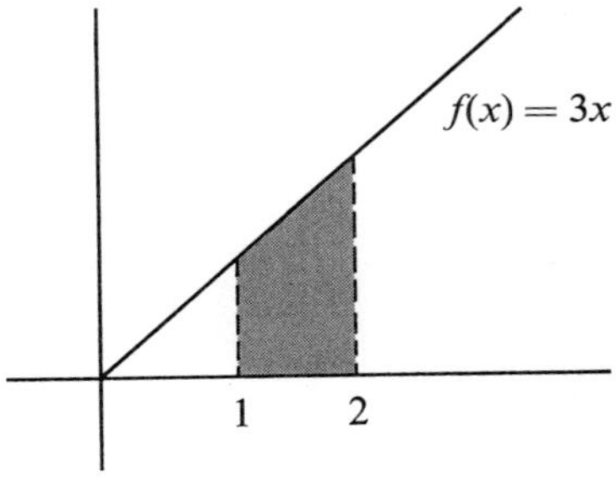

$$\int_1^2 3x\,dx = \left.\frac{3x^2}{2}\right]_1^2 = \frac{3(2)^2}{2} - \frac{3(1)^2}{2} = 6 - \frac{3}{2} = \frac{9}{2}.$$

3. $\int_{-2}^2 x^2\,dx =$ ☐

■ 2.

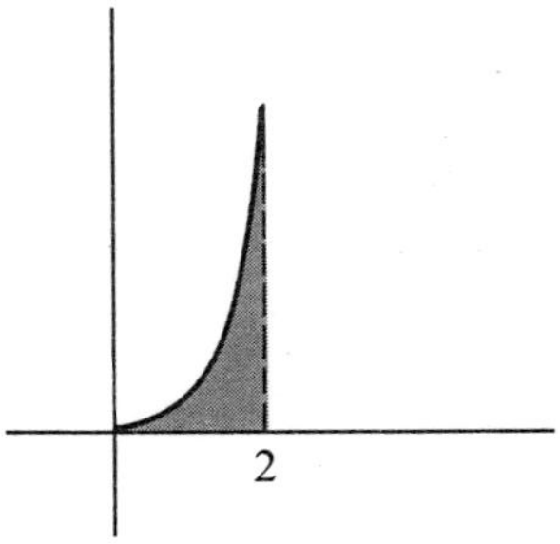

$$\int_0^2 x^3\,dx = \left.\frac{x^4}{4}\right]_0^2 = \frac{2^4}{4} - \frac{0^4}{4} = 4.$$

4. $\int_1^4 x^{5/2}\,dx = \boxed{}$

■ 3.

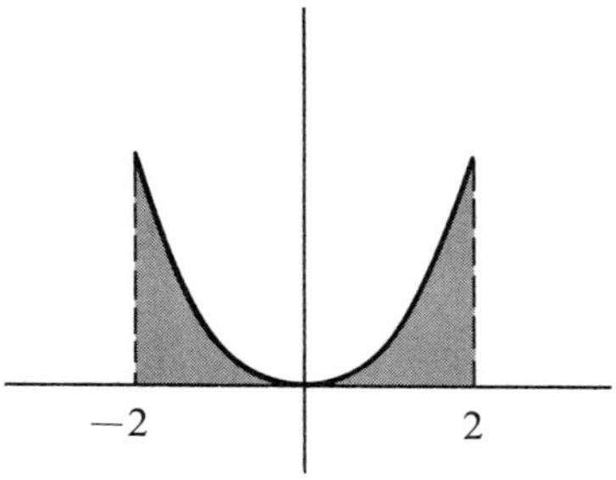

$$\int_{-2}^{2} x^2\,dx = \frac{x^3}{3}\Big]_{-2}^{2} = \frac{(2)^3}{3} - \frac{(-2)^3}{3} = \frac{8}{3} + \frac{8}{3} = \frac{16}{3}.$$

5. $\int_0^2 (x^2 + x + 1)\,dx$

■ 4.

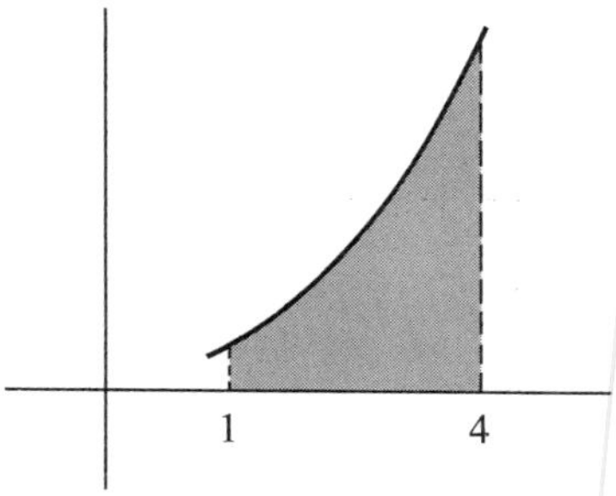

$$\int_1^4 x^{5/2}\,dx = \frac{x^{7/2}}{7/2}\Big]_1^4 = \frac{2x^{7/2}}{7}\Big]_1^4 = \frac{2(4)^{7/2}}{7} - \frac{2(1)^{7/2}}{7}$$

$$= \frac{256}{7} - \frac{2}{7} = \frac{254}{7}.$$

---

6. $\int_1^4 x\sqrt{x}\,dx$

■ 5.

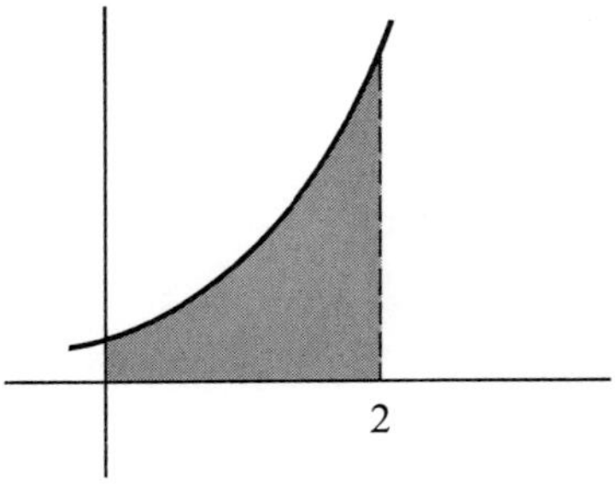

$$\int_0^2 (x^2 + x + 1)\,dx$$

$$= \frac{x^3}{3} + \frac{x^2}{2} + x \Big]_0^2$$

$$= \frac{(2)^3}{3} + \frac{(2)^2}{2} + 2 - 0$$

$$= \frac{8}{3} + 2 + 2 = \frac{20}{3}.$$

---

7. $\int_{-1}^2 (x+1)^2\,dx$

■ 6.

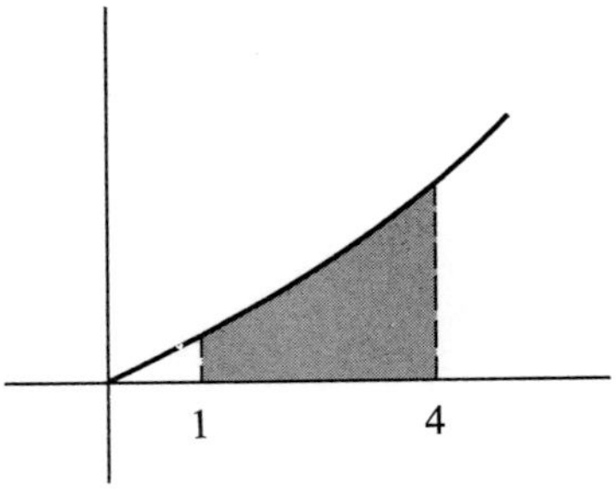

$$\int_1^4 x\sqrt{x}\,dx$$

$$= \int_1^4 x^{3/2}\,dx$$

$$= \frac{2x^{5/2}}{5}\Bigg]_1^4 = \frac{2(4)^{5/2}}{5} - \frac{2(1)^{5/2}}{5}$$

$$= \frac{64}{5} - \frac{2}{5} = \frac{62}{5}.$$

---

8. The **average ordinate** of the graph of a function f in the interval from a to b is

$$y_{(av)} = \frac{\int_a^b f(x)\,dx}{b-a}.$$

Find the average ordinate of the intervals designated in Exercises 1 and 6.

■ 7.

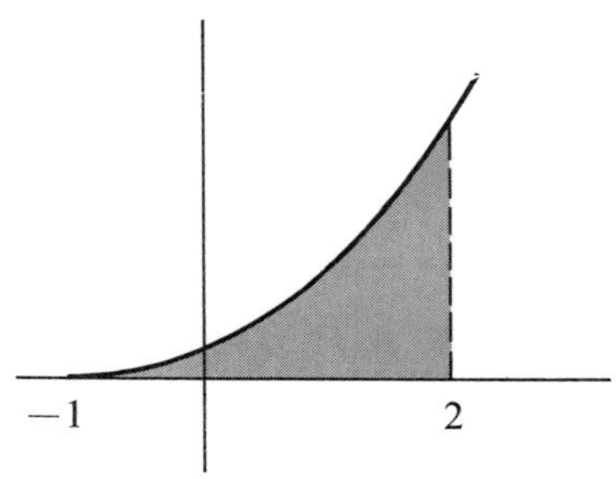

$$\int_{-1}^{2} (x+1)^2\,dx$$

$$= \int_{-1}^{2} (x^2 + 2x + 1)\,dx$$

$$= \frac{x^3}{3} + x^2 + x\Bigg]_{-1}^{2}$$

$$= \left(\frac{(2)^3}{3} + (2)^2 + 2\right) - \left(\frac{(-1)^3}{3} + (-1)^2 - 1\right)$$

$$= \frac{8}{3} + 4 + 2 + \frac{1}{3} - 1 + 1$$

$$= 9.$$

■ 8. (1). $y_{(av)} = \frac{9/2}{2-1} = \frac{9}{2}$.

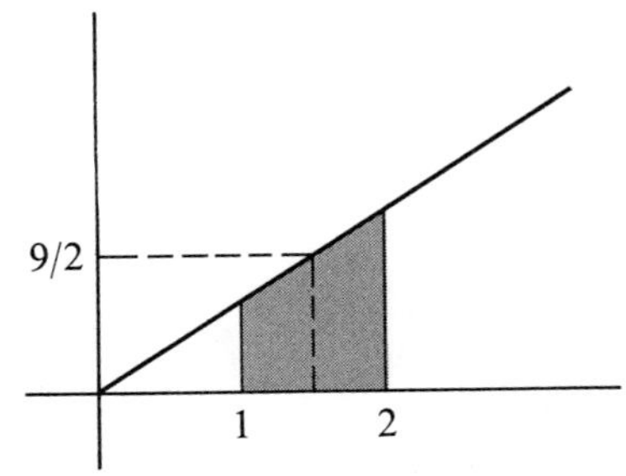

(6). $y_{(av)} = \frac{62/5}{4-1} = \frac{62}{15}$.

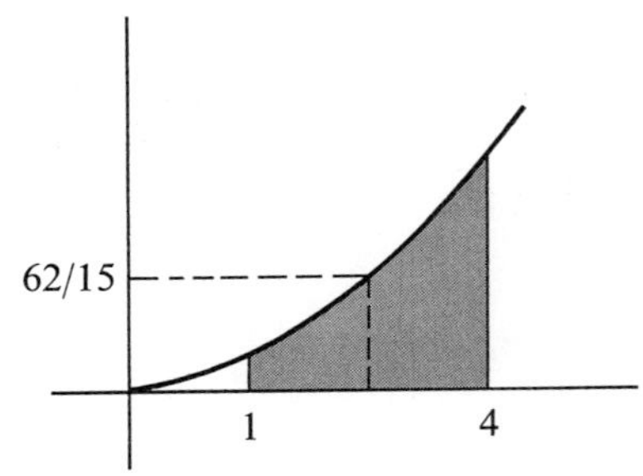

12-3 Problems

1. The energy (work) required to pump oil from the top side of the cylindrical tank is $E = \int_0^{20} 1300\, x\, dx$. Evaluate E.

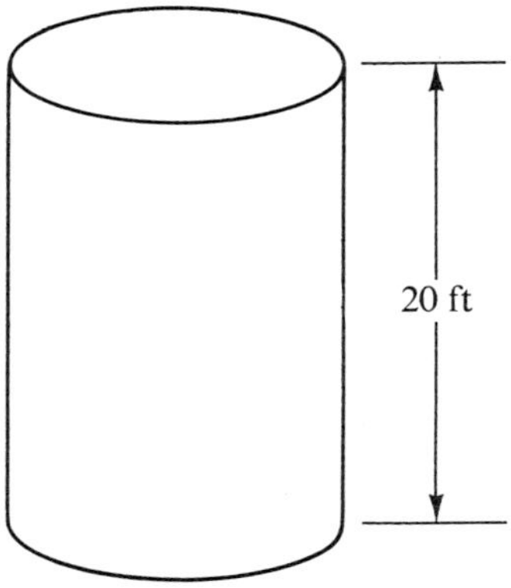

2. Find the shaded area between the graphs of $f(x) = 0.5x^2$ and $g(x) = 3x$.

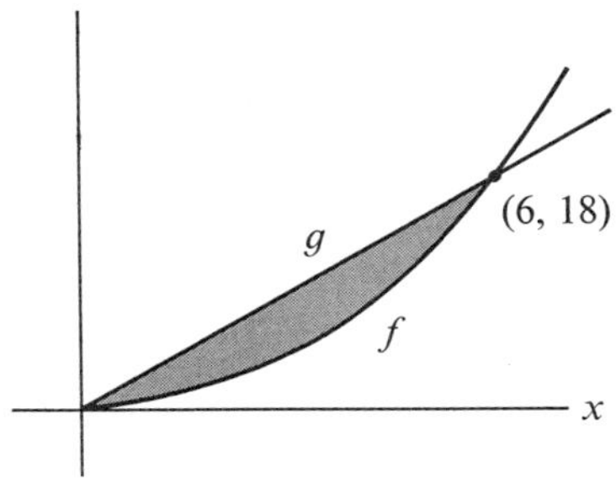

3. The thrust (T) of a piston at different positions (x) in its four-inch stroke is $T = 1.8x^{2.2}$. What is the total energy (work) required to move the piston through one stroke? What is the energy required in the last inch of the stroke?

$$(\text{Work}) = \int_a^b (T)\,dx.$$

4. The area designated in Fig. 12-3h is $\int_4^{10} (-2x^2 + 20x + 100)\,dx$. Find the abscissa, x_m, that divides the area into halves.

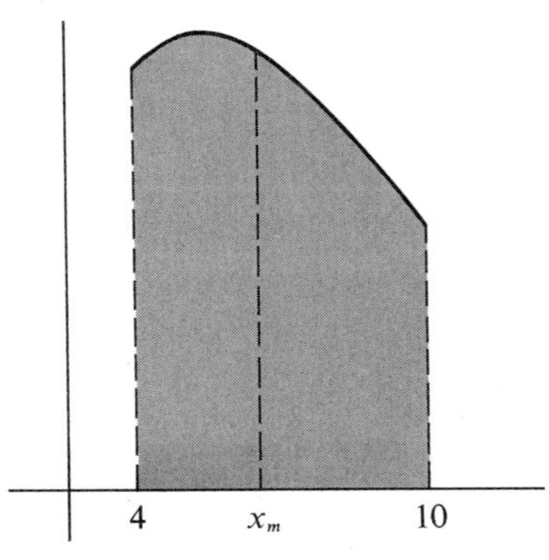

Fig. 12-3h

5. Curve $P_1P_2P_3$ is parabolic. What is the area of the shaded portion of the figure?

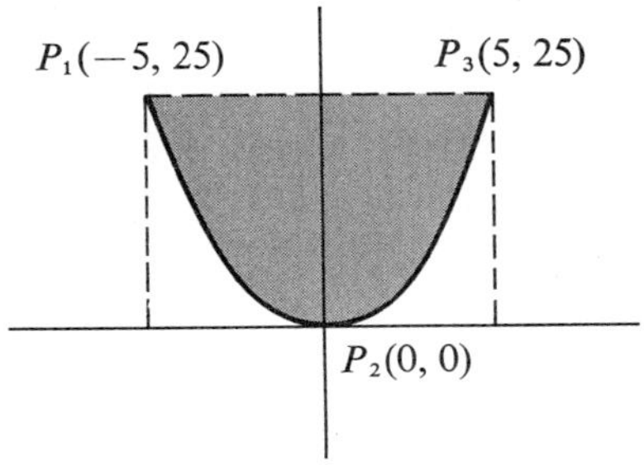

6. The equation of an impulse of voltage is $v = 14t^2 + 1$ from $t = 0$ to $t = 0.25$ seconds. Find the average voltage over the 0.25-second interval.
7. Examine a table of integrals. Notice how the integrals are grouped. Try to use the tables to evaluate the following definite integrals.

   a. $\int_1^3 6e^x\,dx$ (e is the base of the natural logarithms),

   b. $\int_1^2 (2 + 3x)^{1/2}\,dx$,

   c. $\int_0^3 \sqrt{16^2 - x^2}\,dx$,

   d. $\int_0^\pi \sin x\,dx$,

   e. $\int_0^\pi \sin^2 x\,dx$.

8. What is the average ordinate over the interval of integration in Problems 7d and 7e?

12-4 Numerical Methods

In general, it is more difficult to find the integral of a function than to find the derivative of a function. Because of this fact, it is often quite practical to use approximate methods for evaluating definite integrals.

A numerical method of approximation is illustrated in Fig. 12-4a. The area under the curve of the function f is approximated by four rectangles. The

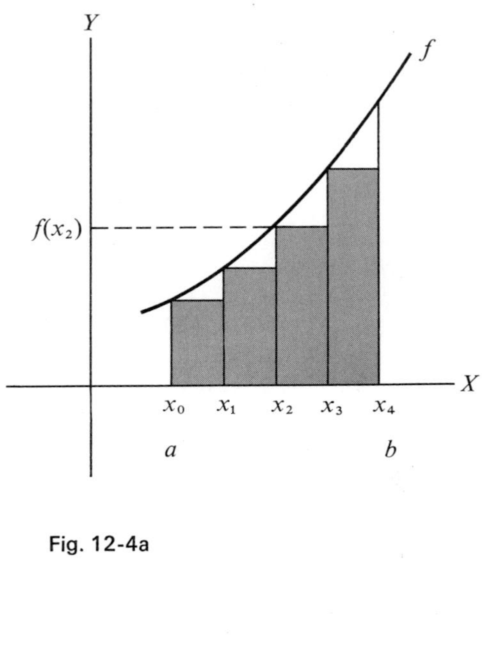

Fig. 12-4a

width of each rectangle is $\Delta x = (b - a)/4$. The length of each rectangle is $f(x_i)$, $i = 0, 1, 2$, and 3. The expression for the area under the curve from x_0 to x_4 is

$$(\text{area})_4 \approx f(x_0)\,\Delta x + f(x_1)\,\Delta x + f(x_2)\,\Delta x + f(x_3)\,\Delta x. \quad \text{(Eq. 12-4a)}$$

This equation is frequently abbreviated by using what is called *sigma notation.*

$$(\text{Area})_4 \approx \sum_{i=0}^{3} f(x_i)\,\Delta x.$$

$\sum$ is the uppercase Greek letter sigma standing for a summation of terms which are alike in form. In our example, the terms are numbered consecutively from $i = 0$ to $i = 3$.

The graph in Fig. 12-4a is repeated again in Fig. 12-4b. This time the area is divided into n rectangles. The width of each rectangle is $\Delta x =$ [ ],

■ $\dfrac{b-a}{n}$,

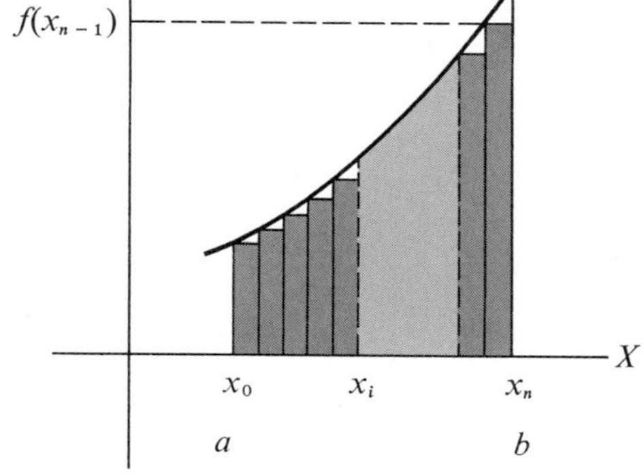

Fig. 12-4b

and the length of each rectangle is $f(x_i)$, $i = 0, 1, 2, \ldots, n-1$. The equation for the approximate area may be written

$$(\text{area})_n \approx f(x_0)\,\Delta x + f(x_1)\,\Delta x + \cdots + f(x_i)\,\Delta x + \cdots + f(x_{n-1})\,\Delta x. \quad \text{(Eq. 12-4b)}$$

In sigma notation, Eq. 12-4b is abbreviated to [ ].

■ $(\text{area})_n = \sum_{i=0}^{n-1} f(x_i)\,\Delta x.$

The method of approximation represented by Eq. 12-4b has the advantage of producing answers to any *degree of accuracy* desired as the value of Δx approaches zero, which is equivalent to n approaching a very large number. So that we may understand and appreciate this claim, let us make some comparisons that come from approximating the area under the curve of $f(x) = x^2$ from $x = 0$ to $x = 1$ (Fig. 12-4c), using Eq. 12-4b. Table 12-4a gives the numbers to be compared.

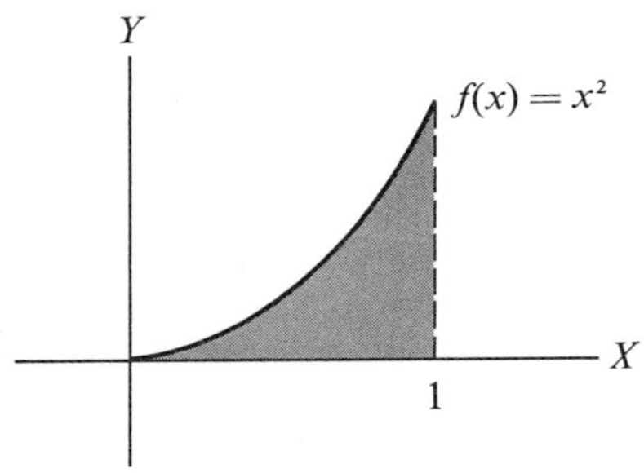

Fig. 12-4c

TABLE 12-4a

Number of rectangles, (n)	Sum of areas of rectangles, $(\text{area})_n$
2	.1250000000
4	.2187500000
10	.2850000000
40	.3209375000
130	.3294970414
220	.3310640496
310	.3317221644
400	.3320843750

As you might expect, as n approaches a very large number, say an infinitely large number, causing Δx to approach zero, the sum approaches the actual area under the curve. This we know to be the definite integral of f. Taking advantage of sigma notation.and the notation of limits, we write

$$\lim_{n\to\infty} \sum_{i=0}^{n-1} f(x_i)\,\Delta x = \int_a^b f(x)\,dx, \quad \text{where} \quad x_0 = a \quad \text{and} \quad x_n = b.$$

In this particular problem, it is not difficult to evaluate precisely the definite integral of $f(x) = x^2$ from $x = 0$ to $x = 1$.

$$\int_0^1 x^2\,dx = \boxed{}.$$

■ $\left.\dfrac{x^3}{3}\right]_0^1 = \dfrac{1}{3} = .333\ldots.$

The method of approximating $\int_a^b f(x)\,dx$ represented by Eq. 12-4b may have the disadvantage of requiring a large number of computations to obtain sufficient accuracy. This is not a problem if computing equipment is used. All of the values in Table 12-4a were computed in just a few seconds using a computer programmed to follow the sequence of steps outlined in the flow chart immediately following.

Read a, b, and n.

Calculate the width of the rectangles.

Start with i, s, and p zero.

Determine the length of a particular rectangle.

Calculate the area of a particular rectangle.

Add on the area last calculated.

Increase i by one.

Return to calculating the area of the next rectangle unless we have completed the last calculation.

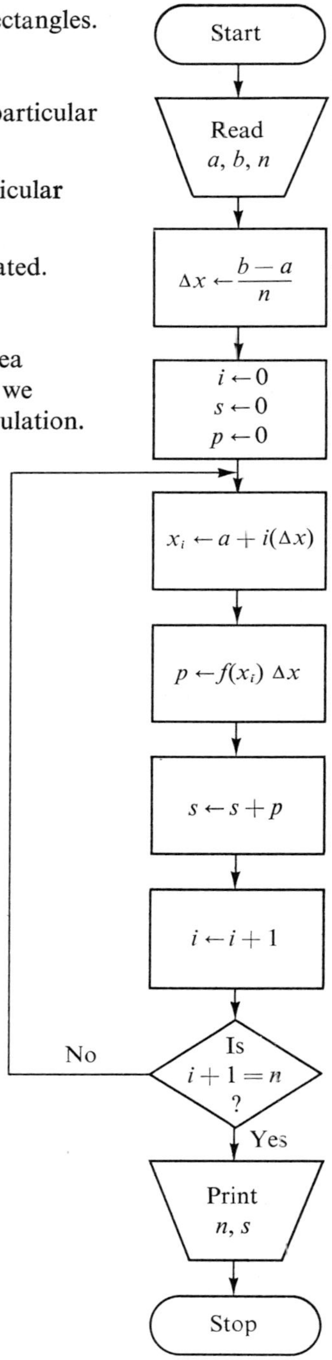

12-4 Exercises

Use a calculator if possible.

1. Approximate $\int_2^7 \frac{1}{x^2}\,dx$ using Eq. 12-4b. Let $n = 5$.

2. Find the precise value of the integral in Problem 1. What would improve the accuracy of the calculation in Problem 1?

■ 1.

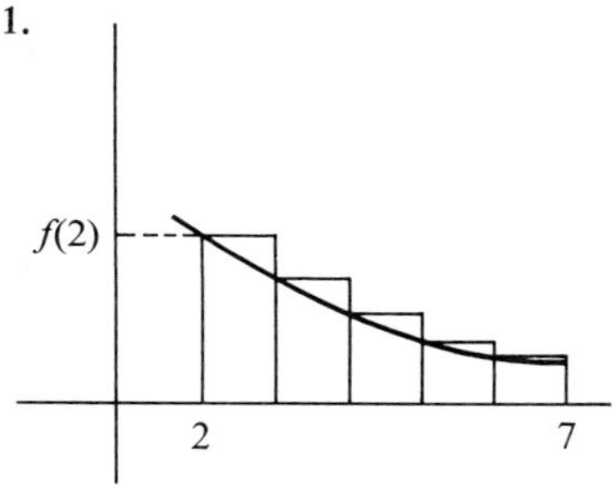

x_i	$f(x_i)$	$f(x_i)\ \Delta x$
2	.2500	.2500
3	.1111	.1111
4	.0625	.0625
5	.0400	.0400
6	.0277	.0277
		.4914
		≈ (Area)$_5$

3. Approximate $\int_3^6 \frac{1}{\sqrt{x-2}}\,dx$ using Eq. 14-4b. Let $n = 6$.

■ 2. i. $$\int_2^7 x^{-2}\,dx = -x^{-1}\Big]_2^7$$
$$= -\frac{1}{x}\Big]_2^7$$
$$= -\frac{1}{7} + \frac{1}{2} = \frac{5}{14}$$
$$\approx .3571.$$

ii. A larger value of n would improve the accuracy.

4. Approximate $\int_0^4 \sqrt{4+x^3}\,dx$ using Eq. 14-4b. Let $n = 8$.

■ 3.

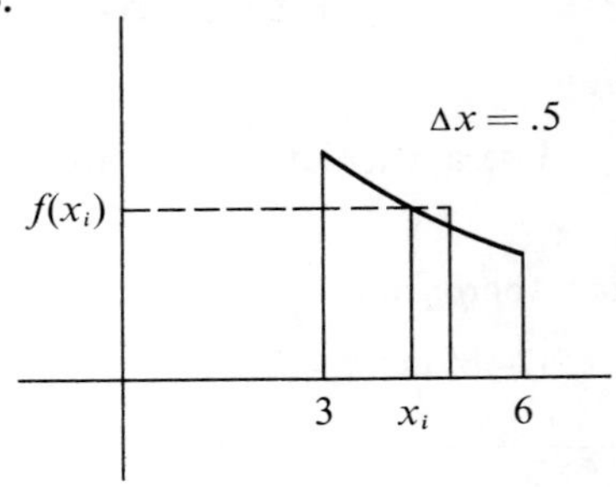

x_i	$f(x_i)$	$f(x_i)\ \Delta x$
3.0	1.0000	.5000
3.5	.8165	.4082
4.0	.7071	.3536
4.5	.6325	.3162
5.0	.5774	.2887
5.5	.5345	.2673
		2.1339 $\approx$ (Area)$_6$

■ 4.

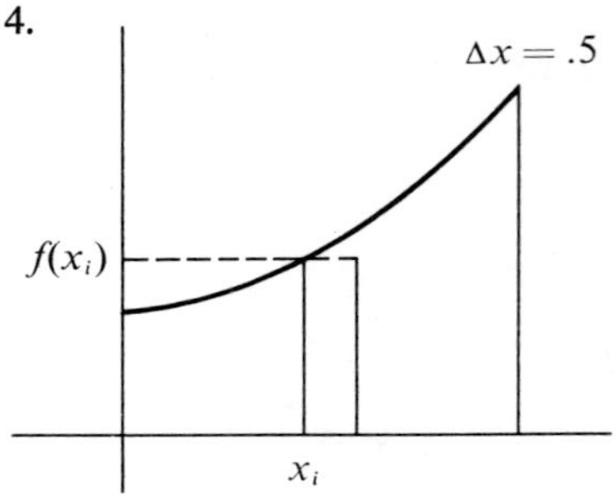

x_i	$f(x_i)$	$f(x_i)\ \Delta x$
0	2.000	1.000
.5	2.031	1.015
1.0	2.236	1.118
1.5	2.716	1.358
2.0	3.464	1.732
2.5	4.430	2.215
3.0	5.568	2.784
3.5	6.846	3.423
		13.645 $\approx$ (Area)$_8$

12-4 Problems

1. Measured data is graphed in Fig. 12-4d. Assume the data corresponds to the graph of a parabola. Approximate the value of $\int_0^{100} f(x)\,dx$.

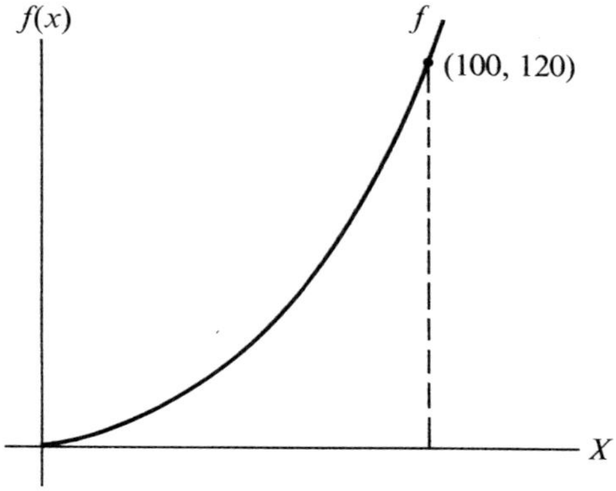

Fig. 12-4d

2. If the area under a curve is approximated with trapezoids, high degrees of accuracy are achieved with considerably fewer divisions of the interval from a to b. (For example, when $n = 40$, the approximate value of $\int_0^1 x^2\,dx$ is .3334375 when

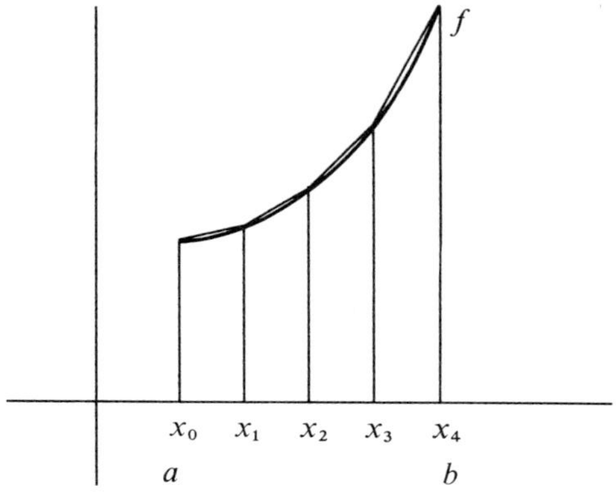

using trapezoidal areas. Compare this with the values of Table 12-4a.) Write a formula analogous to Eq. 12-4a using trapezoidal areas. Apply the formula to the approximation of $\int_0^1 x^2\,dx$.
3. Approximate $\int_0^{\pi/2} \sin^2 x\,dx$ using Eq. 12-4b with $n = 10$.
4. Approximate $\int_0^4 e^{2x}\,dx$ using Eq. 12-4b with $n = 8$.

12 Supplementary Exercises and Problems

1. Find the indefinite integral of $f(x) = 6x^2$. (12-1)
2. Find $\int f(x)\,dx$ when $f(x) = 9x^2$. (12-1)
3. What is the antiderivative of $f(x) = x^{2/3}$? (12-1)
4. Find $\int f(t)\,dt$ when $f(t) = t^{1/3}$. (12-1)
5. Calculate $\int (10x^{-2})\,dx$. (12-1)
6. If $f(w) = 100w - 50$, find $\int f(w)\,dw$. (12-1)
7. Integrate $\int (t^2 + 3t + 1)\,dt$. (12-1)
8. Integrate $\int 5\sqrt{x}\,dx$. (12-2)
9. Integrate $\int 2.06i^{0.8}\,di$. (12-2)
10. If $q = \int i\,dt$ and $i = f(t) = .002t^{1.6}$, find q at $t = 0.1$ when $q = 0$ at $t = 0$. (12-2)
11. If acceleration is represented by the equation $f(t) = -12t$, find the equation for velocity. (12-2)
12. Evaluate $\int_3^5 dx$. (12-3)
13. Evaluate $\int_0^2 (t^2 - 10t)\,dt$. (12-3)
14. Evaluate $\int_0^2 3(x^2 - 2)^2\,dx$. (12-3)
15. Find the area bounded by $f(x) = (-x^2/4) + 5x$ and $g(x) = 10x/6$. (12-3)
16. Evaluate $v = 1/c \int_{t_1}^{t_2} i\,dt$, when $i = 0.06t^{.5} + 0.4$, $c = 0.005$, $t_1 = 0$, and $t_2 = 0.5$. (12-3)
17. a. Evaluate $\bar{x} = \dfrac{\int_0^5 xf(x)dx}{\int_0^5 f(x)dx}$ when $f(x) = \dfrac{x^2}{10} + 4$.

    b. Evaluate $\int_0^{\bar{x}} f(x)\,dx$.

    c. Evaluate $\int_{\bar{x}}^5 f(x)\,dx$. (12-3)

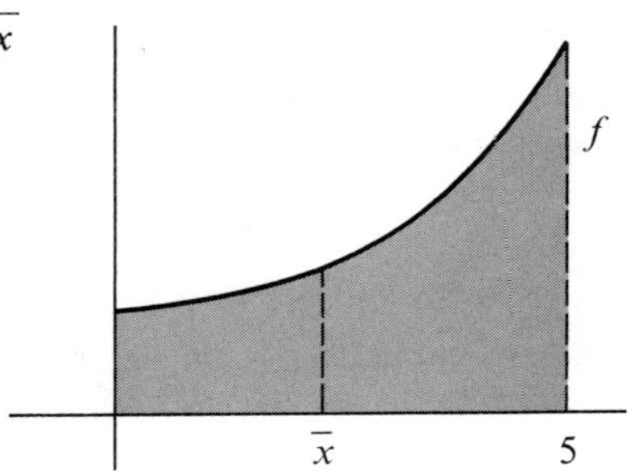

18. Approximate $\int_0^3 \sqrt{2 + 3x}\,dx$ using Eq. 14-4b with $n = 9$. (12-4)

Exercises for Slide Rule or Calculator

1. $6.5 \times 4.2 = 27.30.$
2. $3.7 \times 12 = 44.40.$
3. $.56 \times 26.7 = 14.95^{+}.$
4. $6.25 \times 3.40 = 21.25.$
5. $408 \times 71.2 = 29.050^{-}.$
6. $.00643 \times .0912 = .0005864^{+}.$
7. $3150 \times .207 = 652.1^{-}.$
8. $.0759 \times 63{,}200 \times 20.7 = 99{,}300^{-}.$
9. $.00612 \times 56.7 \times 9.08 = 3.151^{-}.$
10. $364 \times .082 \times \pi = 93.77^{+}.$
11. $(3.81)^2 = 14.52^{-}.$
12. $(.00623)^2 = .00003881^{+}.$
13. $(7.6\pi)^2 = 570.1^{-}.$
14. $(185)^2 = 34{,}230^{-}.$
15. $(.981)^2 = .9624^{-}.$
16. $\dfrac{1065}{230} = 4.630^{+}.$
17. $\dfrac{20.4}{5.04} = 4.048^{-}.$
18. $\dfrac{22.6}{0.125} = 180.8.$
19. $\dfrac{7.59}{102} = .07441^{+}.$
20. $\dfrac{854}{.0023} = 371{,}330^{+}.$
21. $\dfrac{15 \times 3.7}{11 \times 4.9} = 1.030^{-}.$
22. $\dfrac{7.66 \times 4.17 \times 5.8}{62.1 \times 1.3} = 2.295^{-}.$
23. $\left(\dfrac{6.23}{7.21}\right)^2 = .7466^{+}.$
24. $\left(\dfrac{4.9\pi}{.0381}\right)^2 = 163{,}200^{+}.$
25. $\dfrac{1}{3400} = .0002941^{+}.$
26. $\dfrac{1}{.0641} = 15.60^{+}.$
27. $(2.63)^2 + (6.3)^2 = 46.61^{-}.$
28. $(13.4)^2 - (7.41)^2 = 124.7^{-}.$
29. $\dfrac{1}{1 - (0.214)^2} = 1.048^{-}.$
30. $\dfrac{120}{7.82 + \dfrac{16.4}{3.8}} = 9.888^{+}.$
31. $\dfrac{1}{250} + \dfrac{1}{340} = .006941^{+}.$
32. $\left(\dfrac{1}{.432}\right)^2 + 7.32 = 12.68^{-}.$
33. $\dfrac{38.4}{\dfrac{1}{3.71} + \dfrac{1}{4.23}} = 75.90^{-}.$
34. $\dfrac{3000 \times 2500}{3000 + 2500} = 1364^{-}.$
35. $\dfrac{.443(16.2)}{720(1 - (.6)^2)} = .01557.$
36. $\sqrt{7.62} = 2.760^{+}.$
37. $\sqrt{34.6} = 5.882^{+}.$
38. $\sqrt{173.4} = 13.17^{-}.$

39. $\sqrt{42^2 + 19^2} = 46.10^-$.

40. $\sqrt{3 + \dfrac{116}{8.21}} = 4.139^-$.

41. $\dfrac{32.2}{\sqrt{118} - \sqrt{106}} = 56.76^-$.

42. $\sqrt{18.2 + \dfrac{1}{18.2}} = 4.273^-$.

43. $\dfrac{-16.8 + \sqrt{16.8^2 - 4(3)(2.5)}}{6} = -.1530^-$.

Problem Solutions

Section 1-1, page 82

1. (Length) $= f(w) = \dfrac{200}{12.60}\, w.$

$f(w)$	w	$f(w)$	w	$f(w)$	w	$f(w)$	w
0	0	55	3.465	105	6.615	155	9.765
5	.315	60	3.780	110	6.930	160	10.080
10	.630	65	4.095	115	7.245	165	10.395
15	.945	70	4.410	120	7.560	170	10.710
20	1.260	75	4.725	125	7.875	175	11.025
25	1.575	80	5.040	130	8.190	180	11.340
30	1.890	85	5.355	135	8.505	185	11.655
35	2.205	90	5.670	140	8.820	190	11.970
40	2.520	95	5.985	145	9.135	195	12.280
45	2.835	100	6.300	150	9.450	200	12.600

2. a. $W_{Na} = \dfrac{39.3}{60.7} W_{Cl}$, b. $W_{Cl} = .607\, W_{NaCl}$.

3. a. $f(.50) = 41.667$ psi, b. $f(.51) = 42.500$ psi, c. $f(.51) - f(.50) = .833$ psi.

4. a. (Displacement) $= f(F) = (1/2)F$, b. $f(3.5) = 1.75$ cm.

5. (Resistance) $= f(L) = kL$,
$k =$ (coefficient of resistivity);
(velocity) $= f(t) = at$, $a =$ (acceleration);
(force) $= f(a) = ma$, $m =$ (mass).

6. (Degrees) $= f(g) = .90g$.

$f(g)$	g	$f(g)$	g
0	0	49.5	55
4.5	5	54.0	60
9.0	10	58.5	65
13.5	15	63.0	70
18.0	20	67.5	75
22.5	25	72.0	80
27.0	30	76.5	85
31.5	35	81.0	90
36.0	40	85.5	95
40.5	45	90.0	100
45.0	50		

7. (Degrees) $= f(m) = .05625m$.

Section 1-2, page 91

1. a. $g(x) = 4x$;
b. $g(-2) = -8$,
$g(0) = 0$,
$g(1) = 4$,
$g(3) = 12$;
c. $m = \dfrac{12-4}{3-1} = 4$.

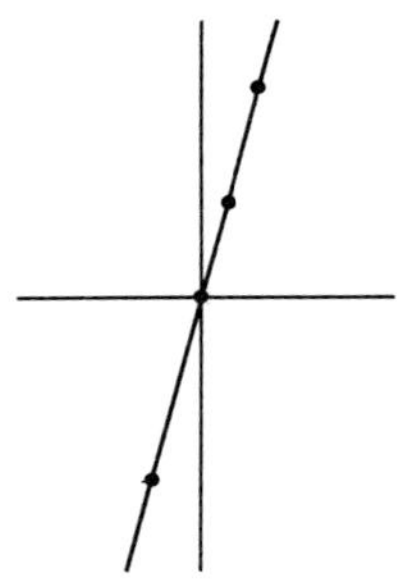

2. a. $m = -1$, b. $f(x) = -x$.
$F = \mu N$,
$F =$ (force in direction of motion),
$\mu =$ (coefficient of friction),
$N =$ (normal force).

4. (Velocity) $= f(t) = at = 2.2t$;
$a =$ (acceleration in miles per hour per second),
$t =$ (time in seconds),
$f(t) =$ (velocity in miles per hour).

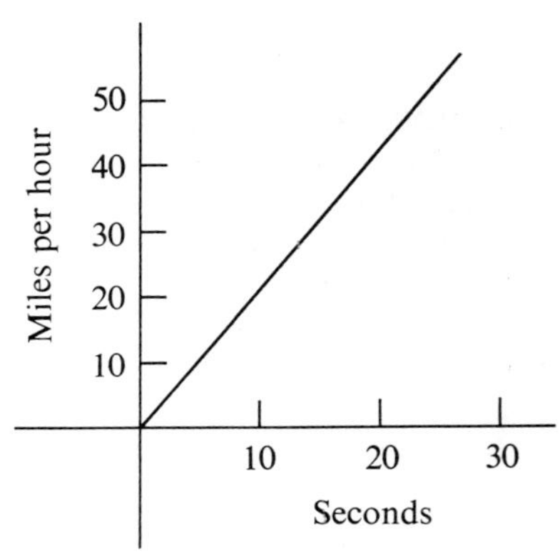

5.

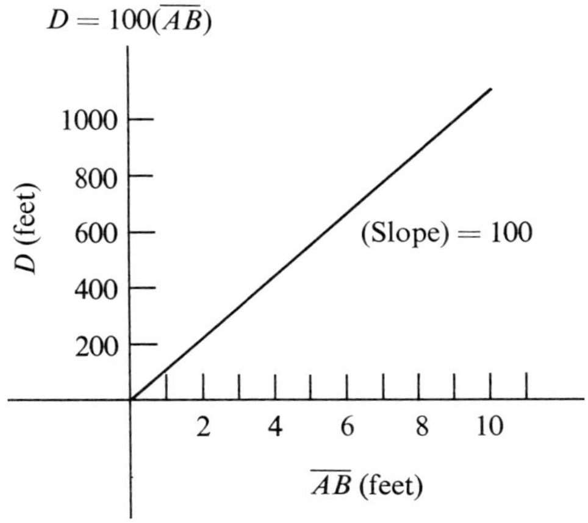

Section 1-3, page 98

1. a. $f(x)$ is R_m; x is E, m is Ω/V, b is $-R_v$; b. $R_m = 10000\,E - 100$.

E	R_m
1	9,900
10	99,900
150	1,499,900

2. (Velocity) $= f(t) = -32t + 88$
$f(2.75) = 0.$

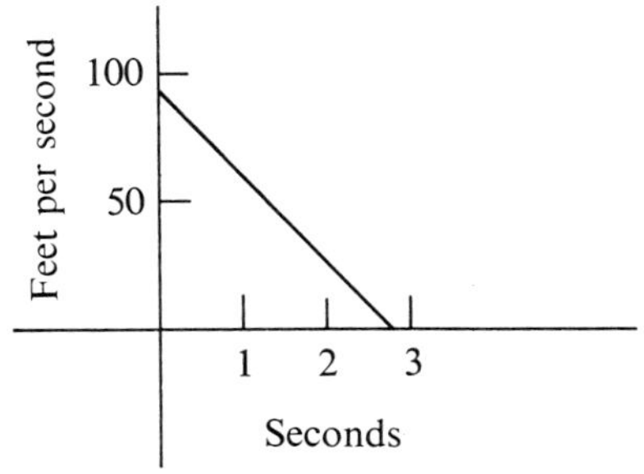

3. $L_t = 2M + 4.4.$

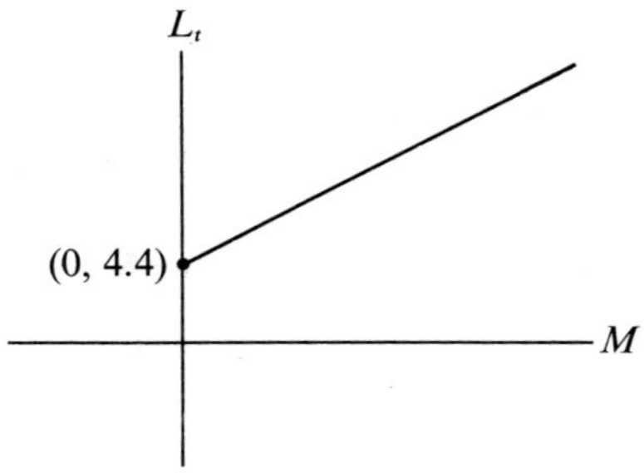

4. Point C is 833 1/3 ft north of point A.

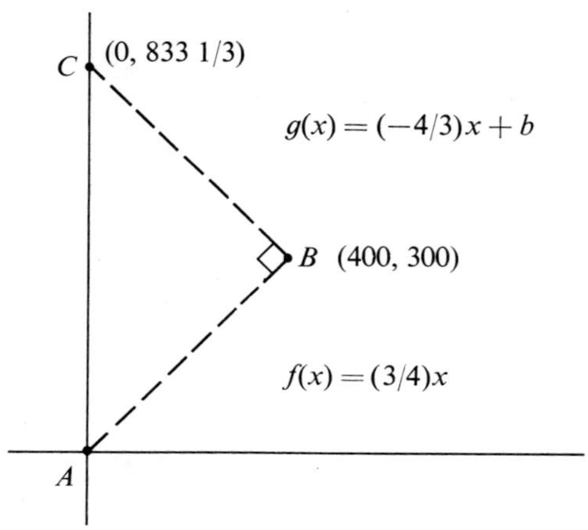

5. a.

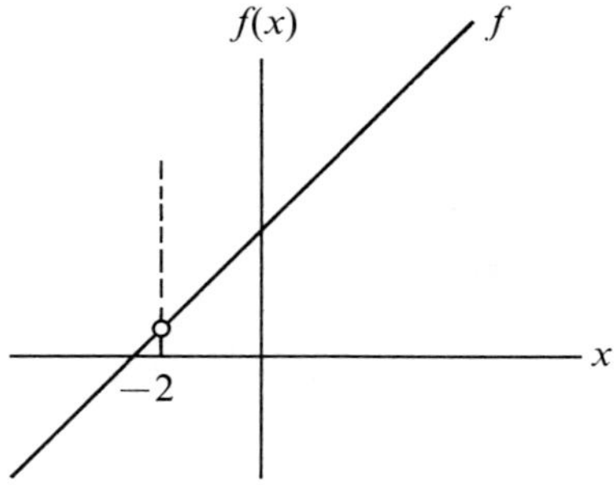

b.

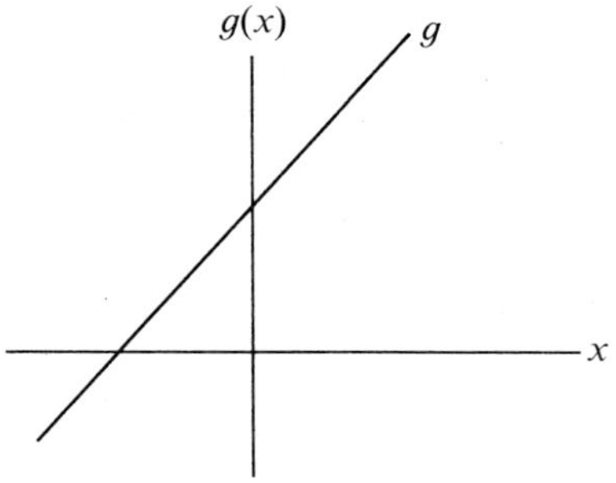

c. no,

d. no.

6. a. All real numbers, b. all real numbers except $x = 1$.

Section 1-4, page 101

1. a. $(t - t_0)a = (v - v_0)$, b. $v_0 = 55.64$ m/sec.

2. $(t - 35)(2.3) = (T - 132.5)$,

$$T = 2.3t + 52,$$

$T =$ (temperature in Fahrenheit),

$t =$ (time in minutes).

3. $T = 190$ F°.

Section 1-5, page 104

1. When time (t) is the independent variable,

$$(w - 120) = \left(\frac{240 - 120}{2.16 - 1.08}\right)(t - 1.08)$$

or

$$w = f(t) = 111.1t.$$

2. $f(1.50) = 166.7$ rps.

3. When frequency (f) is the independent variable,

$$(KE - 2.00) = \left(\frac{2.98 - 2.00}{(7.25 - 5.75)10^{26}}\right)(f - 5.75(10^{26}))$$

or

$$KE = 6.53(10^{-27})f - 1.76.$$

4. $KE = 2.16.$

5. $(I_c - 15) = \left(\dfrac{30 - 15}{.030 - .012}\right)(I_b - .012)$

$$I_c = f(I_b) = 833I_b + 5.00$$

$$f(.021) = 22.5.$$

6. $(F - 32) = \left(\dfrac{212 - 32}{100 - 0}\right)(C - 0)$

$$F = \frac{9}{5}C + 32.$$

7. a. $L = 100.009 = \left(\dfrac{100.035 - 100.009}{106 - 70}\right)(T - 70)$

or

$$L = .00072T + 99.958,$$

b. $L = 100.023$ ft, c. $T = 58.3$ F°.

Section 1-6, page 111

1. $x =$ (number of item A),
$y =$ (number of item B),
$y \leq -\frac{3}{5}x + 24.$

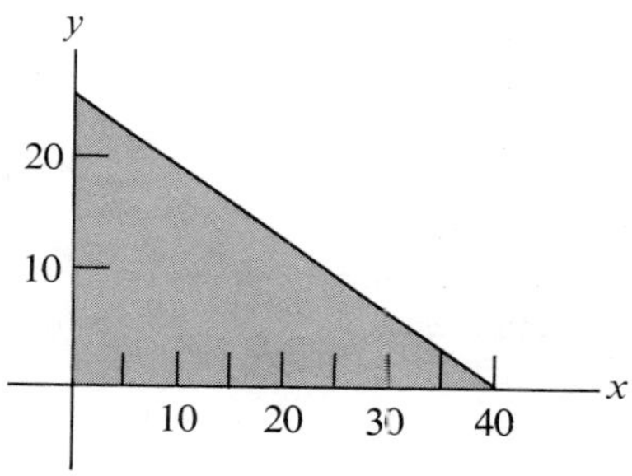

2. $x =$ (number of grade A),
$y =$ (number of grade B),
$y > -1.67x + 41{,}667.$

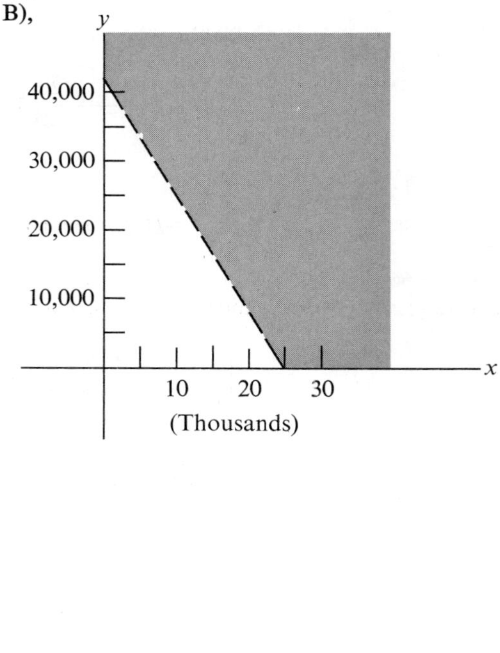

3. Yes.

Section 1-7, page 116

1. a. (Area) = 60 square units,
b.

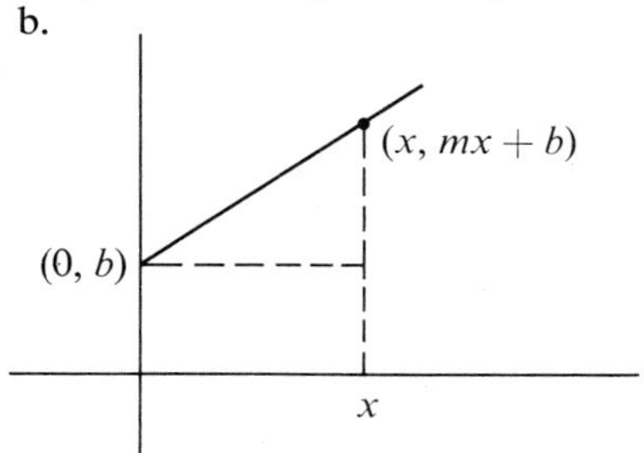

$F(x) = (1/2)mx^2 + bx$

2. i. (Acceleration) $= f'(t) = 15$,
(velocity) $= f(t) = 15t$,
(distance) $= F(t) = 7.5t^2$.
ii. $f(t) = v = 15t$,
$t = v/15 = 180/15 = 12$ sec.
iii. $F(12) = 1080$ ft.

3. $f(s) = 6s$,
$F(s) = 3s^2$,
$F(5) = 75$ in. lb.
$= 6.25$ ft. lb.

4. $F(6) - F(5) = 33$ in. lb.
$= 2.75$ ft. lb.

5. (Current) $= f(t) = .05t$, (Charge) $= F(t) = .025t^2$.

Section 2-1, page 128

1. a.

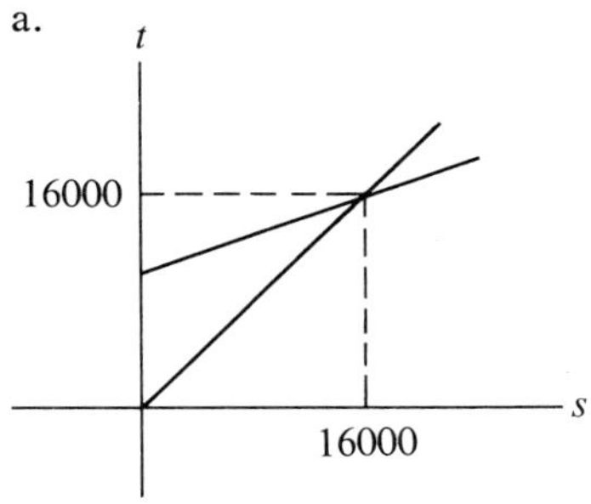

b. Increase in break-even point.

2.

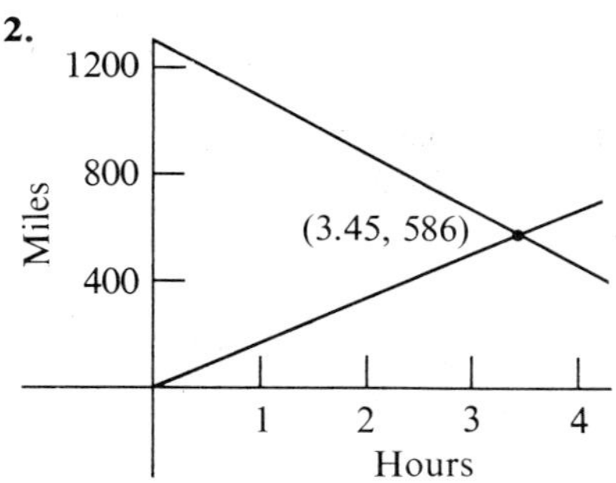

3.

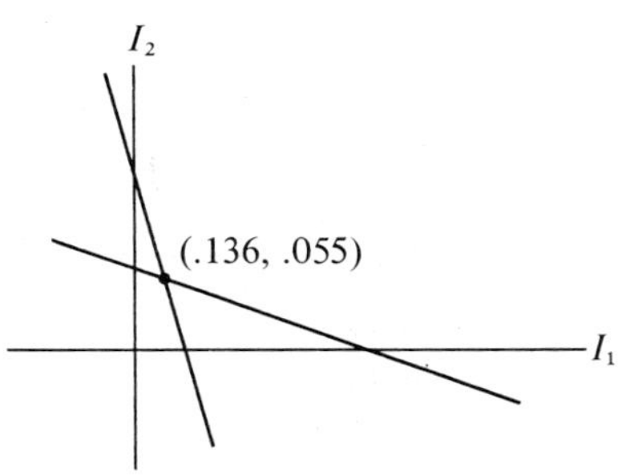

4.

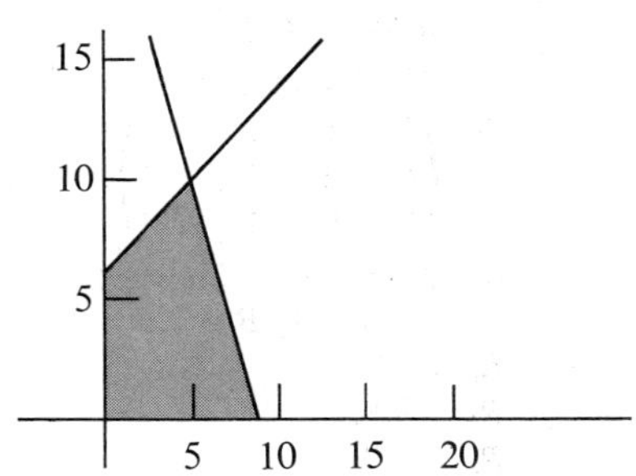

Section 2-2, page 135

1. $i_1 = .012$,
$i_2 = .004$.

2. i. $A + B = 90°$,
$A - B = 18°$.
ii. $A = 54°$,
$B = 36°$.

3. i. $3 = 2m + b$,
$4 = m + b$.
ii. $m = -1$,
$b = 5$.

4. i. $R = kT + b$.
ii. $.42 = k20 + b$,
$.48 = k70 + b$.
iii. $m = -1$,
$b = 5$.

5. i. $W + w = 1000$
$.45W + .25w = .40(1000)$.
ii. $W = 750$ lb (45% zinc),
$w = 250$ lb (25% zinc).

6. i. $5R + 5r = 150$,
$25R - 25r = 150$.
ii. $R = 18$ in./sec $= .12$ rps,
$r = 12$ in./sec $= .08$ rps.

7. i. $I_p = .00429\ P_v - .3$,
$I_p = .00133\ P_v + .4$.
ii. $P_v = 236.5$.

Section 2-3, page 142

1. i. $S + s = 180$,
$.2S + .08s = .15(180)$.
ii. $S = 105$ cc. of 20% solution,
$s = 75$ cc. of 8% solution.

2. i. $5R + 5r = 400$,
$10/3\,R + 10/3\,r = 400$.
ii. $R = 100$ mph (plane),
$r = 20$ mph (wind).

3. $T_1 = 909$,
$T_2 = 2036$.

4. i. $C = 1500 + \dfrac{112(.018)}{.83}\,n$,

$C = 1200 + \dfrac{112(.018)}{.80}\,n$.

ii. $n = 3294$ hr.

5. i. $C = .65L + 550(.0185)$,
$C = .83L + 550(.0125)$.
ii. $L = 18.3$ ft.

Section 2-4, page 155

1. i. $A(-1)^2 + B(-1) + C = -5$,
$A(2)^2 + B(2) + C = 4$,
$A(0)^2 + B(0) + C = -4$.
ii. $\left(\begin{array}{ccc|c} 1 & -1 & 1 & -5 \\ 4 & 2 & 1 & 4 \\ 0 & 0 & 1 & -4 \end{array}\right)$.
iii. $f(x) = x^2 + 2x - 4$;
$\therefore\ A = 1,\ B = 2,\ C = -4$.

2. i. $R_B + R_C = 7$,
$R_A - R_C = 5$,
$R_A - R_B = 4$.
ii. $\left(\begin{array}{ccc|c} 0 & 1 & 1 & 7 \\ 1 & 0 & -1 & 5 \\ 1 & -1 & 0 & 4 \end{array}\right)$.
iii. $R_A = 8,\ R_B = 4,\ R_C = 3$.

3. $R_a = 12/9$,
$R_b = 8/9$,
$R_c = 6/9$.

4. i. $t/365 - d = 0$,
$t/687 - d = -1$.
ii. $t = 778.7$ msec.

5. $I_A = 33/46 = .717,$
$I_B = -(8/23) = -.348,$
$I_C = -(17/46) = -.370.$

6. i. $6R_1 + 6R_2 = 1,$
$8R_2 + 8R_3 = 1,$
$10R_1 + 10R_3 = 1.$
ii. $R_1 = 17/240.$
iii. (time—no. 1) $= 240/17$
$= 14.1$ hr.

Section 3-1, page 169

1.

E	P
0	0
1	.001
2	.004
3	.009
4	.016
5	.025
6	.036
7	.049
8	.064
9	.081
10	.100

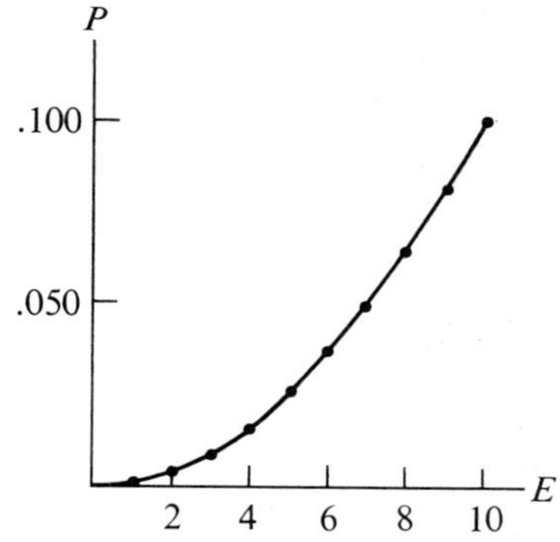

2.

t	E_k
0	0
1	512
2	1024
3	4608
4	8192
5	12800

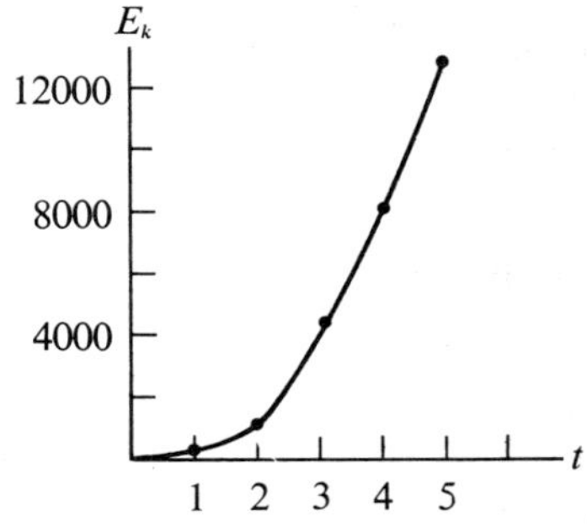

3.

t	s
0	−12
2	180
4	244
6	180
8	−12

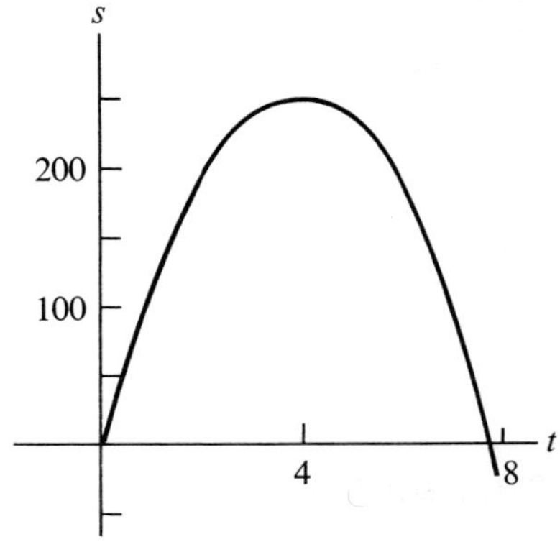

4. i. $F = (1/4)(58.67)^2(3/2) = 1290.7$ poundals,
ii. $F = 322.7$ poundals, that is, 1/4 as much.

5. i. $L = \dfrac{980(1)^2}{4\pi^2} = 24.82$ cm,

ii. $L = \dfrac{980(2)^2}{4\pi^2} = 99.29$ cm.

Section 3-2, page 178

1. $160t - 16t^2 = 0$,
$t = 0$, 10 sec.

2. x | $11 - 2x$ | $2x^2 - 11x + 12 = 0$,

$x = 4$, 3/2 in.

3. $R_2^2 - 300R_2 + 18000 = 0$,
$R_2 = 217.1$ or 82.9,
$R_1 = 82.9$ or 217.1.

4. i. $a = 5\sqrt{3}/24$ in.,
ii. (area) = .3383 sq in.,
iii. (tensile strength) = 10825 lb.

5. i. (time) = $\sqrt{5}/2$ sec,
ii. $x = 6.71$ ft.

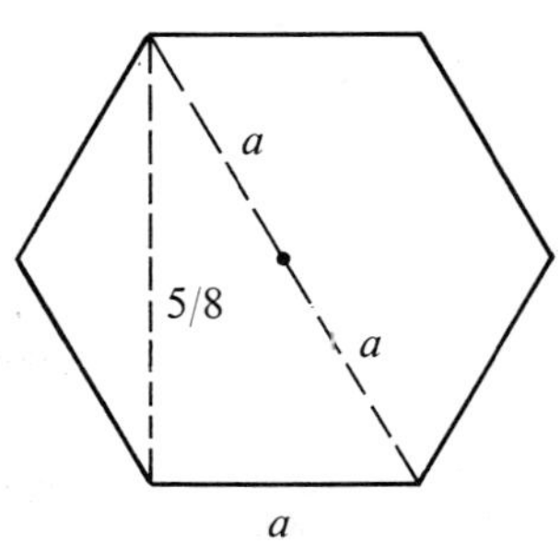

Section 3-3, page 185

1. $x^2 + x = 56$,
$x = 7$ or -8.

2. $t = 2.22$ sec.

3. $f(x) = 2x + 4$,
$F(x) = x^2 + 4x = 96$,
$x = 8$.

4. $f(s) = 6s$,
$F(s) = 6s^2/2 = 6.25(12)$,
$s = 5$ in.

5. i. $\pi r_1^2 + \pi r_2^2 = (4/5)\pi 5^2,$
$r_1 = 5 - r_2.$

ii. $2r_2^2 - 10r_2 + 5 = 0,$
$r_2 = 4.44$ in.,
$r_1 = .56$ in.

6. $\left(r - \frac{1}{3}\right)^2 + \left(r - \frac{1}{2}\right)^2 = r,$
$r = 1.41$ in.

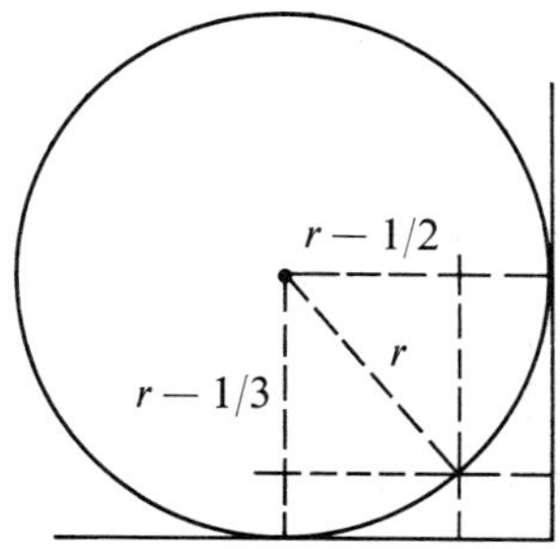

7. $(s - .125)^3 - s^3 = -2.00,$
$s = 2.37$ cm.

8. $x = 23.4$ ft.

Section 3-4, page 190

1. $\bar{t} = 2.5$ sec.

2. $A = w(600 - 2w),$
$\bar{w} = 150,$
$A_{(\max)} = 45{,}000$ sq ft.

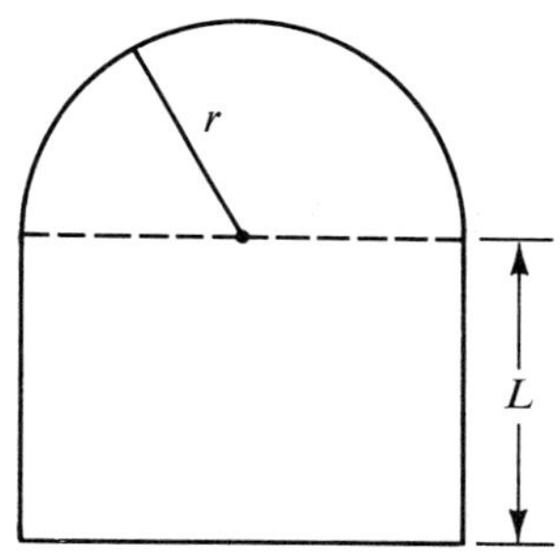

3. $(\text{Area}) = \frac{\pi r^2}{2} + 2r\left(\frac{p - \pi r - 2r}{2}\right),$
$\bar{r} = 1.68,$
$L = 1.68.$

4. $\bar{s} = L/2.$

5. (Gross Amount) $= (300 + 30n)(12 - .8n),$
$\bar{n} = 2.5$ days.

6. i. (Area) $= (20 - x)(8/20\ x),$
$\bar{x} = 10.$

ii. $\overline{FC} = 10, \quad \overline{FE} = 4.$

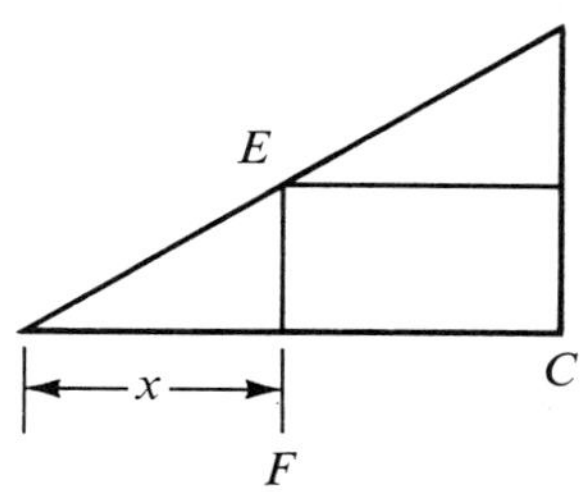

Section 3-5, page 198

1.

s	$f(s)$	$f'(s)$
1	100	0
50	97.5	−.100
100	90.0	−.200

2.

	Middle	Ends
Height	10 ft	35 ft
Slope	0	±1.00

3. $f'(t) = -t + 3 = 0,$
$t = 3$ sec.

4. $f'(x) = -1/2\,x + 1.$

Section 3-6, page 202

1. (Distance) $\approx$ 427 ft.
2. (Bending moment) $\approx$ 40,500 ft lb.
3. (Energy) $\approx$ 333 in. lb = 27.8 ft lb.
4. (Voltage) $\approx \dfrac{.010}{.006} = 1.67$ V.
5. A mechanical device that measures the area of a plane figure by tracing its boundary.

Section 4-1, page 213

1. $V = 43.0.$
2. $r = 5.30(10^{-9})$ cm.
3. $C = 2.89(10^{-8}).$
4. $Q = 1.402(10^{-5}).$

Section 4-2, page 221

1. $i = .0176.$
2. $v = 268.4.$
3.

H	Q
0	0
.5	11.8
1.0	33.3
1.5	61.2
2.0	94.2

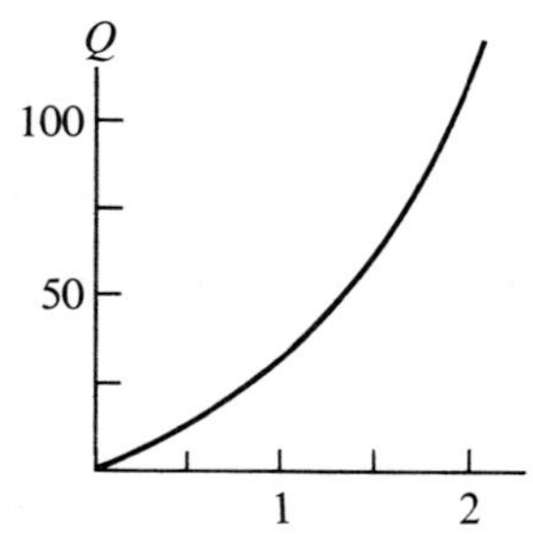

4. $(hp) = 1.03.$

Section 4-3, page 227

1. $F_0 = 14.5.$
2. $V = 16.9$ ft/sec.
3. $I = .0514$ A.
4. $H_c = 82.7.$
5. $z = 89{,}450.$

Section 4-4, page 232

1. $r = \sqrt{v/\pi h}.$
2. $C = L/z_0^2.$
3. $L = \dfrac{1}{4\pi^2 F_0^2 C}.$
4. $\lambda = \dfrac{2\pi V^2}{g}.$

5. $R = \sqrt{\dfrac{24k^2 + a^2}{6}}$.

6. $R = \sqrt{\dfrac{E^2}{I^2} - X^2}$.

7. $C = \dfrac{Q}{L(H + h_r)^{3/2}} = .01288$.

8. $d = 13.42$.

Section 5-1, page 242

1.

N	$\log_2 N$	$\log_8 N$	$\log_{16} N$
2	1	1/3	1/4
4	2	2/3	2/4
8	3	3/3	3/4
16	4	4/3	4/4
32	5	5/3	5/4
64	6	6/3	6/4
128	7	7/3	7/4
256	8	8/3	8/4

2. $n + 2 = 8^{5/3}, \quad n = 30.$

3. $\log_e P = kt.$

4. $\dfrac{P_0}{P_i} = 10^{N/10}$ or $\dfrac{P_0^{10}}{P_i} = 10^N$.

5. $\log_{10}\left(\dfrac{p_2}{p_1}\right) = \dfrac{w}{4883V}$.

6. $(H^+) = 10^{-(pH)}$.

7. $\log_e (R/R_0) = b/T$.

Section 5-2, page 247

1. $\log_{10} 1 = 0$, $\log_5 1 = 0$, $\log_e 1 = 0$.
2. $\log_{10}(1/A) = \log_{10} 1 - \log_{10} A = -\log_{10} A$.
3. $3 \log_{10} A + 4 \log_{10} B = \log_{10}(A^3 B^4)$.
4. $\log_{10} A - \log_{10} B = \log_{10}(A/B)$.
5. $\log_b (AB) = 5$,
   $\log_b (A/B) = -1$,
   $\log_b (B/A) = 1$,
   $\log_b \sqrt{A} = 1$,
   $\log_b \sqrt{B} = 3/2$.
6. $\log_b y^2 = .96$.
7. $\log_b y = -(1/4)$.

Section 5-3, page 254

(No written solutions.)

Section 5-4, page 261

1. $Q = 4.875$ cu ft/sec.
   $= 17{,}550$ cu ft/hr.

2. $N_{db} = 33.01$.

3. $P = 170{,}500$ lb.

4. $P = \dfrac{5600}{1.489} = \$3761.$

5. $H = 0.38(16.40)(2.378)$
$= 14.83.$

6. $F = 2^{12(60)/40} = 2^{18}$
$= 262{,}144.$

Section 5-5, page 270

1. $E = 14{,}030.$

2.

t	i
0	1.000
.2	.819
.4	.670
.6	.549
.8	.449
1.0	.368
1.2	.301
1.4	.247
1.6	.202
1.8	.165
2.0	.135
2.2	.111
2.4	.091
2.6	.074
2.8	.061
3.0	.050

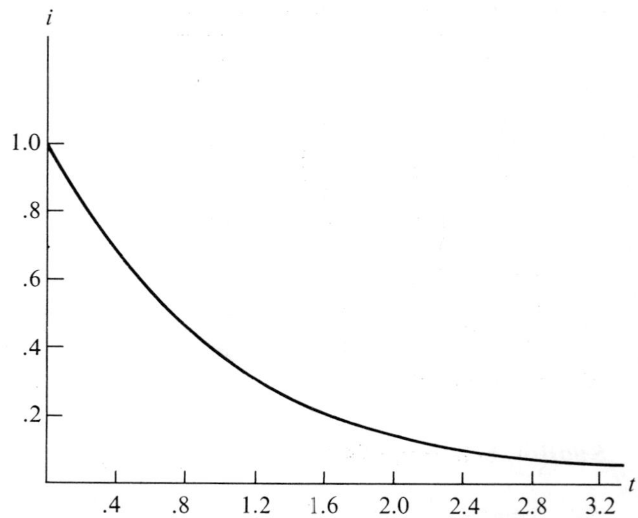

3. $L = 2150(1.0848 - .9218)$
$= 350.3866.$

4. $i = .5(1 - .40657)$
$= .297.$

5.

x	I/I_0
.00	1.000
.02	.773
.04	.597
.06	.461
.08	.356
.10	.275
.12	.213
.14	.164
.16	.127
.18	.098
.20	.076

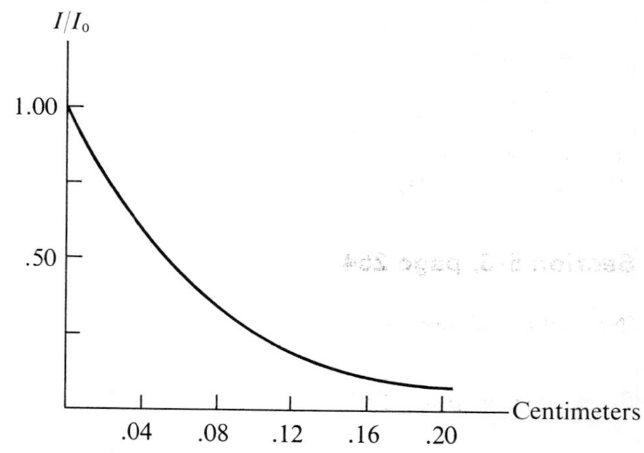

Section 5-6, page 276

1. $k = -\dfrac{.6932}{1800} = -.000385.$

2. $\dfrac{1}{20}(2.3026 - 2.1972) = .00527).$

3. $\dfrac{1}{2}(3.896 - .257) = 1.820.$

4. $\dfrac{\ln 5.24}{\ln .375} = -1.689.$

5. (No written solution.)

Section 5-7, page 283

1. $t = \dfrac{\log(A/P)}{\log(1+r)}$
$= 11.00$ yr.

2. $t = \dfrac{\log(A/P)}{4\log(1+r/4)}$
$= 10.75$ yr.

3. $t = 86.6.$

4. $k = \dfrac{\ln(14.11/14.72)}{-1122}$
$= 3.78(10^{-5}).$

5. $\log a = \log .04 + 300/276$
$a = .4887.$

6. $k = -.275,$
$t = 10.9$ hr.

7. $B = -43342,$
$i = 2.87(10^2).$

8. $(H^+) = 10^{-7}.$

9. $\ln i = -(Rt/L) + \ln E/R,$
$\ln(iR/E) = -(Rt/L),$
$iR/E = e^{-Rt/L},$
$i = (E/R)\, e^{-Rt/L}$

10. $i = E/R(1 - e^{-Rt/L}),$
$iR/E - 1 = -e^{-Rt/L},$
$1 - iR/E = e^{-Rt/L},$
$\ln(1 - iR/E) = -(Rt/L),$
$t = \dfrac{-L\ln(1 - iR/E)}{R}.$

Section 6-1, page 296

1. $y/x = 3 = \tan 71.6°.$

2. $y = 120 \tan 37°$
$= 90.4.$

3. (Altitude) $= 9 \tan 60° = 15.6.$

4. $\tan\theta = 10/14,$
$\theta = 35.5°.$

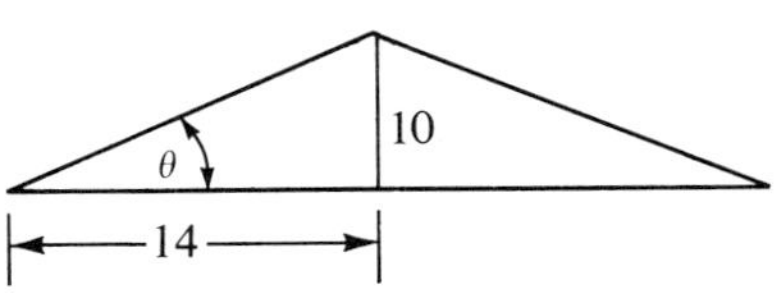

5. $\tan\theta = \dfrac{28}{1000}$
$\theta = 1.6°.$

6. $\overline{BC} = \dfrac{30{,}000}{\tan 11°\, 20'} = 149{,}680$ ft.

7. $\left(\dfrac{\text{riser}}{\text{tread}}\right) = \tan 22° = .404,$
(tread) $= 19.8$ in.

Section 6-2, page 306

1. a. $\theta = \arctan 10/25 = 21.8°$, b. $z = 26.93$.

2. $\overline{AC} = 2.5/\cos 40° = 3.264$, $\theta = \arccos 3.264/4.2 = 39.0°$.

3. $\theta = \arcsin 7.2/9.2 = 50.7°$,
$x = 3.2/\cos 50.7° = 5.06$.

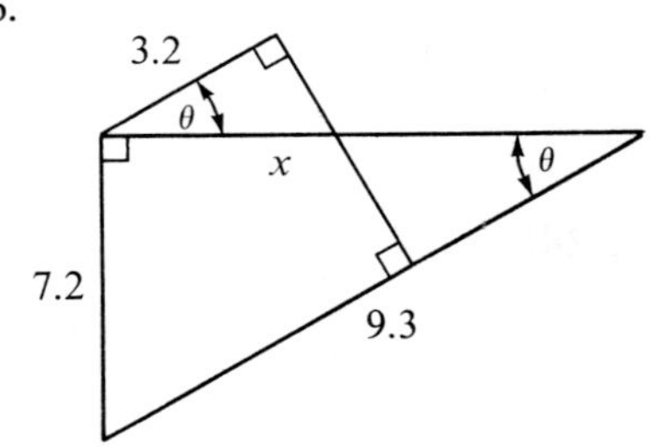

4. $\theta = \arctan 8.5/15 = 29.5°$,
$\overline{OA} = 5.5/\sin 29.5° = 11.2$,
$\overline{OC} = \overline{OA} + 5.5 = 16.7$.
$\therefore \quad \overline{BC} = 1.7$ in. (above rim).

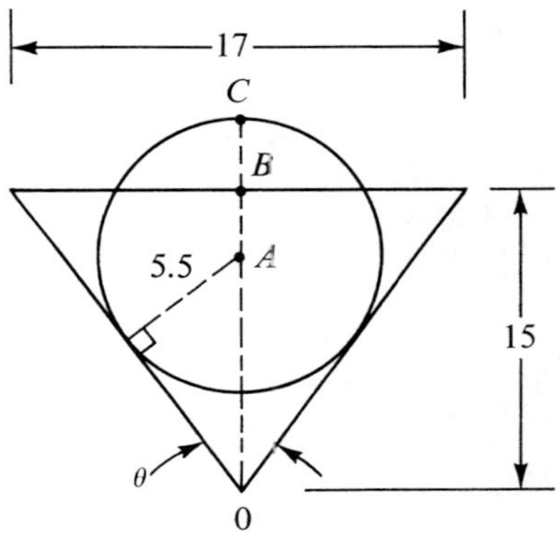

5. $\sin 27.5° = \dfrac{R}{(R+20)}$,
$R = 17.15$ ft.

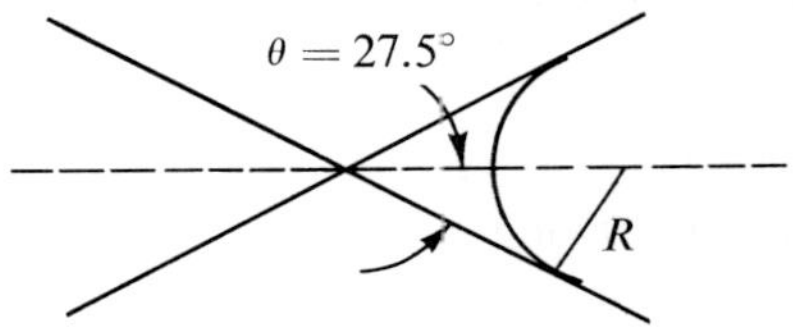

6. $M = T + D + \dfrac{D}{\tan(A/2)}$.

7. $\sin 17.5° = \dfrac{2.5 - r}{2.5 + r}$,
$r = 1.75$ in.

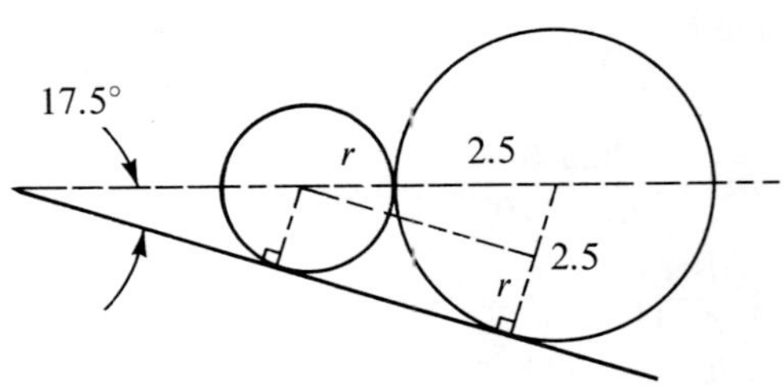

8. i. Let $\overline{D'B'} = 1$;
$\overline{DB} = 1/\cos 42° = 1.3456.$

ii. $\dfrac{\overline{D_1C_1}}{\overline{D_1B_1}} = \tan 65°,$

$\overline{D_1C_1} = \overline{DC} = 2.1445.$

iii. $\theta/2 = \arctan \dfrac{1.3456}{2.1445},$

$\theta = 64.2°.$

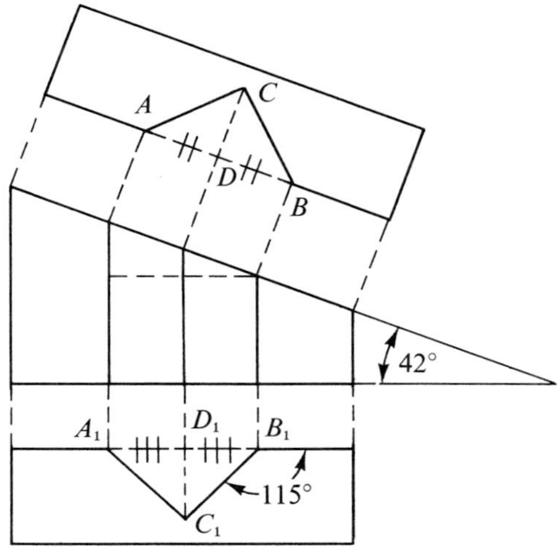

9. i. Let $\overline{OC} = 2$;
$\therefore\ \overline{CB} = 1,\ \overline{OB} = \sqrt{3}.$

ii. $\overline{AB}$ and $\overline{PC}$ are equal and parallel.

iii. $\tan 20° = \dfrac{\overline{AB}}{\sqrt{3}} = \dfrac{\overline{PC}}{\sqrt{3}}$;

$\overline{PC} = .63041.$

iv. $\angle POC = \arctan \overline{PC}/2$
$= 17.5°.$

10. i. $\angle CBA = \arctan 43/36$
$= 50.06°$;

ii. $\overline{BC} = \dfrac{36}{\sin 50.06°} = 46.95$;

iii. $\dfrac{x}{10} = \dfrac{46.95 - x}{5},$

$x = 31.30$;

iv. $\angle OBS = \arccos 10/31.30$
$= 71.37°$;

v. $\theta/2 = 31.43$; $\theta = 62.86°.$

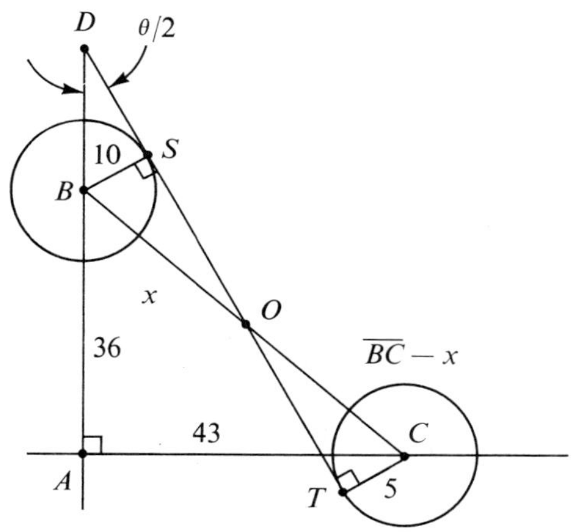

Section 6-3, page 321

1. i. $\theta = 40°$;

ii. $\dfrac{\sin 40°}{450} = \dfrac{\sin 28°}{120 + w},$

$w = 208.7$;

iii. $\dfrac{\sin 17°}{208.7} = \dfrac{\sin 123°}{x},$

$x = 598.6.$

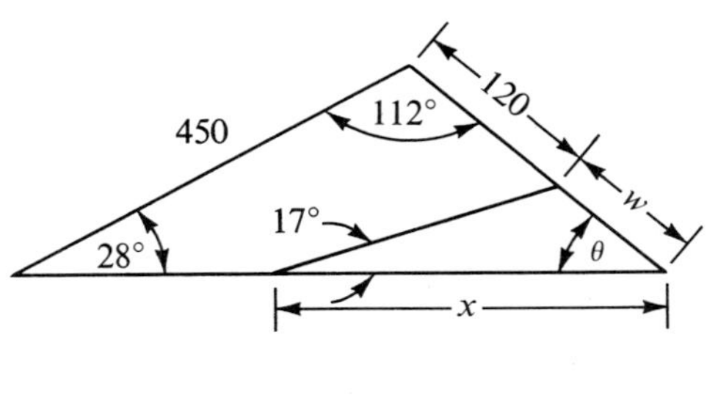

2. i. $\dfrac{248}{\sin 28^\circ} = \dfrac{470}{\sin \phi}$,

$\phi = 62.8^\circ, \quad 117.2^\circ.$

ii. $\beta = 109.2^\circ, \quad 34.8^\circ.$

iii. $\dfrac{248}{\sin 28^\circ} = \dfrac{F_1}{\sin 109.2^\circ}, \quad F_1 = 498.9.$

$\dfrac{248}{\sin 28^\circ} = \dfrac{F_1}{\sin 34.8^\circ}, \quad F_1 = 301.5.$

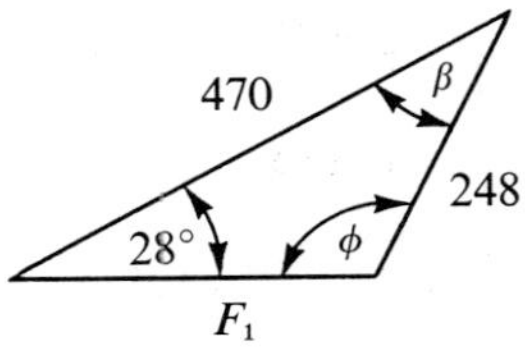

3. i. $\dfrac{\overline{AB}}{\sin 95^\circ} = \dfrac{12}{\sin 42.5^\circ}$,

$\overline{AB} = 17.69.$

ii. (Diameter) $= 2R$
$= 17.69 \tan 21.25$
$= 6.88$ in.

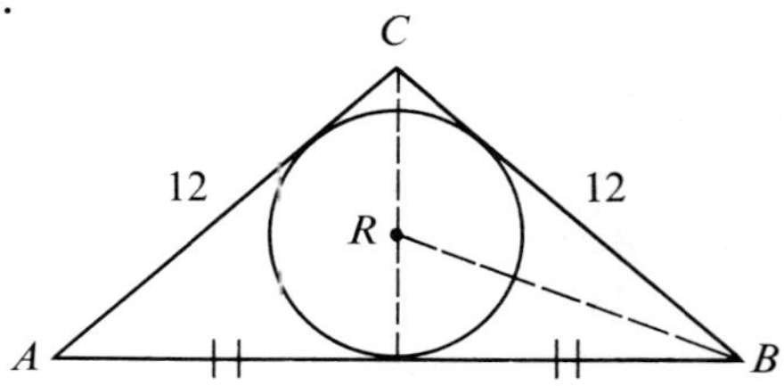

4. $\overline{DE} = 13.48; \qquad \theta = 115^\circ.$

5. $\angle AOB = 15^\circ$,
$\overline{OB} = \overline{BC} = 5/\cos 15^\circ = 5.1764.$

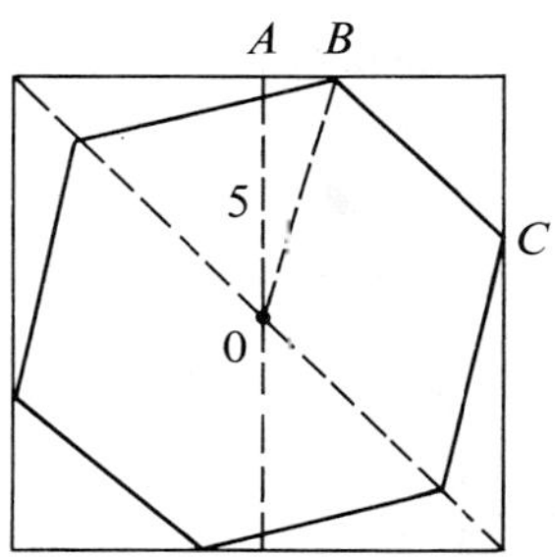

Section 6-4, page 330

1. i. $\overline{BC} = \dfrac{500 \sin 43^\circ}{\sin 24^\circ} = 838.4;$

ii. $\overline{AC} = \dfrac{500 \sin 129^\circ}{\sin 19^\circ} = 1193.5;$

iii. $\overline{AB}^2 = (1193.5)^2 + (838.4)^2 - 2(1193.5)(838.4) \cos 81^\circ,$
$\overline{AB} = 1346.9.$

2. $d = 2.48$ in.

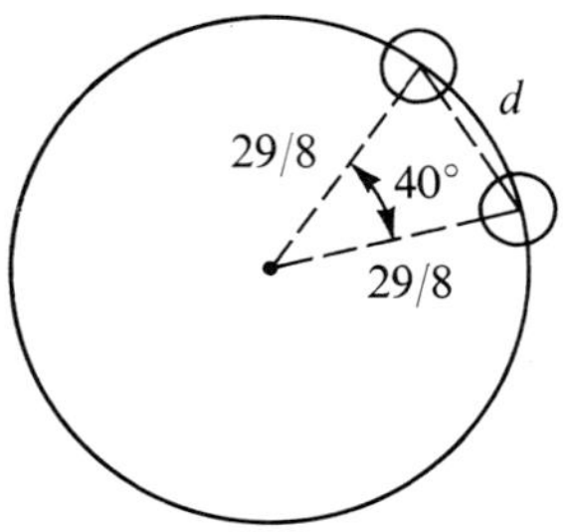

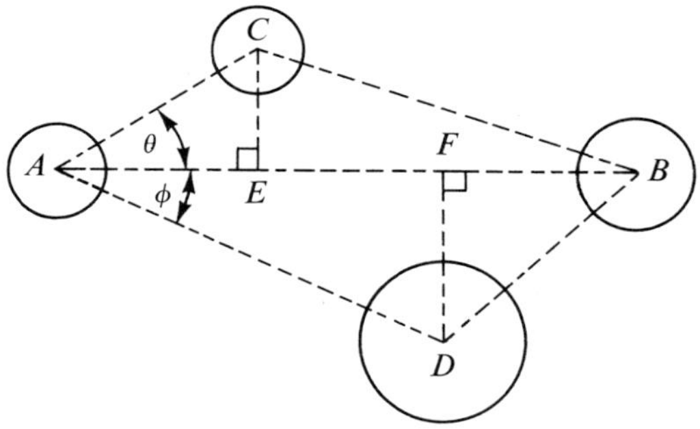

3. i. $\cos\theta = \dfrac{(2.200)^2 - (1.250)^2 - (3.000)^2}{-2(1.250)(3.000)}$

$= .7630,$

$\theta = 40.3°;$

$\overline{CE} = 1.250 \sin 40.3° = .8085$ in.

$\overline{AE} = 1.250 \cos 40.3° = .9538$ in.

ii. $\cos\phi = \dfrac{(1.312)^2 - (2.125)^2 - (3.000)^2}{-2(2.125)(3.000)}$

$= .9250,$

$\phi = 22.3°;$

$\overline{AF} = 2.125 \cos 22.3° = 1.966$ in.,

$\overline{DF} = 2.125 \sin 22.3° = .8064$ in.

4. i. $\angle A = 180° - (105° + 33°) = 42°.$

ii. $\dfrac{\overline{AD}}{\sin 33°} = \dfrac{118 + 254}{\sin 42°}$; $\overline{AD} = 302.78.$

iii. $\overline{AB} = 352.3$ (by the Law of Cosines).

5. (Belt Length) $= 12.0 + 16.0 + 18.5 + \pi 8$

$= 71.6$ in.

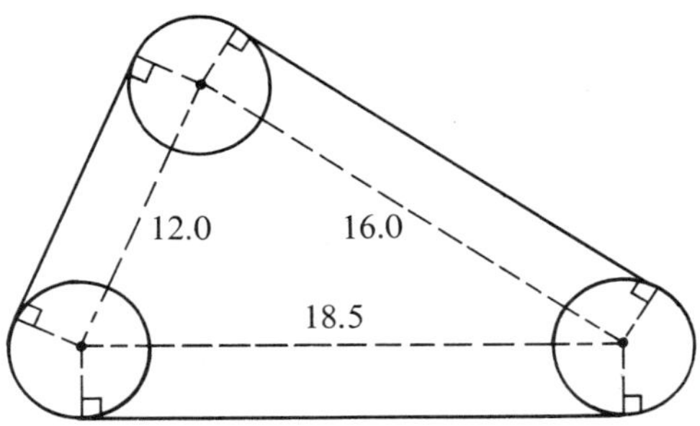

6. $\overline{AB} = 9.5$ ft,
$\overline{BC} = 9.5$ ft,
$\overline{AG} = 11.3$ ft,
$\overline{BG} = 5.11$ ft,
$\overline{GC} = 10.23$ ft.
(Symmetrical)

7. i. $\arctan .3 = 16.7°$,
ii. $d = \dfrac{25 \sin 106.7°}{\sin 38.3°} = 38.6$ ft,
iii. $u = \dfrac{25 \sin 73.3°}{\sin 71.7°} = 25.2$ ft.

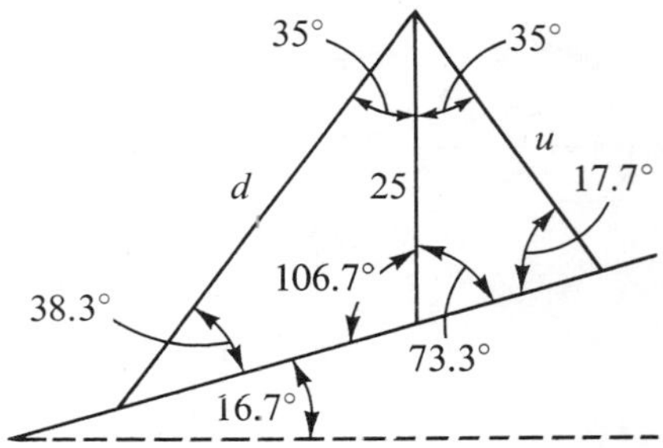

Section 6-5, page 337

1. i. $x = h/\tan 50°$,
$x = 6 - h/\tan 75°$.
ii. $h/\tan 50° = 6 - h/\tan 75°$,
$h = 5.42$.
iii. $(\text{Area}) = \dfrac{5.42}{2}(20 + 14) = 92.2$.

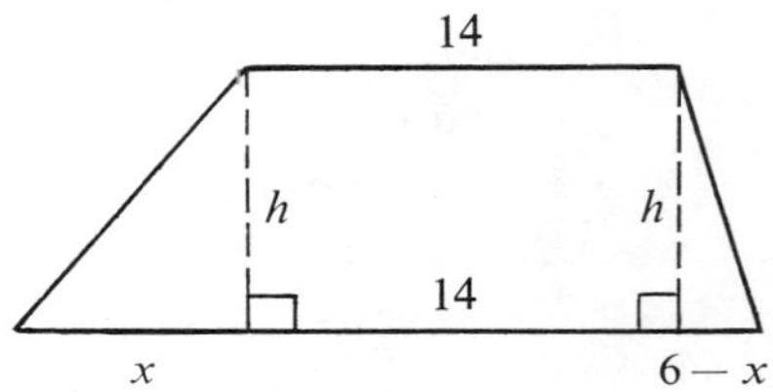

2. (Area) = 1410 sq ft.

3.

Depth (ft)	Volume (cu ft)
0	0
1	31.1
2	82.4
3	141.3
4	200.2
5	251.6
6	282.7

4. i. $A = (40/360)\pi(6)^2 - 6^2(\sin 70°)(\cos 70°)$
$= .9968$ sq in.
ii. (volume) = (.9968)(3.00) = 2.9905 cu in.
iii. (weight) = (2.9905)(.28) = .837 lb.

5. (Area) = 30 cos 50°
= 19.28 sq in.

6. (Area)′ = 120/cos 50°
= 186.7 sq in.

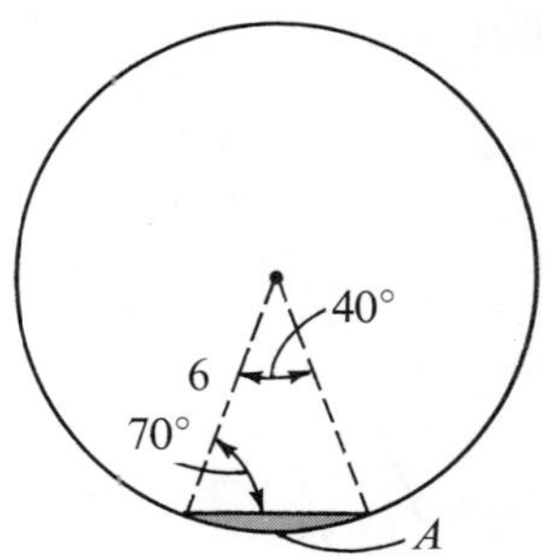

Section 6-6, page 347

1. $|\mathbf{r}| = \sqrt{(-4.7)^2 + (8.1)^2} = 9.36$.

2. $-\mathbf{r} = (-5.2, -8.7)$.

3. $\mathbf{r} = (8,6)$, (rectangular form)
$= (10, 36.9°)$, (polar form).

4. (8.6, 10.4).

5. $\mathbf{v} = (350, 135°)$
$= (-247.5, 247.5)$.

6. $\mathbf{t}_1 = (1000, 135°)$
$= (-707, 707)$,
$\mathbf{t}_2 = (1000, 45°)$
$= (707, 707)$,
$\boldsymbol{\omega} = (1414, 270°)$
$= (0, -1414)$.

Section 6-7, page 354

1. $\mathbf{r} = (35, 20°) = (32.89, 11.97)$,
$\mathbf{s} = (40, 60°) = (20.00, 34.64)$,
$\mathbf{r} + \mathbf{s} = (52.89, 46.61) = (70.50, 41.4°)$.

2. $\theta = \arctan 8/3 = 69.4°$,
$v = \sqrt{8^2 + 3^2} = 8.54$,
$\mathbf{v} = (8.54, 69.4°)$, or 8.54 mph, 20.6° east of north.

3. $\mathbf{A} = (38, 121°) = (-19.57, 32.57)$,
$\mathbf{B} = (76, 261°) = (-11.89, -75.06)$,
$\mathbf{C} = (57, 317°) = (41.69, -38.87)$,
$\mathbf{A} + \mathbf{B} + \mathbf{C} = (10.23, -81.37) = (82.01, 277.2°)$.

4. $\mathbf{A} = (20, 47°) = (13.64, 14.63)$,
$\mathbf{B} = (30, 121°) = (-15.45, 25.72)$,
$\mathbf{C} = (50, 180°) = (-50.00, 0.00)$,
$\mathbf{D} = (70, 300°) = (35.00, -60.62)$,
$\mathbf{A} + \mathbf{B} + \mathbf{C} + \mathbf{D} = (-16.81, -20.28)$
$= (26.34, 230.3°)$.

5. i. $\mathbf{r}_1 = (210, 65°) = (88.75, 190.3)$,
$\mathbf{r}_2 = (340, 325°) = (278.5, -195.0)$,
$\mathbf{r}_3 = (200, 218°) = (-157.6, -123.1)$.
ii. $\mathbf{r}_1 + \mathbf{r}_2 + \mathbf{r}_3 + \mathbf{r}_4 = (0, 0)$,
$\mathbf{r}_4 = -(\mathbf{r}_1 + \mathbf{r}_2 + \mathbf{r}_3)$
$= (-209.7, 127.8) = (245.5, 148.6°)$.

6. i. $\mathbf{A} = (10, 180°) = (-10, 0)$,
$\mathbf{s}_v = (8, 90°) = (0, 25)$,
$\mathbf{s}_L = (25, 270°) = (0, -8)$.
ii. $\mathbf{A} + \mathbf{s}_v + \mathbf{s}_L + \mathbf{s}_{Lv} = (0, 0)$,
$\mathbf{s}_{Lv} = -(\mathbf{A} + \mathbf{s}_v + \mathbf{s}_L)$
$= (10, -17) = (19.72, 300.5°)$.
iii. $\theta = 30.5°$.

7. i. $T = 2200 \cos 35° = 1802$,
$\mathbf{T} = (1802, 55°) = (1034, 1476)$,
$\mathbf{W} = (2200, 270°) = (0, -2200)$,
$\mathbf{T} + \mathbf{W} = (1034, -723.8) = (1262, 325°)$.

8. i. $L = \sqrt{6^2 + 15^2 - 2(6)(15) \cos 70°}$
$= 14.122$.
ii. $\phi = 4.061°$.
iii. $\dfrac{6}{\sin \theta} = \dfrac{14.122}{\sin 70°}$,
$\theta = 23.531°$.

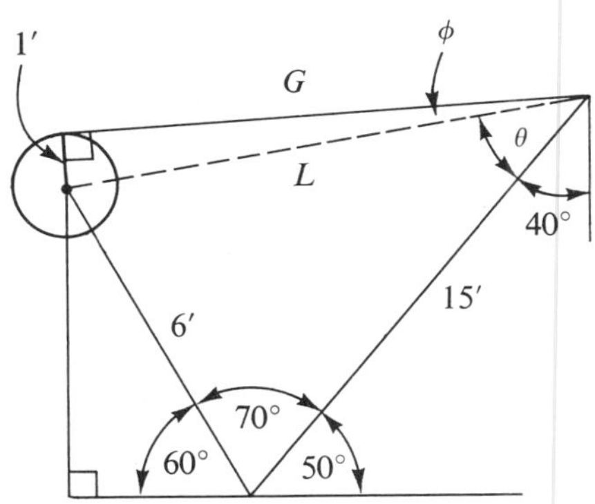

iv. $\mathbf{B} = (7.4, 50°) = (4.757, 5.669)$,
$\mathbf{G} = (5.3, 202.4°) = (-4.900, -2.020)$.

v. $\mathbf{B} + \mathbf{G} = (-.1432, 3.6484)$
$= (3.65, 92.24°)$.

vi. (Max. safe load) $= (3.651 \cos 2.24°, 270°)$
$= (3.648, 270°)$.

vii. Boom will fail first.

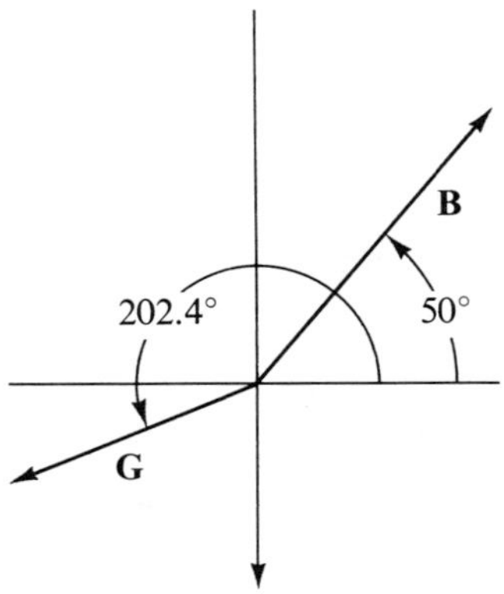

Section 7-1, page 369

1. (Arc length) $= 4.00\pi = 12.57$ in.
(radius) $= 16.0/1.55 = 10.32$ in.
(central angle) $= 12.5/5.52 = 2.265$ rad.

2. $\theta = 0.4(12) = 4.80$ rad.,
$t = 6.4/.5\pi = 4.07$ sec.,
$\omega = 60\pi/.5 = 120\pi$ rad/sec.

3. $\omega r = (1/\text{time})(\text{length}/1) =$ (velocity of a point on the end of r).

4. 1130 deg/sec; 3.14 rev.

5. i. $L = \sqrt{7^2 - 4^2}$
$= 5.745$ in.,
$\theta = \arcsin 4/7$
$= .6082$ rad,
$\beta = \pi/2 + .6082$
$= 2.1790$ rad,
$\alpha = \pi/2 - .6082$
$= .9626$ rad.

ii. (Belt length) $= 2(\alpha + 5\beta + L)$
$= 35.21$ in.

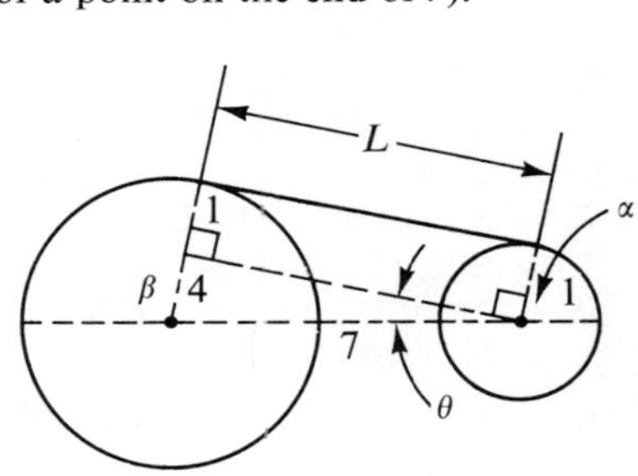

6. $2\pi R_Q = 4\pi/5\ R_P$;
$R_Q = 1$;
$R_P = 5/2$ in.

7. (Velocity) $= \omega r$
$= (15\pi)(17)$
$= 801.1$ in./sec.

8. 111.19 km.

9. $C_{(40\text{th})} = 2\pi(6371 \cos 40°)$
$= 30{,}665$ km.

Section 7-2, page 378

1. Piston of a motor,
Pendulum,
Balance wheel of a watch,
Weight on a spring,
Voltage from a generator.
. . .

3.

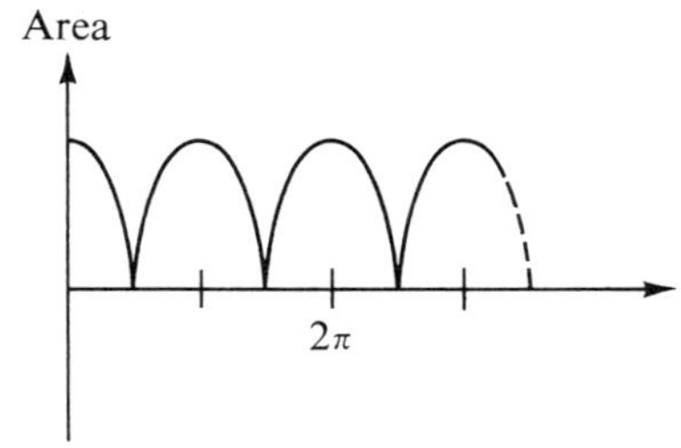

4.

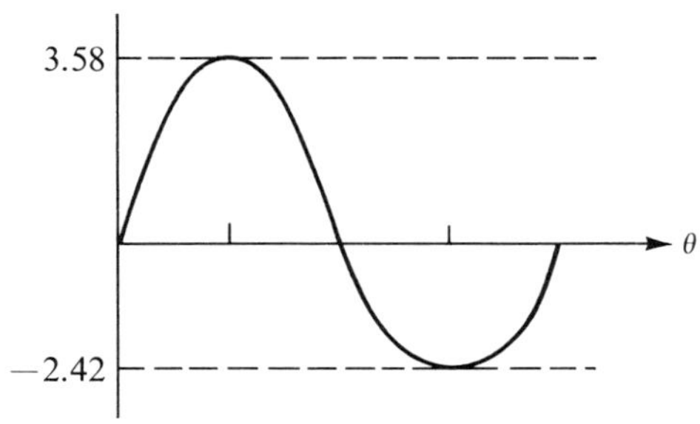

Section 7-3, page 387

1. 1/60 sec; 100 cps.
2. Cycles; radians per second; radians.
3.

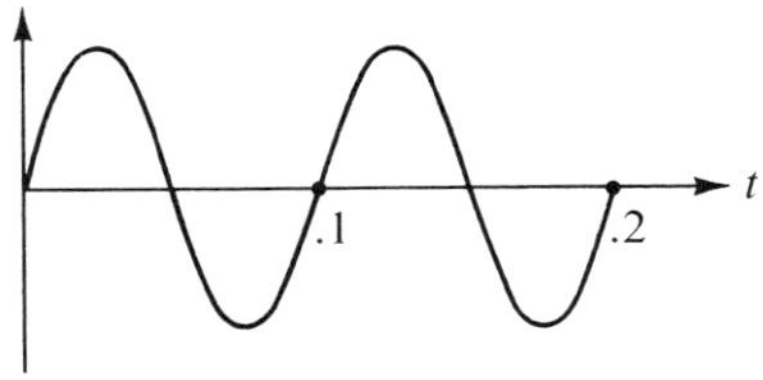

4. (Period) = 0.1 sec.
5. (Frequency) = 1/(Period) = 1/.01 = 100 cps.
6. $\omega = 2\pi 400 = 800\pi$ rad/sec.
   (Period) = 1/400 = .0025 sec.

7. $(1740/60)(2\pi r) = (22)(2\pi R)$,

$\frac{R}{r} = \frac{29}{22} = 1.318.$

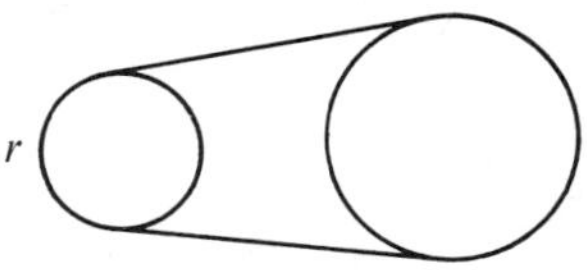

8. $v = 18 \sin 44\pi t.$

Section 7-4, page 398

1. i. $c = \arctan 11/15 = .6327;$
ii. (period) $= 2\pi/b,\ b = 120;$
iii. (phase shift) $= c/b = .6327/120 = .00527$ (lagging).

2. i. $c = \arctan 3500/5000 = .6109,$
ii. $c/b = .6109/120\pi = .00162$ (leading).

3. i. $c = \arctan x_c/R$
$= \arctan 800/1000 = .6749,$
ii. $i = f(t) = 15 \sin(120t + .6749)$
$f(.01) = 14.31.$

4. i. $c = \arctan 2200/2000 = .8334,$
ii. $i = f(t) = I \sin(800t - .8334),$
iii. $800t - .8334 = \pi/2$
$t = .003005.$

5. a. $\omega = 2\pi f = 2\pi(60) = 120\pi,$
b. $v = 10 \sin(120\pi(.01) + \pi/2)$
$= -8.09.$

Section 7-5, page 407

1. $d = s \sin\theta + \sqrt{r^2 - (s\cos\theta)^2} - \sqrt{r^2 - s^2}$
(crank shaft, $s = 3$; rod, $r = 8$).

2. $f_1 = 1/2\ f_2.$

3. $f_3 = 1/0.03(10^{-6}) = 3.33(10^7)$ cps,
$f_1 = (1/3)(3.33)(10^7) = 1.11(10^7)$ cps.

4.

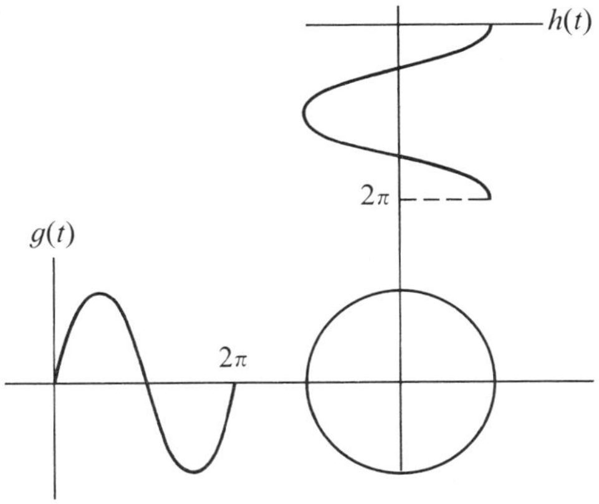

5.

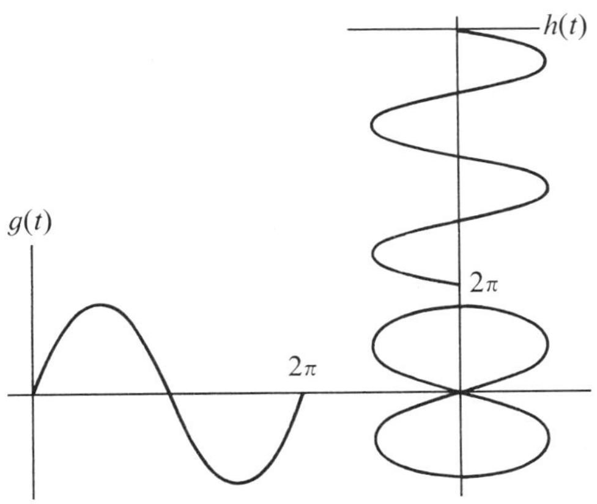

Section 7-6, page 411

1.

t	$f(t)$, (in.)	$f'(t)$, (in./sec)
0	0	12.75
.2	1.90	3.88
.4	1.18	−10.17
.6	−1.18	−10.17
.8	−1.90	3.88
1.0	0	12.57

2. When $f(t) = r$, $f'(t) = 0$.

3.
$$f'(t) = 6.25(3000)(2\pi) = 1178 \text{ m/min} = 19.63 \text{ m/sec.}$$

Section 8-1, page 422

1. a. $\sqrt{1 - \cos^2 .38} = \sin .38 = .3709$,

b. $\sqrt{\sec^2 .74 - 1} = \tan .74 = .9131$,

c. $\dfrac{\sec .42}{\tan .42 + \cot .42} = \sin .42 = .4078.$

2. $\sin \theta = b$,

$\cos \theta = \sqrt{1 - b^2}$,

$\tan \theta = b/\sqrt{1 - b^2}$,

$\csc \theta = 1/b$,

$\sec \theta = 1/\sqrt{1 - b^2}$,

$\cot \theta = \sqrt{1 - b^2}/b$.

3. $\sin .6283 = .5878$,

$\cos .6283 = .8090$.

$\tan .6283 = .7265$,

$\csc .6283 = 1.7014$,

$\sec .6283 = 1.2361$,

$\cot .6283 = 1.3764$.

4. $\sec 152° = -1.1326$,

$\csc 152° = 2.1301$,

$\tan 152° = -1.8807$.

5. $(\text{Area}) = \pi m^2 \tan^2 \theta/2$.

7.

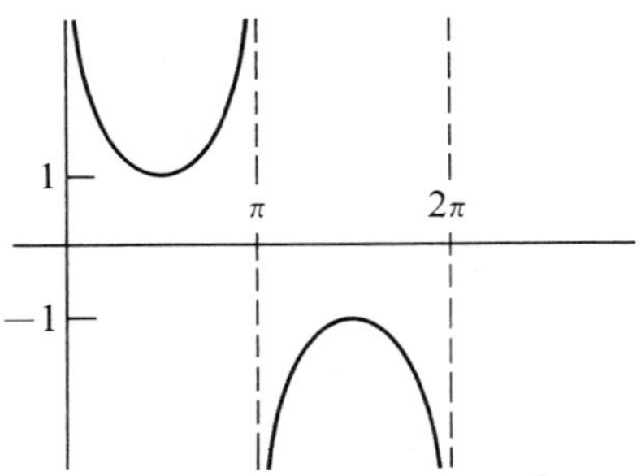

Section 8-2, page 431

1.
$$f(t) + g(t) = A\left(\sin\left(2\pi \frac{t}{T} - 2\pi \frac{a}{\lambda}\right) + \sin\left(2\pi \frac{t}{T} + 2\pi \frac{a}{\lambda}\right)\right)$$
$$= A\left(\sin \frac{2\pi t}{T} \cos \frac{2\pi a}{\lambda} - \cos \frac{2\pi t}{T} \sin \frac{2\pi a}{\lambda}\right.$$
$$\left. + \sin \frac{2\pi t}{T} \cos \frac{2\pi a}{\lambda} + \cos \frac{2\pi t}{T} \sin \frac{2\pi a}{\lambda}\right)$$
$$= \left(2A \cos \frac{2\pi a}{\lambda}\right)\left(\sin \frac{2\pi t}{T}\right).$$

2.
$$a^2 = (a_1 \cos c_1 + a_2 \cos c_2)^2 + (a_1 \sin c_1 + a_2 \sin c_2)^2$$
$$= a_1^2 \cos^2 c_1 + 2a_1a_2 \cos c_1 \cos c_2 + a_2^2 \cos^2 c_2 + a_1^2 \sin^2 c_1$$
$$+ 2a_1a_2 \sin c_1 \sin c_2 + a_2^2 \sin^2 c_2$$
$$= a_1^2 + a_2^2 + 2a_1a_2 \cos(c_1 - c_2)$$
$$= a_1^2 + a_2^2 - 2a_1a_2 \cos(\pi - (c_1 - c_2)).$$

3. a. $F_1 = 22.49$, b. $F_2 = \dfrac{\omega \sin A_1}{\sin(A_1 + A_2)}$.

$F_2 = 37.62$.

4.
$$\begin{aligned} v_1 + v_2 + v_3 &= 3(\sin 120\pi t \cos 0 + \cos 120\pi t \sin 0) \\ &\quad + 3\left(\sin 120\pi t \cos \frac{\pi}{9} + \cos 120\pi t \sin \frac{\pi}{9}\right) \\ &\quad + 3\left(\sin 120\pi t \cos \frac{2\pi}{9} + \cos 120\pi t \sin \frac{2\pi}{9}\right) \\ &= 3\left(1 + \cos \frac{\pi}{9} + \cos \frac{2\pi}{9}\right) \sin 120\pi t \\ &\quad + 3\left(\sin \frac{\pi}{9} + \sin \frac{2\pi}{9}\right) \cos 120\pi t \\ &= \left(8.638 \cos \frac{\pi}{9}\right) \sin 120\pi t + \left(8.638 \sin \frac{\pi}{9}\right) \cos 120\pi t \\ &= 8.638 \sin\left(120\pi t + \frac{\pi}{9}\right). \end{aligned}$$

Section 8-3, page 437

1. i. $\omega = 2\pi f = 120\pi$.

ii. $v = 10 \sin 120\pi t$,

$i = 7.54 \cos 120\pi t$.

iii. $p = iv = (V \sin \omega t)(CV\omega \cos \omega t) = \left(\dfrac{CV^2\omega}{2}\right) \sin 2\omega t = 37.7 \sin 240\pi t$.

iv.

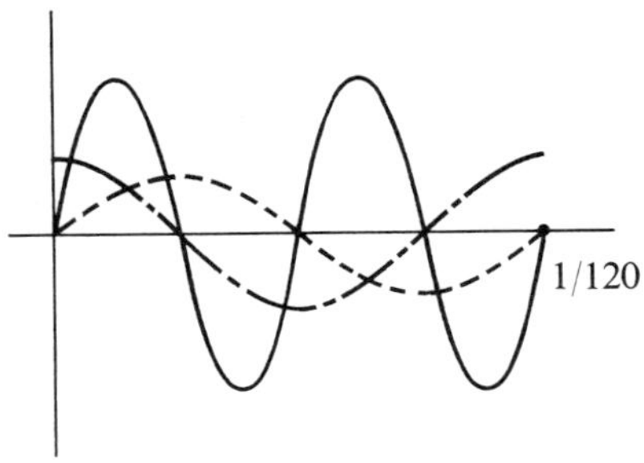

2. $Y = 2A$; $n = 2\pi/\lambda$; $m = 2\pi/P$.

3. a. $\dfrac{Mv^2}{RMg} = \dfrac{(P+Q)\sin\theta}{(P+Q)\cos\theta}$,

$\theta = \arctan \dfrac{v^2}{Rg}$.

b. $v^2 = Rg \tan\theta$,

$$\begin{aligned} v &= \sqrt{2500(32) \tan 15^\circ} \\ &= 146.4 \text{ ft/sec} \\ &= 99.8 \text{ mph}. \end{aligned}$$

Section 8-4, page 446

1. a. $v_1/v_2 = \sin 30°/\sin 60° = \sqrt{3}/3 = .577.$ b. $n_2 = nv_1/v_2 = .7698.$

2. i. Let r be the radius of circle STU.

ii. $\overline{OP}^2 = \overline{QO}^2 + \overline{QP}^2 - 2(\overline{QO})^2(\overline{QP})^2 \cos\theta,$
$(5-r)^2 = 4 + (r+3)^2 - 2(2)(r+3)\cos\theta,$

$$\cos\theta = \frac{4r-3}{r+3}.$$

iii. $\overline{RP}^2 = \overline{QR}^2 + \overline{QP}^2 - 2(\overline{QR})^2(\overline{QP})^2 \cos\theta,$
$(r+2)^2 = 5^2 + (r+3)^2 - 2(5)(r+3)\cos\theta,$

$$\cos\theta = \frac{r+15}{5r+15}.$$

iv. $$\frac{4r-3}{r+3} = \frac{r+15}{5r+15},$$

$19r^2 + 27r - 90 = 0; \quad r = 30/19, \ -3.$

v. $$\theta = \arccos\frac{4(30/19)-3}{(30/19)+3} = \arccos .7241 = 43.6°,$$

$x = -(\overline{PQ}\cos\theta - \overline{OQ}) = -1.316,$
$y = \overline{QP}\sin\theta = 3.158.$

3. a. $F(2\pi) = \pi,$ b. $(\text{mean ordinate}) = \dfrac{F(2\pi)}{2\pi - 0} = \dfrac{1}{2},$

c. $\sqrt{1/2} = .707.$

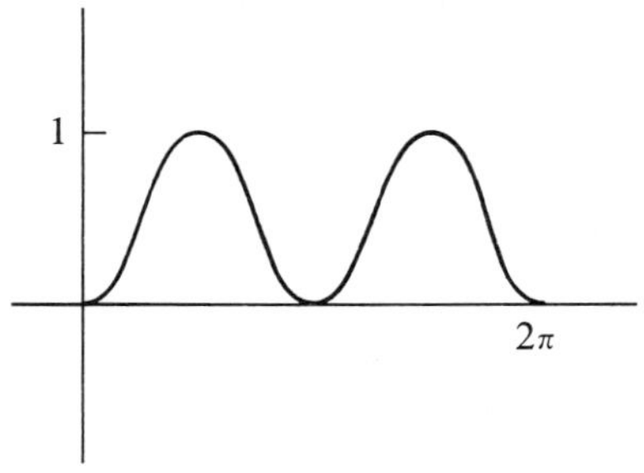

4. i. $(\theta/2\pi)\pi r^2 - r^2 \sin(\theta/2)\cos(\theta/2) = \pi r^2/4,$
$\sin\theta = \theta - \pi/2.$

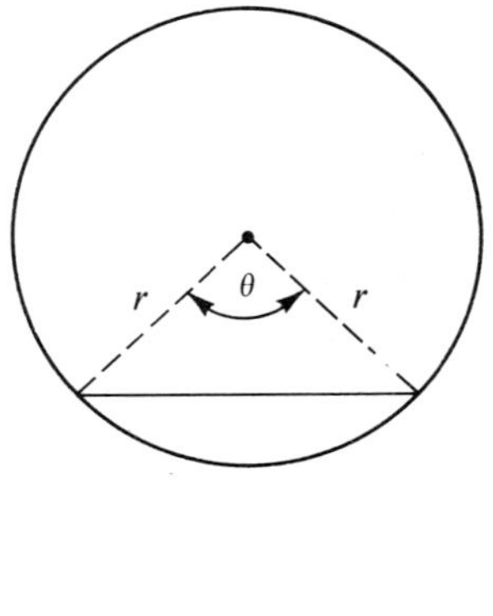

ii.

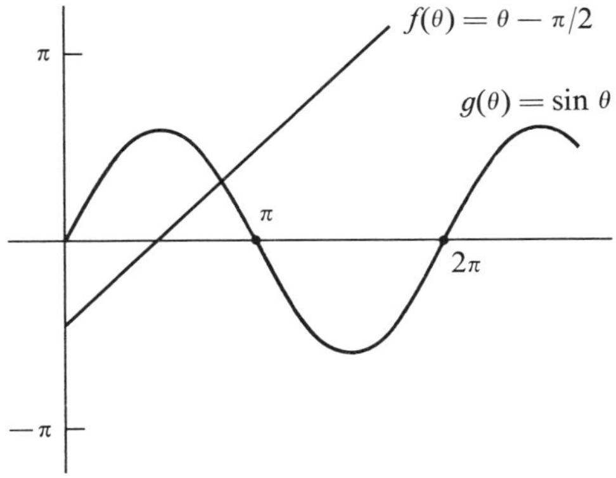

iii. $\theta \approx 2.31$ rad.

Section 9-1, page 455

1. $j250 - j100 = j150.$

2. $j1131 - j1326 = -j195.$

Section 9-2, page 461

1. $Z_1 + Z_2 = 175 - j150.$

2. $Z = (Z_1 Z_2)/(Z_1 + Z_2).$

3. $Z_t = \dfrac{(25 + j35)(30 - j65)}{(25 + j35) + (30 - j65)} = 46.78 + j15.06.$

4. $Z_2 = \dfrac{Z_1 + Z_3}{2} = \dfrac{(480 - j500) + (200 + j350)}{2} = 340 - j75.$

5. a. $f(2 + j3) = 4(2 + j3)^2 + (2 + j3) = -18 + j51;$
b. $f(-1 - j) = 4(-1 - j)^2 + (-1 - j) = -1 + j7.$

Section 9-3, page 468

1. $Z_3 = -(Z_1 + Z_2)$
$= -[(1000 + j2000) + (-800 + j2200)]$
$= -200 - j4200.$

2. $Z_1 = \left(\dfrac{5.3}{3.4}\right)(0 - j42) = 0 - j65.5.$

3. $Z = (0 + j200) + (0 - j300) + (50 + j0) = 50 - j100.$

4. $Z_2 = -\frac{a}{b} Z_1$

$= \left(-\frac{5.0}{1.5}\right)(100(\cos 30° + j \sin 30°))$

$= 333(\cos 210° + j \sin 210°).$

5. $Z_1 Z_2 = 19.2(\cos 7\pi/12 + j \sin 7\pi/12).$

Section 9-4, page 471

1. $Z = 0.9(\cos 36° + j \sin 36°).$

2. $I = \dfrac{12(\cos 0° + j \sin 0°)}{1000(\cos 20° + j \sin 20°)} = 0.012(\cos 340° + j \sin 340°).$

3. $Z_t = \dfrac{8000(\cos 20° + j \sin 20°)}{159.19 + j36.65}$

$= \dfrac{8000(\cos 20° + j \sin 20°)}{163.4(\cos 13.0° + j \sin 13.0°)}$

$= 49.0(\cos 7° + j \sin 7°).$

4. $P = 250(\cos 13\pi/40 + j \sin 13\pi/40).$

5. $P = 200\left(\cos 4\pi ft + \frac{\pi}{12} + j \sin 4\pi ft + \frac{\pi}{12}\right).$

Section 9-5, page 477

1. $Z^{1/3} = \begin{cases} 4(\cos 0° + j \sin 0°), \\ 4(\cos 120° + j \sin 120°), \\ 4(\cos 240° + j \sin 240°). \end{cases}$

2. $P_{\text{app}} = 10(\cos 60° + j \sin 60°).$

3. $E = \begin{cases} 3.87(\cos 26.5° + j \sin 26.5°), \\ 3.87(\cos 206.5° + j \sin 206.5°). \end{cases}$

4. $E = \dfrac{0.12(\cos 15° + j \sin 15°)}{82.0(\cos 37.6° + j \sin 37.6°)}$

$= 0.00146(\cos 337.4° + j \sin 337.4°).$

5. i. $Z^2 + 4 = 0$

$Z^2 = -4$

$= 4(\cos 180° + j \sin 180°).$

ii. $Z = \begin{cases} 2(\cos 90° + j \sin 90°), \\ 2(\cos 270° + j \sin 270°). \end{cases}$

Section 9-6, page 481

1. $Z = \begin{cases} 10 + j7, \\ 12.2(\cos 35.0° + j \sin 35.0°). \end{cases}$

Voltage leads current by 35.0 degrees.

2. $Z = \frac{V}{I} = \frac{18 + j15}{9 + j4} = \frac{23.4(\cos 39.8° + j \sin 39.8°)}{9.8(\cos 24.0° + j \sin 24.0°)}$

$= 2.38(\cos 15.8° + j \sin 15.8°).$

Voltage leads current by 15.8°.

3. The current and voltage are in phase; in other words, the phase constant is zero.

4. $Z_1 = 7 - j4 = 8.06(\cos 330.3° + j \sin 330.3°).$

$Z_2 = 6 + j7 = 9.22(\cos 49.4° + j \sin 49.4°).$

5. a. $Z = \frac{Z_1 Z_2}{Z_1 + Z_2}.$ b. $Z = \frac{74.31(\cos 19.7° + j \sin 19.7°)}{13.34(\cos 13.0° + j \sin 13.0°)}$

$= 5.57(\cos 6.7° + j \sin 6.7°)$

$= 5.53 + j0.650.$

Section 10-1, page 490

1. i.

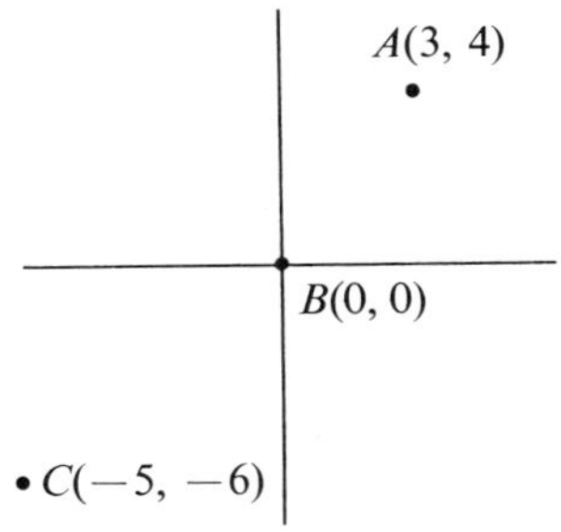

ii. $\overline{AC} = \sqrt{(3 + 5)^2 + (4 + 6)^2}$

$= 12.8.$

2.

Solution 1	*Solution 2*
$v_1(-a, b)$,	$v_1(-b, a)$,
$v_2(a, b)$,	$v_2(b, a)$,
$v_3(a, -b)$,	$v_3(b, -a)$,
$v_4(-a, -b)$.	$v_4(-b, -a)$.

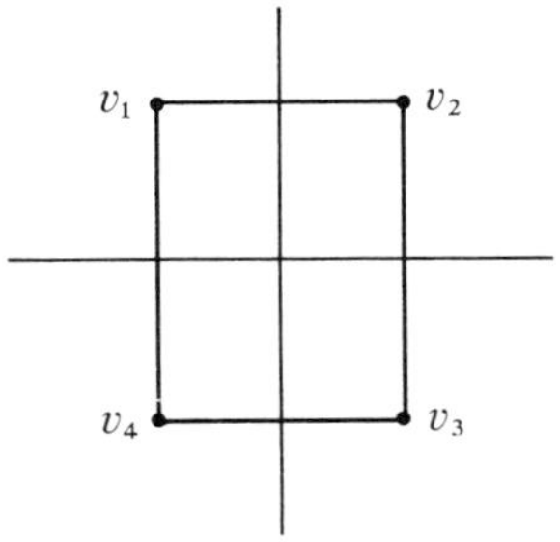

3. $P_1(|a\sqrt{2}|, 0)$,
$P_2(0, |a\sqrt{2}|)$,
$P_3(-|a\sqrt{2}|, 0)$,
$P_4(0, -|a\sqrt{2}|)$.

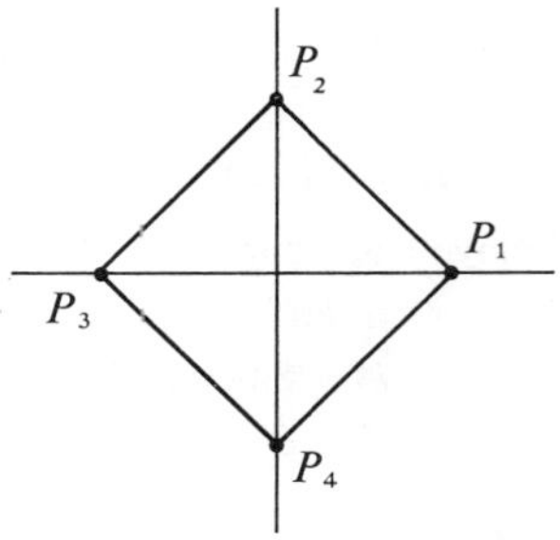

4. i. $(\text{Radius}) = \sqrt{(5-2)^2 + (-2-1)^2}$
$= \sqrt{18}$.

ii. $(\text{Area}) = \pi r^2 = 56.5$.

5. $(10/3, 14/3)$.

6. $(\text{Area}) = (1/2)(3+4)(5+2)$
$= 24.5$.

7. 4.438 miles.

8. i. $M(\overline{P_1P_3}) = (-3, 1)$,
$M(\overline{P_1P_2}) = (1/2, -3/2)$.

ii. Equation of L_1: $y = 1$.

iii. Equation of L_2: $y + x = -1$.

iv. Simultaneous solution of equations for L_1 and L_2:
$x = -2, \quad y = 1$.
$\therefore$ Center $(-2, 1)$.

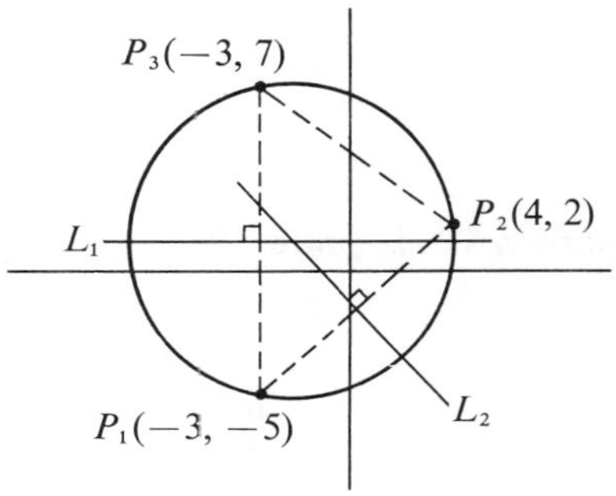

Section 10-2, page 498

1. i. $A(0, 0)$, $B(9, 3)$, $C(7, 6)$, $D(0, 7)$.

ii. $(\text{Area})_{ABCD} = (1/2)(0(3-7) + 9(6-0) + 7(7-3) + 0(0-6))$
$= 41$ sq miles.

2. i. $A(0, 0)$, $B(67.7, 24.6)$, $C(81.7, 73.6)$, $D(-10.4, 59.1)$.

ii. $(\text{Area})_{ABCD} = 4283$ sq ft.

3. i. $A(0, 0)$, $D(350.4, -156.6)$, $C(300.3, 199.9)$, $B(95.4, 152.6)$.

ii. $(\text{Area})_{ABCD} = 71{,}913.6$ sq ft.

Section 10-3, page 506

1. i. $h = -D/2 = 24/2 = 12$,
$k = -E/2 = -8/2 = -4$.

ii. $r = \dfrac{\sqrt{D^2 + E^2 - 4F}}{2} = 7$.

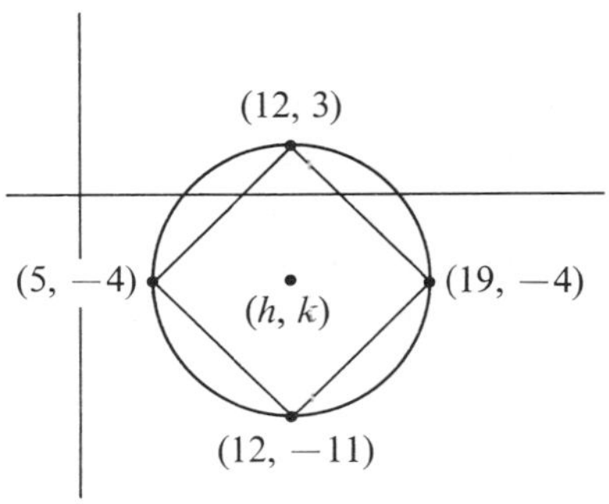

2. i. $\left.\begin{aligned} x^2 + y^2 &= 25, \\ y &= 2/3\,x - 4 \end{aligned}\right\}$ system.

ii. $\{(4.95, -0.70), (-1.26, -4.84)\}$.

iii. $M(\overline{P_1P_2}) = \left(\dfrac{4.95 - 1.26}{2}, \dfrac{-4.84 - .70}{2}\right)$

$= (1.85, -2.77)$.

3.

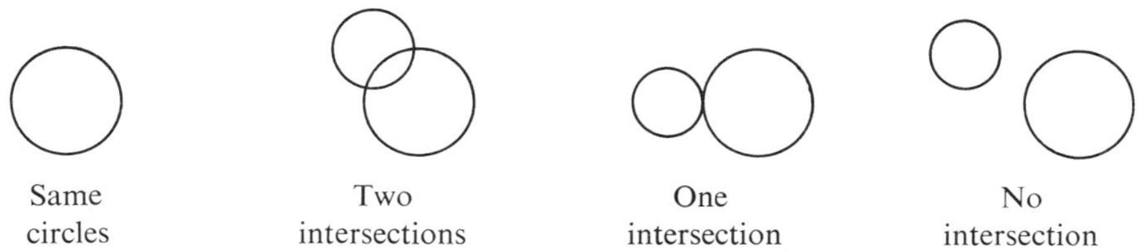

4. i. $x^2 + y^2 - 2x - 4y - 4 = 0$; center: (1, 2), radius: 3.
$x^2 + y^2 - 6x - 8y + 9 = 0$; center: (3, 4), radius: 4.

ii. (Distance from center to center) $= \sqrt{(3-1)^2 + (4-2)^2} = 6.32$.

iii. $\{(3.24, .01), (-.993, 4.24)\}$.

5. i. Center: (0, 0), radius: 280;
$x^2 + y^2 = 78400$.

ii. Center: (300, 0), radius: 120;
$x^2 + y^2 - 600x + 75600 = 0$.

iii. $\left.\begin{aligned} x^2 + y^2 - 78400 &= 0, \\ x^2 + y^2 - 600x + 75600 &= 0 \end{aligned}\right\}$ system;

$\{(256.7, 111.9)(256.7, -111.9)\}$.

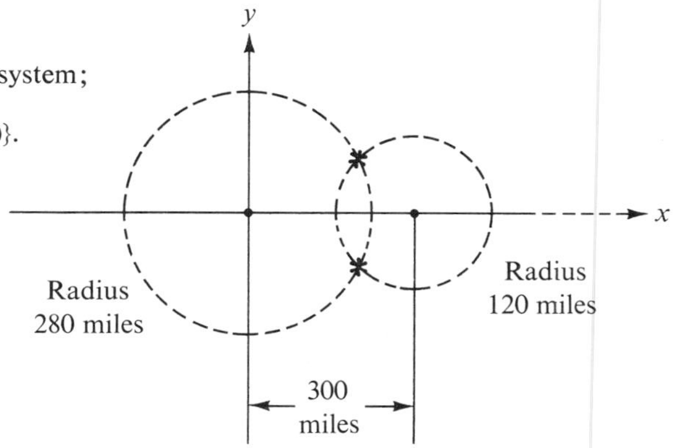

6. i. $\left.\begin{array}{l} 12^2 + 9^2 + 12D + 9E + F = 0, \\ (-15)^2 + 0 - 15D + 0E + F = 0, \\ (-12)^2 + (-20)^2 - 12D - 20E + F = 0 \end{array}\right\}$ system.

ii. $D = -5.06$,
$E = 15.19$.

iii. $h = -D/2 = 2.53$,
$k = -E/2 = -7.59$.

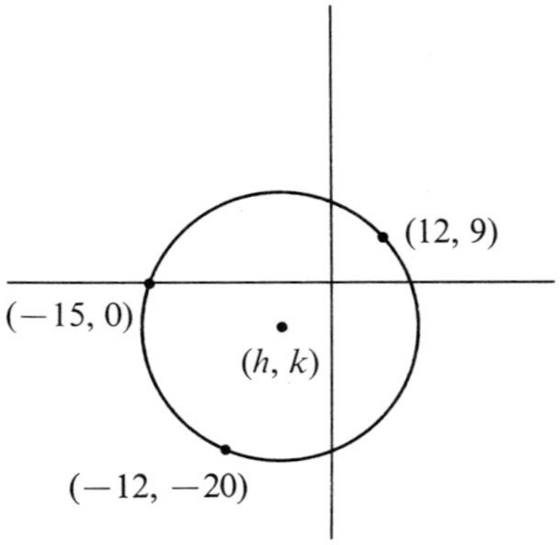

Section 10-4, page 516

1. i. $(20)^2 - 4p(1) = 0$
$p = 100$.

ii. $y = x^2/400$ (where (0, 0) is the center of the cable).

2. $x = \pm\sqrt{400(2)} = \pm 28.28$.

3. i. $x^2 - 4py = 0$
$p = 36/28$.

ii. $|\overline{FV}| = 1.29$ in.

4. $|\overline{FV}|$ increases.

5. a. i. (Slope of T) $= f'(4) = 2$; ii. (slope of N) $= -1/2$;
iii. $\theta = 63.4°$; iv. (slope of L_2) $= \tan 36.8° = .7481$.

b. $(y - 4) = .7481(x - 0)$,
$y_{\text{int}} = 1.00$.

c. $p = 1$ and focus is (0, 1).

6. i. Vertex: (2.5, −4.25).

ii. Let (h, k) be (2.5, −4.25) so that
$x = 2.5 + x^*$,
$y = -4.25 + y^*$.

iii. $-4.25 + y^* = (2.5 + x^*)^2 - 5(2.5 + x) + 2$
$y^* = (x^*)^2$.

iv. In the (x^*, y^*) system:
$p = 1/4$.
Vertex: (0, 0),
Focus: (0, 1/4),
Directrix: $y^* = -1/4$.

v. In the (x, y) system:
Vertex: $(2.5, -4.25)$,
Focus: $(2.5, -4.00)$,
Directrix: $y = -4.50$.

Section 10-5, page 526

1. i. $a = 24$ in.,
$b = 18$ in.,
$c = \sqrt{24^2 - 18^2} = 15.87$.

ii. (String length) $= 2a$
$= 48$ in.

2. i. $\frac{A_1}{A_2} = \frac{3}{4} = \frac{s^2}{48^2}$, $s = 41.57$ in. (string length).

ii. $\frac{3}{4} = \frac{f^2}{15.87^2}$, $f = 13.74$ in.

iii. $\frac{3}{4} = \frac{M^2}{48^2}$, $M = 41.57$ in.

iv. $\frac{3}{4} = \frac{m^2}{36^2}$, $m = 31.18$ in.

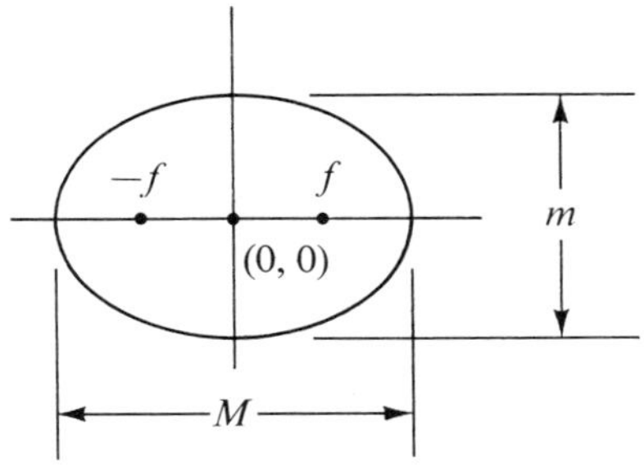

3. $\frac{x^2}{2500} + \frac{y^2}{900} = 1$.

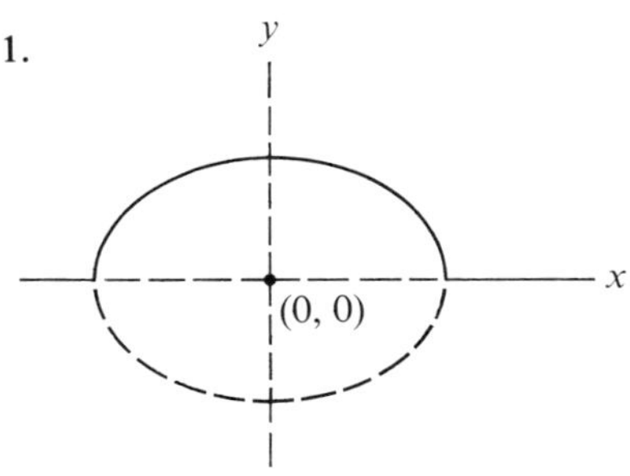

4. Earth's diameter: 7920 miles,
major axis: 32,890 miles,
minor axis: 22,086 miles,
foci: $(\pm 12, 185, 0)$.

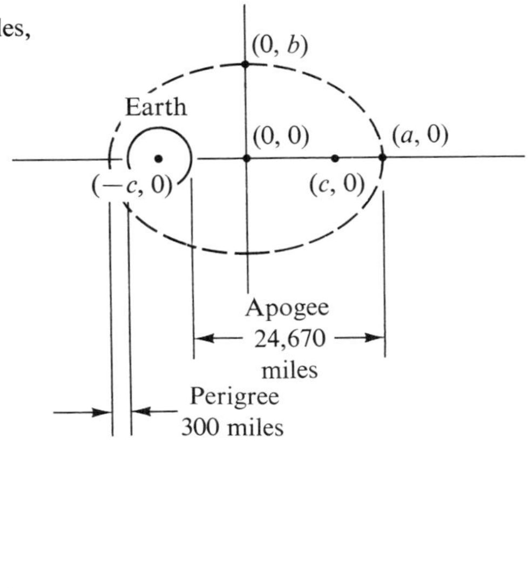

5. i. Major axis: $\dfrac{.75}{\cos 40^\circ} = .979$ in.

ii. minor axis: .75 in.

iii. $\dfrac{x^2}{.240} + \dfrac{y^2}{.141} = 1.$

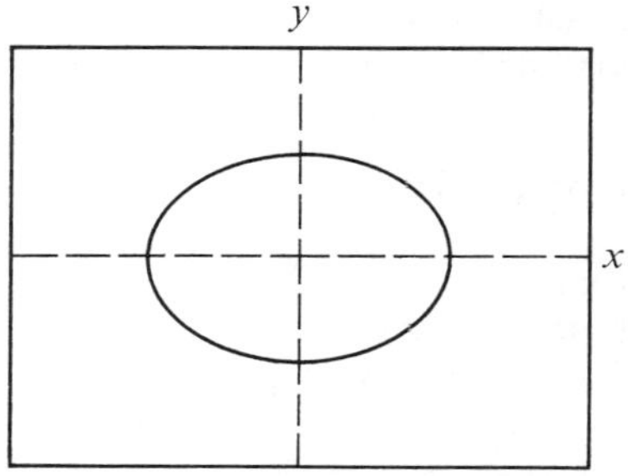

Section 10-6, page 537

1. i. $x = x^* \cos 45^\circ - y^* \sin 45^\circ = \dfrac{x^* - y^*}{\sqrt{2}},$

$y = x^* \sin 45^\circ + y^* \cos 45^\circ = \dfrac{x^* + y^*}{\sqrt{2}}.$

ii. $\left(\dfrac{x^* - y^*}{\sqrt{2}}\right)^2 - \left(\dfrac{x^* + y^*}{\sqrt{2}}\right)^2 = 16,$

$x^* y^* = -8.$

3.

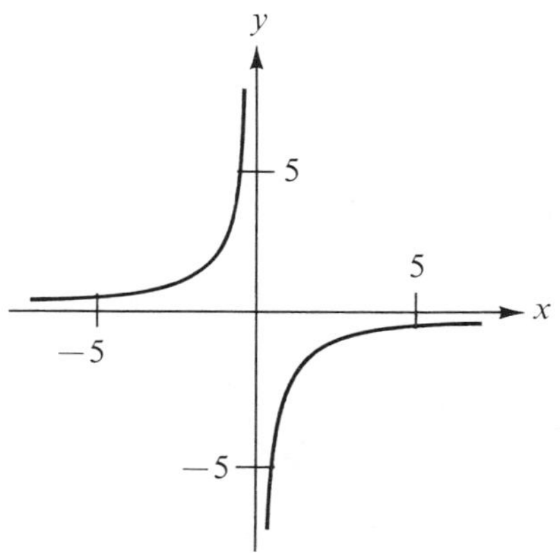

4. a. i. $d_1 = 2a_1 = 186{,}000(.0043)$

$a_1 = 400,$

ii. $c_1 = 1000,$

iii. $b_1^2 = 1000^2 - 400^2 = 840{,}000,$

iv. $x^2/160{,}000 - y^2/840{,}000 = 1.$

(The ship is on the curve of this hyperbola).

b. i. $d_2 = 2a_2 = 186{,}000(.0032),$

$a_2 = 300.$

ii. $c_2 = 1200$.
iii. $b_2^2 = 1200^2 - 300^2 = 1{,}350{,}000$.
iv. $y^2/90{,}000 - x^2/1{,}350{,}000 = 1$.
v. When the equations of parts (a) and (b) are solved simultaneously, the solutions give four possible positions of the ship.

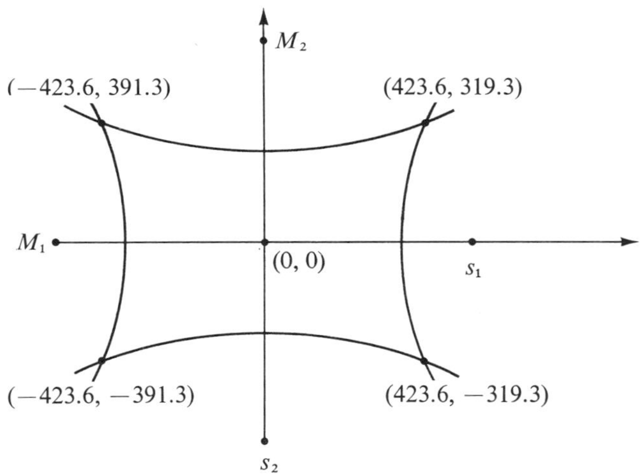

Section 11-1, page 550

1. i. 1st Hour: $1200(.8) + 60(.2) = 972°$ F,
2nd Hour: $1200(.8)^2 + 60(.2) = 789.6°$ F,
3rd Hour: $1200(.8)^3 + 60(.2) = 643.7°$ F.
ii. indefinitely.
iii. 60° F.

2.

Time (seconds)	0	.01	.02	.03
Current	0	.9 E/R	.99 E/R	.999 E/R

3. $m = 16$; $(1 + 1/16)^{16} \approx 2.6378 \ldots$,
$m = 32$; $(1 + 1/32)^{32} \approx 2.6769 \ldots$,
$\lim_{m\to\infty} (1 + 1/m)^m = e = 2.71828 \ldots$ (base of natural logarithms).

4. $\lim_{\theta\to 0} \dfrac{\overline{PM}}{\overset{\frown}{NM}} = \lim_{\theta\to 0} \dfrac{\tan\theta}{\theta} = 1$ (θ in radians).

Section 11-2, page 558

1. i. $f'(x) = \lim_{\Delta x\to 0} \dfrac{2.2(10^{-3})((x + \Delta x)^2 - x^2)}{x + \Delta x - x} = 4.4(10^{-3})x$,

ii. $f'(100) = 0.44$, $\theta = \arctan 0.44 = 23.7°$.

2. $f(x) = x^2/3 - 5,$
$f'(x) = 2x/3.$

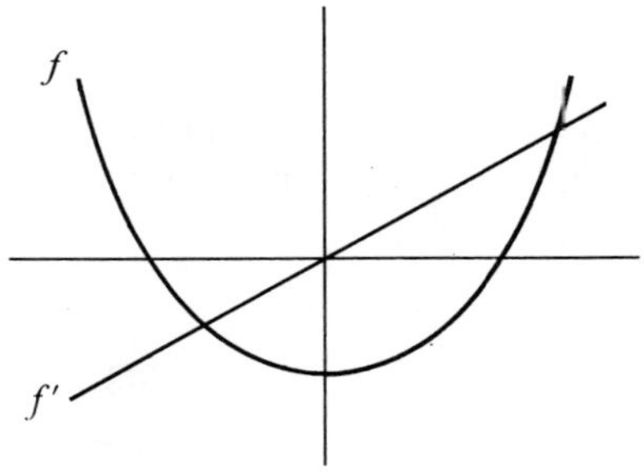

3.

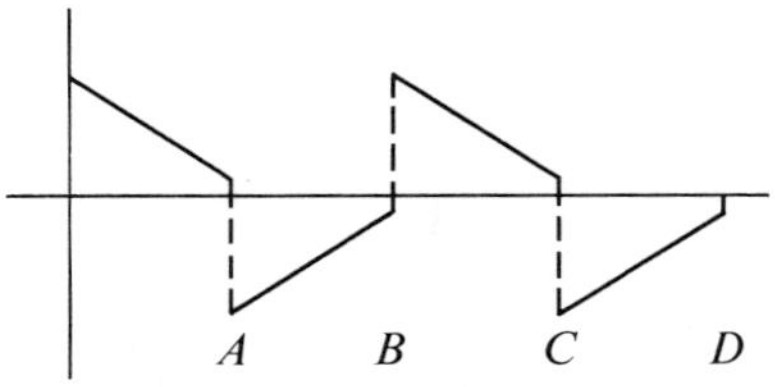

4.

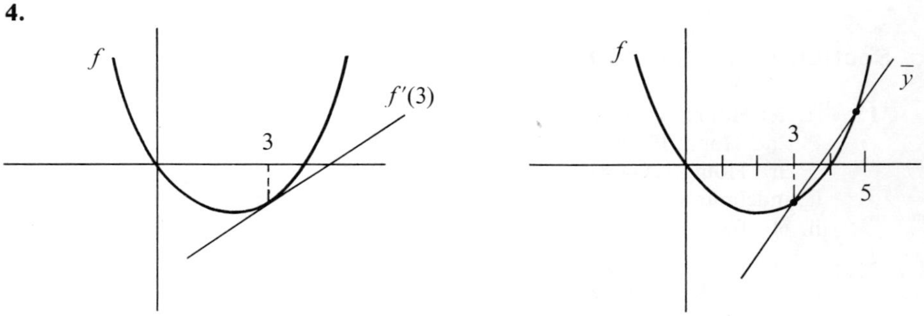

Section 11-3, page 563

1. $\omega = f'(t) = 2\alpha t + \omega_0.$

2. $f'(P) = \dfrac{.00285}{P^{.06}}.$

3. $f'(r) = \dfrac{\mu_0 I}{2\pi r^2}.$

4. $f'(T) = \dfrac{1.2\, R_1}{T_1} T^{.2}.$

5. i. $D(4, f(4)) \cong D(4, -4);$
ii. $f'(x) = x,$
(slope of $\overleftrightarrow{CE}) = f'(4) = 4;$
iii. Equation of $\overleftrightarrow{CE}$, $g(x) = 4x - 20;$
iv. $c(0, g(0)) \approx c(0, -20);$
v. $|\overline{CD}| = \sqrt{(4-0)^2 + (-4-(-20))^2} \approx 16.49.$

6. $V = f(h) = \pi rh^2 - \pi h^3/3,$
$f'(h) = 2\pi rh - \pi h^2.$

7. i. $r = (3/2)t,$
ii. $A = \pi r^2 = \pi(3t/2)^2 = (9\pi/4)t^2, \quad f(t) = A,$
iii. $f'(t) = 18\pi t/4 = 9\pi t/2.$

Section 11-4, page 569

1. a. i. $x^2 + y^2 = 16$
$f(x) = y = \sqrt{16 - x^2};$

ii. $f'(x) = -\dfrac{x}{\sqrt{16 - x^2}}, \quad f'(3) = -1.1339;$

iii. $\angle CEB = \arctan|-1.1339| = 48.6°.$

b. i. (Equation of $\overleftrightarrow{CE}$) $= g(x) = -1.13x + 6.04;$
ii. $g(0) = 6.04, \quad g(5.35) = 0;$
iii. $|\overline{CE}| = \sqrt{(6.04)^2 + (5.35)^2} = 8.07.$

2. i. $f(x) = y = \sqrt{(4x^2/9) + 4}$ (upper half of hyperbola),

ii. $f'(x) = \dfrac{4x}{9\sqrt{\dfrac{4x^2}{9} + 4}},$

iii. $f'(4) = 0.533,$
iv. $\angle ABC = 2(90° - \arctan 0.5333) = 123.8°.$

3. i. $\dfrac{di}{dr} = -\dfrac{5.5}{r^2},$

ii. $\dfrac{dr}{dt} = -\dfrac{6(10^3)}{t^2},$

iii. $\dfrac{di}{dt} = \dfrac{di}{dr} \cdot \dfrac{dr}{dt} = \dfrac{3.3(10^4)}{r^2t^2}.$

4. i. $f'(x) = W/6EI((L - x)^3 - L^4),$
ii. $f'(24) = 100/(6)30(10^6)(72)((48 - 24)^3 - (48)^4)$
$= -.0409.$

5. $f(t) = r = (4t^2 + d^2)^{1/2},$

$f'(t) = \dfrac{4t}{\sqrt{4t^2 + d^2}}.$

Section 11-5, page 577

1. i. $f(t) = r = \dfrac{t^2 - 1}{t^3 + t};$

ii. $g(t) = t^2 - 1, \quad g'(t) = 2t;$
iii. $h(t) = t^3 + t, \quad h'(t) = 3t^2 + 1;$

iv. $f'(t) = \dfrac{(t^3 + t)(2t) - (t^2 - 1)(3t^2 + 1)}{(t^3 + t)^2} = \dfrac{-t^4 + 4t^2 + 1}{(t^3 + t)^2}.$

2. i. $f(h) = v = \left(\dfrac{2GMh}{Rh + R^2}\right)^{1/2}$;

ii. $g(h) = 2GMh, \quad g'(h) = 2GM$;

iii. $h(h) = Rh + R^2, \quad h'(h) = R$;

iv. $$f'(h) = \frac{1}{2}\left(\frac{2GMh}{Rh + R^2}\right)^{-1/2}\left[\frac{(Rh + R^2)(2GM) - 2GMhR}{(Rh + R^2)^2}\right]$$
$$= \left(\frac{GMR^2}{\sqrt{2GMh}}\right)\left(\frac{1}{(Rh + R^2)^{3/2}}\right).$$

3. i. $f(L) = P = 2\pi\sqrt{\dfrac{k^2 + L^2}{Lg}} = 2\pi\sqrt{\dfrac{k^2}{Lg} + \dfrac{L}{g}}$;

ii. $$f'(L) = 2\pi\left[\left(\frac{1}{2}\right)\cdot\left(\frac{k^2}{Lg} + \frac{L}{g}\right)^{-1/2}\cdot\left(\frac{1}{g} - \frac{k^2}{gL^2}\right)\right]$$
$$= \pi\left(\frac{Lg}{k^2 + L^2}\right)^{1/2}\cdot\left(\frac{L^2 - k^2}{L^2 g}\right).$$

4. i. $y = f(t) = 2t^2 + 5$,
$v_y = f'(t) = 4t$.

ii. $v = \sqrt{(v_x)^2 + (v_y)^2} = \sqrt{(7.5/t)^2 + (4t)^2}$.

iii. $v_{(t=2.5)} = 10.44$.

5. i. $x = f(t) = 3(t^2 + 1)^{-1/2}$,
$v_x = f'(t) = -3t/(t^2 + 1)^{3/2}$.

ii. $y = g(t) = 3t(t^2 + 1)^{-1/2}$,
$v_y = g'(t) = 3/(t^2 + 1)^{3/2}$.

iii. $\theta = \arctan(v_y/v_x) = \arctan(1/-t)$,
$$v = \sqrt{(v_y)^2 + (v_x)^2} = \sqrt{\frac{9 + 9t^2}{(t^2 + 1)^3}}.$$

iv. At $t = 1$, $\theta = \arctan(-1) = 135°$, $v = \sqrt{18/8} = 1.50$.

6. i. Let $x_1 = 2.5$, $x_2 = x_1 - f(x_1)/f'(x_1) \approx 2.164$;

ii. let $x_1 = 2.164$, $x_2 = x_1 - f(x_1)/f'(x_1) \approx 2.0971$;

iii. let $x_1 = 2.0971$, $x_2 = x_1 - f(x_1)/f'(x_1) \approx 2.0946$;

iv. etc.

Section 11-6, page 583

1. i. (Volume) $= f(s) = (12 - 2s)^2 s$
$= 144s - 48s^2 + 4s^3$;

ii. $f'(s) = 144 - 96s + 12s^2$;

iii. $12s^2 - 96s + 144 = 0$
$s = 2$ (for max. volume).

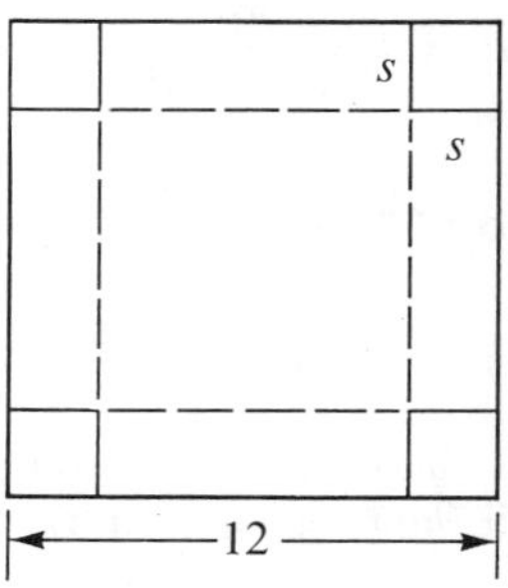

2. i. (Total surface area) $= 2A + L$
$= 2\pi r^2 + 2\pi rh.$

ii. (Volume) $= \pi r^2 h$; $h = V/\pi r^2$.

iii. $\therefore$ (Surface) $= f(r) = 2\pi r^2 + 2\pi r(V/\pi r^2)$.

iv. $f'(r) = 4\pi r - 2V/r^2$.

v. $4\pi r - 2V/r^2 = 0$; $r = h/2$.

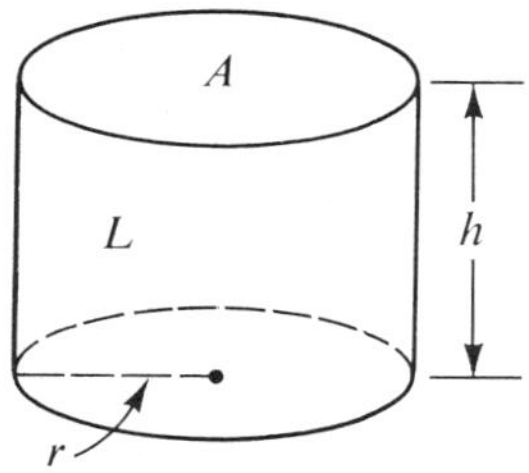

3. i. $|d| = g(x) = ((x-1)^2 + (y-2)^2)^{1/2}$
$= ((x-1)^2 + (x^2/4 - 2)^2)^{1/2}.$

ii. $g'(x) = \dfrac{(x^3/4) - 2}{2\sqrt{x^4/16 - 2x + 5}} = 0$; $x = 2$.

iii. $Q(2, 1)$.

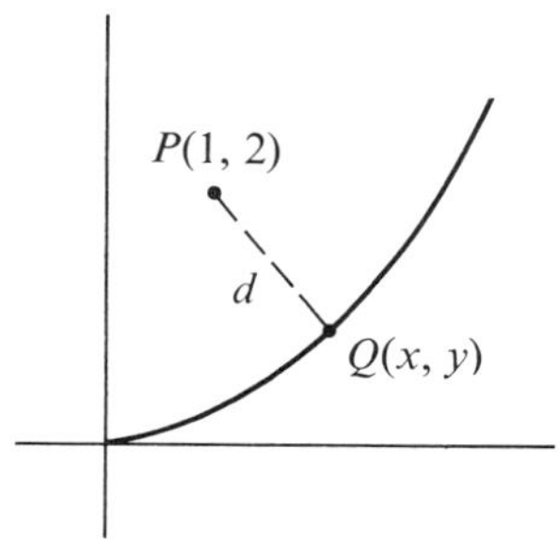

4. i. $|d| = (x^2 + y^2)^{1/2}$
$= ((4/y)^2 + y^2)^{1/2} = f(y).$

ii. $f'(y) = \dfrac{-32y^{-3} + 2y}{2\sqrt{16y^{-2} + y^2}} = 0$; $y = 2$.

iii. $P(2, 2)$.

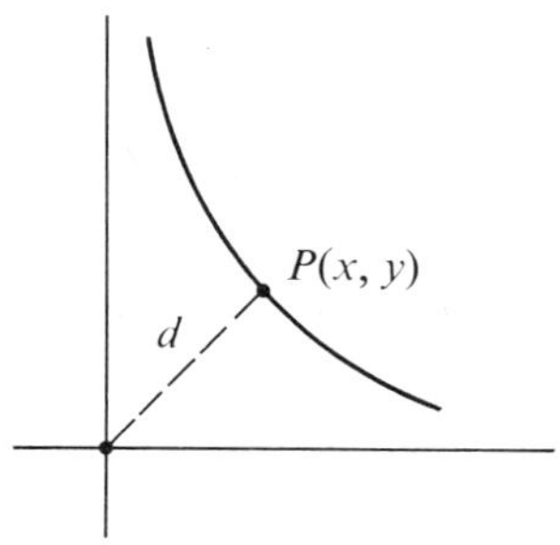

5. i. (Length of road) $= 16 - h + 2\sqrt{h^2 + 6^2} = f(h)$.

ii. $f'(h) = -1 + 2h/\sqrt{h^2 + 6^2} = 0$; $h = 2\sqrt{3}$.

iii. (From A to junction) $= 16 - 2\sqrt{3} \approx 12.54$.

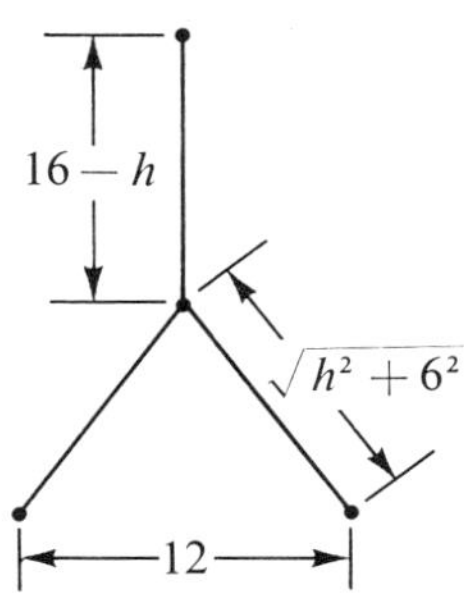

6. i. $r^2 = R^2 - h^2$.

ii. $(\text{Volume}) = (1/3)\pi r^2 h$

$$= \pi/3(R^2 - h^2)h = f(h).$$

iii. $f'(h) = \pi/3(R^2 - 3h^2) = 0, \quad h = R/\sqrt{3}$.

iv. $\theta = a/R = (2\pi R - 2\pi r)/R$

$$= \frac{2\pi R - 2\pi(R^2 - h^2)^{1/2}}{R}$$

$$= \frac{2\pi R - 2\pi(R^2 - R^2/3)^{1/2}}{R}$$

$$= 2\pi(1 - \sqrt{2/3}) \approx .367\pi(\text{rad}) \approx 66°.$$

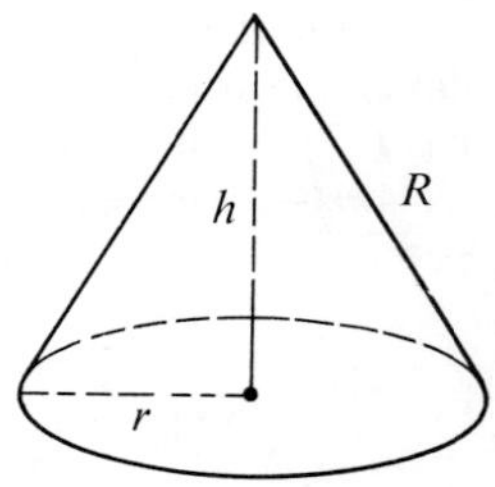

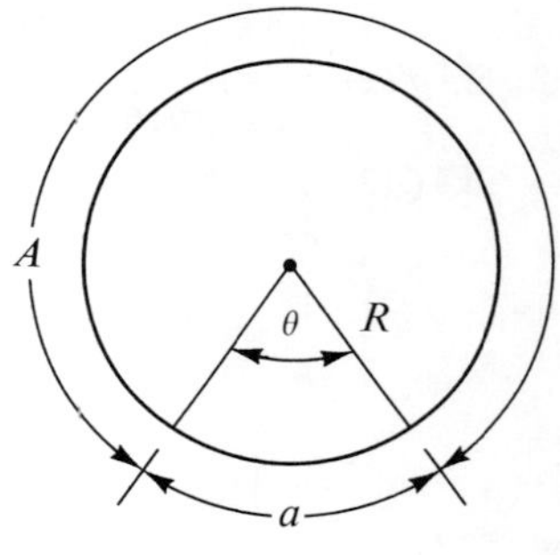

7. i. $d^2 = D^2 - \omega^2$.

ii. $(\text{Strength}) = k\omega d^2$

$$= k\omega(D^2 - \omega^2) = f(\omega).$$

iii. $f'(\omega) = k(D^2 - 3\omega^2) = 0$

$$\omega = D/\sqrt{3} \approx .577D.$$

iv. $d = \sqrt{2/3}D \approx .816D$.

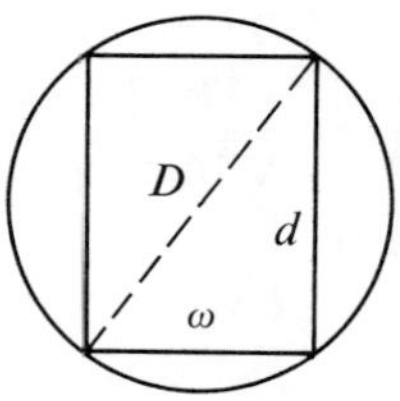

8. i. $f(t) = s = ((500 - 300t)^2 + (250 - 220t)^2)^{1/2}$;

ii. $$f'(t) = \frac{2(500 - 300t)(-300) + 2(250 - 220t)(-220)}{2((500 - 300t)^2 + (250 - 220t)^2)^{1/2}} = 0.$$

iii. $t = 1.481$.

iv. $f(1.481) = 94.08$ miles.

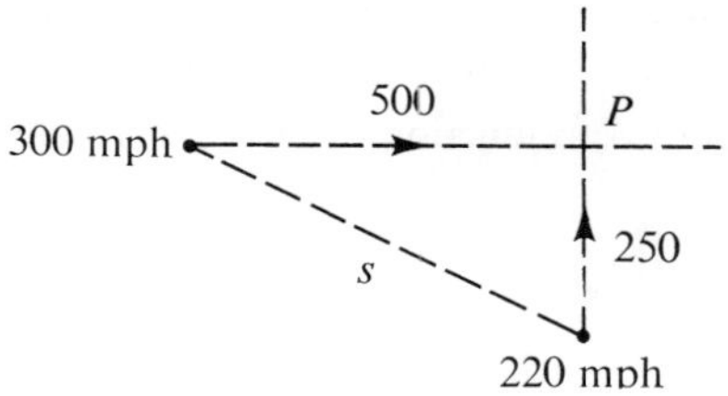

9. i. $R = \dfrac{6 - 8I}{I}$.

ii. $f(I) = P = I^2R = I^2\left(\dfrac{6 - 8I}{I}\right)$.

iii. $f'(I) = 6 - 16I = 0, \quad I = 3/8.$

iv. $\therefore \quad R = 8$ (ohms).

10. i. $f(x) = c = bn/x + pn + ix/2;$

ii. $f'(x) = -bn/x^2 + i/2 = 0;$

iii. $\therefore \quad x = \sqrt{2bn/i}$

$$= \sqrt{\frac{2(3{,}000)(12{,}000)}{4}} \approx 4243.$$

Section 11-7, page 592

1. $f'(x) = \cos x,$
$f'(x) = -\sin x,$
$f'(x) = \sec^2 x,$
$f'(x) = -\csc x \cot x,$
$f'(x) = \sec x \tan x,$
$f'(x) = -\csc^2 x.$

2. i. $g'(x) = \cos x = 1, \quad x = 0,$

ii. one.

3. i. $|f'(t)| = \omega|\cos \omega t| = \omega, \quad t = 0, \quad \pi/\omega;$

ii. ω.

4. i. $f'(t) = 48\pi \cos 4\pi t = 0,$
$\cos 4\pi t = 0.$
ii. $4\pi t = \pi/2,$
$t = 1/8.$

5. i. $f'(t) = -800\pi \sin(50\pi t + \pi/12) = 0,$
$\sin(50\pi t + \pi/12) = 0.$

ii. $50\pi t + \pi/12 = \pi,$
$t = .01833.$

iii. $50\pi t + \pi/12 = 2\pi,$
$t = .03833.$

6. i. $f'(t) = -130\pi \sin 20\pi t,$ ii. $f'(.02) = -388.4.$

7. i. $f'(t) = -\csc^2 t,$ ii. $f'(\pi/6) = -4.000.$

8. i. $e = L(24 \cos 120t),$ ii. $e_{(t=.01)} = (0.6)(24) \cos 1.2$
$= 5.218.$

9. $f'(t) = \dfrac{-s^2 \sin t \cos t}{\sqrt{r^2 - (s \sin t)^2}}.$

10. $f'(t) = s\omega \cos \omega t + \dfrac{s^2\omega \cos \omega t \sin \omega t}{\sqrt{r^2 - s^2 \cos^2 \omega t}}.$

Section 12-1, page 600

1. i. (Distance) $= \int f(t)\,dt = 10t^2 + c = F(t)$.

ii. $F(0) = 0; \quad \therefore\ c = 0,$

$F(10) = 10(10)^2 = 1000$ ft.

2. $F(t) = q = -t^{-1} + c$.

3. i. $C = .00004x^3 + .06x^2 + 10x + c$.

ii. Since $C = 20{,}000$ when $x = 0$, $c = 20{,}000$.

iii. $C_{1000} = .00004(1000)^3 + .06(1000)^2 + 10(1000) + 20{,}000$

$= \$130{,}000$.

4. $E = -(ks^2/2)$.

Section 12-2, page 605

1. i. $v = F(t) = \int 9.8\,dt = 9.8t + c$.

ii. $F(0) = 2.0, \quad \therefore\ c = 2.0$.

iii. $F(10) = 9.8(10) + 2.0 = 100$ m/sec.

2. i. $f(x) = \int f'(x)\,dx$

$= \int (1/2\,x - 3)\,dx = (x^2/4) - 3x + c$.

ii. Since $f(2) = 4, \quad c = 9$.

iii. $f(x) = (x^2/4) - 3x + 9$.

3. i. (Velocity) $= \int 12\,dt = 12t + c, \quad c = 0$.

ii. (Distance) $= \int (12t)\,dt = 6t^2 + c, \quad c = 0$.

iii. $12(t) = 150; \quad t = 12.5$ sec.

iv. (Distance) $= 6(12.5)^2 = 937.5$ ft.

4.

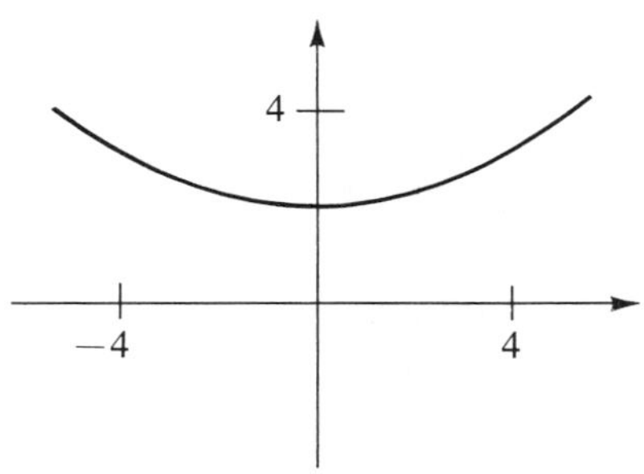

5. $f(x) = x/4; \quad \int f(x)\,dx = (x^2/8) + 2$.

6.

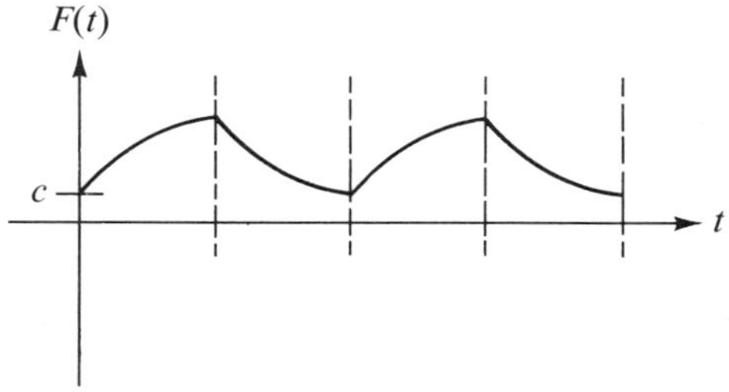

7.

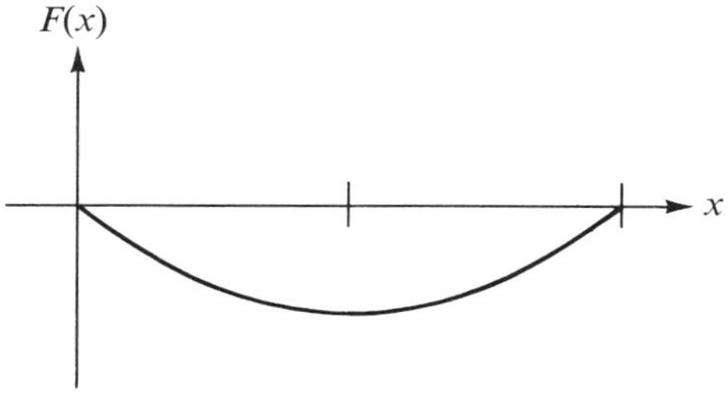

8. $F(x) = \dfrac{wx^2}{2} - kx + c, \quad c = 0.$

Section 12-3, page 616

1. $E = \displaystyle\int_0^{20} 1300\pi x\,dx = \frac{1300\pi x^2}{2}\Big]_0^{20} = 816{,}816.$

2. $(\text{Area}) = \displaystyle\int_0^6 3x\,dx - \int_0^6 0.5x^2\,dx = \frac{3x^2}{2}\Big]_0^6 - \frac{x^3}{6}\Big]_0^6 = 54 - 36 = 18.$

3. i. $(\text{Energy}) = \displaystyle\int_0^4 (1.8x^{2.2})\,dx = \frac{1.8x^{3.2}}{3.2}\Big]_0^4 = 47.50.$

ii. $(\text{Energy}) = \displaystyle\int_3^4 (1.8x^{2.2})\,dx = \frac{1.8x^{3.2}}{3.2}\Big]_3^4 = 47.50 - 18.92 = 28.58.$

4. $\displaystyle\int_4^{x_m} (-2x^2 + 20x + 100)\,dx = \int_{x_m}^{10} (-2x^2 + 20x + 100)\,dx.$

$\dfrac{-2x_m{}^3}{3} + 10x_m^2 + 100x_m = 925.3, \quad x_m = 6.82.$

5. $(\text{Area}) = 2\left(5(25) - \int_0^5 x^2\, dx\right) = 2(125 - 41.67) = 166.7.$

6. $V_{\text{av}} = \dfrac{\int_0^{.25} (14t^2 + 1)\, dt}{.25} = \dfrac{.3229}{.25} = 1.292.$

7. a. $\int_1^3 6e^x\, dx = 6e^x \Big]_1^3 = 104.2.$

b. $\int_1^2 (2 + 3x)^{1/2}\, dx = \dfrac{2}{3(3)} \sqrt{(2 + 3x)^3} \Big]_1^2 = 2.544.$

c. $\int_0^3 (\sqrt{16^2 - x^2})\, dx = \dfrac{x}{2} \sqrt{16^2 - x^2} + \dfrac{16^2}{2} \sin^{-1} \dfrac{x}{16} \Big]_0^3 = 47.70.$

d. $\int_0^\pi \sin x\, dx = -\cos x \Big]_0^\pi = 2.$

e. $\int_0^\pi \sin^2 x\, dx = \dfrac{1}{2} x - \dfrac{1}{4} \sin 2x \Big]_0^\pi = \pi/2.$

8. (7d) $2/\pi = .637.$

(7e) $(\pi/2)/\pi = .500.$

Section 12-4, page 625

1. i. $f(x) = ax^2, \quad 120 = ax^2, \quad a = 0.012.$

ii. $(\text{Area})_{10} \approx \sum_{n=0}^{9} (0.012)(10n)^2(10) = 3420.$

Note: $\int_0^{100} (0.012\, x^2)\, dx = 4000.$

2. i. $(\text{Area})_4 \approx \dfrac{1}{2} (f(x_0) + f(x_1))\, \Delta x + \dfrac{1}{2} (f(x_1) + f(x_2))\, \Delta x$

$$+ \frac{1}{2} (f(x_2) + f(x_3))\, \Delta x + \frac{1}{2} (f(x_3) + f(x_4))\, \Delta x$$

$$\approx \Delta x/2(f(x_0) + 2f(x_1) + 2f(x_2) + 2f(x_3) + f(x_4)).$$

ii. $(\text{Area})_4 \approx .25/2(0 + 2(.0625) + 2(.2500) + 2(.5625) + 1.000)$

$\approx 0.3437.$

3. $(\text{Area})_{10} \approx \sum_{n=0}^{9} \left(\sin^2 \dfrac{n\pi}{20}\right)\left(\dfrac{\pi}{20}\right) = .7069.$

Note: $\int_0^{\pi/2} \sin^2 x\, dx = \dfrac{1}{2} x - \dfrac{1}{4} \sin 2x \Big]_0^{\pi/2} = .7854.$

4. $(\text{Area})_8 \approx \sum_{n=0}^{7} e^{(.2)(.5)n}(.5) = 5.83.$

Note: $\displaystyle\int_0^4 e^{.2x}\,dx = \frac{e^{.2x}}{.2}\Big]_0^4 = 6.128.$

TABLE 1
Square Roots ($\sqrt{N}$)

Number	0	1	2	3	4	5	6	7	8	9
1.0	1.000	$1.00\bar{5}$	1.010	$1.01\bar{5}$	1.020	$1.02\bar{5}$	1.030	1.034	1.039	1.044
1.1	1.049	1.054	1.058	1.063	1.068	1.072	1.077	1.082	1.086	1.091
1.2	1.095	1.100	$1.10\bar{5}$	1.109	1.114	1.118	1.122	1.127	1.131	1.136
1.3	1.140	$1.14\bar{5}$	1.149	1.153	1.158	1.162	1.166	1.170	$1.17\bar{5}$	1.179
1.4	1.183	1.187	1.192	1.196	1.200	1.204	1.208	1.212	1.217	1.221
1.5	$1.22\bar{5}$	1.229	1.233	1.237	1.241	$1.24\bar{5}$	1.249	1.253	1.257	1.261
1.6	$1.26\bar{5}$	1.269	1.273	1.277	1.281	$1.28\bar{5}$	1.288	1.292	1.296	1.300
1.7	1.304	1.308	1.311	1.315	1.319	1.323	1.327	1.330	1.334	1.338
1.8	1.342	1.345	1.349	1.353	1.356	1.360	1.364	1.367	1.371	$1.37\bar{5}$
1.9	1.378	1.382	1.386	1.389	1.393	1.396	1.400	1.404	1.407	1.411
2.0	1.414	1.418	1.421	$1.42\bar{5}$	1.428	1.432	1.435	1.439	1.442	1.446
2.1	1.449	1.453	1.456	1.459	1.463	1.466	1.470	1.473	1.476	1.480
2.2	1.483	1.487	1.490	1.493	1.497	1.500	1.503	1.507	1.510	1.513
2.3	1.517	1.520	1.523	1.526	1.530	1.533	1.536	1.539	1.543	1.546
2.4	1.549	1.552	1.556	1.559	1.562	1.565	1.568	1.572	$1.57\bar{5}$	1.578
2.5	1.581	1.584	1.587	1.591	1.594	1.597	1.600	1.603	1.606	1.609
2.6	1.612	1.616	1.619	1.622	$1.62\bar{5}$	1.628	1.631	1.634	1.637	1.640
2.7	1.643	1.646	1.649	1.652	1.655	1.658	1.661	1.664	1.667	1.670
2.8	1.673	1.676	1.679	1.682	1.685	1.688	1.691	1.694	1.697	1.700
2.9	1.703	1.706	1.709	1.712	$1.71\bar{5}$	1.718	1.720	1.723	1.726	1.729
3.0	1.732	$1.73\bar{5}$	1.738	1.741	1.744	1.746	1.749	1.752	$1.75\bar{5}$	1.758
3.1	1.761	1.764	1.766	1.769	1.772	$1.77\bar{5}$	1.778	1.780	1.783	1.786
3.2	1.789	1.792	1.794	1.797	1.800	1.803	1.806	1.808	1.811	1.814
3.3	1.817	1.819	1.822	$1.82\bar{5}$	1.828	1.830	1.833	1.836	1.838	1.841
3.4	1.844	1.847	1.849	1.852	$1.85\bar{5}$	1.857	1.860	1.863	1.865	1.868
3.5	1.871	1.873	1.876	1.879	1.881	1.884	1.887	1.889	1.892	$1.89\bar{5}$
3.6	1.897	1.900	1.903	1.905	1.908	1.910	1.913	1.916	1.918	1.921
3.7	1.924	1.926	1.929	1.931	1.934	1.936	1.939	1.942	1.944	1.947
3.8	1.949	1.952	1.954	1.957	1.960	1.962	$1.96\bar{5}$	1.967	1.970	1.972
3.9	$1.97\bar{5}$	1.977	1.980	1.982	$1.98\bar{5}$	1.987	1.990	1.992	$1.99\bar{5}$	1.997
4.0	2.000	2.002	$2.00\bar{5}$	2.007	2.010	2.012	$2.01\bar{5}$	2.017	2.020	2.022
4.1	$2.02\bar{5}$	2.027	2.030	2.032	$2.03\bar{5}$	2.037	2.040	2.042	$2.04\bar{5}$	2.047
4.2	2.049	2.052	2.054	2.057	2.059	2.062	2.064	2.066	2.069	2.071
4.3	2.074	2.076	2.078	2.081	2.083	2.086	2.088	2.090	2.093	2.095
4.4	2.098	2.100	2.102	$2.10\bar{5}$	2.107	2.110	2.112	2.114	2.117	2.119

TABLE 1 (Continued)

Number	0	1	2	3	4	5	6	7	8	9
4.5	2.121	2.124	2.126	2.128	2.131	2.133	2.135	2.138	2.140	2.142
4.6	$2.14\overline{5}$	2.147	2.149	2.152	2.154	2.156	2.159	2.161	2.163	2.166
4.7	2.168	2.170	2.173	$2.17\overline{5}$	2.177	2.179	2.182	2.184	2.186	2.189
4.8	2.191	2.193	2.195	2.198	2.200	2.202	$2.20\overline{5}$	2.207	2.209	2.211
4.9	2.214	2.216	2.218	2.220	2.223	$2.22\overline{5}$	2.227	2.229	2.232	2.234
5.0	2.236	2.238	2.241	2.243	$2.24\overline{5}$	2.247	2.249	2.252	2.254	2.256
5.1	2.258	2.261	2.263	$2.26\overline{5}$	2.267	2.269	2.272	2.274	2.276	2.278
5.2	2.280	2.283	$2.28\overline{5}$	2.287	2.289	2.291	2.293	2.296	2.298	2.300
5.3	2.302	2.304	2.307	2.309	2.311	2.313	2.315	2.317	2.319	2.322
5.4	2.324	2.326	2.328	2.330	2.332	$2.33\overline{5}$	2.337	2.339	2.341	2.343
5.5	2.345	2.347	2.349	2.352	2.354	2.356	2.358	2.360	2.362	2.364
5.6	2.366	2.369	2.371	2.373	$2.37\overline{5}$	2.377	2.379	2.381	2.383	2.385
5.7	2.387	2.390	2.392	2.394	2.396	2.398	2.400	2.402	2.404	2.406
5.8	2.408	2.410	2.412	$2.41\overline{5}$	2.417	2.419	2.421	2.423	$2.42\overline{5}$	2.427
5.9	2.429	2.431	2.433	2.435	2.437	2.439	2.441	2.443	2.445	2.447
6.0	2.449	2.452	2.454	2.456	2.458	2.460	2.462	2.464	2.466	2.468
6.1	2.470	2.472	2.474	2.476	2.478	2.480	2.482	2.484	2.486	2.488
6.2	2.490	2.492	2.494	2.496	2.498	2.500	2.502	2.504	2.506	2.508
6.3	2.510	2.512	2.514	2.516	2.518	2.520	2.522	2.524	2.526	2.528
6.4	2.530	2.532	2.534	2.536	2.538	2.540	2.542	2.544	2.546	2.548
6.5	$2.55\overline{0}$	2.551	2.553	2.555	2.557	2.559	2.561	2.563	2.565	2.567
6.6	2.569	2.571	2.573	$2.57\overline{5}$	2.577	2.579	2.581	2.583	$2.58\overline{5}$	2.587
6.7	2.588	2.590	2.592	2.594	2.596	2.598	2.600	2.602	2.604	2.606
6.8	2.608	2.610	2.612	2.613	2.615	2.617	2.619	2.621	2.623	2.625
6.9	2.627	2.629	2.631	2.632	2.634	2.636	2.638	2.640	2.642	2.644
7.0	2.646	2.648	2.650	2.651	2.653	2.655	2.657	2.659	2.661	2.663
7.1	2.665	2.666	2.668	2.670	2.672	2.674	2.676	2.678	2.680	2.681
7.2	2.683	2.685	2.687	2.689	2.691	2.693	2.694	2.696	2.698	2.700
7.3	2.702	2.704	2.076	2.707	2.709	2.711	2.713	$2.71\overline{5}$	2.717	2.178
7.4	2.720	2.722	2.724	2.726	2.728	2.729	2.731	2.733	$2.73\overline{5}$	2.737
7.5	2.739	2.740	2.742	2.744	2.746	2.748	$2.75\overline{0}$	2.751	2.753	$2.75\overline{5}$
7.6	2.757	2.759	2.760	2.762	2.764	2.766	2.768	2.769	2.771	2.773
7.7	$2.77\overline{5}$	2.777	2.778	2.780	2.782	2.784	2.786	2.787	2.789	2.791
7.8	2.793	$2.79\overline{5}$	2.796	2.798	2.800	2.802	2.804	2.805	2.807	2.809
7.9	2.811	2.812	2.814	2.816	2.818	2.820	2.821	2.823	$2.82\overline{5}$	2.827

TABLE 1 (Continued)

Number	0	1	2	3	4	5	6	7	8	9
8.0	2.828	2.830	2.832	2.834	2.835	2.837	2.839	2.841	2.843	2.844
8.1	2.846	2.848	2.850	2.851	2.853	2.855	2.857	2.858	2.860	2.862
8.2	2.864	2.865	2.867	2.869	2.871	2.872	2.874	2.876	2.877	2.879
8.3	2.881	2.883	2.884	2.886	2.888	2.890	2.891	2.893	2.895	2.897
8.4	2.898	2.900	2.902	2.903	2.905	2.907	2.909	2.910	2.912	2.914
8.5	2.915	2.917	2.919	2.921	2.922	2.924	2.926	2.927	2.929	2.931
8.6	2.933	2.934	2.936	2.938	2.939	2.941	2.943	2.944	2.946	2.948
8.7	2.950	2.951	2.953	2.955	2.956	2.958	2.960	2.961	2.963	2.965
8.8	2.966	2.968	2.970	2.972	2.973	2.975	2.977	2.978	2.980	2.982
8.9	2.983	2.985	2.987	2.988	2.990	2.992	2.993	2.995	2.997	2.998
9.0	3.000	3.002	3.003	3.005	3.007	3.008	3.010	3.012	3.013	3.015
9.1	3.017	3.018	3.020	3.022	3.023	3.025	3.027	3.028	3.030	3.032
9.2	3.033	3.035	3.036	3.038	3.040	3.041	3.043	3.045	3.046	3.048
9.3	3.050	3.051	3.053	3.055	3.056	3.058	3.059	3.061	3.063	3.064
9.4	3.066	3.068	3.069	3.071	3.072	3.074	3.076	3.077	3.079	3.081
9.5	3.082	3.084	3.085	3.087	3.089	3.090	3.092	3.094	3.095	3.097
9.6	3.098	3.100	3.102	3.103	3.105	3.106	3.108	3.110	3.111	3.113
9.7	3.114	3.116	3.118	3.119	3.121	3.122	3.124	3.126	3.127	3.129
9.8	3.130	3.132	3.134	3.135	3.137	3.138	3.140	3.142	3.143	3.145
9.9	3.146	3.148	3.150	3.151	3.153	3.154	3.156	3.158	3.159	3.161
10.	3.162	3.178	3.194	3.209	3.225	3.240	3.256	3.271	3.286	3.302
11.	3.317	3.332	3.347	3.362	3.376	3.391	3.406	3.421	3.435	3.450
12.	3.464	3.479	3.493	3.507	3.521	3.536	3.550	3.564	3.578	3.592
13.	3.606	3.619	3.633	3.647	3.661	3.674	3.688	3.701	3.715	3.728
14.	3.742	3.755	3.768	3.782	3.795	3.808	3.821	3.834	3.847	3.860
15.	3.873	3.886	3.899	3.912	3.924	3.937	3.950	3.962	3.975	3.987
16.	4.000	4.012	4.025	4.037	4.050	4.062	4.074	4.087	4.099	4.111
17.	4.123	4.135	4.147	4.159	4.171	4.183	4.195	4.207	4.219	4.231
18.	4.243	4.254	4.266	4.278	4.290	4.301	4.313	4.324	4.336	4.347
19.	4.359	4.370	4.382	4.393	4.405	4.416	4.427	4.438	4.450	4.461
20.	4.472	4.483	4.494	4.506	4.517	4.528	4.539	4.550	4.561	4.572
21.	4.583	4.593	4.604	4.615	4.626	4.637	4.648	4.658	4.669	4.680
22.	4.690	4.701	4.712	4.722	4.733	4.743	4.754	4.764	4.775	4.785
23.	4.796	4.806	4.817	4.827	4.837	4.848	4.858	4.868	4.879	4.889
24.	4.899	4.909	4.919	4.930	4.940	4.950	4.960	4.970	4.980	4.990

TABLE 1 (Continued)

Number	0	1	2	3	4	5	6	7	8	9
25.	5.000	5.010	5.020	5.030	5.040	$5.05\overline{0}$	5.060	5.070	5.079	5.089
26.	5.099	5.109	5.119	5.128	5.138	5.148	5.158	5.167	5.177	5.187
27.	5.196	5.206	5.215	$5.22\overline{5}$	$5.23\overline{5}$	5.244	5.254	5.263	5.273	5.282
28.	5.292	5.301	5.310	5.320	5.329	5.339	5.348	5.357	5.367	5.376
29.	5.385	5.394	5.404	5.413	5.422	5.431	5.441	$5.45\overline{0}$	5.459	5.468
30.	5.477	5.486	5.495	$5.50\overline{5}$	5.514	5.523	5.532	5.541	$5.55\overline{0}$	5.559
31.	5.568	5.577	5.586	$5.59\overline{5}$	5.604	5.612	5.621	5.630	5.639	5.648
32.	5.657	5.666	$5.67\overline{5}$	5.683	5.692	5.701	5.710	5.718	5.727	5.736
33.	$5.74\overline{5}$	5.753	5.762	5.771	5.779	5.788	5.797	5.805	5.814	5.822
34.	5.831	5.840	5.848	5.857	5.865	5.874	5.882	5.891	5.899	5.908
35.	5.916	$5.92\overline{5}$	5.933	5.941	$5.95\overline{0}$	5.958	5.967	$5.97\overline{5}$	5.983	5.992
36.	6.000	6.008	6.017	$6.02\overline{5}$	6.033	6.042	$6.05\overline{0}$	6.058	6.066	$6.07\overline{5}$
37.	6.083	6.091	6.099	6.107	6.116	6.124	6.132	6.140	6.148	6.156
38.	6.164	6.173	6.181	6.189	6.197	$6.20\overline{5}$	6.213	6.221	6.229	6.237
39.	$6.24\overline{5}$	6.253	6.261	6.269	6.277	$6.28\overline{5}$	6.293	6.301	6.309	6.317
40.	$6.32\overline{5}$	6.332	6.340	6.348	6.356	6.364	6.372	6.380	6.387	6.395
41.	6.403	6.411	6.419	6.427	6.434	6.442	$6.45\overline{0}$	6.458	6.465	6.473
42.	6.481	6.488	6.496	6.504	6.512	6.519	6.527	$6.53\overline{5}$	6.542	$6.55\overline{0}$
43.	6.557	6.565	6.573	6.580	6.588	6.595	6.603	6.611	6.618	6.626
44.	6.633	6.641	6.648	6.656	6.663	6.671	6.678	6.686	6.693	6.701
45.	6.708	6.716	6.723	6.731	6.738	6.745	6.753	6.760	6.768	$6.77\overline{5}$
46.	6.782	6.790	6.797	6.804	6.812	6.819	6.826	6.834	6.841	6.848
47.	6.856	6.863	6.870	6.877	$6.88\overline{5}$	6.892	6.899	6.907	6.914	6.921
48.	6.928	6.935	6.943	$6.95\overline{0}$	6.957	6.964	6.971	6.979	6.986	6.993
49.	7.000	7.007	7.014	7.021	7.029	7.036	7.043	$7.05\overline{0}$	7.057	7.064
50.	7.071	7.078	7.085	7.092	7.099	7.106	7.113	7.120	7.127	7.134
51.	7.141	7.148	7.155	7.162	7.169	7.176	7.183	7.190	7.197	7.204
52.	7.211	7.218	$7.22\overline{5}$	7.232	7.239	7.246	7.253	7.259	7.266	7.273
53.	7.280	7.287	7.294	7.301	7.308	7.314	7.321	7.328	$7.33\overline{5}$	7.342
54.	7.348	7.355	7.362	7.369	7.376	7.382	7.389	7.396	7.403	7.409
55.	7.416	7.423	7.430	7.436	7.443	$7.45\overline{0}$	7.457	7.463	7.470	7.477
56.	7.483	7.490	7.497	7.503	7.510	7.517	7.523	7.530	7.537	7.543
57.	$7.55\overline{0}$	7.556	7.563	7.570	7.576	7.583	7.589	7.596	7.603	7.609
58.	7.616	7.622	7.629	7.635	7.642	7.649	7.655	7.662	7.668	$7.67\overline{5}$
59.	7.681	7.688	7.694	7.701	7.707	7.714	7.720	7.727	7.733	7.740

TABLE 1 (Continued)

Number	0	1	2	3	4	5	6	7	8	9
60.	7.746	7.752	7.759	7.765	7.772	7.778	$7.78\overline{5}$	7.791	7.797	7.804
61.	7.810	7.817	7.823	7.829	7.836	7.842	7.849	$7.85\overline{5}$	7.861	7.868
62.	7.874	7.880	7.887	7.893	7.899	7.906	7.912	7.918	$7.92\overline{5}$	7.931
63.	7.937	7.944	$7.95\overline{0}$	7.956	7.962	7.969	$7.97\overline{5}$	7.981	7.987	7.994
64.	8.000	8.006	8.012	8.019	$8.02\overline{5}$	8.031	8.037	8.044	$8.05\overline{0}$	8.056
65.	8.062	8.068	$8.07\overline{5}$	8.081	8.087	8.093	8.099	8.106	8.112	8.118
66.	8.124	8.130	8.136	8.142	8.149	$8.15\overline{5}$	8.161	8.167	8.173	8.179
67.	8.185	8.191	8.198	8.204	8.210	8.216	8.222	8.228	8.234	8.240
68.	8.246	8.252	8.258	8.264	8.270	8.276	8.283	8.289	$8.29\overline{5}$	8.301
69.	8.307	8.313	8.319	$8.32\overline{5}$	8.331	8.337	8.343	8.349	$8.35\overline{5}$	8.361
70.	8.367	8.373	8.379	8.385	8.390	8.396	8.402	8.408	8.414	8.420
71.	8.426	8.432	8.438	8.444	$8.45\overline{0}$	8.456	8.462	8.468	8.473	8.479
72.	8.485	8.491	8.497	8.503	8.509	$8.51\overline{5}$	8.521	8.526	8.532	8.538
73.	8.544	$8.55\overline{0}$	8.556	8.562	8.567	8.573	8.579	$8.58\overline{5}$	8.591	8.597
74.	8.602	8.608	8.614	8.620	8.626	8.631	8.637	8.643	8.649	8.654
75.	8.660	8.666	8.672	8.678	8.683	8.689	$8.69\overline{5}$	8.701	8.706	8.712
76.	8.718	8.724	8.729	$8.73\overline{5}$	8.741	8.746	8.752	8.758	8.764	8.769
77.	$8.77\overline{5}$	8.781	8.786	8.792	8.798	8.803	8.809	$8.81\overline{5}$	8.820	8.826
78.	8.832	8.837	8.843	8.849	8.854	8.860	8.866	8.871	8.877	8.883
79.	8.888	8.894	8.899	8.905	8.911	8.916	8.922	8.927	8.933	8.939
80.	8.944	$8.95\overline{0}$	8.955	8.961	8.967	8.972	8.978	8.983	8.989	8.994
81.	9.000	9.006	9.011	9.017	9.022	9.028	9.033	9.039	9.044	$9.05\overline{0}$
82.	9.055	9.061	9.066	9.072	9.077	9.083	9.088	9.094	9.099	$9.10\overline{5}$
83.	9.110	9.116	9.121	9.127	9.132	9.138	9.143	9.149	9.154	9.160
84.	9.165	9.171	9.176	9.182	9.187	9.192	9.198	9.203	9.209	9.214
85.	9.220	$9.22\overline{5}$	9.230	9.236	9.241	9.247	9.252	9.257	9.263	9.268
86.	9.274	9.279	9.284	9.290	9.295	9.301	9.306	9.311	9.317	9.322
87.	9.327	9.333	9.338	9.343	9.349	9.354	9.359	$9.36\overline{5}$	9.370	9.375
88.	9.381	9.386	9.391	9.397	9.402	9.407	9.413	9.418	9.423	9.429
89.	9.434	9.439	$9.44\overline{5}$	$9.45\overline{0}$	9.455	9.460	9.466	9.471	9.476	9.482
90.	9.487	9.492	9.497	9.503	9.508	9.513	9.518	9.524	9.529	9.534
91.	9.539	$9.54\overline{5}$	$9.55\overline{0}$	9.555	9.560	9.566	9.571	9.576	9.581	9.586
92.	9.592	9.597	9.602	9.607	9.612	9.618	9.623	9.628	9.633	9.638
93.	9.644	9.649	9.654	9.659	9.664	9.670	$9.67\overline{5}$	9.680	9.685	9.690
94.	9.695	9.701	9.706	9.711	9.716	9.721	9.726	9.731	9.737	9.742

TABLE 1 (Continued)

Number	0	1	2	3	4	5	6	7	8	9
95.	9.747	9.752	9.757	9.762	9.767	9.772	9.778	9.783	9.788	9.793
96.	9.798	9.803	9.808	9.813	9.818	9.823	9.829	9.834	9.839	9.844
97.	9.849	9.854	9.859	9.864	9.869	9.874	9.879	9.884	9.889	9.894
98.	9.899	$9.90\overline{5}$	9.910	$9.91\overline{5}$	9.920	$9.92\overline{5}$	9.930	$9.93\overline{5}$	9.940	$9.94\overline{5}$
99.	$9.95\overline{0}$	$9.95\overline{5}$	9.960	$9.96\overline{5}$	9.970	$9.97\overline{5}$	9.980	$9.98\overline{5}$	9.990	$9.99\overline{5}$

Source: *Basic Mathematics Simplified,* C. Thomas Olivo, Delmar Publisher, Inc., Albany, New York, 1953

TABLE 2
Common Logarithms ($Log_{10} N$)

N	0	1	2	3	4	5	6	7	8	9	PP 1	PP 2	PP 3	PP 4	PP 5	PP 6	PP 7	PP 8	PP 9
10	0000	0043	0086	0128	0170	0212	0253	0294	0334	0374	4	8	12	17	21	25	29	33	37
11	0414	0453	0492	0531	0569	0607	0645	0682	0719	0755	4	8	11	15	19	23	26	30	34
12	0792	0828	0864	0899	0934	0969	1004	1038	1072	1106	3	7	10	14	17	21	24	28	31
13	1139	1173	1206	1239	1271	1303	1335	1367	1399	1430	3	6	10	13	16	19	23	26	29
14	1461	1492	1523	1553	1584	1614	1644	1673	1703	1732	3	6	9	12	15	18	21	24	27
15	1761	1790	1818	1847	1875	1903	1931	1959	1987	2014	3	6	8	11	14	17	20	22	25
16	2041	2068	2095	2122	2148	2175	2201	2227	2253	2279	3	5	8	11	13	16	18	21	24
17	2304	2330	2355	2380	2405	2430	2455	2480	2504	2529	2	5	7	10	12	15	17	20	22
18	2553	2577	2601	2625	2648	2672	2695	2718	2742	2765	2	5	7	9	12	14	16	19	21
19	2788	2810	2833	2856	2878	2900	2923	2945	2967	2989	2	4	7	9	11	13	16	18	20
20	3010	3032	3054	3075	3096	3118	3139	3160	3181	3201	2	4	6	8	11	13	15	17	19
21	3222	3243	3263	3284	3304	3324	3345	3365	3385	3404	2	4	6	8	10	12	14	16	18
22	3424	3444	3464	3483	3502	3522	3541	3560	3579	3598	2	4	6	8	10	12	14	15	17
23	3617	3636	3655	3674	3692	3711	3729	3747	3766	3784	2	4	6	7	9	11	13	15	17
24	3802	3820	3838	3856	3874	3892	3909	3927	3945	3962	2	4	5	7	9	11	12	14	16
25	3979	3997	4014	4031	4048	4065	4082	4099	4116	4133	2	3	5	7	9	10	12	14	15
26	4150	4166	4183	4200	4216	4232	4249	4265	4281	4298	2	3	5	7	8	10	11	13	15
27	4314	4330	4346	4362	4378	4393	4409	4425	4440	4456	2	3	5	6	8	9	11	13	14
28	4472	4487	4502	4518	4533	4548	4564	4579	4594	4609	2	3	5	6	8	9	11	12	14
29	4624	4639	4654	4669	4683	4698	4713	4728	4742	4757	1	3	4	6	7	9	10	12	13
30	4771	4786	4800	4814	4829	4843	4857	4871	4886	4900	1	3	4	6	7	9	10	11	13
31	4914	4928	4942	4955	4969	4983	4997	5011	5024	5038	1	3	4	6	7	8	10	11	12
32	5051	5065	5079	5092	5105	5119	5132	5145	5159	5172	1	3	4	5	7	8	9	11	12
33	5185	5198	5211	5224	5237	5250	5263	5276	5289	5302	1	3	4	5	6	8	9	10	12
34	5315	5328	5340	5353	5366	5378	5391	5403	5416	5428	1	3	4	5	6	8	9	10	11
35	5441	5453	5465	5478	5490	5502	5514	5527	5539	5551	1	2	4	5	6	7	9	10	11
36	5563	5575	5587	5599	5611	5623	5635	5647	5658	5670	1	2	4	5	6	7	8	10	11
37	5682	5694	5705	5717	5729	5740	5752	5763	5775	5786	1	2	3	5	6	7	8	9	10
38	5798	5809	5821	5832	5843	5855	5866	5877	5888	5899	1	2	3	5	6	7	8	9	10
39	5911	5922	5933	5944	5955	5966	5977	5988	5999	6010	1	2	3	4	5	7	8	9	10
40	6021	6031	6042	6053	6064	6075	6085	6096	6107	6117	1	2	3	4	5	6	8	9	10
41	6128	6138	6149	6160	6170	6180	6191	6201	6212	6222	1	2	3	4	5	6	7	8	9
42	6232	6243	6253	6263	6274	6284	6294	6304	6314	6325	1	2	3	4	5	6	7	8	9
43	6335	6345	6355	6365	6375	6385	6395	6405	6415	6425	1	2	3	4	5	6	7	8	9
44	6435	6444	6454	6464	6474	6484	6493	6503	6513	6522	1	2	3	4	5	6	7	8	9

Proportional Parts = PP columns 1–9.

TABLE 2 (Continued)

N	0	1	2	3	4	5	6	7	8	9	Proportional Parts								
											1	2	3	4	5	6	7	8	9
45	6532	6542	6551	6561	6571	6580	6590	6599	6609	6618	1	2	3	4	5	6	7	8	9
46	6628	6637	6646	6656	6665	6675	6684	6693	6702	6712	1	2	3	4	5	6	7	7	8
47	6721	6730	6739	6749	6758	6767	6776	6785	6794	6803	1	2	3	4	5	5	6	7	8
48	6812	6821	6830	6839	6848	6857	6866	6875	6884	6893	1	2	3	4	4	5	6	7	8
49	6902	6911	6920	6928	6937	6946	6955	6964	6972	6981	1	2	3	4	4	5	6	7	8
50	6990	6998	7007	7016	7024	7033	7042	7050	7059	7067	1	2	3	3	4	5	6	7	8
51	7076	7084	7093	7101	7110	7118	7126	7135	7143	7152	1	2	3	3	4	5	6	7	8
52	7160	7168	7177	7185	7193	7202	7210	7218	7226	7235	1	2	2	3	4	5	6	7	7
53	7243	7251	7259	7267	7275	7284	7292	7300	7308	7316	1	2	2	3	4	5	6	6	7
54	7324	7332	7340	7348	7356	7364	7372	7380	7388	7396	1	2	2	3	4	5	6	6	7
55	7404	7412	7419	7427	7435	7443	7451	7459	7466	7474	1	2	2	3	4	5	5	6	7
56	7482	7490	7497	7505	7513	7520	7528	7536	7543	7551	1	2	2	3	4	5	5	6	7
57	7559	7566	7574	7582	7589	7597	7604	7612	7619	7627	1	2	2	3	4	5	5	6	7
58	7634	7642	7649	7657	7664	7672	7679	7686	7694	7701	1	1	2	3	4	4	5	6	7
59	7709	7716	7723	7731	7738	7745	7752	7760	7767	7774	1	1	2	3	4	4	5	6	7
60	7782	7789	7796	7803	7810	7818	7825	7832	7839	7846	1	1	2	3	4	4	5	6	6
61	7853	7860	7868	7875	7882	7889	7896	7903	7910	7917	1	1	2	3	4	4	5	6	6
62	7924	7931	7938	7945	7952	7959	7966	7973	7980	7987	1	1	2	3	3	4	5	6	6
63	7993	8000	8007	8014	8021	8028	8035	8041	8048	8055	1	1	2	3	3	4	5	5	6
64	8062	8069	8075	8082	8089	8096	8102	8109	8116	8122	1	1	2	3	3	4	5	5	6
65	8129	8136	8142	8149	8156	8162	8169	8176	8182	8189	1	1	2	3	3	4	5	5	6
66	8195	8202	8209	8215	8222	8228	8235	8241	8248	8254	1	1	2	3	3	4	5	5	6
67	8261	8267	8274	8280	8287	8293	8299	8306	8312	8319	1	1	2	3	3	4	5	5	6
68	8325	8331	8338	8344	8351	8357	8363	8370	8376	8382	1	1	2	3	3	4	4	5	6
69	8388	8395	8401	8407	8414	8420	8426	8432	8439	8445	1	1	2	2	3	4	4	5	6
70	8451	8457	8463	8470	8476	8482	8488	8494	8500	8506	1	1	2	2	3	4	4	5	6
71	8513	8519	8525	8531	8537	8543	8549	8555	8561	8567	1	1	2	2	3	4	4	5	5
72	8573	8579	8585	8591	8597	8603	8609	8615	8621	8627	1	1	2	2	3	4	4	5	5
73	8633	8639	8645	8651	8657	8663	8669	8675	8681	8686	1	1	2	2	3	4	4	5	5
74	8692	8698	8704	8710	8716	8722	8727	8733	8739	8745	1	1	2	2	3	4	4	5	5
75	8751	8756	8762	8768	8774	8779	8785	8791	8797	8802	1	1	2	2	3	3	4	5	5
76	8808	8814	8820	8825	8831	8837	8842	8848	8854	8859	1	1	2	2	3	3	4	5	5
77	8865	8871	8876	8882	8887	8893	8899	8904	8910	8915	1	1	2	2	3	3	4	4	5
78	8921	8927	8932	8938	8943	8949	8954	8960	8965	8971	1	1	2	2	3	3	4	4	5
79	8976	8982	8987	8993	8998	9004	9009	9015	9020	9025	1	1	2	2	3	3	4	4	5

TABLE 2 (Continued)

N	0	1	2	3	4	5	6	7	8	9	Proportional Parts								
											1	2	3	4	5	6	7	8	9
80	9031	9036	9042	9047	9053	9058	9063	9069	9074	9079	1	1	2	2	3	3	4	4	5
81	9085	9090	9096	9101	9106	9112	9117	9122	9128	9133	1	1	2	2	3	3	4	4	5
82	9138	9143	9149	9154	9159	9165	9170	9175	9180	9186	1	1	2	2	3	3	4	4	5
83	9191	9196	9201	9206	9212	9217	9222	9227	9232	9238	1	1	2	2	3	3	4	4	5
84	9243	9248	9253	9258	9263	9269	9274	9279	9284	9289	1	1	2	2	3	3	4	4	5
85	9294	9299	9304	9309	9315	9320	9325	9330	9335	9340	1	1	2	2	3	3	4	4	5
86	9345	9350	9355	9360	9365	9370	9375	9380	9385	9390	1	1	2	2	3	3	4	4	5
87	9395	9400	9405	9410	9415	9420	9425	9430	9435	9440	0	1	1	2	2	3	3	4	4
88	9445	9450	9455	9460	9465	9469	9474	9479	9484	9489	0	1	1	2	2	3	3	4	4
89	9494	9499	9504	9509	9513	9518	9523	9528	9533	9538	0	1	1	2	2	3	3	4	4
90	9542	9547	9552	9557	9562	9566	9571	9576	9581	9586	0	1	1	2	2	3	3	4	4
91	9590	9595	9600	9605	9609	9614	9619	9624	9628	9633	0	1	1	2	2	3	3	4	4
92	9638	9643	9647	9652	9657	9661	9666	9671	9675	9680	0	1	1	2	2	3	3	4	4
93	9685	9689	9694	9699	9703	9708	9713	9717	9722	9727	0	1	1	2	2	3	3	4	4
94	9731	9736	9741	9745	9750	9754	9759	9763	9768	9773	0	1	1	2	2	3	3	4	4
95	9777	9782	9786	9791	9795	9800	9805	9809	9814	9818	0	1	1	2	2	3	3	4	4
96	9823	9827	9832	9836	9841	9845	9850	9854	9859	9863	0	1	1	2	2	3	3	4	4
97	9868	9872	9877	9881	9886	9890	9894	9899	9903	9908	0	1	1	2	2	3	3	4	4
98	9912	9917	9921	9926	9930	9934	9939	9943	9948	9952	0	1	1	2	2	3	3	4	4
99	9956	9961	9965	9969	9974	9978	9983	9987	9991	9996	0	1	1	2	2	3	3	3	4

Source: *Mathematical Handbook of Formulas and Tables,* M. R. Spiegel, Schaum's Outline Series, McGraw-Hill Book Company. 1968

TABLE 3
Natural or Napierian Logarithms (Ln x)

x	0	1	2	3	4	5	6	7	8	9
1.0	.00000	.00995	.01980	.02956	.03922	.04879	.05827	.06766	.07696	.08618
1.1	.09531	.10436	.11333	.12222	.13103	.13976	.14842	.15700	.16551	.17395
1.2	.18232	.19062	.19885	.20701	.21511	.22314	.23111	.23902	.24686	.25464
1.3	.26236	.27003	.27763	.28518	.29267	.30010	.30748	.31481	.32208	.32930
1.4	.33647	.34359	.35066	.35767	.36464	.37156	.37844	.38526	.39204	.39878
1.5	.40547	.41211	.41871	.42527	.43178	.43825	.44469	.45108	.45742	.46373
1.6	.47000	.47623	.48243	.48858	.49470	.50078	.50682	.51282	.51879	.52473
1.7	.53063	.53649	.54232	.54812	.55389	.55962	.56531	.57098	.57661	.58222
1.8	.58779	.59333	.59884	.60432	.60977	.61519	.62058	.62594	.63127	.63658
1.9	.64185	.64710	.65233	.65752	.66269	.66783	.67294	.67803	.68310	.68813
2.0	.69315	.69813	.70310	.70804	.71295	.71784	.72271	.72755	.73237	.73716
2.1	.74194	.74669	.75142	.75612	.76081	.76547	.77011	.77473	.77932	.78390
2.2	.78846	.79299	.79751	.80200	.80648	.81093	.81536	.81978	.82418	.82855
2.3	.83291	.83725	.84157	.84587	.85015	.85442	.85866	.86289	.86710	.87129
2.4	.87547	.87963	.88377	.88789	.89200	.89609	.90016	.90422	.90826	.91228
2.5	.91629	.92028	.92426	.92822	.93216	.93609	.94001	.94391	.94779	.95166
2.6	.95551	.95935	.96317	.96698	.97078	.97456	.97833	.98208	.98582	.98954
2.7	.99325	.99695	1.00063	1.00430	1.00796	1.01160	1.01523	1.01885	1.02245	1.02604
2.8	1.02962	1.03318	1.03674	1.04028	1.04380	1.04732	1.05082	1.05431	1.05779	1.06126
2.9	1.06471	1.06815	1.07158	1.07500	1.07841	1.08181	1.08519	1.08856	1.09192	1.09527
3.0	1.09861	1.10194	1.10526	1.10856	1.11186	1.11514	1.11841	1.12168	1.12493	1.12817
3.1	1.13140	1.13462	1.13783	1.14103	1.14422	1.14740	1.15057	1.15373	1.15688	1.16002
3.2	1.16315	1.16627	1.16938	1.17248	1.17557	1.17865	1.18173	1.18479	1.18784	1.19089
3.3	1.19392	1.19695	1.19996	1.20297	1.20597	1.20896	1.21194	1.21491	1.21788	1.22083
3.4	1.22378	1.22671	1.22964	1.23256	1.23547	1.23837	1.24127	1.24415	1.24703	1.24990
3.5	1.25276	1.25562	1.25846	1.26130	1.26413	1.26695	1.26976	1.27257	1.27536	1.27815
3.6	1.28093	1.28371	1.28647	1.28923	1.29198	1.29473	1.29746	1.30019	1.30291	1.30563
3.7	1.30833	1.31103	1.31372	1.31641	1.31909	1.32176	1.32442	1.32708	1.32972	1.33237
3.8	1.33500	1.33763	1.34025	1.34286	1.34547	1.34807	1.35067	1.35325	1.35584	1.35841
3.9	1.36098	1.36354	1.36609	1.36864	1.37118	1.37372	1.37624	1.37877	1.38128	1.38379
4.0	1.38629	1.38879	1.39128	1.39377	1.39624	1.39872	1.40118	1.40364	1.40610	1.40854
4.1	1.41099	1.41342	1.41585	1.41828	1.42070	1.42311	1.42552	1.42792	1.43031	1.43270
4.2	1.43508	1.43746	1.43984	1.44220	1.44456	1.44692	1.44927	1.45161	1.45395	1.45629
4.3	1.45862	1.46094	1.46326	1.46557	1.46787	1.47018	1.47247	1.47476	1.47705	1.47933
4.4	1.48160	1.48387	1.48614	1.48840	1.49065	1.49290	1.49515	1.49739	1.49962	1.50185

TABLE 3 (Continued)

x	0	1	2	3	4	5	6	7	8	9
4.5	1.50408	1.50630	1.50851	1.51072	1.51293	1.51513	1.51732	1.51951	1.52170	1.52388
4.6	1.52606	1.52823	1.53039	1.53256	1.53471	1.53687	1.53902	1.54116	1.54330	1.54543
4.7	1.54756	1.54969	1.55181	1.55393	1.55604	1.55814	1.56025	1.56235	1.56444	1.56653
4.8	1.56862	1.57070	1.57277	1.57485	1.57691	1.57898	1.58104	1.58309	1.58515	1.58719
4.9	1.58924	1.59127	1.59331	1.59534	1.59737	1.59939	1.60141	1.60342	1.60543	1.60744
5.0	1.60944	1.61144	1.61343	1.61542	1.61741	1.61939	1.62137	1.62334	1.62531	1.62728
5.1	1.62924	1.63120	1.63315	1.63511	1.63705	1.63900	1.64094	1.64287	1.64481	1.64673
5.2	1.64866	1.65058	1.65250	1.65441	1.65632	1.65823	1.66013	1.66203	1.66393	1.66582
5.3	1.66771	1.66959	1.67147	1.67335	1.67523	1.67710	1.67896	1.68083	1.68269	1.68455
5.4	1.68640	1.68825	1.69010	1.69194	1.69378	1.69562	1.69745	1.69928	1.70111	1.70293
5.5	1.70475	1.70656	1.70838	1.71019	1.71199	1.71380	1.71560	1.71740	1.71919	1.72098
5.6	1.72277	1.72455	1.72633	1.72811	1.72988	1.73166	1.73342	1.73519	1.73695	1.73871
5.7	1.74047	1.74222	1.74397	1.74572	1.74746	1.74920	1.75094	1.75267	1.75440	1.75613
5.8	1.75786	1.75958	1.76130	1.76302	1.76473	1.76644	1.76815	1.76985	1.77156	1.77326
5.9	1.77495	1.77665	1.77834	1.78002	1.78171	1.78339	1.78507	1.78675	1.78842	1.79009
6.0	1.79176	1.79342	1.79509	1.79675	1.79840	1.80006	1.80171	1.80336	1.80500	1.80665
6.1	1.80829	1.80993	1.81156	1.81319	1.81482	1.81645	1.81808	1.81970	1.82132	1.82294
6.2	1.82455	1.82616	1.82777	1.82938	1.83098	1.83258	1.83418	1.83578	1.83737	1.83896
6.3	1.84055	1.84214	1.84372	1.84530	1.84688	1.84845	1.85003	1.85160	1.85317	1.85473
6.4	1.85630	1.85786	1.85942	1.86097	1.86253	1.86408	1.86563	1.86718	1.86872	1.87026
6.5	1.87180	1.87334	1.87487	1.87641	1.87794	1.87947	1.88099	1.88251	1.88403	1.88555
6.6	1.88707	1.88858	1.89010	1.89160	1.89311	1.89462	1.89612	1.89762	1.89912	1.90061
6.7	1.90211	1.90360	1.90509	1.90658	1.90806	1.90954	1.91102	1.91250	1.91398	1.91545
6.8	1.91692	1.91839	1.91986	1.92132	1.92279	1.92425	1.92571	1.92716	1.92862	1.93007
6.9	1.93152	1.93297	1.93442	1.93586	1.93730	1.93874	1.94018	1.94162	1.94305	1.94448
7.0	1.94591	1.94734	1.94876	1.95019	1.95161	1.95303	1.95445	1.95586	1.95727	1.95869
7.1	1.96009	1.96150	1.96291	1.96431	1.96571	1.96711	1.96851	1.96991	1.97130	1.97269
7.2	1.97408	1.97547	1.97685	1.97824	1.97962	1.98100	1.98238	1.98376	1.98513	1.98650
7.3	1.98787	1.98924	1.99061	1.99198	1.99334	1.99470	1.99606	1.99742	1.99877	2.00013
7.4	2.00148	2.00283	2.00418	2.00553	2.00687	2.00821	2.00956	2.01089	2.01223	2.01357
7.5	2.01490	2.01624	2.01757	2.01890	2.02022	2.02155	2.02287	2.02419	2.02551	2.02683
7.6	2.02815	2.02946	2.03078	2.03209	2.03340	2.03471	2.03601	2.03732	2.03862	2.03992
7.7	2.04122	2.04252	2.04381	2.04511	2.04640	2.04769	2.04898	2.05027	2.05156	2.05284
7.8	2.05412	2.05540	2.05668	2.05796	2.05924	2.06051	2.06179	2.06306	2.06433	2.06560
7.9	2.06686	2.06813	2.06939	2.07065	2.07191	2.07317	2.07443	2.07568	2.07694	2.07819

TABLE 3 (Continued)

x	0	1	2	3	4	5	6	7	8	9
8.0	2.07944	2.08069	2.08194	2.08318	2.08443	2.08567	2.08691	2.08815	2.08939	2.09063
8.1	2.09186	2.09310	2.09433	2.09556	2.09679	2.09802	2.09924	2.10047	2.10169	2.10291
8.2	2.10413	2.10535	2.10657	2.10779	2.10900	2.11021	2.11142	2.11263	2.11384	2.11505
8.3	2.11626	2.11746	2.11866	2.11986	2.12106	2.12226	2.12346	2.12465	2.12585	2.12704
8.4	2.12823	2.12942	2.13061	2.13180	2.13298	2.13417	2.13535	2.13653	2.13771	2.13889
8.5	2.14007	2.14124	2.14242	2.14359	2.14476	2.14593	2.14710	2.14827	2.14943	2.15060
8.6	2.15176	2.15292	2.15409	2.15524	2.15640	2.15756	2.15871	2.15987	2.16102	2.16217
8.7	2.16332	2.16447	2.16562	2.16677	2.16791	2.16905	2.17020	2.17134	2.17248	2.17361
8.8	2.17475	2.17589	2.17702	2.17816	2.17929	2.18042	2.18155	2.18267	2.18380	2.18493
8.9	2.18605	2.18717	2.18830	2.18942	2.19054	2.19165	2.19277	2.19389	2.19500	2.19611
9.0	2.19722	2.19834	2.19944	2.20055	2.20166	2.20276	2.20387	2.20497	2.20607	2.20717
9.1	2.20827	2.20937	2.21047	2.21157	2.21266	2.21375	2.21485	2.21594	2.21703	2.21812
9.2	2.21920	2.22029	2.22138	2.22246	2.22354	2.22462	2.22570	2.22678	2.22786	2.22894
9.3	2.23001	2.23109	2.23216	2.23324	2.23431	2.23538	2.23645	2.23751	2.23858	2.23965
9.4	2.24071	2.24177	2.24284	2.24390	2.24496	2.24601	2.24707	2.24813	2.24918	2.25024
9.5	2.25129	2.25234	2.25339	2.25444	2.25549	2.25654	2.25759	2.25863	2.25968	2.26072
9.6	2.26176	2.26280	2.26384	2.26488	2.26592	2.26696	2.26799	2.26903	2.27006	2.27109
9.7	2.27213	2.27316	2.27419	2.27521	2.27624	2.27727	2.27829	2.27932	2.28034	2.28136
9.8	2.28238	2.28340	2.28442	2.28544	2.28646	2.28747	2.28849	2.28950	2.29051	2.29152
9.9	2.29253	2.29354	2.29455	2.29556	2.29657	2.29757	2.29858	2.29958	2.30058	2.30158

$\ln 10 = 2.30259$ $\qquad 4 \ln 10 = 9.21034$ $\qquad 7 \ln 10 = 16.11810$

$2 \ln 10 = 4.60517$ $\qquad 5 \ln 10 = 11.51293$ $\qquad 8 \ln 10 = 18.42068$

$3 \ln 10 = 6.90776$ $\qquad 6 \ln 10 = 13.81551$ $\qquad 9 \ln 10 = 20.72327$

Source: *Mathematical Handbook of Formulas and Tables*, M. R. Spiegel, Schaum's Outline Series, McGraw-Hill Book Company. 1968

TABLE 4
Trigonometric Functions – Degrees, Radians

Rad	Deg	Sin	Tan	Cot	Cos	Deg	Rad
0.0000	0.0	0.00000	0.00000	∞	1.0000	90.0	1.5708
0.0017	0.1	0.00175	0.00175	573.0	1.0000	89.9	1.5691
0.0035	0.2	0.00349	0.00349	286.5	1.0000	89.8	1.5673
0.0052	0.3	0.00524	0.00524	191.0	1.0000	89.7	1.5656
0.0070	0.4	0.00698	0.00698	143.24	1.0000	89.6	1.5638
0.0087	0.5	0.00873	0.00873	114.59	1.0000	89.5	1.5621
0.0105	0.6	0.01047	0.01047	95.49	0.9999	89.4	1.5603
0.0122	0.7	0.01222	0.01222	81.85	0.9999	89.3	1.5586
0.0139	0.8	0.01396	0.01396	71.62	0.9999	89.2	1.5568
0.0157	0.9	0.01571	0.01571	63.66	0.9999	89.1	1.5551
0.0175	1.0	0.01745	0.01746	57.29	0.9998	89.0	1.5533
0.0192	1.1	0.01920	0.01920	52.08	0.9998	88.9	1.5516
0.0209	1.2	0.02094	0.02095	47.74	0.9998	88.8	1.5499
0.0227	1.3	0.02269	0.02269	44.07	0.9997	88.7	1.5481
0.0244	1.4	0.02443	0.02444	40.92	0.9997	88.6	1.5464
0.0262	1.5	0.02618	0.02619	38.19	0.9997	88.5	1.5446
0.0279	1.6	0.02792	0.02793	35.80	0.9996	88.4	1.5429
0.0297	1.7	0.02967	0.02968	33.69	0.9996	88.3	1.5411
0.0314	1.8	0.03141	0.03143	31.82	0.9995	88.2	1.5394
0.0332	1.9	0.03316	0.03317	30.14	0.9995	88.1	1.5376
0.0349	2.0	0.03490	0.03492	28.64	0.9994	88.0	1.5359
0.0367	2.1	0.03664	0.03667	27.27	0.9993	87.9	1.5341
0.0384	2.2	0.03839	0.03842	26.03	0.9993	87.8	1.5324
0.0401	2.3	0.04013	0.04016	24.90	0.9992	87.7	1.5301
0.0419	2.4	0.04188	0.04191	23.86	0.9991	87.6	1.5289
0.0436	2.5	0.04362	0.04366	22.90	0.9990	87.5	1.5272
0.0454	2.6	0.04536	0.04541	22.02	0.9990	87.4	1.5254
0.0471	2.7	0.04711	0.04716	21.20	0.9989	87.3	1.5237
0.0489	2.8	0.04885	0.04891	20.45	0.9988	87.2	1.5219
0.0506	2.9	0.05059	0.05066	19.74	0.9987	87.1	1.5202
0.0524	3.0	0.05234	0.05241	19.081	0.9986	87.0	1.5184
0.0541	3.1	0.05408	0.05416	18.464	0.9985	86.9	1.5167
0.0559	3.2	0.05582	0.05591	17.886	0.9984	86.8	1.5149
0.0576	3.3	0.05756	0.05766	17.343	0.9983	86.7	1.5132
0.0593	3.4	0.05931	0.05941	16.832	0.9982	86.6	1.5115
Rad	Deg	Cos	Cot	Tan	Sin	Deg	Rad

TABLE 4 (Continued)

Rad	Deg	Sin	Tan	Cot	Cos	Deg	Rad
0.0611	3.5	0.06105	0.06116	16.350	0.9981	86.5	1.5097
0.0628	3.6	0.06279	0.06291	15.895	0.9980	86.4	1.5080
0.0645	3.7	0.06453	0.06467	15.464	0.9979	86.3	1.5062
0.0663	3.8	0.06627	0.06642	15.056	0.9978	86.2	1.5045
0.0681	3.9	0.06802	0.06817	14.669	0.9977	86.1	1.5027
0.0698	4.0	0.06976	0.06993	14.301	0.9976	86.0	1.5010
0.0716	4.1	0.07150	0.07168	13.951	0.9974	85.9	1.4992
0.0733	4.2	0.07324	0.07344	13.617	0.9973	85.8	1.4975
0.0750	4.3	0.07498	0.07519	13.300	0.9972	85.7	1.4957
0.0768	4.4	0.07672	0.07695	12.996	0.9971	85.6	1.4940
0.0785	4.5	0.07846	0.07870	12.706	0.9969	85.5	1.4923
0.0803	4.6	0.08020	0.08046	12.429	0.9968	85.4	1.4905
0.0820	4.7	0.08194	0.08221	12.163	0.9966	85.3	1.4888
0.0838	4.8	0.08368	0.08397	11.909	0.9965	85.2	1.4870
0.0855	4.9	0.08542	0.08573	11.664	0.9963	85.1	1.4853
0.0873	5.0	0.08716	0.08749	11.430	0.9962	85.0	1.4835
0.0890	5.1	0.08889	0.08925	11.205	0.9960	84.9	1.4818
0.0908	5.2	0.09063	0.09101	10.988	0.9959	84.8	1.4800
0.0925	5.3	0.09237	0.09277	10.780	0.9957	84.7	1.4783
0.0942	5.4	0.09411	0.09453	10.579	0.9956	84.6	1.4765
0.0960	5.5	0.09585	0.09629	10.385	0.9954	84.5	1.4748
0.0977	5.6	0.09758	0.09805	10.199	0.9952	84.4	1.4731
0.0995	5.7	0.09932	0.09981	10.019	0.9951	84.3	1.4713
0.1012	5.8	0.10106	0.10158	9.845	0.9949	84.2	1.4696
0.1030	5.9	0.10279	0.10334	9.677	0.9947	84.1	1.4678
0.1047	6.0	0.10453	0.10510	9.514	0.9945	84.0	1.4661
0.1065	6.1	0.10626	0.10687	9.357	0.9943	83.9	1.4643
0.1082	6.2	0.10800	0.10863	9.205	0.9942	83.8	1.4626
0.1100	6.3	0.10973	0.11040	9.058	0.9940	83.7	1.4608
0.1117	6.4	0.11147	0.11217	8.915	0.9938	83.6	1.4591
0.1134	6.5	0.11320	0.11394	8.777	0.9936	83.5	1.4574
0.1152	6.6	0.11494	0.11570	8.643	0.9934	83.4	1.4556
0.1169	6.7	0.11667	0.11747	8.513	0.9932	83.3	1.4539
0.1187	6.8	0.11840	0.11924	8.386	0.9930	83.2	1.4521
0.1204	6.9	0.12014	0.12101	8.264	0.9928	83.1	1.4504
Rad	Deg	Cos	Cot	Tan	Sin	Deg	Rad

TABLE 4 (Continued)

Rad	Deg	Sin	Tan	Cot	Cos	Deg	Rad
0.1222	7.0	0.12187	0.12278	8.144	0.9925	83.0	1.4486
0.1239	7.1	0.12360	0.12456	8.028	0.9923	82.9	1.4469
0.1257	7.2	0.12533	0.12633	7.916	0.9921	82.8	1.4451
0.1274	7.3	0.12706	0.12810	7.806	0.9919	82.7	1.4434
0.1292	7.4	0.12880	0.12988	7.700	0.9917	82.6	1.4416
0.1309	7.5	0.13053	0.13165	7.596	0.9914	82.5	1.4399
0.1326	7.6	0.13226	0.13343	7.495	0.9912	82.4	1.4382
0.1344	7.7	0.13399	0.13521	7.396	0.9910	82.3	1.4364
0.1361	7.8	0.13572	0.13698	7.300	0.9907	82.2	1.4367
0.1379	7.9	0.13744	0.13876	7.207	0.9905	82.1	1.4329
0.1396	8.0	0.13917	0.14054	7.115	0.9903	82.0	1.4312
0.1414	8.1	0.14090	0.14232	7.026	0.9900	81.9	1.4294
0.1431	8.2	0.14263	0.14410	6.940	0.9898	81.8	1.4277
0.1449	8.3	0.14436	0.14588	6.855	0.9895	81.7	1.4259
0.1466	8.4	0.14608	0.14767	6.772	0.9893	81.6	1.4242
0.1484	8.5	0.14781	0.14945	6.691	0.9890	81.5	1.4224
0.1501	8.6	0.14954	0.15124	6.612	0.9888	81.4	1.4207
0.1518	8.7	0.15126	0.15302	6.535	0.9885	81.3	1.4190
0.1536	8.8	0.15299	0.15481	6.460	0.9882	81.2	1.4172
0.1553	8.9	0.15471	0.15660	6.386	0.9880	81.1	1.4155
0.1571	9.0	0.15643	0.15838	6.314	0.9877	81.0	1.4137
0.1588	9.1	0.15816	0.16017	6.243	0.9874	80.9	1.4120
0.1606	9.2	0.15988	0.16196	6.174	0.9871	80.8	1.4102
0.1623	9.3	0.16160	0.16376	6.107	0.9869	80.7	1.4085
0.1641	9.4	0.16333	0.16555	6.041	0.9866	80.6	1.4067
0.1658	9.5	0.16505	0.16734	5.976	0.9863	80.5	1.4050
0.1676	9.6	0.16677	0.16914	5.912	0.9860	80.4	1.4032
0.1693	9.7	0.16849	0.17093	5.850	0.9857	80.3	1.4015
0.1710	9.8	0.17021	0.17273	5.789	0.9854	80.2	1.3998
0.1728	9.9	0.17193	0.17453	5.730	0.9851	80.1	1.3980
0.1745	10.0	0.1736	0.1763	5.671	0.9848	80.0	1.3963
0.1763	10.1	0.1754	0.1781	5.614	0.9845	79.9	1.3945
0.1780	10.2	0.1771	0.1799	5.558	0.9842	79.8	1.3928
0.1798	10.3	0.1788	0.1817	5.503	0.9839	79.7	1.3910
0.1815	10.4	0.1805	0.1835	5.449	0.9836	79.6	1.3893
Rad	Deg	Cos	Cot	Tan	Sin	Deg	Rad

TABLE 4 (Continued)

Rad	Deg	Sin	Tan	Cot	Cos	Deg	Rad
0.1833	10.5	0.1822	0.1853	5.396	0.9833	79.5	1.3875
0.1850	10.6	0.1840	0.1871	5.343	0.9829	79.4	1.3858
0.1868	10.7	0.1857	0.1890	5.292	0.9826	79.3	1.3840
0.1885	10.8	0.1874	0.1908	5.242	0.9823	79.2	1.3823
0.1902	10.9	0.1891	0.1926	5.193	0.9820	79.1	1.3806
0.1920	11.0	0.1908	0.1944	5.145	0.9816	79.0	1.3788
0.1937	11.1	0.1925	0.1962	5.097	0.9813	78.9	1.3771
0.1955	11.2	0.1942	0.1980	5.050	0.9810	78.8	1.3753
0.1972	11.3	0.1959	0.1998	5.005	0.9806	78.7	1.3736
0.1990	11.4	0.1977	0.2016	4.959	0.9803	78.6	1.3718
0.2007	11.5	0.1994	0.2035	4.915	0.9799	78.5	1.3701
0.2025	11.6	0.2011	0.2053	4.872	0.9796	78.4	1.3683
0.2042	11.7	0.2028	0.2071	4.829	0.9792	78.3	1.3666
0.2059	11.8	0.2045	0.2089	4.787	0.9789	78.2	1.3648
0.2077	11.9	0.2062	0.2107	4.745	0.9785	78.1	1.3641
0.2094	12.0	0.2079	0.2126	4.705	0.9781	78.0	1.3614
0.2112	12.1	0.2096	0.2144	4.665	0.9778	77.9	1.3596
0.2129	12.2	0.2113	0.2162	4.625	0.9774	77.8	1.3579
0.2147	12.3	0.2130	0.2180	4.586	0.9770	77.7	1.3561
0.2164	12.4	0.2147	0.2199	4.548	0.9767	77.6	1.3544
0.2182	12.5	0.2164	0.2217	4.511	0.9763	77.5	1.3526
0.2199	12.6	0.2181	0.2235	4.474	0.9759	77.4	1.3509
0.2217	12.7	0.2198	0.2254	4.437	0.9755	77.3	1.3491
0.2234	12.8	0.2215	0.2272	4.402	0.9751	77.2	1.3474
0.2251	12.9	0.2233	0.2290	4.366	0.9748	77.1	1.3456
0.2269	13.0	0.2250	0.2309	4.331	0.9744	77.0	1.3439
0.2286	13.1	0.2267	0.2327	4.297	0.9740	76.9	1.3422
0.2304	13.2	0.2284	0.2345	4.264	0.9736	76.8	1.3404
0.2321	13.3	0.2300	0.2364	4.230	0.9732	76.7	1.3387
0.2339	13.4	0.2317	0.2382	4.198	0.9728	76.6	1.3369
0.2356	13.5	0.2334	0.2401	4.165	0.9724	76.5	1.3352
0.2374	13.6	0.2351	0.2419	4.134	0.9720	76.4	1.3334
0.2391	13.7	0.2368	0.2438	4.102	0.9715	76.3	1.3317
0.2409	13.8	0.2385	0.2456	4.071	0.9711	76.2	1.3299
0.2426	13.9	0.2402	0.2475	4.041	0.9707	76.1	1.3282
Rad	Deg	Cos	Cot	Tan	Sin	Deg	Rad

TABLE 4 (Continued)

Rad	Deg	Sin	Tan	Cot	Cos	Deg	Rad
0.2443	14.0	0.2419	0.2493	4.011	0.9703	76.0	1.3265
0.2461	14.1	0.2436	0.2512	3.981	0.9699	75.9	1.3247
0.2478	14.2	0.2453	0.2530	3.952	0.9694	75.8	1.3230
0.2496	14.3	0.2470	0.2549	3.923	0.9690	75.7	1.3212
0.2513	14.4	0.2487	0.2568	3.895	0.9686	75.6	1.3195
0.2531	14.5	0.2504	0.2586	3.867	0.9681	75.5	1.3177
0.2548	14.6	0.2521	0.2605	3.839	0.9677	75.4	1.3160
0.2566	14.7	0.2538	0.2623	3.812	0.9673	75.3	1.3142
0.2583	14.8	0.2554	0.2642	3.785	0.9668	75.2	1.3125
0.2600	14.9	0.2571	0.2661	3.758	0.9664	75.1	1.3107
0.2618	15.0	0.2588	0.2679	3.732	0.9659	75.0	1.3090
0.2635	15.1	0.2605	0.2698	3.706	0.9655	74.9	1.3073
0.2653	15.2	0.2622	0.2717	3.681	0.9650	74.8	1.3055
0.2670	15.3	0.2639	0.2736	3.655	0.9646	74.7	1.3038
0.2688	15.4	0.2656	0.2754	3.630	0.9641	74.6	1.3020
0.2705	15.5	0.2672	0.2773	3.606	0.9636	74.5	1.3003
0.2723	15.6	0.2689	0.2792	3.582	0.9632	74.4	1.2985
0.2740	15.7	0.2706	0.2811	3.558	0.9627	74.3	1.2968
0.2758	15.8	0.2723	0.2830	3.534	0.9622	74.2	1.2950
0.2775	15.9	0.2740	0.2849	3.511	0.9617	74.1	1.2933
0.2793	16.0	0.2756	0.2867	3.487	0.9613	74.0	1.2915
0.2810	16.1	0.2773	0.2886	3.465	0.9608	73.9	1.2898
0.2827	16.2	0.2790	0.2905	3.442	0.9603	73.8	1.2881
0.2845	16.3	0.2807	0.2924	3.420	0.9598	73.7	1.2863
0.2862	16.4	0.2823	0.2943	3.398	0.9593	73.6	1.2846
0.2880	16.5	0.2840	0.2962	3.376	0.9588	73.5	1.2828
0.2897	16.6	0.2857	0.2981	3.354	0.9583	73.4	1.2811
0.2915	16.7	0.2874	0.3000	3.333	0.9578	73.3	1.2793
0.2932	16.8	0.2890	0.3019	3.312	0.9573	73.2	1.2776
0.2950	16.9	0.2907	0.3038	3.291	0.9568	73.1	1.2758
0.2967	17.0	0.2924	0.3057	3.271	0.9563	73.0	1.2741
0.2985	17.1	0.2940	0.3076	3.251	0.9558	72.9	1.2723
0.3002	17.2	0.2957	0.3096	3.230	0.9553	72.8	1.2706
0.3019	17.3	0.2974	0.3115	3.211	0.9548	72.7	1.2689
0.3037	17.4	0.2990	0.3134	3.191	0.9542	72.6	1.2671
Rad	Deg	Cos	Cot	Tan	Sin	Deg	Rad

TABLE 4 (Continued)

Rad	Deg	Sin	Tan	Cot	Cos	Deg	Rad
0.3054	17.5	0.3007	0.3153	3.172	0.9537	72.5	1.2654
0.3072	17.6	0.3024	0.3172	3.152	0.9532	72.4	1.2636
0.3089	17.7	0.3040	0.3191	3.133	0.9527	72.3	1.2619
0.3107	17.8	0.3057	0.3211	3.115	0.9521	72.2	1.2601
0.3124	17.9	0.3074	0.3230	3.096	0.9516	72.1	1.2584
0.3142	18.0	0.3090	0.3249	3.078	0.9511	72.0	1.2566
0.3159	18.1	0.3107	0.3269	3.060	0.9505	71.9	1.2549
0.3177	18.2	0.3123	0.3288	3.042	0.9500	71.8	1.2531
0.3194	18.3	0.3140	0.3307	3.024	0.9494	71.7	1.2514
0.3211	18.4	0.3156	0.3327	3.006	0.9489	71.6	1.2497
0.3229	18.5	0.3173	0.3346	2.989	0.9483	71.5	1.2479
0.3246	18.6	0.3190	0.3365	2.971	0.9478	71.4	1.2462
0.3264	18.7	0.3206	0.3385	2.954	0.9472	71.3	1.2444
0.3281	18.8	0.3223	0.3404	2.937	0.9466	71.2	1.2427
0.3299	18.9	0.3239	0.3424	2.921	0.9461	71.1	1.2409
0.3317	19.0	0.3256	0.3443	2.904	0.9455	71.0	1.2392
0.3334	19.1	0.3272	0.3463	2.888	0.9449	70.9	1.2374
0.3351	19.2	0.3289	0.3482	2.872	0.9444	70.8	1.2357
0.3368	19.3	0.3305	0.3502	2.856	0.9438	70.7	1.2339
0.3386	19.4	0.3322	0.3522	2.840	0.9432	70.6	1.2322
0.3403	19.5	0.3338	0.3541	2.824	0.9426	70.5	1.2305
0.3421	19.6	0.3355	0.3561	2.808	0.9421	70.4	1.2287
0.3438	19.7	0.3371	0.3581	2.793	0.9415	70.3	1.2270
0.3456	19.8	0.3387	0.3600	2.778	0.9409	70.2	1.2252
0.3473	19.9	0.3404	0.3620	2.762	0.9403	70.1	1.2235
0.3491	20.0	0.3420	0.3640	2.747	0.9397	70.0	1.2217
0.3508	20.1	0.3437	0.3659	2.733	0.9391	69.9	1.2200
0.3526	21.2	0.3453	0.3679	2.718	0.9385	69.8	1.2182
0.3543	21.3	0.3469	0.3699	2.703	0.9379	69.7	1.2165
0.3560	21.4	0.3486	0.3719	2.689	0.9373	69.6	1.2147
0.3578	21.5	0.3502	0.3739	2.675	0.9367	69.5	1.2130
0.3595	21.6	0.3518	0.3759	2.660	0.9361	69.4	1.2113
0.3613	21.7	0.3535	0.3779	2.646	0.9354	69.3	1.2095
0.3630	21.8	0.3551	0.3799	2.633	0.9348	69.2	1.2078
0.3648	21.9	0.3567	0.3819	2.619	0.9342	69.1	1.2060
Rad	Deg	Cos	Cot	Tan	Sin	Deg	Rad

TABLE 4 (Continued)

Rad	Deg	Sin	Tan	Cot	Cos	Deg	Rad
0.3665	21.0	0.3584	0.3839	2.605	0.9336	69.0	1.2043
0.3683	21.1	0.3600	0.3859	2.592	0.9330	68.9	1.2025
0.3700	21.2	0.3616	0.3879	2.578	0.9323	68.8	1.2008
0.3718	21.3	0.3633	0.3899	2.565	0.9317	68.7	1.1990
0.3735	21.4	0.3649	0.3919	2.552	0.9311	68.6	1.1973
0.3752	21.5	0.3665	0.3939	2.539	0.9304	68.5	1.1956
0.3770	21.6	0.3681	0.3959	2.526	0.9298	68.4	1.1938
0.3787	21.7	0.3697	0.3979	2.513	0.9291	68.3	1.1921
0.3805	21.8	0.3714	0.4000	2.500	0.9285	68.2	1.1903
0.3822	21.9	0.3730	0.4020	2.488	0.9278	68.1	1.1886
0.3840	22.0	0.3746	0.4040	2.475	0.9272	68.0	1.1868
0.3857	22.1	0.3762	0.4061	2.463	0.9265	67.9	1.1851
0.3875	22.2	0.3778	0.4081	2.450	0.9259	67.8	1.1833
0.3892	22.3	0.3795	0.4101	2.438	0.9252	67.7	1.1816
0.3910	22.4	0.3811	0.4122	2.426	0.9245	67.6	1.1798
0.3927	22.5	0.3827	0.4142	2.414	0.9239	67.5	1.1781
0.3944	22.6	0.3843	0.4163	2.402	0.9232	67.4	1.1764
0.3962	22.7	0.3859	0.4183	2.391	0.9225	67.3	1.1746
0.3979	22.8	0.3875	0.4204	2.379	0.9219	67.2	1.1729
0.3997	22.9	0.3891	0.4224	2.367	0.9212	67.1	1.1711
0.4014	23.0	0.3907	0.4245	2.356	0.9205	67.0	1.1694
0.4032	23.1	0.3923	0.4265	2.344	0.9198	66.9	1.1676
0.4049	23.2	0.3939	0.4286	2.333	0.9191	66.8	1.1659
0.4067	23.3	0.3955	0.4307	2.322	0.9184	66.7	1.1641
0.4084	23.4	0.3971	0.4327	2.311	0.9178	66.6	1.1624
0.4102	23.5	0.3987	0.4348	2.300	0.9171	66.5	1.1606
0.4119	23.6	0.4003	0.4369	2.289	0.9164	66.4	1.1589
0.4136	23.7	0.4019	0.4390	2.278	0.9157	66.3	1.1572
0.4154	23.8	0.4035	0.4411	2.267	0.9150	66.2	1.1554
0.4171	23.9	0.4051	0.4431	2.257	0.9143	66.1	1.1537
0.4189	24.0	0.4067	0.4452	2.246	0.9135	66.0	1.1519
0.4206	24.1	0.4083	0.4473	2.236	0.9128	65.9	1.1502
0.4224	24.2	0.4099	0.4494	2.225	0.9121	65.8	1.1484
0.4241	24.3	0.4115	0.4515	2.215	0.9114	65.7	1.1467
0.4259	24.4	0.4131	0.4536	2.204	0.9107	65.6	1.1449
Rad	Deg	Cos	Cot	Tan	Sin	Deg	Rad

TABLE 4 (Continued)

Rad	Deg	Sin	Tan	Cot	Cos	Deg	Rad
0.4276	24.5	0.4147	0.4557	2.194	0.9100	65.5	1.1432
0.4294	24.6	0.4163	0.4578	2.184	0.9092	65.4	1.1414
0.4311	24.7	0.4179	0.4599	2.174	0.9085	65.3	1.1397
0.4328	24.8	0.4195	0.4621	2.164	0.9078	65.2	1.1379
0.4346	24.9	0.4210	0.4642	2.154	0.9070	65.1	1.1362
0.4363	25.0	0.4226	0.4663	2.145	0.9063	65.0	1.1345
0.4381	25.1	0.4242	0.4684	2.135	0.9056	64.9	1.1337
0.4398	25.2	0.4258	0.4706	2.125	0.9048	64.8	1.1310
0.4416	25.3	0.4274	0.4727	2.116	0.9041	64.7	1.1292
0.4433	25.4	0.4289	0.4748	2.106	0.9033	64.6	1.1275
0.4451	25.5	0.4305	0.4770	2.097	0.9026	64.5	1.1257
0.4468	25.6	0.4321	0.4791	2.087	0.9018	64.4	1.1240
0.4485	25.7	0.4337	0.4813	2.078	0.9011	64.3	1.1222
0.4503	25.8	0.4352	0.4834	2.069	0.9003	64.2	1.1205
0.4520	25.9	0.4368	0.4856	2.059	0.8996	64.1	1.1188
0.4538	26.0	0.4384	0.4877	2.050	0.8988	64.0	1.1170
0.4555	26.1	0.4399	0.4899	2.041	0.8980	63.9	1.1153
0.4573	26.2	0.4415	0.4921	2.032	0.8973	63.8	1.1135
0.4590	26.3	0.4431	0.4942	2.023	0.8965	63.7	1.1118
0.4608	26.4	0.4446	0.4964	2.014	0.8957	63.6	1.1100
0.4625	26.5	0.4462	0.4986	2.006	0.8949	63.5	1.1083
0.4643	26.6	0.4478	0.5008	1.997	0.8942	63.4	1.1065
0.4660	26.7	0.4493	0.5029	1.988	0.8934	63.3	1.1048
0.4677	26.8	0.4509	0.5051	1.980	0.8926	63.2	1.1030
0.4695	26.9	0.4524	0.5073	1.971	0.8918	63.1	1.1013
0.4712	27.0	0.4540	0.5095	1.963	0.8910	63.0	1.0996
0.4730	27.1	0.4555	0.5117	1.954	0.8902	62.9	1.0978
0.4747	27.2	0.4571	0.5139	1.946	0.8894	62.8	1.0961
0.4765	27.3	0.4586	0.5161	1.937	0.8886	62.7	1.0943
0.4782	27.4	0.4602	0.5184	1.929	0.8878	62.6	1.0926
0.4800	27.5	0.4617	0.5206	1.921	0.8870	62.5	1.0908
0.4817	27.6	0.4633	0.5228	1.913	0.8862	62.4	1.0891
0.4835	27.7	0.4648	0.5250	1.905	0.8854	62.3	1.0873
0.4852	27.8	0.4664	0.5272	1.897	0.8846	62.2	1.0856
0.4869	27.9	0.4679	0.5295	1.889	0.8838	62.1	1.0838
Rad	Deg	Cos	Cot	Tan	Sin	Deg	Rad

TABLE 4 (Continued)

Rad	Deg	Sin	Tan	Cot	Cos	Deg	Rad
0.4887	28.0	0.4695	0.5317	1.881	0.8829	62.0	1.0821
0.4904	28.1	0.4710	0.5340	1.873	0.8821	61.9	1.0804
0.4922	28.2	0.4726	0.5362	1.865	0.8813	61.8	1.0786
0.4939	28.3	0.4741	0.5384	1.857	0.8805	61.7	1.0769
0.4957	28.4	0.4756	0.5407	1.849	0.8796	61.6	1.0751
0.4974	28.5	0.4772	0.5430	1.842	0.8788	61.5	1.0734
0.4992	28.6	0.4787	0.5452	1.834	0.8780	61.4	1.0716
0.5009	28.7	0.4802	0.5475	1.827	0.8771	61.3	1.0699
0.5027	28.8	0.4818	0.5498	1.819	0.8763	61.2	1.0681
0.5044	28.9	0.4833	0.5520	1.811	0.8755	61.1	1.0664
0.5061	29.0	0.4848	0.5543	1.804	0.8746	61.0	1.0647
0.5079	29.1	0.4863	0.5566	1.797	0.8738	60.9	1.0629
0.5096	29.2	0.4879	0.5589	1.789	0.8729	60.8	1.0612
0.5114	29.3	0.4894	0.5612	1.782	0.8721	60.7	1.0594
0.5131	29.4	0.4909	0.5635	1.775	0.8712	60.6	1.0577
0.5149	29.5	0.4924	0.5658	1.767	0.8704	60.5	1.0559
0.5166	29.6	0.4939	0.5681	1.760	0.8695	60.4	1.0542
0.5184	29.7	0.4955	0.5704	1.753	0.8686	60.3	1.0524
0.5201	29.8	0.4970	0.5727	1.746	0.8678	60.2	1.0507
0.5219	29.9	0.4985	0.5750	1.739	0.8669	60.1	1.0489
0.5236	30.0	0.5000	0.5774	1.7321	0.8660	60.0	1.0472
0.5253	30.1	0.5015	0.5797	1.7251	0.8652	59.9	1.0455
0.5271	30.2	0.5030	0.5820	1.7182	0.8643	59.8	1.0437
0.5288	30.3	0.5045	0.5844	1.7113	0.8634	59.7	1.0420
0.5306	30.4	0.5060	0.5867	1.7045	0.8625	59.6	1.0402
0.5323	30.5	0.5075	0.5890	1.6977	0.8616	59.5	1.0385
0.5341	30.6	0.5090	0.5914	1.6909	0.8607	59.4	1.0367
0.5358	30.7	0.5105	0.5938	1.6842	0.8599	59.3	1.0350
0.5376	30.8	0.5120	0.5961	1.6775	0.8590	59.2	1.0332
0.5393	30.9	0.5135	0.5985	1.6709	0.8581	59.1	1.0315
0.5411	31.0	0.5150	0.6009	1.6643	0.8572	59.0	1.0297
0.5428	31.1	0.5165	0.6032	1.6577	0.8563	58.9	1.0280
0.5445	31.2	0.5180	0.6056	1.6512	0.8554	58.8	1.0263
0.5463	31.3	0.5195	0.6080	1.6447	0.8545	58.7	1.0245
0.5480	31.4	0.5210	0.6104	1.6383	0.8536	58.6	1.0228
Rad	Deg	Cos	Cot	Tan	Sin	Deg	Rad

TABLE 4 (Continued)

Rad	Deg	Sin	Tan	Cot	Cos	Deg	Rad
0.5498	31.5	0.5225	0.6128	1.6319	0.8526	58.5	1.0210
0.5515	31.6	0.5240	0.6152	1.6255	0.8517	58.4	1.0193
0.5533	31.7	0.5255	0.6176	1.6191	0.8508	58.3	1.0175
0.5550	31.8	0.5270	0.6200	1.6128	0.8499	58.2	1.0158
0.5568	31.9	0.5284	0.6224	1.6066	0.8490	58.1	1.0140
0.5585	32.0	0.5299	0.6249	1.6003	0.8480	58.0	1.0123
0.5603	32.1	0.5314	0.6273	1.5941	0.8471	57.9	1.0105
0.5620	32.2	0.5329	0.6297	1.5880	0.8462	57.8	1.0088
0.5637	32.3	0.5344	0.6322	1.5818	0.8453	57.7	1.0071
0.5655	32.4	0.5358	0.6346	1.5757	0.8443	57.6	1.0053
0.5672	32.5	0.5373	0.6371	1.5697	0.8434	57.5	1.0036
0.5690	32.6	0.5388	0.6395	1.5637	0.8425	57.4	1.0018
0.5707	32.7	0.5402	0.6420	1.5577	0.8415	57.3	1.0000
0.5725	32.8	0.5417	0.6445	1.5517	0.8406	57.2	0.9983
0.5742	32.9	0.5432	0.6469	1.5458	0.8396	57.1	0.9966
0.5760	33.0	0.5446	0.6494	1.5399	0.8387	57.0	0.9948
0.5777	33.1	0.5461	0.6519	1.5340	0.8377	56.9	0.9931
0.5794	33.2	0.5476	0.6544	1.5282	0.8368	56.8	0.9913
0.5812	33.3	0.5490	0.6569	1.5224	0.8358	56.7	0.9896
0.5829	33.4	0.5505	0.6594	1.5166	0.8348	56.6	0.9879
0.5847	33.5	0.5519	0.6619	1.5108	0.8339	56.5	0.9861
0.5864	33.6	0.5534	0.6644	1.5051	0.8329	56.4	0.9844
0.5882	33.7	0.5548	0.6669	1.4994	0.8320	56.3	0.9826
0.5899	33.8	0.5563	0.6694	1.4938	0.8310	56.2	0.9809
0.5917	33.9	0.5577	0.6720	1.4882	0.8300	56.1	0.9791
0.5934	34.0	0.5592	0.6745	1.4826	0.8290	56.0	0.9774
0.5952	34.1	0.5606	0.6771	1.4770	0.8281	55.9	0.9756
0.5969	34.2	0.5621	0.6796	1.4715	0.8271	55.8	0.9739
0.5986	34.3	0.5635	0.6822	1.4659	0.8261	55.7	0.9721
0.6004	34.4	0.5650	0.6847	1.4605	0.8251	55.6	0.9704
0.6021	34.5	0.5664	0.6873	1.4550	0.8241	55.5	0.9687
0.6039	34.6	0.5678	0.6899	1.4496	0.8231	55.4	0.9669
0.6056	34.7	0.5693	0.6924	1.4442	0.8221	55.3	0.9652
0.6074	34.8	0.5707	0.6950	1.4388	0.8211	55.2	0.9634
0.6091	34.9	0.5721	0.6976	1.4335	0.8202	55.1	0.9617
Rad	Deg	Cos	Cot	Tan	Sin	Deg	Rad

TABLE 4 (Continued)

Rad	Deg	Sin	Tan	Cot	Cos	Deg	Rad
0.6109	35.0	0.5736	0.7002	1.4281	0.8192	55.0	0.9599
0.6126	35.1	0.5750	0.7028	1.4229	0.8181	54.9	0.9582
0.6144	35.2	0.5764	0.7054	1.4176	0.8171	54.8	0.9564
0.6161	35.3	0.5779	0.7080	1.4124	0.8161	54.7	0.9547
0.6178	35.4	0.5793	0.7107	1.4071	0.8151	54.6	0.9530
0.6196	35.5	0.5807	0.7133	1.4019	0.8141	54.5	0.9512
0.6213	35.6	0.5821	0.7159	1.3968	0.8131	54.4	0.9495
0.6231	35.7	0.5835	0.7186	1.3916	0.8121	54.3	0.9477
0.6248	35.8	0.5850	0.7212	1.3865	0.8111	54.2	0.9460
0.6266	35.9	0.5864	0.7239	1.3814	0.8100	54.1	0.9442
0.6283	36.0	0.5878	0.7265	1.3764	0.8090	54.0	0.9425
0.6301	36.1	0.5892	0.7292	1.3713	0.8080	53.9	0.9407
0.6318	36.2	0.5906	0.7319	1.3663	0.8070	53.8	0.9390
0.6336	36.3	0.5920	0.7346	1.3613	0.8059	53.7	0.9372
0.6353	36.4	0.5934	0.7373	1.3564	0.8049	53.6	0.9355
0.6370	36.5	0.5948	0.7400	1.3514	0.8039	53.5	0.9338
0.6388	36.6	0.5962	0.7427	1.3465	0.8028	53.4	0.9320
0.6405	36.7	0.5976	0.7454	1.3416	0.8018	53.3	0.9303
0.6423	36.8	0.5990	0.7481	1.3367	0.8007	53.2	0.9285
0.6440	36.9	0.6004	0.7508	1.3319	0.7997	53.1	0.9268
0.6458	37.0	0.6018	0.7536	1.3270	0.7986	53.0	0.9250
0.6475	37.1	0.6032	0.7563	1.3222	0.7976	52.9	0.9233
0.6493	37.2	0.6046	0.7590	1.3175	0.7965	52.8	0.9215
0.6510	37.3	0.6060	0.7618	1.3127	0.7955	52.7	0.9198
0.6528	37.4	0.6074	0.7646	1.3079	0.7944	52.6	0.9180
0.6545	37.5	0.6088	0.7673	1.3032	0.7934	52.5	0.9163
0.6562	37.6	0.6101	0.7701	1.2985	0.7923	52.4	0.9146
0.6580	37.7	0.6115	0.7729	1.2938	0.7912	52.3	0.9128
0.6597	37.8	0.6129	0.7757	1.2892	0.7902	52.2	0.9111
0.6615	37.9	0.6143	0.7785	1.2846	0.7891	52.1	0.9093
0.6632	38.0	0.6157	0.7813	1.2799	0.7880	52.0	0.9076
0.6650	38.1	0.6170	0.7841	1.2753	0.7869	51.9	0.9058
0.6667	38.2	0.6184	0.7869	1.2708	0.7859	51.8	0.9041
0.6685	38.3	0.6198	0.7898	1.2662	0.7848	51.7	0.9023
0.6702	38.4	0.6211	0.7926	1.2617	0.7837	51.6	0.9006
Rad	Deg	Cos	Cot	Tan	Sin	Deg	Rad

TABLE 4 (Continued)

Rad	Deg	Sin	Tan	Cot	Cos	Deg	Rad
0.6720	38.5	0.6225	0.7954	1.2572	0.7826	51.5	0.8988
0.6737	38.6	0.6239	0.7983	1.2527	0.7815	51.4	0.8971
0.6754	38.7	0.6252	0.8012	1.2482	0.7804	51.3	0.8954
0.6772	38.8	0.6266	0.8040	1.2437	0.7793	51.2	0.8936
0.6789	38.9	0.6280	0.8069	1.2393	0.7782	51.1	0.8919
0.6807	39.0	0.6293	0.8098	1.2349	0.7771	51.0	0.8901
0.6824	39.1	0.6307	0.8127	1.2305	0.7760	50.9	0.8884
0.6842	39.2	0.6320	0.8156	1.2261	0.7749	50.8	0.8866
0.6859	39.3	0.6334	0.8185	1.2218	0.7738	50.7	0.8849
0.6877	39.4	0.6347	0.8214	1.2174	0.7727	50.6	0.8831
0.6894	39.5	0.6361	0.8243	1.2131	0.7716	50.5	0.8814
0.6912	39.6	0.6374	0.8273	1.2088	0.7705	50.4	0.8796
0.6929	39.7	0.6388	0.8302	1.2045	0.7694	50.3	0.8779
0.6946	39.8	0.6401	0.8332	1.2002	0.7683	50.2	0.8762
0.6964	39.9	0.6414	0.8361	1.1960	0.7672	50.1	0.8744
0.6981	40.0	0.6428	0.8391	1.1918	0.7660	50.0	0.8727
0.6999	40.1	0.6441	0.8421	1.1875	0.7649	49.9	0.8709
0.7016	40.2	0.6455	0.8451	1.1833	0.7638	49.8	0.8691
0.7034	40.3	0.6468	0.8481	1.1792	0.7627	49.7	0.8674
0.7051	40.4	0.6481	0.8511	1.1750	0.7615	49.6	0.8656
0.7069	40.5	0.6494	0.8541	1.1708	0.7604	49.5	0.8639
0.7086	40.6	0.6508	0.8571	1.1667	0.7593	49.4	0.8622
0.7103	40.7	0.6521	0.8601	1.1626	0.7581	49.3	0.8604
0.7121	40.8	0.6534	0.8632	1.1585	0.7570	49.2	0.8587
0.7138	40.9	0.6547	0.8662	1.1544	0.7559	49.1	0.8570
0.7156	41.0	0.6561	0.8693	1.1504	0.7547	49.0	0.8552
0.7173	41.1	0.6574	0.8724	1.1463	0.7536	48.9	0.8535
0.7191	41.2	0.6587	0.8754	1.1423	0.7524	48.8	0.8517
0.7208	41.3	0.6600	0.8785	1.1383	0.7513	48.7	0.8500
0.7226	41.4	0.6613	0.8816	1.1343	0.7501	48.6	0.8482
0.7243	41.5	0.6626	0.8847	1.1303	0.7490	48.5	0.8465
0.7261	41.6	0.6639	0.8878	1.1263	0.7478	48.4	0.8447
0.7278	41.7	0.6652	0.8910	1.1224	0.7466	48.3	0.8430
0.7295	41.8	0.6665	0.8941	1.1184	0.7455	48.2	0.8412
0.7313	41.9	0.6678	0.8972	1.1145	0.7443	48.1	0.8395
Rad	Deg	Cos	Cot	Tan	Sin	Deg	Rad

TABLE 4 (Continued)

Rad	Deg	Sin	Tan	Cot	Cos	Deg	Rad
0.7330	42.0	0.6691	0.9004	1.1106	0.7431	48.0	0.8378
0.7348	42.1	0.6704	0.9036	1.1067	0.7420	47.9	0.8360
0.7365	42.2	0.6717	0.9067	1.1028	0.7408	47.8	0.8343
0.7383	42.3	0.6730	0.9099	1.0990	0.7396	47.7	0.8325
0.7400	42.4	0.6743	0.9131	1.0951	0.7385	47.6	0.8303
0.7418	42.5	0.6756	0.9163	1.0913	0.7373	47.5	0.8290
0.7435	42.6	0.6769	0.9195	1.0875	0.7361	47.4	0.8273
0.7453	42.7	0.6782	0.9228	1.0837	0.7349	47.3	0.8255
0.7470	42.8	0.6794	0.9260	1.0799	0.7337	47.2	0.8238
0.7487	42.9	0.6807	0.9293	1.0761	0.7325	47.1	0.8221
0.7505	43.0	0.6820	0.9325	1.0724	0.7314	47.0	0.8203
0.7522	43.1	0.6833	0.9358	1.0686	0.7302	46.9	0.8186
0.7540	43.2	0.6845	0.9391	1.0649	0.7290	46.8	0.8168
0.7558	43.3	0.6858	0.9424	1.0612	0.7278	46.7	0.8151
0.7575	43.4	0.6871	0.9457	1.0575	0.7266	46.6	0.8133
0.7592	43.5	0.6884	0.9490	1.0538	0.7254	46.5	0.8116
0.7610	43.6	0.6896	0.9523	1.0501	0.7242	46.4	0.8098
0.7627	43.7	0.6909	0.9556	1.0464	0.7230	46.3	0.8081
0.7645	43.8	0.6921	0.9590	1.0428	0.7218	46.2	0.8063
0.7662	43.9	0.6934	0.9623	1.0392	0.7206	46.1	0.8046
0.7679	44.0	0.6947	0.9657	1.0355	0.7193	46.0	0.8029
0.7697	44.1	0.6959	0.9691	1.0319	0.7181	45.9	0.8011
0.7714	44.2	0.6972	0.9725	1.0283	0.7169	45.8	0.7994
0.7732	44.3	0.6984	0.9759	1.0247	0.7157	45.7	0.7976
0.7749	44.4	0.6997	0.9793	1.0212	0.7145	45.6	0.7959
0.7767	44.5	0.7009	0.9827	1.0176	0.7133	45.5	0.7941
0.7784	44.6	0.7022	0.9861	1.0141	0.7120	45.4	0.7924
0.7802	44.7	0.7034	0.9896	1.0105	0.7108	45.3	0.7907
0.7819	44.8	0.7046	0.9930	1.0070	0.7096	45.2	0.7889
0.7837	44.9	0.7059	0.9965	1.0035	0.7083	45.1	0.7871
0.7854	45.0	0.7071	1.0000	1.0000	0.7071	45.0	0.7854
Rad	Deg	Cos	Cot	Tan	Sin	Deg	Rad

Index

Symbols

$\lvert x \rvert$	Absolute value of x
$\angle a$	Angle a
$\angle ABC$	Angle with B the vertex and A and C points on the rays or sides of the angle
ω	Angular velocity (Greek letter omega)
$F(x)$	Antiderivative, or area of F at x
$\approx$	Approximately equal to
$\widehat{AB}$	Arc from point A to B
$\bar{x}$	Average x
$\{\ \}$	Braces, to specify a set
$\cong$	Congruent to
csc	Cosecant
cos	Cosine
cot	Cotangent
$^\circ$	Degree
Δx	Delta x, small positive increment of x
f'	Derivative of the function f
$\dfrac{d\,f(x)}{dx}$	Derivative of $f(x)$ with respect to x
$F(x)\Big]_b^a$	$F(b) - F(a)$
$f(x), g(x)$	The value at x of the function f and g
$>$	Is greater than
$\geq$	Is greater than or equal to
$\leftrightarrow$	If and only if (implication in both directions)
$\rightarrow$	Implies, or is equivalent to
$\int_b^a f(x)\,dx$	The definite integral of f with respect to x between the limits of a and b
$\int f(x)\,dx$	The indefinite integral or antiderivative of f with respect to x
j	Square root of (-1) used in imaginary and complex number notation